THE CYTOSKELETAL BASIS
OF PLANT GROWTH AND FORM

THE CYTOSKELETAL BASIS OF PLANT GROWTH AND FORM

Edited by

Clive W. Lloyd

Department of Cell Biology, John Innes Institute, Norwich, UK

ACADEMIC PRESS

Harcourt Brace Jovanovich, Publishers

London San Diego New York
Boston Sydney Tokyo Toronto

ACADEMIC PRESS LIMITED
24–28 Oval Road
London NW1 7DX

US edition published by
ACADEMIC PRESS INC
San Diego, CA 92101

This book is printed on acid-free paper

A catalogue record for this book is available from
the British Library

ISBN 0–12–453770–7

Typeset by Columns Design and Production Services Ltd, Reading, UK
Printed and bound in Great Britain at The Bath Press, Avon

CONTRIBUTORS

R.C. Brown   Department of Biology, University of Southwestern Louisiana, Lafayette, LA 70504–2451 USA

B. Buchen   Botanisches Institut, Universität Bonn, Venusbergweg 22, D-5300 Bonn 1, Germany

R.J. Cyr   Department of Biology, Pennsylvania State University, 208 Mueller Laboratory, University Park, PA 16802, USA

D.P. Delmer   Department of Botany, Institute of Life Sciences, The Hebrew University, Jerusalem 91904, Israel

J.H. Doonan   Department of Cell Biology, John Innes Institute, Colney Lane, Norwich NR4 7UH, UK

D.J. Fairbairn   Department of Cell Biology, John Innes Institute, Colney Lane, Norwich NR4 7UH, UK

M.M. Falconer   Department of Biology, University of Ottawa, Ottawa, Ontario, Canada K1N 6N5

T.H. Giddings   Department of Molecular, Cellular and Developmental Biology, University of Colorado, Boulder, CO 80309–0347, USA

P.B. Green   Department of Biological Sciences, Stanford University, Stanford, CA 94305, USA

T. Hayashi   Wood Research Institute, Kyoto University, Gokasho, Uji, Kyoto 611, Japan

Z. Hejnowicz   Botanisches Institut, Universität Bonn, Venusbergweg 22, D-5300 Bonn 1, Germany

P.J. Hussey   Biochemistry Department, Royal Holloway and Bedford New College, Egham Hill, Egham, Surrey TW20 0EX, UK

U. Kutschera   Botanisches Institut, Universität Bonn, Venusbergweg 22, D-5300 Bonn 1, Germany

A.-M. Lambert   Institut de Biologie Moléculaire des Plantes, Centre National de la Recherche Scientifique, 12 Rue du Général Zimmer, 67084 Strasbourg Cedex, France

B.E. Lemmon   Department of Biology, University of Southwestern Louisiana, Lafayette, LA 70504–2451, USA

C.W. Lloyd   Department of Cell Biology, John Innes Institute, Colney Lane, Norwich NR4 7UH, UK

M.C. McCann   Department of Cell Biology, John Innes Institute, Colney Lane, Norwich NR4 7UH, UK

D.W. McCurdy   Plant Cell Biology Group, Research School of Biological Sciences, The Australian National University, PO Box 475, Canberra City, ACT 2601, Australia

L.C. Morejohn   Department of Botany, University of Texas, Austin, TX 78713, USA

T. Murata   Department of Biology, Faculty of Science, Tokyo Metropolitan University, Fukazawa, Tokyo 158, Japan

B.A. Palevitz   Department of Botany, University of Georgia, Athens, GA 30602, USA

K. Roberts   Department of Cell Biology, John Innes Institute, Colney Lane, Norwich NR4 7UH, UK

A.-C. Schmidt   Institut de Biologie Moléculaire des Plantes, Centre National de la Recherche Scientifique, 12 Rue de Général Zimmer, 67084 Strasbourg Cedex, France

R.W. Seagull   USDA/ARS, Southern Regional Research Center, 1100 Robert E. Lee Boulevard, PO Box 19687, New Orleans, LA 70179, USA

J.M.L. Selker   Department of Biology, University of Oregon, Eugene, OR 97403–1210, USA

P.J. Shaw   Department of Cell Biology, John Innes Institute, Colney Lane, Norwich NR4 7UH, UK

H. Shibaoka   Department of Biology, Osaka University, Toyonaka, Osaka 560, Japan

A. Sievers   Botanisches Institut, Universität Bonn, Venusbergweg 22, D-5300 Bonn 1, Germany

C.D. Silflow   Department of Genetics and Cell Biology, 250 BioScience Center, 1445 Gortner Avenue, St Paul, MN 55108, USA

D.P. Snustad   Department of Genetics and Cell Biology, 250 BioScience Center, 1445 Gortner Avenue, St Paul, MN 55108, USA

L.A. Staehelin   Department of Molecular, Cellular and Developmental Biology, University of Colorado, Boulder, CO 80309–0347, USA

H. Stoeckel   Institut de Biologie Moléculaire des

Plantes, Centre National de la Recherche Scientifique, 12 Rue du Général Zimmer, 67084 Strasbourg Cedex, France

M. Wada   Department of Biology, Faculty of Science, Tokyo Metropolitan University, Fukazawa, Tokyo 158, Japan

S.M. Wick   Department of Plant Biology, University of Minnesota, 220 BSC, 1445 Gortner Avenue, St Paul, MN 55108, USA

R.E. Williamson   Plant Cell Biology Group, Research School of Biological Sciences, The Australian National University, PO Box 475, Canberra City, ACT 2601, Australia

S.M. Wolniak   Department of Botany and Center for Agricultural Biotechnology, University of Maryland, College Park, MD 20742, USA

M. Vantard   Institut de Biologie Moléculaire des Plantes, Centre National de la Recherche Scientifique, 12 Rue du Général Zimmer, 67084 Strasbourg Cedex, France

D. Volkmann   Botanisches Institut, Universität Bonn, Venusbergweg 22, D-5300 Bonn 1, Germany

PREFACE

This book is about the shape of plant cells and the factors that affect it during development. A common strand in most of the chapters is how the cytoskeleton and the cell wall combine to direct the growth of plant cells.

If we use a low power microscope to look at sections of plants which have been cleared of cytoplasm the pattern of cells can still be seen outlined in the form of the wall polymers. But although walls can retain complex organic shapes as 'dead wood', it is the living cytoplasm that plays an important architectural role in moulding new growth. Stripped of its wall, the protoplast is spherical and is unable to maintain or develop an asymmetric shape but as the wall regenerates, the scaffolding proteins of the cytoskeleton help to direct the alignment of cellulose microfibrils in the wall, thereby establishing a non-spherical cell morphology. It is this partnership between filamentous cytoplasmic proteins and extracellular polymers that is the key to plant form.

During interphase, environmental factors and plant growth regulators restructure the cytoskeleton in order to remodel the direction in which the walls expand or, in the case of tip-growing cells, extend. Cytoskeletal proteins form other complex scaffolding assemblies during division and these are involved in orienting the cell plate. In this way the envelope of cellular space is defined by the yielding properties of side walls and the placement of cross walls. Plant form is the history of asymmetries developed out of the partnership between wall and cytoskeleton. The nature of this relationship is now known in broad outline and we are beginning to find out how it is co-ordinated as part of the developmental programme.

A previous book, *The Cytoskeleton in Plant Growth and Development* also covered this area. That book was largely a product of the previous 20 years of ultrastructural research, almost exclusively on microtubules. The subsequent decade has seen advances in the biochemical and molecular biological studies on plant tubulin but more is gradually being discovered about the other cytoskeletal systems. However, the greatest shift of emphasis came as a result of applying immunofluorescence microscopy to plant cells. The change in scale allowed entire cytoskeletons to be seen in many cells in a single preparation; this new perspective is therefore one of complex, integral, and often physically very large assemblies, that are undoubtedly dynamic. In addition, the earlier studies on protoplasts, single suspension cells and dissociated tissue cells have now moved on to larger developmental units such as stomatal complexes, entire filamentous plants and even apices of higher plants. The developmental relationships between cells can now be visualized more clearly and we can see how internal agents of morphogenetic change (such as plant growth regulators) and external influences (such as light and gravity) impinge upon the cytoskeleton. It is as a result of these advances that the role of the cytoskeleton in plant development is re-examined.

In discussing plant morphology the cell wall cannot be ignored for there is no point in discussing the cytoskeleton in isolation. There is a section on the chemistry, synthesis and architecture of the wall but later sections on division, expansion and development – even though the wall is not described in molecular detail – are still involved with the larger issues of wall placement.

The chapters are arranged from the more

molecular, though cellular, to developmental but because of the enormity of understanding how cell behaviour is integrated into development, the emphasis is somewhere in the middle of this scale. This book is not intended to provide a seamless collection of definitive reviews for the field is still developing on several fronts and ideas are far from fixed. Instead, a larger than usual number of rather shorter chapters was invited, enabling the authors to concentrate on recent developments or to summarize their own subject areas. This format should provide a better feel for recent progress and for how the field might develop. Brief linking pieces are provided before each group of chapters to help unite them and to underline the wider frame of reference.

I would like to thank my colleagues who provided pictures used throughout the book: Kim Goodbody, Paul Green, Anne-Marie Lambert, Paul Linstead and Maureen McCann. I am also grateful to Andy Richford, Sarah Robertson and Carol Parr of Academic Press.

CLIVE LLOYD

CONTENTS

The cytoskeleton in plant development

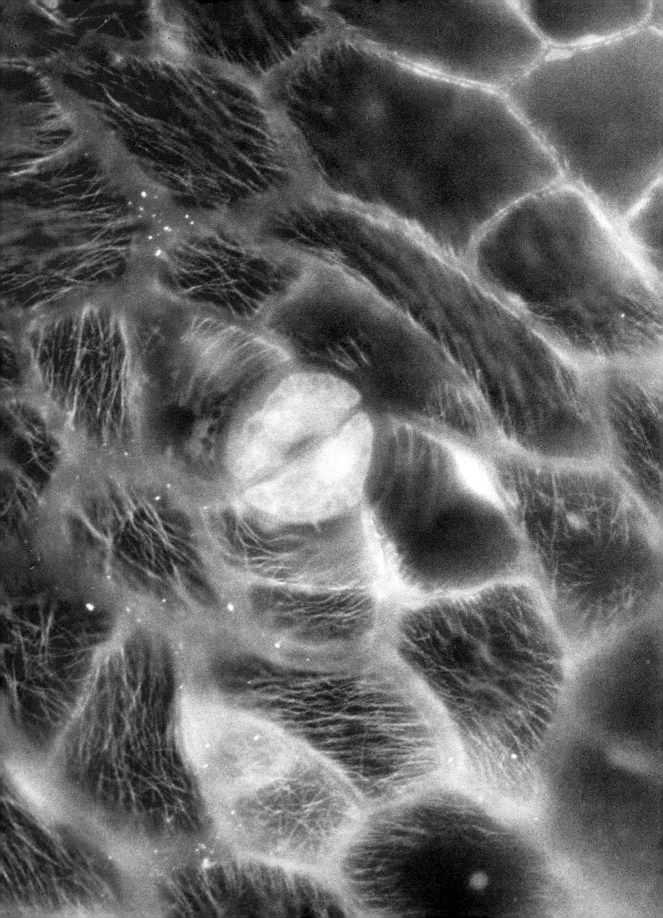

MOLECULES OF THE CYTOSKELETON

It has only recently become apparent that plants contain all three major filamentous classes throughout the cell cycle. The microtubule cycle was already quite well understood in the 1960s but it was only in the latter part of the 1980s that actin filaments were found to remain beyond interphase, during mitosis and cytokinesis. Similarly, the third cytoskeletal element, intermediate filaments, have also recently been detected in plants. The major filament-forming proteins have been identified but very few accessory proteins have been pinpointed, let alone characterized. It has been said elsewhere that the biochemistry of the plant cytoskeleton would have advanced more rapidly if, like animals, plants possessed brains, hooves and muscles to enable the major filamentous proteins to be isolated in bulk. The plant field is therefore still trying to name the parts, particularly the accessory proteins which will modify and regulate the behaviour of the filaments.

The characean algae have played an important part in understanding cytoplasmic streaming for their actin filaments can be exposed to experimentation by simply cutting the large filamentous cells. In Chapter 1, McCurdy and Williamson review the components of the actin cytoskeleton and their localization in algae and higher plants. One of the most interesting aspects for morphogenesis has been the realization – largely through the use of fluorescent phallotoxins – that actin filaments do not depolymerize upon entry into mitosis. It is at this stage that cytoplasmic streaming stops but actin filaments persist and, as will be discussed in chapters on dividing cells, are implicated in setting the division plane.

The major microtubule polypeptide, tubulin, is well conserved throughout phylogeny, allowing anti-fungal or anti-vertebrate tubulin antibodies to be used for immunostaining plant cells. For a long time the difficulty of concentrating plant tubulin to a level sufficient to initiate self-assembly *in vitro*, hindered its characterization. Fortunately, methods are now available for assembling plant tubulin – the most successful involving the use of taxol to lower the critical concentration for polymerization. From biochemical and immunoblotting studies it can be seen that plant and animal tubulins are subtly different. One of the most pertinent differences between plant and brain tubulin lies in the relative insensitivity of the plant protein to the depoly-merizing agent colchicine, and enhanced sensitivity to a range of herbicides. This is discussed by Morejohn in Chapter 3.

Like other cytoskeletal proteins, tubulins are part of a multigene family. Their multiplicity is enhanced by post-translational modification and it is important to know whether this diversity of isotypes is reflected in functional specialization. Phosphorylation, detyrosination, acetylation of tubulins have all been considered as potential modifiers of microtubule dynamics in animal cells. At the time of writing, microinjection studies using cytoskeletal probes have only just begun on plant cells and so nothing is known about cytoskeletal dynamics, but post-translational modification of tubulins will probably figure highly in future research and such modifications are discussed by Hussey, Snustad and Silflow in Chapter 2. They also review the molecular biological aspects of tubulin multiplicity, particularly whether there is developmental or functional significance to be attached to the expression of particular sets of isotypes in particular tissues.

A very important aspect of cell cycle progression is the relationship between the microtubule arrays. The reorganization of one cytoskeletal structure as it is supplanted by the next implies reutilization of subunits although there is little firm evidence to prove that all arrays are expressions of the same tubulin pool. Recent microinjection studies by Zhang, Wadsworth and Hepler (*Proc. Natl Acad. Sci. U.S.A.*, **87**, 8820–8824, 1990) have shown that it is possible to follow microtubule dynamics in living plant cells and they report that exogenous tubulin incorporated into the spindle is re-used by the phragmoplast. This shows the way ahead but it also underlines another major requirement: to establish with certainty *where* microtubules (MTs) are nucleated. Without tools that identify MT nucleation sites to universal satisfaction the location of material, which is so important to cellular morphogenesis, remains a mystery. This whole area of MT rearrangement and reutilization is discussed by Palevitz in Chapter 4, and further discussion, in other contexts, can be found in Chapters 17, 18 and 20 by Wick, Lloyd, and Wada and Murata, respectively.

One area which will yield great dividends is the study of microtubule-associated proteins (MAPs). MAPs affect the stability of microtubules *in vitro* and – as described by Cyr in Chapter 5 – undoubtedly exist in plants. They must constitute the cross-bridges that are seen between cortical MTs in ultrastructural studies and must also provide the links between MTs and the plasma membrane. Whether these two types of link belong to the same class is unknown but such proteins will almost certainly turn out to be involved in the lateral registration of MTs and hence play an architectural part in forming the cytoskeletons. But as will be seen in succeeding chapters on cell shape, MTs are dynamic, reorganizing in response to growth regulators, and it is essential to find out if and how these MAPs modulate cytoskeletal activity. It is a reasonably safe prediction that knowledge of MT dynamics and motors will revolutionize the way in which we think of plant cells.

In Chapter 6 by Shaw *et al.*, the last chapter in this section, evidence is reviewed that plants possess intermediate filament antigens in nucleus and cytoplasm. As in animal cells their function is not yet known but this at least constitutes another homology between the two kingdoms!

1. ACTIN AND ACTIN-ASSOCIATED PROTEINS

David W. McCurdy and Richard E. Williamson

Plant Cell Biology Group, Research School of Biological Sciences,
The Australian National University, PO Box 475, Canberra City, ACT 2601, Australia

Actin filaments are recognized as widespread components of plant cells (Staiger and Schliwa, 1987; Lloyd, 1988). This chapter summarizes knowledge of plant actin and of the proteins which associate with it and briefly considers the organization, function and regulation of the actin cytoskeleton in selected cell types.

Component proteins of the actin cytoskeleton

Actin

Sequenced actin genes encode proteins with substantial homologies with animal actins (reviewed by Meagher and McLean, 1990). Single plant species have more diverse actin families than do individual animal species and the diversity in a single plant is comparable with that between highly diverged species of animals (Meagher and McLean, 1990). Plants may have individual isovariants of actin that are specialized for particular functions and/or expression in certain cell types (Meagher and McLean, 1990; McLean *et al.*, 1990). More distantly related proteins may also exist, sharing DNase-binding activity and some epitopes with actins (McCurdy and Williamson, 1987; Williamson *et al.*, 1987).

The protein products of some of these genes have been purified and partially character-ized (Kato and Tonomura, 1977; Vahey and Scordilis, 1980; Vahey *et al.*, 1982; Turkina *et al.*, 1987; Ma and Yen, 1989; Villaneuva *et al.*, 1990). They can be recognized without purifica-tion by immunoblotting, partial proteolysis and their affinity for pancreatic DNase I (Williamson *et al.*, 1985; McCurdy and Williamson, 1987). No functional properties that set plant actins

apart from animal actins have been recognized. While this is not unexpected given the high sequence conservation, the physicochemical properties of plant actins require more detailed characterization.

All actins form 6-nm filaments that can be identified by their ability to bind proteolytic fragments of myosin to form arrowhead fila-ments (Palevitz *et al.*, 1974), to bind actin-specific antibodies (Williamson and Toh, 1979) and to bind fluorescent phallotoxins (Nothnagel *et al.*, 1981). Myosin molecules move towards the barbed end of the actin filament (Huxley, 1963) and the underlying polarity gives the two filament ends different assembly properties; the barbed ends of actin filaments grow faster than the pointed ends (Woodrum *et al.*, 1975). Animals, slime moulds (Pollard and Cooper, 1986) and yeast (Drubin, 1990) have actin-binding proteins that regulate filament extension, bundle or sever filaments or impart Ca^{2+} sensitivity. Comparable proteins have yet to be reliably identified in plants.

Myosins

Myosins are actin-activated ATPases that can be consigned by physicochemical properties to two families. The myosin I family has a catalytically active heavy chain of approxi-mately 100–140 kD. Its conserved region is homologous to the head portion of the larger heavy chain of myosin IIs and encodes an ATP-sensitive actin-binding site. The remainder of the myosin I molecule can encode different features in different myosins (Korn and Hammer, 1988; Adams and Pollard, 1989a; Jung *et al.*, 1989; Titus *et al.*, 1989; Jung and Hammer, 1990). These can include: a second,

The Cytoskeletal Basis of Plant Growth and Form   ISBN 0–12–453770–7

ATP-insensitive actin-binding site that allows such myosin Is to cross-link and cause relative movements of actin filaments (Korn and Hammer, 1988; Korn et al., 1988); lipid-binding domains that are believed responsible for myosin Is association with membranes (Adams and Pollard, 1989b), and calmodulin-binding sites (Mooseker and Coleman, 1989). Myosin IIs (reviewed by Korn and Hammer, 1988) have an enzymically active subunit of approximately 200 kD that is divided into a globular head and a long tail. The globular head has one actin-binding site from which actin is released by ATP while the tails self-associate to a stable dimer. Further assembly into larger aggregates varies between different gene products and can be controlled by phosphorylation.

Plant myosin genes have not been cloned and their gene products have been difficult to study biochemically or to localize in situ. Several have been purified (Kato and Tonomura, 1977; Ohsuka and Inoue, 1979; Vahey et al., 1982; Turkina et al., 1987; Ma and Yen, 1989) and more recently, two antibodies to animal myosins have been found to react with plant polypeptides on immunoblots (Parke et al., 1986; Grolig et al., 1988; Tang et al., 1989; Lin et al., 1989). By the criterion of heavy chain M_r, most putative plant myosins fall within the myosin II range. Several, however, are smaller than the heavy chains of the majority of animal myosin IIs and lie rather close to the upper end of the size range of myosin Is. Self-association may therefore be a better criterion as to family; myosins from Nitella (Kato and Tonomura, 1977), Egaria (Ohsuka and Inoue, 1979) and Lycopersicon (Vahey and Scordilis, 1980) form bipolar filaments and estimates of the native molecular weight of pea tendril myosin indicate a dimer (Ma and Yen, 1989). Two putative myosins fall within the size range of myosin Is. Lycopersicon myosin has a 100-kD heavy chain but its reported formation of bipolar filaments suggests that the 100-kD polypeptide is a proteolysed myosin II (Vahey et al., 1982; compare with Vahey and Scordilis, 1980). The 110-kD polypeptide from Nitella that is recognized by the monoclonal antibody to the myosin II of 3T3 cells (Grolig et al., 1988) remains to be confirmed as a myosin by further criteria.

This lack of firm evidence for myosin Is and the presence of myosin IIs has to be considered in the context of the longstanding view (see below) that plant myosins bind to and move organelles. Myosin Is would be favoured to move organelles (Sinard and Pollard, 1989; Adams and Pollard, 1989a; Spudich, 1990) with their affinity for lipids (Adams and Pollard, 1989b) and demonstrated ability to associate with organelles and support their movements (Adams and Pollard, 1986). Myosin IIs are expected to generate tension by cross-linking anti-parallel actin filaments but have yet to be implicated in organelle movements except possibly in the case of the Saccharomyces nucleus (Watts et al., 1987). Documented bundles of plant actin are unipolar when decorated with arrowheads (Palevitz et al., 1974; Condeelis, 1974) but bidirectional organelle movements within single cytoplasmic strands (Mahlberg, 1964) suggest that actin filaments with both polarities exist in close proximity. These could be cross-linked by self-associated myosin IIs or by myosin Is with a second actin-binding site. The cytochalasin-sensitive deformation of protoplasts (Hahne and Hoffman, 1984) is consistent with plant actomyosin generating tension as well as generating a shearing force to drive cytoplasmic streaming. Clearly much remains to be learnt about plant myosins by further biochemical and molecular studies.

Ca²⁺-dependent protein kinase (CDPK)

CDPK, a 52-kD calcium-dependent, but calmodulin and phospholipid independent, protein kinase purified from suspension-cultured soybean cells (Harmon et al., 1987; Putnam-Evans et al., 1990), is the only actin-associated protein other than myosin to be characterized in plants. CDPK co-distributes with the F-actin network in Allium roots and Tradescantia pollen tubes but in vitro binding studies show that the purified protein does not bind either G- or F-actin (Putnam-Evans et al., 1989). Its in vivo association with plant actin is therefore probably via an unidentified actin-binding protein(s) (Putnam-Evans et al., 1989). CDPK is stimulated up to 100-fold by micromolar amounts of free Ca^{2+} ($K_{0.5} = 1.5$ μM free Ca^{2+}) and can efficiently phosphorylate gizzard myosin light chains (Putnam-Evans *et al.*, 1990), two features which suggest that CDPK may be an important regulator of the plant cytoskeleton (Putnam-Evans *et al.*, 1989).

Localization and function of actin and associated proteins

Algae

Algal cells, particularly giant cells, provide some of the best opportunities for cell biological studies of the actin cytoskeleton (Menzel, 1991). We will confine our attention to the characean algae that have been influential in formulating ideas regarding the function and regulation of actomyosin in plant cells.

Internodal cells of the characean algae show vigorous cytoplasmic streaming (reviewed by Kuroda, 1990; Williamson, 1991). Actin filaments form large bundles at the interface between streaming endoplasm and stationary cortical cytoplasm with all filaments having the polarity to move an endoplasmic myosin in the direction of the streaming endoplasm (Kersey *et al.*, 1976). Myosin(s) are localized on endoplasmic organelles (Grolig *et al.*, 1988) supporting the hypothesis that myosin-coated organelles show ATP-dependent movements along the actin bundles and form rigor complexes with them in the absence of ATP (Williamson, 1975). Endoplasmic reticulum may help to mobilize the entire mass of endoplasm (Kachar and Reese, 1988), much of which will not physically contact the actin bundles. The regulation of actomyosin interaction in streaming may be controlled by CDPK. Its characean homologue is also localized on endoplasmic organelles and reticulum, as well as along the actin bundles (Fig. 1.1), and a physical association between CDPK and myosin is likely since both are extracted by perfusion with ATP or high Ca^{2+} (Grolig *et al.*, 1988; Harmon and McCurdy, 1990).

The actin bundles in characean cells, like other components of the cortical cytoplasm (Green, 1964), are strain-aligned parallel to the cell's major axis of growth. They may also

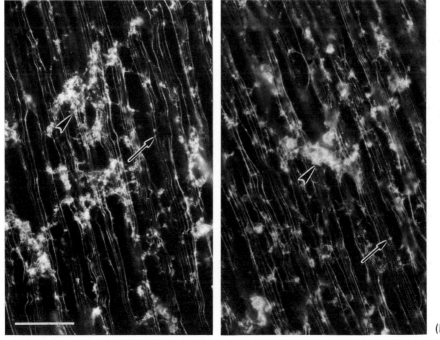

(a)   (b)

Fig. 1.1. Immunofluorescence microscopy using antimyosin (a) and anti-CDPK (b) monoclonal antibodies in internodal cells of *Chara*. Both antibodies label the bundles of actin as well as small organelles associated with these bundles (arrows). Large aggregates of endoplasmic material are also labelled (open arrows). Scale bar = 20 μm (× 812).

contribute to their own alignment by the cytoplasmic flow that they generate (Williamson *et al.*, 1984). Actin bundle assembly in growing cells is sensitive to lower concentrations of cytochalasin than are required to inhibit streaming (Williamson and Hurley, 1986). Extension of severed bundles into wounded regions is predominantly from bundles exposing barbed filament ends (Williamson *et al.*, 1984), raising the possibility that the pointed ends may be blocked in some way. Cytochalasin completely inhibits bundle extension into wounds at very low concentrations (Williamson and Hurley, 1986), consistent with its activity as a barbed-end capping agent.

Non-dividing cells of higher plants

Actin
Individual filaments of actin aggregate to form large bundles and these have been observed in a wide variety of differentiated cell types from many species of vascular plants. Bundles of actin were first observed by electron microscopy (reviewed by Hepler and Palevitz, 1974; Seagull, 1989) but their three-dimensional organization is best appreciated by fluorescence microscopy (e.g. Parthasarathy, 1985). The orientation of the actin cytoskeleton is dependent largely on the shape of the cell (Parthasarathy, 1985), so that in elongated cells that display vigorous cytoplasmic streaming, such as those from root vascular tissue or epidermal layers of leaves or coleoptiles, the actin cytoskeleton consists of large subcortical bundles connected to a peripheral network of finer elements, both of which show net longitudinal orientation. More spherical cells, such as those in suspension culture (see next section), endosperm (Schmit and Lambert, 1987) or fruit tissue (Parthasarathy *et al.*, 1985) contain three-dimensional networks of actin but with no preferred orientation.

The dominant function of the large subcortical bundles of actin in elongated cells, as in probably most cells, is to drive 'cytoplasmic' or organelle streaming. Of the 18 cell types containing filamentous actin examined from 17 species of vascular plants, 16 displayed cytoplasmic streaming (Parthasarathy *et al.*, 1985).

Direct evidence comes from observations that longitudinal bundles shown to contain F-actin support a unidirectional movement of different organelles that is only inhibited by concentrations of cytochalasin that physically disrupt the bundles themselves (Parthasarathy, 1985).

While the role of subcortical actin is reasonably clear, the function(s) of the network of finer, peripheral bundles seen in most cells is less so. Individual organelle movements can be supported by peripheral actin (Parthasarathy, 1985), but organized and extensive streaming is not a common feature in cortical regions of most cells. Peripheral actin may have a role in the spatial organization of the components of the parietal cytoplasm. Electron microscopy has revealed close associations of cortical actin with peripheral endoplasmic reticulum (ER) (reviewed by Hepler *et al.*, 1990), and suggestions have been made that the organization of this peripheral ER is influenced by actin (Quader *et al.*, 1987; Lichtscheidl and Weiss, 1988; Hepler *et al.*, 1990). In other instances, actin filaments may anchor chloroplasts in cortical locations that are favourable to maximize light-absorption (Cox *et al.*, 1987; see also Menzel and Elsner-Menzel, 1989), as well as provide the motive force for light-dependent movements of chloroplasts in leaf cells (Witztum and Parthasarathy, 1985). An indirect function of cortical actin may be to add mass to the total actin cytoskeleton so that, in the absence of specific anchorages to the plasma membrane, its increased inertia provides a stable base against which the forces that generate cytoplasmic streaming can work (Pesacreta and Parthasarathy, 1984). Actin has also been implicated in the polar distribution of organelles in plant cells, specifically during statolith movements in root-cap cells and the establishment of cell polarity in *Fucus* eggs. The former is discussed later by Sievers *et al.*, Chapter 13 this volume, and will not be covered here.

Higher plant cells that expand by tip growth provide valuable material to study actin organization and function. Pollen tubes prepared by chemical fixation show thick undulating bundles of actin which have a net axial organization and are prominent in subapical regions but also extend along the length of the tube (reviewed by Heslop-Harrison and Heslop-Harrison, 1989a). Alternative methods of tissue preparation have revealed a more complex assembly of very fine bundles and single filaments which are also predominantly axial (Lancelle *et al.*, 1987; Pierson, 1988; Tiwari and Polito, 1988; Heslop-Harrison and Heslop-Harrison, 1989a). Freeze-substitution reveals bundles of actin closely associated with cortical microtubules which are themselves cross-bridged to the plasma membrane (Lancelle *et al.*, 1987). Tubular elements of ER and vacuoles are linked to this cortical cytoskeleton. Bundles of actin are seen deeper in the cytoplasm, and at the tip region, fine bundles and individual filaments form a complex meshwork (Tiwari and Polito, 1988).

Bundles of actin filaments support uni-directional streaming of organelles in older regions of pollen tubes which show extensive vacuolation (Pierson, 1988; Heslop-Harrison and Heslop-Harrison, 1989a). That actin bundles act as tracks for streaming organelles is supported by an earlier study showing that individual bundles are composed of numerous 6-nm filaments having uniform polarity as indicated by labelling patterns with heavy meromyosin (Condeelis, 1974). Streaming in younger, subapical regions of pollen tubes, however, is more complex with adjacent channels of oppositely flowing organelles. In this case the streaming organelles presumably follow closely packed actin bundles that have either uniform acropetal or basipetal polarity (Heslop-Harrison and Heslop-Harrison, 1989a). Actin filaments have also been implicated in the control of pollen tube growth (reviewed by Steer and Steer, 1989). The vectoral delivery of wall precursor-containing vesicles to the tip expansion zone requires actin filaments, and Picton and Steer (1982) have proposed that an actin filament network stabilizes the tip region and that growth driven by turgor is regulated by a Ca^{2+}-controlled weakening of this network. The freeze-substitution studies mentioned above confirm the structural basis of this model.

Actin-associated proteins

Pollen tubes provide the non-dividing higher-plant cell type in which the localization of myosin is best understood. Myosin is associated with the outer surfaces of organelles such as amyloplasts, mitochondria, wall-precursor vesicles, and the vegetative nucleus and generative cells (Heslop-Harrison and Heslop-Harrison, 1989b; Tang *et al.*, 1989). Unidirectional movement of the smaller organelles probably occurs via interactions between the myosin coating these structures and the polar axial bundles of F-actin, while the slower tip-tracking movements of the vegetative nucleus and generative cell, and the accompanying distortions in their shape, is explained by a dynamic balance of forces created by multiple contacts with actin bundles of different polarities (Heslop-Harrison and Heslop-Harrison, 1989b).

CDPK co-localizes with axial bundles of F-actin in pollen tubes (Putnam-Evans *et al.*, 1989). Unlike myosins in this system, however, CDPK does not show a punctate distribution indicative of an association with organelles. Thus, the co-localization of CDPK and myosin seen in *Chara* is not apparent in pollen tubes. Whether this difference is real or reflects different methods used to prepare pollen tubes for immunofluorescence microscopy (compare Tang *et al.*, 1989 with Putnam-Evans *et al.*, 1989) remains to be established.

Dividing cells of higher plants

Actin

We will concentrate on the organization of the actin cytoskeleton in two dividing cell types that show quite different architectural features: highly vacuolated cells, typically from suspension cultures, and meristematic cells from root tips. The actin cytoskeleton during mitosis in wall-less endosperm cells will be discussed by Lambert *et al.* in Chapter 15, this volume.

During interphase the actin cytoskeleton in both cell types is similar to that described for differentiated cells; bundles of actin filaments are prominent in subcortical regions and these give rise to finer elements that typically are random but in some cases are arranged in transverse arrays similar in orientation to cortical microtubules (MTs) (see references listed below). This predominantly similar organization of actin is not apparent, however, when these two cell types perform mitosis. In highly vacuolated cells a filamentous network of actin persists throughout mitosis and is prominent in the central raft of cytoplasm (the phragmosome) that suspends the mitotic nucleus and spindle (Kakimoto and Shibaoka, 1987; Traas et al., 1987; Lloyd and Traas, 1988; and see Lloyd, Chapter 18, this volume). Radial filaments of actin in the phragmosome connect the centrifugally expanding phragmoplast to the adjacent side wall and Lloyd and Traas (1988) believe that this actin, by guiding the expanding phragmoplast to the correct division site in the cortex, performs an essential structural role in division plane alignment in plant cells. The disruption of oblique cell plate alignment by cytochalasin supports this concept (Palevitz, 1980; Gunning, 1982; Cho and Wick, 1990).

In contrast, however, the available evidence from root-tip cells indicates that the entire actin cytoskeleton is disassembled during prophase (Clayton and Lloyd, 1985; Palevitz, 1987a, 1988; McCurdy and Gunning, 1990). In wheat, this disassembly is preceded by a complete re-organization of the actin cytoskeleton during preprophase into highly parallel arrays of transverse cortical filaments (McCurdy et al., 1989; McCurdy and Gunning, 1990). Once these arrays are disassembled, however, usually prior to metaphase, the cells then complete mitosis in the absence of a detectable actin cytoskeleton (McCurdy and Gunning, 1990). The absence of filamentous actin provides a structural explanation for the observation that cytoplasmic streaming ceases during mitosis (see Gunning, 1982).

A consensus in actin filament organiza-tion returns during cytokinesis where actin is clearly present in the phragmoplast (Clayton and Lloyd, 1985; Gunning and Wick, 1985; Kakimoto and Shibaoka, 1987; Palevitz, 1987b; Seagull et al., 1987; McCurdy and Gunning, 1990). Isolation of phragmosomal complexes from suspension-cultured cells has shown that the actin filaments in the two halves of this structure have a net polarity consistent with providing unidirectional transport of vesicles towards the cell plate (Kakimoto and Shibaoka,

1988; see next section). Actin filaments are also associated with the mitotic spindle in some cells (reviewed by Baskin and Cande, 1990). The weight of available evidence, however, clearly discounts a role for actin in providing the motive force for chromosome separation (Baskin and Cande, 1990). Actin may have a role in maintaining the orientation and struc-tural integrity of the spindle (Schmit and Lambert, 1987; Traas et al., 1989; Cho and Wick, 1990).

The evidence reviewed above points towards different uses of actin during mitosis based on the different structural features of highly vacuolated cells versus those from the meri-stem of root tips. In vacuolated cells, for example, actin filaments are involved in both the premitotic migration and positioning of the nucleus within the phragmosome, as well as maintaining the structural integrity of the phragmoplast (Thomas et al., 1977; Gunning, 1982; Venverloo and Libbenga, 1987; Katsuta and Shibaoka, 1988). In contrast, nuclear migration and phragmosomal formation in root-tip cells are typically not observed, and so actin filaments in this case may not be required. Actin filaments in many plant cells are sensitive to chemical fixation (for discussion see McCurdy and Gunning, 1990), however, and so progress in understanding their organi-zation during mitosis will only be achieved by development of preparative techniques that visualize the true *in vivo* arrangement of the actin cytoskeleton. The permeabilization (Traas et al., 1987) and stabilization (Kakimoto and Shibaoka, 1987) techniques used to visualize filaments in mitotic suspension culture cells are not suitable for multicellular tissues. Micro-injection of labelled actin probes (Schmit and Lambert, 1990), preferably derivatized actin, offers a promising method to study actin filament distribution and dynamics in single cells, whereas rapid freeze-fixation methods, in combination with careful electron microscopy or possibly immunofluorescence microscopy, may provide a satisfactory method to determine actin filament organization in root-tip cells.

Actin-associated proteins
Only the heavy chain of myosin (Parke et al.,

1986) and CDPK (Putnam-Evans *et al.*, 1989) have been localized in dividing higher plant cells. Immunofluorescence in both studies identified an accumulation of the relevant protein in the phragmoplast. The presence of both polarized actin filaments and now myosin in this structure clearly invites the conclusion that they provide an actomyosin-based system, possibly regulated by Ca^{2+} via CDPK activity, for delivering vesicles to the expanding cell plate. Evidence against such a mechanism comes from observations that Ca^{2+} deficiency causes inhibition of cell plate formation (Gunning, 1982), and that in some instances cytochalasin does not inhibit migration of vesicles into the division plane of the cell plate (Palevitz, 1980). Thus, while both results might be explained by invoking a higher level of complexity in the operation of each mechanism, they serve to illustrate the problems associated with assigning function based solely on location (see Staiger and Schliwa, 1987).

An epitope recognized by a monoclonal antibody to troponin T, an actin-associated protein that helps regulate actomyosin function in muscle, co-distributes with MTs in the cortical array, preprophase band, mitotic spindle and phragmoplast in *Allium* root-tip cells (Lim *et al.*, 1986). However, the molecular characteristics of the protein carrying this epitope are not known, and thus the significance of its association with MTs remains unclear.

Regulation by Ca^{2+}

Actomyosin can be either activated or inhibited by rises in the concentration of free Ca^{2+} in plant cells and both actin and myosin may be targets for these effects. The comparable diversity of responses to Ca^{2+} in animal cells arises from the existence of multiple Ca^{2+}-binding proteins with diverse effects on the cytoskeleton (Pollard and Cooper, 1986).

Cytoplasmic streaming in characean algae ceases during an action potential within a few hundred milliseconds of peak membrane depolarization (Tazawa and Kishimoto, 1968). It resumes gradually, reaching full velocity after a delay ranging up to several minutes. The action potential raises cytoplasmic free Ca^{2+} from approximately 0.2 µM to 7 µM at about the time of maximum membrane depolarization (Williamson and Ashley, 1982). Micromolar Ca^{2+} concentrations completely inhibit ATP-dependent organelle movements in cells whose plasma membranes have been permeabilized (Tominaga *et al.*, 1983) and, dependent on preparation methods, completely (Tominaga *et al.*, 1987) or significantly (Williamson, 1975) inhibit them in cells whose tonoplasts have been removed or permeabilized.

Indirect evidence supports the view that myosin is inhibited by a Ca^{2+}-stimulated protein kinase while actin remains potentially active. Characean actin bundles transport myosin-coated beads (Sheetz and Spudich, 1983; Shimmen and Yano, 1984; Shimmen, 1988). These movements are only Ca^{2+} sensitive when the exogenous myosin is Ca^{2+} sensitive (e.g. *Physarum* or scallop myosin), leading to the conclusion that characean actin bundles are insensitive to Ca^{2+}. By default, therefore, myosin must be inhibited (reviewed by Shimmen, 1988; Williamson, 1991). A role for protein phosphorylation in inhibiting streaming is supported by experiments on ATP-dependent movements in tonoplast-free cells. Motility is arrested by inhibitors of protein phosphatases and irreversibly inhibited by ATP-γ-S, a substrate for protein kinases that produces thiophosphorylated proteins that cannot be dephosphorylated by protein phosphatases (Tominaga *et al.*, 1987). CDPK (Harmon and McCurdy, 1990) and myosin (Grolig *et al.*, 1988) occur in the actin bundles and on endoplasmic organelles, making CDPK an attractive candidate for the inhibitory kinase. Dephosphorylation of the kinase substrate(s) was postulated to involve a calmodulin-activated phosphatase (Tominaga *et al.*, 1985) since calmodulin inhibitors block the recovery of cytoplasmic streaming that has been inhibited by Ca^{2+}. Calmodulin is widely distributed in the endoplasm but not in the actin bundles. In contrast to CDPK, however, it is readily extracted by vacuolar perfusion (Jablonsky *et al.*, 1990). Streaming is irreversibly inhibited by Ca^{2+} in cells depleted of calmodulin by perfusion (Williamson, 1979), a finding consistent with a role for calmodulin in reactivation.

In other algae, Ca^{2+}-activation of processes involving actomyosin occurs. Organelle movements in *Acetabularia* probably occur along actin bundles (Koop and Kiermayer, 1980; Koop, 1981; Nagai and Fukui, 1981). These movements are activated by ≥ 1 µM Ca^{2+} in tonoplast-permeabilized cells (Fukui and Nagai, 1985), the reverse of the reaction in characean cells. Actin, myosin and calmodulin are also implicated in the cytoplasmic contractions elicited by wounding cells of the Siphonocladales (reviewed by La Claire, 1991). External Ca^{2+} is required if wounding is to elicit contractions and these can be triggered by 0.5–1.0 µM Ca^{2+} in detergent-extracted cells (La Claire, 1984). Organization of the contractile proteins as well as their activation appears to be Ca^{2+}-regulated.

In higher plants, the ionophore A23187 inhibits cytoplasmic streaming (Dorée and Picard, 1980; Woods *et al.*, 1984a; Takagi and Nagai, 1986; Kohno and Shimmen, 1988b) and ionophoretic injection of Ca^{2+} inhibits streaming in *Tradescantia* stamen-hair cells (Hepler and Wayne, 1985). In *Elodea*, there is circumstantial evidence that phytochrome controls cytoplasmic Ca^{2+} concentrations and through them, the presence or absence of streaming (Takagi and Nagai, 1983, 1985). Filament bundles remain intact in cells where streaming is arrested (Takagi and Nagai, 1983). While inhibition by Ca^{2+} appears therefore widespread, the physiological agents eliciting rises in free Ca^{2+} concentration are not well known. Rises may occur during stress (Woods *et al.*, 1984a, b) but it is not yet known whether rises sufficient to inhibit streaming occur under other physiological conditions.

Ca^{2+} may regulate both the organization of actin and the activity of myosin in pollen tubes. Ca^{2+}-ionophores inhibit *in situ* organelle movements and fragment actin cables when external Ca^{2+} concentrations are ≥ 10 µM (Kohno and Shimmen, 1988b). The failure of fragmentation to reverse when ionophore is washed out leaves open the possibility that pathological changes occur although assembly of bundles could, for example, require some factor present only in the extending zone of the cell. There may also be a myosin-based site of action for Ca^{2+} since ATP-dependent movements of isolated pollen tube organelles along characean actin bundles are inhibited when cells are homogenized in high Ca^{2+} solutions (Kohno and Shimmen, 1988a). The physiological relevance of these observations depends on the free Ca^{2+} concentration *in vivo*. It is attractive to think that Ca^{2+} regulates bundle assembly and vesicle movement in the growing tip but sufficiently elevated concentrations of Ca^{2+} have yet to be demonstrated (Nobiling and Reiss, 1987).

References

Adams, R.J. and Pollard, T.D. (1986). Propulsion of organelles isolated from *Acanthamoeba* along actin filaments by myosin-I. *Nature (Lond.)* **322**, 754–756.

Adams, R.J. and Pollard, T.D. (1989a). Membrane-bound myosin-I provides new mechanisms in cell motility. *Cell Motil. Cytoskeleton* **14**, 178–182.

Adams, R.J. and Pollard, T.D. (1989b). Binding of myosin I to membrane lipids. *Nature (Lond.)* **340**, 565–568.

Baskin, T.I. and Cande, W.Z. (1990). The structure and function of the mitotic spindle in flowering plants. *Annu. Rev. Plant Physiol. Plant Mol. Biol.* **41**, 277–315.

Cho, S.-O. and Wick, S.M. (1990). Distribution and function of actin in the developing stomatal complex of winter rye (*Secale cereale* cv. Puma). *Protoplasma* **157**, 154–164.

Clayton, L. and Lloyd, C.W. (1985). Actin organisation during the cell cycle in meristematic plant cells. Actin is present in the cytokinetic phragmoplast. *Exp. Cell Res.* **156**, 231–238.

Condeelis, J.S. (1974). The identification of F-actin in the pollen tube and protoplast of *Amaryllis belladonna*. *Exp. Cell Res.* **88**, 435–439.

Cox, G., Hawes, C.R., van der Lubbe, L. and Juniper, B.E. (1987). High-voltage electron microscopy of whole, critical-point dried plant cells. 2. Cytoskeletal structures and plastid motility in *Selaginella*. *Protoplasma* **140**, 173–186.

Dorée, M. and Picard, A. (1980). Release of Ca^{2+} from intracellular pools stops cytoplasmic streaming in *Tradescantia* staminal hairs. *Experientia* **36**, 1291–1292.

Drubin, D.G. (1990). Actin and actin-binding proteins in yeast. *Cell Motil. Cytoskeleton* **15**, 7–11.

Fukui, S. and Nagai, R. (1985). Reactivation of cytoplasmic streaming in a tonoplast-permeabilized cell model of *Acetabularia*. *Plant Cell Physiol.* **26**, 737–744.

Green, P.B. (1964). Cinematic observations of the growth and division of chloroplasts in *Nitella*. *Am. J. Bot.* **51**, 334–342.

Grolig, F., Williamson, R.E., Parke, J., Miller, C. and Anderton, B.H. (1988). Myosin and Ca^{2+}-sensitive streaming in the alga *Chara*: two polypeptides reacting with a monoclonal anti-myosin and their localization in the streaming endoplasm. *Eur. J. Cell Biol.* **47**, 22–31.

Gunning, B.E.S. (1982). The cytokinetic apparatus: its development and spatial regulation. In *The Cytoskeleton in Plant Growth and Development* (ed. C.W. Lloyd), pp. 229–292. Academic Press, New York.

Gunning, B.E.S. and Wick, S.M. (1985). Preprophase bands, phragmoplasts and spatial control of cytokinesis. *J. Cell Sci. Suppl.* **2**, 157–179.

Hahne, G. and Hoffman, F. (1984). The effect of laser microsurgery on cytoplasmic strands and cytoplasmic streaming in isolated plant protoplasts. *Eur. J. Cell Biol.* **33**, 175–179.

Harmon, A.C. and McCurdy, D.W. (1990). Calcium-dependent protein kinase and its possible role in the regulation of the cytoskeleton. In *Current Topics in Plant Biochemistry*, vol. 9 (eds D.D. Randall and D.G. Blevins), pp. 119–128. University of Missouri Press, Columbia.

Harmon, A.C., Putnam-Evans, C. and Cormier, M.J. (1987). A calcium-dependent but calmodulin-independent protein kinase from soybean. *Plant Physiol.* **83**, 830–837.

Hepler, P.K. and Palevitz, B.A. (1974). Microtubules and microfilaments. *Annu. Rev. Plant Physiol.* **25**, 309–362.

Hepler, P.K. and Wayne, R.O. (1985). Calcium and plant development. *Annu. Rev. Plant Physiol.* **36**, 397–439.

Hepler, P.K., Palevitz, B.A., Lancelle, S.A., McCauley, M.M. and Lichtscheidl, I. (1990). Cortical endoplasmic reticulum in plants. *J. Cell Sci.* **96**, 355–373.

Heslop-Harrison, J. and Heslop-Harrison, Y. (1989a). Actomyosin and movement in the angiosperm pollen tube: an interpretation of some recent results. *Sex. Plant Reprod.* **2**, 199–207.

Heslop-Harrison, J. and Heslop-Harrison, Y. (1989b). Myosin associated with the surface of organelles, vegetative nuclei and generative cells in angiosperm pollen grains and tubes. *J. Cell Sci.* **94**, 319–325.

Huxley, H.E. (1963). Electron microscope studies on the structure of natural and synthetic protein filaments from striated muscle. *J. Mol. Biol.* **7**, 281–308.

Jablonsky, P.P., Hagan, R.P., Grolig, F. and Williamson, R.E. (1990). Immunolocalization of *Chara* calmodulin and the reversibility of the inhibition of cytoplasmic streaming by Ca^{2+}. In *Calcium in Plant Growth and Development, Curr. Topics Plant Physiol.*, vol. 4 (eds R.T. Leonard and P.K. Hepler), pp. 79–85. American Society of Plant Physiology, Rockville.

Jung, G. and Hammer, J.A. III (1990). Generation and characterization of *Dictyostelium* cells deficient in a myosin I heavy chain isoform. *J. Cell Biol.* **110**, 1955–1964.

Jung, G., Schmidt, C.J. and Hammer, J.A. III (1989). Myosin I heavy-chain genes of *Acanthamoeba castellanii*: cloning of a second gene and evidence for the existence of a third isoform. *Gene* **82**, 269–280.

Kachar, B. and Reese, T.S. (1988). The mechanism of cytoplasmic streaming in characean algal cells: sliding of endoplasmic reticulum along actin filaments. *J. Cell Biol.* **106**, 1545–1552.

Kakimoto, T. and Shibaoka, H. (1987). Actin filaments and microtubules in the preprophase band and phragmoplast of tobacco cells. *Protoplasma* **140**, 151–156.

Kakimoto, T. and Shibaoka, H. (1988). Cytoskeletal ultrastructure of phragmoplast-nuclei complexes isolated from cultured tobacco cells. *Protoplasma* (Suppl. 2), 95–103.

Kato, T. and Tonomura, Y. (1977). Identification of myosin in *Nitella flexilis*. *J. Biochem. (Tokyo)* **82**, 777–782.

Katsuta, J. and Shibaoka, H. (1988). The roles of the cytoskeleton and the cell wall in nuclear positioning in tobacco BY-2 cells. *Plant Cell Physiol.* **29**, 403–413.

Kersey, Y.M., Hepler, P.K., Palevitz, B.A. and Wessels, N.K. (1976). Polarity of actin filaments in characean algae. *Proc. Natl Acad. Sci. U.S.A.* **73**, 165–167.

Kohno, T. and Shimmen, T. (1988a). Accelerated sliding of pollen tube organelles along Characeae actin bundles regulated by Ca^{2+}. *J. Cell Biol.* **106**, 1539–1543.

Kohno, T. and Shimmen, T. (1988b). Mechanism of Ca^{2+} inhibition of cytoplasmic streaming in lily pollen tubes. *J. Cell Sci.* **91**, 501–509.

Koop, H.U. (1981). Protoplasmic streaming in *Acetabularia*. *Protoplasma* **109**, 143–157.

Koop, H.U. and Kiermayer, O. (1980). Protoplasmic streaming in the giant unicellular green alga *Acetabularia mediterranea*. II. Differential sensitivity of movement systems to substances acting on microfilaments and microtubuli. *Protoplasma* **102**, 295–306.

Korn, E.D. and Hammer, J.A. III (1988). Myosins of nonmuscle cells. *Annu. Rev. Biophys. Biophys. Chem.* **17**, 23–45.

Korn, E.D., Atkinson, M.A.L., Brzeska, H., Hammer, J.A. III, Jung, G. and Lynch, T.J. (1988). Structure–function studies on *Acanthamoeba* myosins IA, IB and II. *J. Cell Biochem.* **36**, 37–50.

Kuroda, K. (1990). Cytoplasmic streaming in plant cells. *Int. Rev. Cytol.* **121**, 267–307.

La Claire II, J.W. (1984). Cell motility during wound healing in giant algal cells: contraction in detergent-permeabilized cells models of *Ernodesmis*. *Eur. J. Cell Biol.* **33**, 180–189.

La Claire II, J.W. (1991). Contractile movements in the algae: the Siphonocladales as model systems. In *The Cytoskeleton of the Algae* (ed. D. Menzel). CRC Press, in press.

Lancelle, S.A., Cresti, M. and Hepler, P.K. (1987). Ultrastructure of the cytoskeleton in freeze-substituted pollen tubes of *Nicotiana alata*. *Protoplasma* **140**, 141–150.

Lichtscheidl, I.K. and Weiss, D.G. (1988). Visualization of submicroscopic structures in the cytoplasm of *Allium cepa* inner epidermal cells by video-enhanced contrast light microscopy. *Eur. J. Cell Biol.* **46**, 376–382.

Lim, S.-S., Hering, G.E. and Borisy, G.G. (1986). Widespread occurrence of anti-troponin T cross-reactive components in non-muscle cells. *J. Cell Sci.* **85**, 1–19.

Lin, Q., Grolig, F., Jablonsky, P.P. and Williamson, R.E. (1989). Myosin heavy chains: detection by immunoblotting in higher plants and localization by immunofluorescence in the alga *Chara*. *Cell Biol. Int. Rep.* **13**, 107–117.

Lloyd, C.W. (1988). Actin in plants. *J. Cell Sci.* **90**, 185–188.

Lloyd, C.W. and Traas, J.A. (1988). The role of F-actin in determining the division plane of carrot suspension cells. Drug studies. *Development* **102**, 211–221.

Ma, Y.-Z. and Yen, L.-F. (1989). Actin and myosin in pea tendrils. *Plant Physiol.* **89**, 586–589.

Mahlberg, P.G. (1964). Rates of organelle movement in streaming cytoplasm of plant tissue culture cells. In *Primitive Motile Systems in Cell Biology* (eds R.D. Allen and N. Kamiya), pp. 43–68. Academic Press, New York.

McCurdy, D.W. and Gunning, B.E.S. (1990). Reorganization of cortical actin microfilaments and microtubules at preprophase and mitosis in wheat root-tip cells: a double label immunofluorescence study. *Cell Motil. Cytoskeleton* **15**, 76–87.

McCurdy, D.W. and Williamson, R.E. (1987). An actin-related protein inside pea chloroplasts. *J. Cell Sci.* **87**, 449–456.

McCurdy, D.W., Sammut, M. and Gunning, B.E.S. (1989). Immunofluorescent visualization of arrays of transverse cortical actin microfilaments in wheat root-tip cells. *Protoplasma* **147**, 204–206.

McLean, B.G., Eubanks, S. and Meagher, R.B. (1990). Tissue-specific expression of divergent actins in soybean root. *Plant Cell* **2**, 335–344.

Meagher, R.B. and McLean, B.G. (1990). Diversity of plant actins. *Cell Motil. Cytoskeleton* **16**, 164–166.

Menzel, D. (1991). *The Cytoskeleton of the Algae* (ed. D. Menzel). CRC Press, in press.

Menzel, D. and Elsner-Menzel, C. (1989). Actin-based chloroplast rearrangements in the cortex of the giant coenocytic alga *Caulerpa*. *Protoplasma*, **150**, 1–8.

Mooseker, M.S. and Coleman, T.R. (1989). The 110-kD protein–calmodulin complex of the intestinal microvillus (brush border myosin I) is a mechanoenzyme. *J. Cell Biol.* **108**, 2395–2400.

Nagai, R. and Fukui, S. (1981). Differential treatment of *Acetabularia* with cytochalasin B and N-ethylmaleimide with special reference to their effects on cytoplasmic streaming. *Protoplasma* **109**, 79–89.

Nobiling, R. and Reiss, H.-D. (1987). Quantitative analysis of calcium gradients and activity in growing pollen tubes of *Lilium longiflorum*. *Protoplasma* **139**, 20–24.

Nothnagal, E.A., Barak, L.S., Sanger, J.W. and Webb, W.W. (1981). Fluorescence studies on modes of cytochalasin B and phallotoxin action on cytoplasmic streaming in *Chara*. *J. Cell Biol.* **88**, 364–372.

Ohsuka, K. and Inoue, A. (1979). Identification of myosin in a flowering plant, *Egeria densa*. *J. Biochem.* **85**, 375–378.

Palevitz, B.A. (1980). Comparative effects of phalloidin and cytochalasin B on motility and morphogenesis in *Allium*. *Can. J. Bot.* **58**, 773–785.

Palevitz, B.A. (1987a). Actin in the preprophase band of *Allium cepa*. *J. Cell Biol.* **104**, 1515–1519.

Palevitz, B.A. (1987b). Accumulation of F-actin during cytokinesis in *Allium*. Correlation with microtubule distribution and the effects of drugs. *Protoplasma* **141**, 24–32.

Palevitz, B.A. (1988). Cytochalasin-induced reorganization of actin in *Allium* root cells. *Cell Motil. Cytoskeleton* **9**, 283–298.

Palevitz, B.A., Ash, J.F. and Hepler, P.K. (1974). Actin in the green alga, *Nitella*. *Proc. Natl Acad. Sci. U.S.A.* **71**, 363–366.

Parke, J., Miller, C. and Anderson, B.H. (1986).

Higher plant myosin heavy-chain identified using a monoclonal antibody. *Eur. J. Cell Biol.* **41**, 9–13.

Parthasarathy, M.V. (1985). F-actin architecture in coleoptile epidermal cells. *Eur. J. Cell Biol.* **39**, 1–12.

Parthasarathy, M.V., Perdue, T.D., Witztum, A. and Alvernaz, J. (1985). An actin network as a normal component of the cytoskeleton in many vascular plants. *Am. J. Bot.* **72**, 1318–1323.

Pesacreta, T.C. and Parthasarathy, M.V. (1984). Microfilament bundles in the roots of a conifer *Chamaecyparis obtusa*. *Protoplasma* **121**, 54–64.

Picton, J.M. and Steer, M.W. (1982). A model for the mechanism of tip extention in pollen tubes. *J. Theor. Biol.* **98**, 15–20.

Pierson, E.S. (1988). Rhodamine–phalloidin staining of F-actin in pollen after dimethylsulphoxide permeabilization. A comparison with the conventional formaldehyde preparation. *Sex. Plant Reprod.* **1**, 83–87.

Pollard, T.D. and Cooper, J.A. (1986). Actin and actin-binding proteins. A critical evaluation of mechanisms and functions. *Annu. Rev. Biochem.* **55**, 987–1035.

Putnam-Evans, C., Harmon, A.C., Palevitz, B.A., Fechheimer, M. and Cormier, M.J. (1989). Calcium-dependent protein kinase is localized wth F-actin in plant cells. *Cell Motil. Cytoskeleton* **12**, 12–22.

Putnam-Evans, C.L., Harmon, A.C. and Cormier, M.J. (1990). Purification and characterization of a novel protein kinase from soybean. *Biochemistry* **29**, 2488–2495.

Quader, H., Hofmann, A. and Schnepf, E. (1987). Shape and movement of the endoplasmic reticulum in onion bulb cells: possible involvement of actin. *Eur. J. Cell Biol.* **44**, 17–26.

Schmit, A.-C. and Lambert, A.-M. (1987). Characterization of dynamics of cytoplasmic F-actin in higher plant endosperm cells during interphase, mitosis, and cytokinesis. *J. Cell Biol.* **105**, 2157–2166.

Schmit, A.-C. and Lambert, A.-M. (1990). Microinjected fluorescent phalloidin *in vivo* reveals the F-actin dynamics and assembly in higher plant mitotic cells. *Plant Cell* **2**, 129–138.

Seagull, R.W. (1989). The plant cytoskeleton. *CRC Crit. Rev. Plant Sci.* **8**, 131–167.

Seagull, R.W., Falconer, M.M. and Weerdenburg, C.A. (1987). Microfilaments: dynamic arrays in plant cells. *J. Cell Biol.* **104**, 995–1004.

Sheetz, M.P. and Spudich, J.A. (1983). Movement of myosin-coated fluorescent beads on actin cables *in vitro*. *Nature (Lond.)* **303**, 31–35.

Shimmen, T. (1988). Characean actin bundles as a tool for studying actomyosin-based motility. *Bot.*

Mag. Tokyo **101**, 533–544.

Shimmen, T. and Yano, M. (1984). Active sliding movement of latex beads coated with skeletal muscle myosin on *Chara* actin bundles. *Protoplasma* **121**, 132–137.

Sinard, J.H. and Pollard, T.D. (1989). Microinjection into *Acanthamoeba castellanii* of monoclonal antibodies to myosin-II slows but does not stop cell locomotion. *Cell Motil. Cytoskeleton* **12**, 42–52.

Spudich, J.A. (1990). In pursuit of myosin function. *Cell Regulation* **1**, 1–11.

Staiger, C.J. and Schliwa, M. (1987). Actin localization and function in higher plants. *Protoplasma* **141**, 1–12.

Steer, M.W. and Steer, J.M. (1989). Pollen tube tip growth. *New Phytol.* **111**, 323–358.

Takagi, S. and Nagai, R. (1983). Regulation of cytoplasmic streaming in *Vallisneria* mesophyll cells. *J. Cell Sci.* **62**, 385–405.

Takagi, S. and Nagai, S. (1985). Light-controlled cytoplasmic streaming in *Vallisneria* mesophyll cells. *Plant Cell Physiol.* **26**, 941–951.

Takagi, S. and Nagai, R. (1986). Intracellular Ca^{2+} concentration and cytoplasmic streaming in *Vallisneria* mesophyll cells. *Plant Cell Physiol.* **27**, 953–959.

Tang, X., Hepler, P.K. and Scordilis, S.P. (1989). Immunochemical and immunocytochemical identification of a myosin heavy chain polypeptide in *Nicotiana* pollen tubes. *J. Cell Sci.* **92**, 569–574.

Tazawa, M. and Kishimoto, U. (1968). Cessation of cytoplasmic streaming of *Chara* internodes during action potential. *Plant Cell Physiol.* **9**, 361–368.

Thomas, D.D.S., Dunn, D.M. and Seagull, R.W. (1977). Rapid cytoplasmic responses of oat coleoptiles to cytochalasin B, auxin, and colchicine. *Can. J. Bot.* **55**, 1797–1800.

Titus, M.A., Warrick, H.M. and Spudich, J.A. (1989). Multiple actin-based motor genes in *Dictyostelium*. *Cell Regulation* **1**, 55–63.

Tiwari, S.C. and Polito, V.S. (1988). Organisation of the cytoskeleton in pollen tubes of *Pyrus communis*: a study employing conventional and freeze-substitution electron microscopy, immunofluorescence and rhodamine–phalloidin. *Protoplasma* **147**, 100–112.

Tominaga, Y., Shimmen, T. and Tazawa, M. (1983). Control of cytoplasmic streaming by extracellular Ca^{2+} in permeabilized *Nitella* cells. *Protoplasma* **116**, 75–77.

Tominaga, Y., Muto, S., Shimmen, T. and Tazawa, M. (1985). Calmodulin and Ca^{2+}-controlled cytoplasmic streaming in characean cells. *Cell Struct. Funct.* **10**, 315–325.

Tominaga, Y., Wayne, R., Tung, H.Y.L. and Tazawa, M. (1987). Phosphorylation–dephosphorylation is involved in Ca^{2+}-controlled cytoplasmic streaming of characean cells. *Protoplasma* **136**, 161–169.

Traas, J.A., Doonan, J.H., Rawlins, D.J., Shaw, P.J., Watts, J. and Lloyd, C.W. (1987). An actin network is present in cytoplasm throughout the cell cycle of carrot cells and associates with the dividing nucleus. *J. Cell Biol.* **105**, 387–395.

Traas, J.A., Burgain, S. and Dumas de Vaulx, R. (1989). The organization of the cytoskeleton during meiosis in eggplant (*Solanum melongena* (L.)): microtubules and F-actin are both necessary for coordinated meiotic division. *J. Cell Sci.* **92**, 541–550.

Turkina, M.V., Kulikova, A.L., Sokolov, O.I., Bogatyrev, V.A. and Kursanov, A.L. (1987). Actin and myosin filaments from the conducting tissues of *Heracleum sosnowskyi. Plant Physiol. Biochem.* **25**, 689–696.

Vahey, M. and Scordilis, S. (1980). Contractile proteins from tomato. *Can. J. Bot.* **58**, 797–801.

Vahey, M., Titus, M., Trautwein, R. and Scordilis, S. (1982). Tomato actin and myosin: contractile proteins from a higher land plant. *Cell Motil. Cytoskeleton* **2**, 131–147.

Venverloo, C.J. and Libbenga, K.R. (1987). Regulation of the plane of cell division in vacuolated cells. I. The function of nuclear positioning and phragmosome formation. *J. Plant Physiol.* **131**, 267–284.

Villaneuva, M.A., Ho, S.C. and Wang, J.L. (1990). Isolation and characterization of one isoform of actin from cultured soybean cells. *Arch. Biochem. Biophys.* **277**, 35–41.

Watts, F.Z., Shiels, G. and Orr, E. (1987). The yeast MY01 gene encoding a myosin-like protein required for cell division. *EMBO J.* **6**, 3499–3505.

Williamson, R.E. (1975). Cytoplasmic streaming in *Chara*: a cell model activated by ATP and inhibited by cytochalasin B. *J. Cell Sci.* **17**, 655–668.

Williamson, R.E. (1979). Filaments associated with the endoplasmic reticulum in the streaming cytoplasm of *Chara corallina. Eur. J. Cell Biol.* **20**, 177–183.

Williamson, R.E. (1991). Cytoplasmic streaming in characean algae: mechanism, regulation by Ca^{2+} and organization. In *Algal Cell Motility* (ed. M. Melkonian), Chapman and Hall, New York, in press.

Williamson, R.E. and Ashley, C.C. (1982). Free Ca^{2+} and cytoplasmic streaming in the alga *Chara. Nature (Lond.)* **296**, 647–651.

Williamson, R.E. and Hurley, U.A. (1986). Growth and regrowth of actin bundles in *Chara*: bundle assembly by mechanisms differing in sensitivity to cytochalasin B. *J. Cell Sci.* **85**, 21–32.

Williamson, R.E. and Toh, B.H. (1979). Motile models of plant cells and the immunofluorescent localization of actin in a motile *Chara* cell model. In *Cell Motility: Molecules and Organization* (ed. S. Hatano, H. Ishikawa and H. Sato), pp. 339–346. University of Tokyo Press, Tokyo.

Williamson, R.E., Hurley, U.A. and Perkin, J.L. (1984). Regeneration of actin bundles in *Chara*: polarized growth and orientation by endoplasmic flow. *Eur. J. Cell Biol.* **34**, 221–228.

Williamson, R.E., Perkin, J.L. and Hurley, U.A. (1985). Selective extraction of *Chara* actin bundles: identification of actin and two coextracting proteins. *Cell Biol. Int. Rep.* **89**, 547–554.

Williamson, R.E., McCurdy, D.W., Hurley, U.A. and Perkin, J.L. (1987). Actin of *Chara* giant internodal cells. *Plant Physiol.* **85**, 268–272.

Witztum, A. and Parthasarathy, M.V. (1985). Role of actin in chloroplast clustering and banding in leaves of *Egeria, Elodea* and *Hydrilla. Eur. J. Cell Biol.* **39**, 21–26.

Woodrum, D.T., Rich, S.A. and Pollard, T.D. (1975). Evidence for biased bidirectional polymerization of actin filaments using heavy meromyosin prepared by an improved method. *J. Cell Biol.* **67**, 231–237.

Woods, C.M., Reid, M.S. and Patterson, B.D. (1984a). Response to chilling stress in plant cells. I. Changes in cyclosis and cytoplasmic structure. *Protoplasma*, **121**, 8–16.

Woods, C.M., Polito, V.S. and Reid, M.S. (1984b). Response to chilling stress in plant cells. II. Redistribution of intracellular calcium. *Protoplasma* **121**, 17–24.

2. TUBULIN GENE EXPRESSION IN HIGHER PLANTS

Patrick J. Hussey[1], D. Peter Snustad[2] and Carolyn D. Silflow[2]

[1] Biochemistry Department, Royal Holloway and Bedford New College, University of London, Egham Hill, Egham, Surrey TW20 0EX, UK
[2] Department of Genetics and Cell Biology, University of Minnesota, 250 BioScience Center, 1445 Gortner Avenue, St Paul, MN 55108, USA

Tubulin gene expression in higher plants

Molecular genetic analysis of the structure and expression of tubulin genes in plants has progressed significantly in recent years. We will describe ongoing work in this area and comment on future directions. Extensive reviews have been published on tubulin gene structure and expression in animal systems (see Cleveland and Sullivan, 1985) and two earlier reviews dealt with the subject of plant tubulin genes (Silflow et al., 1987; Fosket, 1989). In this chapter we will refer to animal systems only in order to extend significant comparisons to plant tubulin genes.

Tubulin heterogeneity

It has now been established that both α and β tubulins exist as families of related isotypes in most eukaryotic cells. These multiple tubulin isotypes are differentially expressed between different cell types and tissues within an organism and as a consequence of specific developmental programmes. Evidence of multiple tubulins within individual cells raised the possibility that different microtubule arrays might be composed of distinct tubulins. Before discussing this possibility, we shall first consider the identification of multiple tubulins in higher plants.

Evidence for multiple forms of α and β tubulin in higher plants has come from a number of groups using a variety of polyacrylamide gel electrophoretic (PAGE) techniques (Dawson and Lloyd, 1985; Hussey and Gull, 1985; Mizuno et al., 1985; Cyr et al., 1987; Joyce, 1990; Kerr and Carter, 1990). Two-dimensional PAGE of plant total protein extracts and immunoblotting using well characterized anti-α tubulin (YOL 1/34, Kilmartin et al., 1982) and anti-β tubulin (DM1B, Blose et al., 1984) monoclonal antibodies identified four α tubulin isotypes and four β tubulin isotypes in Phaseolus vulgaris (Hussey and Gull, 1985). The recognition epitope for the antibody YOL 1/34 is between amino acids 414–422 of pig brain α tubulin, whereas the epitope for the antibody DM1B is between amino acids 416–430 of pig brain β tubulin (Breitling and Little, 1986). Both amino acid domains are conserved over a wide range of taxonomically distinct organisms (Little, 1985). Comparison of the gel migration pattern of the tubulins of P. vulgaris with that of mammalian tubulins revealed a significant difference (Hussey and Gull, 1985). Although the plant and animal β tubulins co-migrated on the gel system used, the plant α tubulins migrated faster than the β tubulins; the animal α tubulins migrated more slowly. Inversion of the α and β tubulin subunits on sodium dodecyl sulphate (SDS) gels has been described in several lower eukaryotes (Gull et al., 1986). In its electrophoretic properties, tubulin from plants resembled that of lower eukaryotes rather than mammalian tubulin. Dawson and Lloyd (1985) observed the inverted migration pattern for taxol-isolated carrot tubulins although other groups have not (Cyr et al., 1987; Kerr and Carter, 1990). It is possible that the discrepancy is due to differences in

The Cytoskeletal Basis of Plant Growth and Form   ISBN 0–12–453770–7

commercially available preparations of SDS (Best *et al.*, 1981) and that the SDS used by Hussey and Gull (1985) and Dawson and Lloyd (1985) (Fisons, FSA Laboratory Supplies, Loughborough, England) reveals a significant difference in SDS binding between plant and animal α tubulins. The biochemical differences between plant and animal tubulins revealed by differences in drug and herbicide binding and differences in antibody cross-reactivity were reviewed by Morejohn and Fosket (1986). These differences exist within a background of extensive (85–90%) amino acid sequence homology between animal and plant tubulins (Silflow *et al.*, 1987).

In all plants analysed to date, multiple isotypes of α and β tubulin were detected. Some examples of 2-D gel immunoblots of different plant tissue extracts probed with anti-α tubulin and anti-β tubulin monoclonal antibodies are shown in Fig. 2.1. Multiple electrophoretically-separable α and β tubulin isotypes are identifiable in *Daucus carota*, *Zea mays* and *Arabidopsis thaliana* protein extracts. Although the relative amounts of each isotype are variable in different plant tissues, the overall constellation of tubulin isotypes for each plant is unique and could be used as a 'signature' for that species.

Further analysis of the expression of carrot (Hussey *et al.*, 1988) and maize (Joyce *et al.*, 1991) tubulin isotypes was achieved by comparing 2-D gel immunoblots prepared from different plant organs at various stages of development. A total of six electrophoretically separable β tubulin isotypes were identifiable in carrot protein extracts. Each isotype was found to be expressed in a complex programme of differentiation and development. That is, the β6 tubulin isotype was found to be expressed solely in seedlings, whilst the β tubulin, β5, was identified only in the vegetative organs of the mature plant. Furthermore, β5 had its highest expression in the leaf lamina. The β4 tubulin was the dominant β tubulin isotype of the two forms detected in pollen; the other form, β2, was found to be ubiquitously expressed in carrot tissues. The remaining β tubulins also had specific expression programmes with β1 present in all tissues except

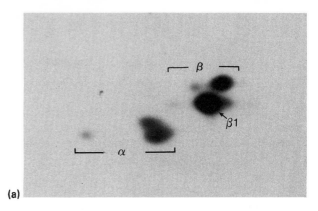

(a)

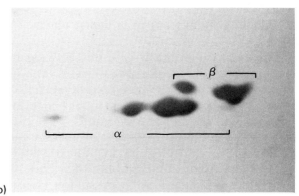

(b)

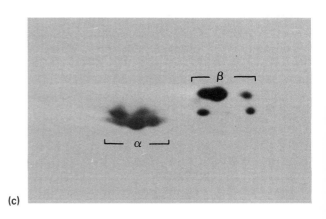

(c)

Fig. 2.1. Tubulin areas of immunoblots of two-dimensional gels of (a) *Daucus carota* suspension culture cells (b) *Zea mays* pollen (c) *Arabidopsis thaliana* bud. The blots were probed with both anti-α and anti-β tubulin monoclonal antibodies. (a) Probed with YOL 1/34, anti-α tubulin (Kilmartin *et al.*, 1982); KMX1 (Birkett *et al.*, 1985) and DM1B (Blose *et al.*, 1984), anti-β tubulins (b) and (c) probed with B512, anti-α tubulin and 2-10-B6, anti-β tubulin (gifts from G. Piperno).

pollen and β3 absent only from pollen and leaf tissue. There are at least four α tubulin isotypes in carrot. The carrot α1 isotype (Dawson and Lloyd, 1985) was found to be absent from protein samples of carrot seedlings (Hussey, 1986).

The ease of dissection of the maize root into separate tissues has allowed an accurate determination of the tubulin isotype composition of tissue within one organ (Joyce et al., 1991). Differences were observed in the relative levels of four different α tubulin isotypes (including one complex isotype) and six different β tubulin isotypes (including one complex isotype) expressed in the root tip, the cortex and the vascular cylinder.

Environmental effects on the tubulin isotype patterns in rye root tips were examined by Kerr and Carter (1990) who detected differences in the α and β tubulins between cold-acclimated and non-acclimated root tissues.

Tubulin genes and tubulin post-translational modification

Multiple electrophoretically-separable tubulin isotypes may arise by three different processes:

(1) Multiple α and β tubulin genes may encode separate polypeptide products.
(2) A single α or β tubulin gene may encode multiple mRNAs by variation in initiation, termination or intron splicing.
(3) The α and β tubulin polypeptides may be post-translationally modified.

Multiple tubulin genes

Genetic and molecular techniques have revealed the presence of families of α and β tubulin genes in all eukaryotes, with the exception of certain unicellular organisms (Cleveland and Sullivan, 1985; Silflow et al., 1987). Southern blot hybridization of genomic DNA to tubulin coding region probes has provided estimates of gene/pseudogene number ranging from two in Chlamydomonas reinhardtii to 20 in humans (Silflow and Rosenbaum, 1981; Lee et al., 1983). Multiple α and β tubulin sequences have been detected in the DNA of all plants analysed, including

Arabidopsis thaliana (Ludwig et al., 1987; Marks et al., 1987), soybean (Guiltanin et al., 1987a), Brassica campestris, Brassica oleracea, carrot, cucumber (Marks et al., 1987), maize (Hussey et al., 1990; Joyce, 1990; Montoliu et al., 1990) and oats (Mendu, 1990).

Large numbers of tubulin pseudogenes have been documented in mammalian genomes, where only seven of the approximately 15 β tubulin sequences seen on Southern blots represent expressed genes (Lee et al., 1983; Wang et al., 1986). The possible presence of pseudogenes among plant tubulin sequences has been investigated completely in only one plant system, Arabidopsis thaliana, where all five α tubulin sequences and all nine β tubulin sequences detected on Southern blots have been cloned. DNA sequences indicated that each of the cloned genes encodes a normal tubulin protein. In addition, the expression of each putative gene has been directly tested by preparing gene-specific probes from the 3' noncoding portion of each gene for hybridization to RNA. Transcripts were detected for each of the gene specific probes, indicating that all the genes are expressed (Ludwig et al., 1987; Marks et al., 1987; Oppenheimer et al., 1988; Kopczak et al., 1991; Snustad et al., 1991).

An alternative method for the identification of expressed tubulin genes involves the isolation of tubulin cDNA clones. By screening libraries prepared from RNA from a variety of plant tissues, it should be possible to obtain clones representing all expressed tubulin genes. Tubulin cDNA clones have been isolated from maize root (Montoliu et al., 1990), shoot, endosperm and pollen cDNA libraries (Hussey et al., 1990; Joyce et al., 1991). Comparison of the results of Southern blot hybridization performed using coding-region probes and gene-specific probes has indicated that the six α tubulin cDNA clones and genomic DNA clones isolated to date represent most of the α tubulin sequences detected in the maize genome by hybridization (Joyce, 1990; Villemur and Silflow, unpublished observations). Pseudogenes for α tubulin thus do not appear to be abundant in the maize genome. Tubulin cDNA clones from pea (Cerff et al., 1986) and oats (Colbert et al., 1990; Mendu, 1990) have also been reported.

The number of expressed tubulin genes is not necessarily indicative of the number of different tubulin isotypes produced. In both *Arabidopsis* and maize, several tubulin gene pairs have been shown to encode identical or nearly identical tubulin isotypes. For example, the *TUA2* and *TUA4* α tubulin genes of *Arabidopsis* encode identical proteins, as do the *TUA3* and *TUA5* genes. Thus, the complement of α tubulin gene products in *Arabidopsis* includes only three different forms (Kopczak *et al.*, 1991). Presumably, these closely related gene pairs have resulted from recent (on the evolutionary time scale) gene duplication events. Evidence for similar recent duplication of α tubulin genes in the maize genome was presented by Montoliu *et al.* (1990). In contrast to the α tubulin genes, the β tubulin genes in *Arabidopsis* encode eight different isotypes; only one linked gene pair has been found to encode an identical protein (Snustad *et al.*, 1991).

Full-length cDNA clones have facilitated the identification of products of individual tubulin genes. For example, Hussey *et al.* (1990) and Joyce (1990) used *in vitro* transcription to synthesize mRNA from maize α and β tubulin cDNA clones that contained complete coding sequences. The mRNA was translated *in vitro* and the translation products were analysed on 2-D gels. The labelled proteins co-migrated with one of the α or β tubulin isotypes found in plant cell extracts.

Post-translational modification

The heterogeneity of α and β tubulins observed by PAGE may also result from post-translational modification of existing polypeptides. The three post-translational modifications that have received the most attention are the acetylation of α tubulin, the phosphorylation of β tubulin and the detyrosination of α tubulin. The first two modifications can be detected as polypeptide shifts on 2-D gels because they result in a change in the pI of tubulin polypeptides. The acetylation of α tubulin was first reported in *Chlamydomonas* flagellar tubulin (L'Hernault and Rosenbaum, 1985). Monoclonal antibodies specific for acetylated tubulin (6-11B-1, Piperno and Fuller, 1985; 1-6.1,

Schulze *et al.*, 1987) have been used to identify the modified protein in subsets of microtubules in some but not all mammalian cultured cell types (Piperno *et al.*, 1987; Schatten *et al.*, 1988) and in certain microtubule arrays in *Physarum polycephalum* (Sasse *et al.*, 1987). The 6-11B-1 antibody has been used to probe Western blots of a variety of higher plant tissue extracts (Hussey, unpublished results), but only one, *Secale cereale*, has an α tubulin showing cross-reactivity to this antibody (Kerr, 1988). A lysine codon at position 40, the acetylation site conserved in animal and protistan α tubulins (LeDizet and Piperno, 1987), has also been found in two of the five *Arabidopsis* α tubulin genes (Kopczak *et al.*, 1991) and three of four *Zea mays* α tubulin genes (Montoliu *et al.*, 1990; Villemur, Haas, Silflow and Snustad, unpublished results). Piperno *et al.* (1987) have correlated an increased resistance to drug-induced depolymerization with acetylated microtubules, suggesting that acetylated rather than non-acetylated microtubules are more stable.

The extent of β tubulin phosphorylation has been reported to vary during the cell cycle (Piras and Piras, 1975) and during differentiation of certain animal tissues (Gard and Kirschner, 1985). Specific tubulin kinases have been identified (Goldenring *et al.*, 1982), but the role of phosphorylation in tubulin biochemistry is still an enigma. Phosphorylated β tubulin can be detected by a decrease in the pI of β tubulin polypeptides. The question of β tubulin phosphorylation has not been examined extensively in plants. However, when comparing 2-D gel immunoblots of β tubulins in cell protein extracts with autoradiographs of β tubulin synthesized *in vitro* from hybridization-selected tubulin mRNA, no polypeptide shifts indicative of a phosphorylation were detected in maize pollen (Hussey *et al.*, 1990).

The detyrosination of α tubulin is dependent on the presence of an encoded tyrosine at the carboxy terminus and the activity of two cellular enzymes: a specific carboxypeptidase (Kumar and Flavin, 1981) and tubulin tyrosine ligase (Barra *et al.*, 1974). All plant α tubulin genes/cDNAs sequenced to date, including five *Arabidopsis* genes and six maize cDNAs encode a carboxy terminal tyrosine (Joyce, 1990; Kopczak *et al.*, 1991; Villemur and Silflow, unpublished data). In addition, a monoclonal antibody specific for tyrosinated α tubulin, YL 1/2 (Kilmartin *et al.*, 1982), detected a constellation of polypeptides similar to that detected by YOL 1/34 when used to probe 2-D gel immunoblots of taxol-purified tubulins from carrot suspension cells (Dawson and Lloyd, 1985). The results suggested that at least the electrophoretically-separable carrot α tubulin isotypes possess a carboxy terminal tyrosine. Whether plant microtubules/tubulin are subject to the well-described detyrosination–tyrosination cycle (Gunderson *et al.*, 1987) remains to be investigated. The presence of the two necessary cellular enzymes has not been reported in plant cells. An interesting observation concerning the distribution of detyrosinated and tyrosinated microtubules was that in *Trypanosoma brucei*, the tyrosinated microtubules are newly polymerized (Sherwin and Gull, 1989).

Expression of α and β tubulin genes

The expression of plant tubulin genes has been investigated by four different approaches:

(1) the tubulin isotypes present in various plant tissues and organs have been examined by 2-D gel electrophoresis and immunoblot analysis using anti-α and anti-β tubulin monoclonal antibodies;
(2) changes in total α and/or β tubulin transcript levels in organs and tissues in response to light or to gibberellic acid (GA₃) treatment have been measured by northern and RNA dot blot analyses using tubulin coding-sequence hybridization probes;
(3) the accumulation of individual tubulin gene transcripts in various tissues and organs has been examined by analysing northern and dot blots of RNAs isolated from these tissues/ organs with gene-specific hybridization probes; and
(4) more refined patterns of tubulin gene expression have been obtained by histochemical analyses of transgenic plants harbouring tubulin gene promoter–reporter gene fusions.

Although direct measurements of rates of

transcript initiation, elongation, and turnover are not available for any plant tubulin gene, the results of studies of the type listed above suggest that some tubulin genes are regulated at the level of transcription in a tissue-specific manner, whereas other tubulin genes are constitutively expressed.

As mentioned earlier, Hussey *et al.* (1988) used 2-D electrophoresis and immunoblot analyses to study the distribution of six separable β tubulin isotypes in various tissues of *Daucus carota*. Similarly, Joyce (1990) examined the distribution in various *Zea mays* tissues of four α and six β tubulin isotypes that could be distinguished by the 2-D gel immunoblotting procedure. Mendu (1990) used 2-D immunoblots to resolve four distinct α tubulin isotypes in elongating internodes of *Avena sativa*. Clearly, the 2-D immunoblotting approach only provides an estimate of the minimum number of tubulin isotypes in any tissue or plant since isotypes that have similar primary structures may not be resolved. Thus, a given protein 'spot' on an immunoblot of a 2-D gel may contain two or more polypeptides rather than a single polypeptide, as demonstrated for the β tubulins of *Physarum polycephalum* by Burland *et al.* (1984). Indeed, it is possible that the β1 tubulin 'spot' in carrot detected with the DM1B antibody on blots of 2-D gels (see Fig. 2.1a) is actually a composite containing two or more β tubulin polypeptides (Hussey *et al.*, 1988). In *Z. mays*, one of the six distinct β tubulin 'spots' on 2-D immunoblots exhibits a complex morphology and clearly contains two or more distinct tubulin isoforms, indicating that maize contains a minimum of seven β tubulin isotypes (Joyce, 1990). Therefore, more precise methods must be employed to elucidate the spatial and temporal patterns of expression of all members of the α and β tubulin gene families.

The accumulation of tubulin gene transcripts in plant tissues/organs has been analysed by northern blot and/or dot blot hybridization procedures in several higher plant species. Cyr *et al.* (1987) used northern blots and tubulin coding-sequence probes to demonstrate that tubulin transcript levels increase during somatic embryogenesis in cell cultures of

D. carota in conjunction with an increase in the number of cortical microtubules. Cortical microtubules are thought to play an important role in cell wall deposition during cell elongation in many plant systems (see discussions by Gunning and Hardham, 1982, and Lloyd, 1984). Several studies have documented changes in tubulin transcript levels that accompany increases or decreases in cell elongation rate. Bustos *et al.* (1989) and Colbert *et al.* (1990) used β tubulin coding-sequence probes and northern blot analyses to demonstrate that β tubulin transcript levels decrease upon exposure of etiolated seedlings of *Glycine max*, *Hordeum vulgare* and *Avena sativa* to red light. Their results correlated well with red-light induced decreases in the rates of cell division and cell elongation in mesocotyls and coleoptiles of monocot seedlings and in internodes of dicot seedlings. Mendu and Silflow (1991) have used RNA dot blots hybridized with α and β tubulin coding-sequence probes to demonstrate seven- to eight-fold increases in α and β tubulin transcript levels in excised *A. sativa* internode sections in response to treatment with GA_3. This increase was closely correlated with GA_3-induced cell elongation within the internode sections.

More detailed information about the programmes of plant tubulin gene expression have been obtained by using gene-specific hybridization probes to detect the transcripts of individual tubulin genes. These gene-specific probes are usually subclones containing 5' or 3' transcribed, but nontranslated, sequences of the individual genes. These noncoding sequences have been found to be highly divergent even in the case of tubulin genes that encode identical or nearly identical tubulin polypeptides (Montoliu *et al.*, 1990; Kopczak *et al.*, 1991; Snustad *et al.*, 1991; Villemur and Silflow, unpublished results). In contrast, the coding sequences of the tubulin genes are highly conserved even across species, such that tubulin coding-sequence probes will cross-hybridize with the transcripts of all the tubulin genes in the family (either all α tubulin gene transcripts or all β tubulin gene transcripts). For example, the β1 and β2 tubulin coding sequences of *Z. mays* exhibit 85% sequence

identity, and the β1 coding sequence of maize shares 76% identity with the coding sequence of the β1 tubulin gene of *A. thaliana* (Hussey *et al.*, 1990).

Such 5' and 3' noncoding gene-specific hybridization probes have been used to begin to investigate the developmentally regulated programmes of expression of all five α and eight of the nine β tubulin genes of *Arabidopsis thaliana* (Ludwig *et al.*, 1987; Silflow *et al.*, 1987; Oppenheimer *et al.*, 1988; Kopczak *et al.*, 1991; Snustad *et al.*, 1991), six α and two β tubulin genes of *Z. mays* (Hussey *et al.*, 1990; Joyce, 1990; Montoliu *et al.*, 1990) and two β tubulin genes of *G. max* (Han *et al.*, 1991). Of the 13 tubulin genes of *A. thaliana* studied to date, three were found to exhibit differential patterns of transcript accumulation in roots, leaves plus petioles, and flowers, whereas transcripts of the other 10 accumulated to significant levels in all tissues/organs examined. A ninth β tubulin gene (*TUB4*) of *A. thaliana* has been shown to be transcribed by S1 nuclease analysis (Marks *et al.*, 1987), but no information is available regarding its pattern of expression. Transcripts of *TUA1*, *TUB1* and *TUB5* genes were shown to accumulate predominantly in flowers, roots, and leaves plus petioles, respectively (Ludwig *et al.*, 1987; Silflow *et al.*, 1987; Oppenheimer *et al.*, 1988). When northern blots of RNAs isolated from parts of dissected flowers were hybridized to gene-specific probes, the *TUA1* transcript was detected only in the male flower tissues, i.e. stamens and pollen (Carpenter *et al.*, 1991). No *TUA1* transcript was detected in RNAs isolated from pistils, petals, or sepals by the northern blot hybridization procedure; however, *TUA1* expression can be detected in other tissues by more sensitive techniques (see below).

In *Z. mays*, both members of the tandem pair of α tubulin genes studied by Montoliu *et al.* (1990) were reported to be preferentially expressed in roots. The accumulation of the transcripts of four other maize α tubulin genes in root tip, root vascular cylinder, root cortex, immature cob, immature embryo, and pollen has been analysed by using dot blots quantitated by densitometry (Joyce, 1990). Although each of the four genes exhibited a unique pattern of transcript accumulation, the α 9.5 gene exhibited the most striking expression pattern with a large accumulation of its transcript in pollen. Similar studies on two β tubulin genes of maize (Hussey *et al.*, 1990; Joyce *et al.*, 1991) have shown that the β1 transcript accumulates predominantly in tissues that are undergoing rapid cell divisions, e.g. root tips, whereas the β2 transcript is present at relatively low levels in most tissues, being most abundant in unfertilized, immature cobs. The expression patterns of two β tubulin genes of *G. max* have been investigated by Han *et al.* (1991). The *sb*-2 gene appears to be constitutively expressed; its transcript was prevalent in all tissues examined. In contrast, the expression of *sb*-1 is developmentally regulated, with transcript levels nearly undetectable in young leaves, but increasing during leaf maturation. The highest levels of *sb*-1 transcript were present in hypocotyl tissue, particularly in elongating hypocotyls of etiolated seedlings.

Additional information about the programmes of tubulin gene expression have been obtained by studying the expression of chimeric genes in transgenic plants. Guiltanin *et al.* (1987b) demonstrated the expression of a chimeric soybean β tubulin gene in transgenic *Nicotiana tabacum* plants. More recently, Carpenter *et al.* (1991) have constructed fusion genes containing the 5' (putative promoter) regions of *A. thaliana* tubulin genes and the coding sequence of the β-glucuronidase (GUS) gene of *Escherichia coli* and introduced these into *Arabidopsis* plants by transformation with *Agrobacterium tumefaciens*. Histochemical assays of GUS activity in the transgenic plants provide sensitive indicators of the levels of expression of these fusion genes in transgenic plants (Jefferson *et al.*, 1986; Jefferson, 1987). GUS assays on organs/tissues of transgenic plants carrying 550 nucleotide-pairs of 5' sequence of the *TUA*1 gene fused to the GUS coding sequence (Fig. 2.2) support the results of northern blot studies in indicating that the *TUA*1 transcript is present at high levels in pollen and anthers (Carpenter *et al.*, 1991). No GUS activity has been detected in the pistils, sepals, or petals of these transgenic plants. Moreover, histochemical studies of developing anthers in transgenic plants showed that GUS activity is first evident in pollen after mitosis when both generative nuclei are present. The increased sensitivity of the GUS assays, as compared to northern blot studies, has permitted the detection of lower levels of putative *TUA*1 expression in other tissues. In particular, low levels (relative to the levels in pollen) of GUS activity were consistently observed in receptacles of young flower buds.

In addition to the transcriptional regulation of tubulin gene expression seen in both plant and animal systems, an unusual feedback regulation of tubulin synthesis at the transcriptional level has been documented in animal systems by Cleveland and co-workers. Increases in the pool of tubulin subunits caused by microtubule destabilizing drugs such as colchicine (Cleveland *et al.*, 1981) or by microinjection of tubulin subunits (Cleveland *et al.*, 1983) resulted in a decrease in the levels of tubulin transcripts. The mechanism of autoregulation appears to involve recognition of the nascent β tubulin amino-terminal tetrapeptide as it emerges from the ribosome and specific degradation of the polysome-associated tubulin mRNA (Yen *et al.*, 1988).

Recent reports of changes in tubulin gene expression in the presence of colchicine have indicated a striking difference between plant and animal cells. When excised oat internode sections or suspension culture cells were treated with colchicine to depolymerize microtubules, tubulin transcript levels showed an increase rather than the expected decrease (Mendu, 1990). Likewise, colchicine treatment of isolated *Zinnia* mesophyll cells in culture did not result in the expected inhibition of tubulin synthesis (Fukuda, 1989). In contrast to the results seen with colchicine, oat culture cells showed a decrease in tubulin transcript levels when treated with either of the anti-microtubule herbicides oryzalin or amiprophos-methyl (Mendu, 1990). These results suggest that plant cells have a tubulin autoregulatory mechanism, but that it cannot be detected using colchicine, possibly because the high concentrations of colchicine needed to depolymerize plant microtubules result in the specific degradation of tubulins (Fukuda, 1989).

(a)

(b)

Fig. 2.2. Histochemical assays of GUS activity in mature flowers of transgenic *Arabidopsis thaliana* plants harbouring a transgene composed of the GUS coding sequence (a) with no promoter and (b) fused to the promoter of the *TUA1* gene of *Arabidopsis*. (a) No blue indigo dye could be detected in flowers of these control plants, indicating that no GUS activity was present. (b) Blue colour and thus GUS activity were present only in the anthers and pollen (dark specks on surfaces of other flower parts) of mature flowers of plants carrying the *TUA1* promoter – GUS coding sequence fusion gene.

Multiple tubulin genes and function?

Microtubules serve a variety of cellular functions that mostly concern organellar movement, cell shape and cell motility. We have seen that both α and β tubulins are encoded by small gene families. This raised the possibility that microtubules performing different functions might be composed of distinct tubulin polypeptides – the 'multi-tubulin hypothesis' (Fulton and Simpson, 1976). A variety of biochemical, genetic, and cell biological approaches have been used in attempts to evaluate the validity of this hypothesis. Until very recently, the majority of

these studies revealed that many, if not all, tubulin isotypes were functionally interchangeable (Lewis *et al.*, 1987; Lopata and Cleveland, 1987). These observations argued against the multi-tubulin hypothesis and supported the view that the tubulin isotypes are all functionally equivalent. The alternative hypothesis as to why multiple tubulin genes are present in most eukaryotes is that these multiple genes have evolved to facilitate the complex patterns of regulatory fine tuning required to meet the differential requirements for tubulins during cell division and growth and morphogenesis of tissues and organs throughout development and in response to different environmental conditions (Raff, 1984). This latter hypothesis is compatible with the view that all tubulin isotypes are functionally equivalent.

Three recent observations have re-kindled interest in the multi-tubulin hypothesis. The first two observations relate to the identification of a subcellular sorting of tubulin isotypes in pheochromocytoma (PC12) cells (Asai and Remolona, 1989; Joshi and Cleveland, 1989) and in *Drosophila melanogaster* wing blade epidermal cells (Kimble *et al.*, 1989). The third observation was the failure of the *Drosophila* β3 tubulin gene to complement a β2 null mutation that results in a male sterile phenotype in homozygous flies (Hoyle and Raff, 1990). The significance of the first two observations is that they provide the first evidence for differential distribution of single tubulin gene products within a cell. The third observation implies that at least the *Drosophila* β3 and β2 isotypes are functionally dissimilar. Hoyle and Raff (1990) demonstrated that the developmentally regulated β3 isotype and the testis-specific β2 isotype of *Drosophila* are not functionally equivalent. A chimeric gene was constructed with the 5' regulatory sequences of the β2 gene fused to the coding and 3' sequences of the β3 gene. *Drosophila* transformed with this hybrid gene synthesized the β3 isotype with the normal β2 isotype pattern in postmitotic male germ cells. However, in the absence of the normal testis-specific β2 isotype (in β2^{null} homozygotes), the transgenic male flies were sterile. In these sterile males, the microtubules associated with mitochondrial derivative elongation formed and appeared to function normally, but axoneme assembly, meiosis, and nuclear shaping did not occur. When β3 was co-expressed with β2 in the male germ line, all classes of cytoplasmic microtubules functioned normally unless the β3 isotype exceeded 20% of the total β tubulin pool. When over 20% of the β tubulin in the testis was the β3 isotype, axoneme assembly was disrupted and doublet microtubules acquired certain morphological features typical of singlet microtubules.

Few experiments have been carried out using plants to test the multi-tubulin hypothesis.

Plants do, however, provide an excellent model system for studies of tubulin utilization in differentiating cells for two reasons:

(1) Plant cells assemble tubulins into one of at least four microtubule arrays; the interphase cortical microtubule array, the preprophase band, the spindle and the phragmoplast.
(2) Plant cells differentiate within cell files, this differentiation being dependent on the balance of cell division and cell elongation.

A root cell file, for example, will originate in the quiescent centre, pass through a region of rapid cell division where the three division related microtubule arrays will predominate, then a region of cell extension where cortical microtubules dominate and finally cell maturation with cessation of growth. From cell extension to maturation there are often cortical microtubule reorientations occurring, until a final transverse form is established.

An experiment has been carried out in which the tubulin isotype composition of one microtubule array was compared with that of whole cells which also contained unpolymerized tubulin in the soluble tubulin pool (Hussey *et al.*, 1987). Plant cells blocked in S-phase were extracted with detergent to produce stable cytoskeletons. These cytoskeletons contained only the interphase cortical microtubule array. The tubulin isotype composition of the interphase cortical array was compared with that of S-phase blocked intact cells (containing interphase cortical microtubules plus the soluble pool of tubulin) using 2-D gel electrophoresis and immunoblotting with anti-α and anti-β tubulin monoclonal antibodies. The results showed that the same tubulin isotypes were present in the cytoskeletal preparations and in intact cells, indicating that no differential usage of tubulin isotypes occurs in the formation of interphase microtubule arrays.

Another approach that is being used in an attempt to identify the functions of individual tubulin genes in higher plants is to investigate the effects of gene-specific anti-sense RNAs on the phenotypes of transgenic plants harbouring anti-sense fusion genes. The original idea was that the anti-sense transcript would anneal with the mRNA and block its translation, resulting in a mutant phenotype due to reduced expression of the gene. Although it is clear that the anti-sense RNA approach can be used to block or lower the level of expression of a given gene, the mechanism by which this effect occurs is still not established. Nevertheless, this approach is being employed in an attempt to gain further insight into the function(s) of the *TUA1* gene of *Arabidopsis*, which is expressed predominantly in anthers and pollen, based on data previously summarized. The preliminary results suggest that the low levels of expression of the *TUA1* gene observed in receptacles of flower buds, and perhaps in other parts of the plant, may have important functions. Regenerated transgenic plantlets carrying the gene-specific 'anti-sense' constructs exhibit a range of phenotypes:

(1) some are grossly deformed with only the formation of bulbous, fused leaf-like structures;
(2) others are smaller than wild-type and produce flower buds with undeveloped internal structures, and
(3) some produce normal-looking flowers, but no pollen (Kopczak, unpublished results).

A range of phenotypic effects of this type is expected since position effects, resulting from the variable chromosomal sites of transgene insertion, are known to yield a spectrum of effects from essentially no reduction to almost complete blockage of the expression of the sense gene (Delauney *et al.*, 1988). Possibly, the more normal-looking plantlets are those with minimal reduction in the levels of *TUA1* expression, whereas those with more extreme defects represent more complete blockage of *TUA1* expression. Alternatively, the effects of the anti-sense transgenes could result from perturbations in the total intracellular tubulin pools and be unrelated to any specific function of the divergent α1 tubulin isotype. Careful measurements of the levels of *TUA1* transcript and α1 tubulin in these transgenic plants must be available before any conclusions are feasible regarding the function of the *TUA1* gene.

Another approach to examine the function of tubulins produced by individual members of a gene family is to determine the distribution of isotypes within various microtubule arrays using immunofluorescence localization techniques. As mentioned above, the use of isotype-specific antibodies in localization studies has indicated that certain isotypes are utilized preferentially in specific microtubule arrays during differentiation processes in animal systems (Asai and Remolona, 1989; Joshi and Cleveland, 1989; Kimble *et al.*, 1989). The cloning and sequencing of plant tubulin genes in *Arabidopsis* and maize have made it possible to develop isotype-specific antibodies for use in immunolocalization studies in these plant systems. Based on predicted amino acid sequences at the divergent, carboxy termini of the tubulin isotypes, peptides have been synthesized and used as antigens to produce polyclonal antibodies for the three different maize α tubulin isotypes and the two different *Arabidopsis* α tubulin isotypes (Villemur, Mendu, Goddard, Wick, Snustad and Silflow, unpublished data). Specificity of the antibodies has been tested with a series of fusion protein constructs containing carboxy terminal sequences from all isotypes in each family. These antibodies are now being used to examine isotype distribution in plant microtubule arrays.

References

Asai, D.J. and Remolona, N.M. (1989). Tubulin isotype usage *in vivo*: A unique spatial distribution of the minor neuronal specific β tubulin isotype in pheochromocytoma cells. *Devel. Biol.* **132**, 398–409.

Barra, H.S., Arce, C.A., Rodriguez, J.A. and Caputto, R. (1974). Some common properties of the protein that incorporates tyrosine as a single unit into microtubule proteins. *Biochem. Biophys. Res. Commun.* **60**, 1384–1390.

Best, D., Warr, P.J. and Gull, K. (1981). Influence of the composition of commercial sodium dodecyl sulfate preparations on the separation of α and β tubulin during polyacrylamide gel electrophoresis. *Anal. Biochem.* **114**, 281–284.

Birkett, C.R., Foster, K.E., Johnson, L. and Gull, K. (1985). Use of monoclonal antibodies to analyse the expression of a multi-tubulin family. *FEBS Lett.* **187**, 211–218.

Blose, S.H., Meltzer, D.I. and Feramisco, J.R. (1984). 10 nm filaments are induced to collapse in living cells microinjected with monoclonal and polyclonal antibodies against tubulin. *J. Cell Biol.* **98**, 847–858.

Breitling, F. and Little, M. (1986). Carboxy-terminal regions on the surface of tubulin and microtubules. Epitope locations of YOL 1/34, DM1A and DM1B. *J. Mol. Biol.* **189**, 367–370.

Burland, T.G., Schedl, T., Gull, K. and Dove, W.F. (1984). Genetic analysis of resistance to benzimidazoles in *Physarum*: Differential expression of β-tubulin genes. *Genetics* **108**, 123–141.

Bustos, M.M., Guiltinan, M.J., Cyr, R.J., Ahdoot, D. and Fosket, D.E. (1989). Light regulation of β-tubulin gene expression during internode development in soybean (*Glycine max* [L.] Merr.). *Plant Physiol.* **91**, 1157–1161.

Carpenter, J., Ploense, S.E., Snustad, D.P. and Silflow, C.D. (1991). Preferential expression of the α1-tubulin gene of *Arabidopsis thaliana* in pollen. In preparation.

Cerff, R., Hundrieser, J. and Friedrich, R. (1986). Subunit B of chloroplast glyceraldehyde-3-phosphate dehydrogenase is related to β tubulin. *Mol. Gen. Genet.* **204**, 44–51.

Cleveland, D.W. and Sullivan, K.F. (1985). Molecular biology and genetics of tubulin. *Annu. Rev. Biochem.* **54**, 331–365.

Cleveland, D.W., Lopata, M.A., Sherline, P. and Kirschner, M.W. (1981). Unpolymerized tubulin modulates the level of tubulin mRNAs. *Cell* **25**, 537–546.

Cleveland, D.W., Pittenger, M.F. and Feramisco, J.R. (1983). Elevation of tubulin levels by micro-injection suppresses new tubulin synthesis. *Nature* **305**, 738–740.

Colbert, J.T., Costigan, S.A. and Zhao, Z. (1990). Photoregulation of β-tubulin mRNA abundance in etiolated oat and barley seedlings. *Plant Physiol.* **93**, 1196–1202.

Cyr, R.J., Bustos, M.M., Guiltinan, M.J. and Fosket, D.E. (1987). Developmental modulation of tubulin protein and mRNA levels during somatic embryogenesis in cultured carrot cells. *Planta* **171**, 365–376.

Dawson, P.J. and Lloyd, C.W. (1985). Identification of multiple tubulins in taxol microtubules purified from carrot suspension cells. *EMBO J.* **4**, 2451–2455.

Delauney, A.J., Tabaeizadeh, Z. and Verma, D.P.S. (1988). A stable bifunctional antisense transcript inhibiting gene expression in transgenic plants. *Proc. Natl Acad. Sci. U.S.A.* **85**, 4300–4304.

Fosket, D.E. (1989). Cytoskeletal proteins and their genes in higher plants. In *The Biochemistry of Plants* (ed. A. Marcus), vol. 15, pp. 393–454. Academic Press, London.

Fukuda, H. (1989). Regulation of tubulin degradation in isolated *Zinnea* mesophyll cells in culture. *Plant Cell Physiol.* **30**, 243–252.

Fulton, C. and Simpson, P.A. (1976). Selective synthesis and utilization of flagellar tubulin. The multitubulin hypothesis. In *Cell Motility* (eds R. Goldman, T. Pollard and J. Rosenbaum), pp. 987–1005. Cold Spring Harbor Laboratory, Cold Spring Harbor, New York.

Gard, D.L. and Kirschner, M.W. (1985). A polymer dependent increase in phosphorylation of β tubulin accompanies differentiation of a mouse neuroblastoma cell line. *J. Cell Biol.* **100**, 764–774.

Goldenring, J.R., Gonzalez, B., McGuire, J.S. and DeLorenzo, R.J. (1982). Purification and characterization of a calmodulin-dependent kinase from rat brain cytosol able to phosphorylate tubulin and microtubule-associated proteins. *J. Biol. Chem.* **258**, 12632–12640.

Guiltinan, M.J., Ma, D.P., Barker, R.F., Bustos, M.M., Cyr, R.J., Yadegari, R. *et al.* (1987a). The isolation, characterization and sequence of two divergent β tubulin genes from soybean (*Glycine max* L.). *Plant Mol. Biol.* **10**, 171–184.

Guiltinan, M.J., Velten, J., Bustos, M.M., Cyr, R.J., Schell, J. and Fosket, D.E. (1987b). The expression of a chimeric soybean beta-tubulin gene in tobacco. *Mol. Gen. Genet.* **207**, 328–334.

Gull, K., Hussey, P.J., Sasse, R., Schneider, A., Seebeck, T. and Sherwin, T. (1986). Tubulin isotypes: Generation of diversity in cells and microtubular organelles. In *The Cytoskeleton Cell Function and Organization* (eds C.W. Lloyd, J.S. Hyams and R.M. Warn), *J. Cell. Sci.* **Suppl. 5**.

Gunderson, G.G., Khawaja, S. and Bulinski, J.C. (1987). Post polymerization detyrosination of alpha tubulin: a mechanism for subcellular differentiation of microtubules. *J. Cell Biol.* **106**, 251–264.

Gunning, B.E.S. and Hardham, A.R. (1982). Microtubules. *Ann. Rev. Plant Physiol.* **33**, 651–691.

Han, I.S., Jongewaard, I. and Fosket, D.E. (1991). Limited expression of a diverged beta-tubulin gene during soybean (*Glycine max* [L.] Merr.) development. *Plant Physiol.* In press.

Hoyle, H.D. and Raff, E.C. (1990). Two *Drosophila* beta tubulin isoforms are not functionally equivalent. *J. Cell Biol.* **111**, 1009–1026.

Hussey, P.J. (1986). Studies on the molecular and cell biology of plant tubulin. PhD Thesis, University of Kent at Canterbury.

Hussey, P.J. and Gull, K. (1985). Multiple isotypes of α and β tubulin in the plant *Phaseolus vulgaris*. *FEBS Lett.* **181**, 113–118.

Hussey, P.J., Traas, J.A., Gull, K. and Lloyd, C.W. (1987). Isolation of cytoskeletons from synchronized plant cells: the interphase microtubule array utilises multiple tubulin isotypes. *J. Cell Sci.* **88**, 225–230.

Hussey, P.J., Lloyd, C.W. and Gull, K. (1988). Differential and developmental expression of β tubulins in a higher plant. *J. Biol. Chem.* **263**, 5474–5479.

Hussey, P.J., Haas, N., Hunsperger, J., Larkin, J., Snustad, D.P. and Silflow, C.D. (1990). The β tubulin gene family in *Zea mays*: two differentially expressed β tubulin genes. *Plant Mol. Biol.* **15**, 957–972.

Jefferson, R.A. (1987). Assaying chimeric genes in plants: The GUS gene fusion system. *Plant Mol. Biol. Rep.* **5**, 387–405.

Jefferson, R.A., Burgess, S.M. and Hirsh, D. (1986). β-Glucuronidase from *Escherichia coli* as a gene fusion marker. *Proc. Natl Acad. Sci. U.S.A.* **83**, 8447–8451.

Joshi, H.C. and Cleveland, D.W. (1989). Differential utilization of β tubulin isotypes in differentiating neurites. *J. Cell Biol.* **109**, 663–673.

Joyce, C.M. (1990). Characterization of alpha-tubulin and beta-tubulin isotype expression in *Zea mays*. PhD Thesis, University of Minnesota.

Joyce, C.M., Hussey, P.J., Haas, N., Snustad, D.P. and Silflow, C.D. (1991). Tubulin isotype expression in the maize seedling root. In preparation.

Kerr, G.P. (1988). Studies of tubulin and microtubules of rye roots with emphasis on effects of low temperature. PhD Thesis, University of Minnesota.

Kerr, G.P. and Carter, J.V. (1990). Tubulin isotypes in rye roots are altered during cold acclimation. *Plant Physiol.* **93**, 83–88.

Kilmartin, J.V., Wright, B. and Milstein, C. (1982). Rat monoclonal antitubulin antibodies derived by using a new non-secreting rat cell line. *J. Cell Biol.* **93**, 576–582.

Kimble, M., Incardona, J.P. and Raff, E.C.D. (1989). A variant β tubulin isoform of *Drosophila melanogaster* is expressed primarily in tissues of mesodermal origin in embryos and pupae, and is utilized in populations of transient microtubules. *Devel. Biol.* **131**, 415–429.

Kopczak, S.D., Silflow, C.D. and Snustad, D.P. (1991). The small genome of *Arabidopsis thaliana* contains five expressed α-tubulin genes. In preparation.

Kumar, N. and Flavin, M. (1981). Preferential action of a brain detyrosinylating carboxypeptidase on polymerised tubulin. *J. Biol. Chem.* **256**, 7678–7686.

LeDizet, M. and Piperno, G. (1987). Identification of an acetylation site of *Chlamydomonas* alpha-tubulin. *Proc. Natl Acad. Sci. U.S.A.* **84**, 5720–5724.

Lee, M.G.S., Lewis, S.A., Wilde, C.D. and Cowan, N.J. (1983). Evolutionary history of a multigene family: an expressed human β tubulin gene and three processed pseudogenes. *Cell* **33**, 477–487.

Lewis, S.A., Gu, W. and Cowan, N.J. (1987). Free intermingling of mammalian β-tubulin isotypes among functionally distinct microtubules. *Cell* **49**, 539–548.

L'Hernault, S.W. and Rosenbaum, J.L. (1985). *Chlamydomonas* α tubulin is post-translationally modified by acetylation on the ε-amino group of a lysine. *Biochemistry* **24**, 473–478.

Little, M. (1985). An evaluation of tubulin as a molecular clock. *Biosystems* **18**, 241–247.

Lloyd, C.W. (1984). Toward a dynamic helical model for the influence of microtubules on wall patterns in plants. *Int. Rev. Cytol.* **86**, 1–51.

Lopata, M.A. and Cleveland, D.W. (1987). *In vivo* microtubules are copolymers of available β-tubulin isotypes: Localization of each of six vertebrate β-tubulin isotypes using polyclonal antibodies elicited by synthetic peptide antigens. *J. Cell Biol.* **105**, 1707–1720.

Ludwig, S.R., Oppenheimer, D.G., Silflow, C.D. and Snustad, D.P. (1987). Characterisation of the α tubulin gene family of *Arabidopsis thaliana*. *Proc. Natl Acad. Sci. U.S.A.* **84**, 5833–5837.

Marks, M.D., West, J. and Weeks, D.P. (1987). The relatively large β tubulin gene family of *Arabidopsis* contains a member with an unusual transcribed 5′ noncoding sequence. *Plant Mol. Biol.* **10**, 91–104.

Mendu, N. (1990). Oat tubulin genes and their expression during internode elongation. PhD Thesis, University of Minnesota.

Mendu, N. and Silflow, C.D. (1991). Elevated levels of tubulin transcripts accompany the GA$_3$-induced elongation of oat internode segments. *Plant Physiol.* In press.

Mizuno, K., Sek, F., Perkin, J., Wick, S., Duniec, J. and Gunning, B. (1985). Monoclonal antibodies specific to plant tubulin. *Protoplasma* **129**, 100–108.

Montoliu, L., Rigau, J. and Puigdomenech, P. (1990). A tandem of α-tubulin genes preferentially expressed in radicular tissues from *Zea mays*. *Plant Mol. Biol.* **14**, 1–15.

Morejohn, L.C. and Fosket, D.E. (1986). Tubulin from plants, fungi and protists. In *Cell and Molecular Biology of the Cytoskeleton* (ed. J.W. Shay), pp. 2257–3229. Plenum, New York.

Oppenheimer, D.G., Haas, N., Silflow, C.D. and Snustad, D.P. (1988). The β tubulin gene family of *Arabidopsis thaliana*: preferential accumulation of the β1 transcript in roots. *Gene* **63**, 87–102.

Piperno, G. and Fuller, M.T. (1985). Monoclonal antibodies specific for an acetylated form of α tubulin recognise the antigen in cilia and flagella from a variety of organisms. *J. Cell Biol.* **101**, 2085–2094.

Piperno, G., LeDizet, M. and Change, X. (1987). Microtubules containing acetylated α tubulin in mammalian cells in culture. *J. Cell Biol.* **104**, 289–302.

Piras, R. and Piras, M.M. (1975). Changes in microtubule phosphorylation during cell cycle of HeLa cells. *Proc. Natl Acad. Sci.* **72**, 1161–1165.

Raff, E.C. (1984). Genetics of microtubule function. *J. Cell Biol.* **99**, 1–10.

Sasse, R., Glyn, M.C.P., Birkett, C.R. and Gull, K. (1987). Acetylated α tubulin in *Physarum*: immunological characterization of the isotype and its usage in particular microtubular organelles. *J. Cell Biol.* **104**, 41–49.

Schatten, G., Simerly, C., Asai, D.J., Szoke, E., Cooke, P. and Schatten, H. (1988). Acetylated α tubulin in microtubules during mouse fertilization and early development. *Devel. Biol.* **130**, 74–86.

Schulze, E., Asai, D.J., Bulinski, J.C. and Kirschner, H.W. (1987). Post translational modification and microtubule stability. *J. Cell Biol.* **105**, 2167–2177.

Sherwin, T. and Gull, K. (1989). Visualization of detyrosination along single microtubules reveals novel mechanisms of assembly during cytoskeletal duplication in trypanosomes. *Cell* **57**, 211–221.

Silflow, C.D. and Rosenbaum, J.L. (1981). Multiple α- and β-tubulin genes in *Chlamydomonas* and regulation of tubulin mRNA levels after deflagellation. *Cell* **24**, 81–88.

Silflow, C.D., Oppenheimer, D.G., Kopczak, S.D., Ploense, S.E., Ludwig, S.R., Haas, N. and Snustad, D.P. (1987). Plant tubulin genes: structure and differential expression during development. *Dev. Genet.* **8**, 435–460.

Snustad, D.P., Kopczak, S.D., Haas, N.A. and Silflow, C.D. (1991). The small genome of *Arabidopsis thaliana* contains at least nine expressed β-tubulin genes. In preparation.

Wang, D., Villasante, A., Lewis, S.A. and Cowan, N.J. (1986). The mammalian β tubulin gene repertoire: hematopoietic expression of a novel heterologous β tubulin isotype. *J. Cell Biol.* **103**, 1903–1909.

Yen, T.J., Machlin, P.S. and Cleveland, D.W. (1988). Autoregulated instability of β-tubulin mRNAs recognition of the nascent amino terminus of β-tubulin. *Nature* **334**, 580–585.

3. THE MOLECULAR PHARMACOLOGY OF PLANT TUBULIN AND MICROTUBULES

Louis C. Morejohn

Department of Botany, University of Texas at Austin,
Austin, TX 78713, USA

Preface

Microtubules are dynamic filamentous poly-mers composed mainly of tubulin, a hetero-dimeric protein with similar α- and β-subunits (MW = 50000). Microtubules are essential structures for the formation and functions of several cytoskeletal arrays in plant cells, including the preprophase band, the mitotic and meiotic spindles, the phragmoplast, and the interphase and differentiation arrays. Treatment of plant cells or organs with drugs that cause the specific loss of microtubules abolishes the normal polar distribution of subcellular compartments and biochemical processes determined by particular cytoskeletal arrays. A most dramatic example of this can be seen in anti-microtubule drug-treated roots where the loss of microtubules ultimately results in isodiametric cell shapes within bulbous 'tumour roots'. Thus, anti-microtubule drugs have played a large part in understanding the functions of the microtubule cytoskeleton in morphogenetic programmes of oriented cell division and differentiation. Because anti-microtubule drugs remain powerful probes of microtubule function, the molecular basis of their interactions with tubulin is discussed here. The most commonly used compounds causing the loss of microtubules include colchicine and a variety of commercially derived herbicidal compounds belonging to at least three different chemical classes: these include the dinitroaniline, phosphoric amide, and N-phenyl carbamate herbicides. Taxol, a secondary plant product, is the only drug found to promote the polymerization of micro-tubules. This chapter summarizes many salient features of anti-microtubule drug action, and focuses on recent biochemical advances in plant systems using anti-microtubule herbicides and taxol. For comprehensive reviews on the biochemistry and molecular biology of plant tubulin and microtubules the reader is referred to Morejohn and Fosket (1986), Dawson and Lloyd (1987), Silflow *et al.* (1987) and Fosket (1989). Early information on anti-microtubule herbicides may be obtained in reviews by Parka and Soper (1977), Ashton and Crafts (1981) and Fedtke (1982).

Biochemistry of plant microtubules

The acquisition of biochemical information on the plant cytoskeleton has lagged significantly behind that on the animal cytoskeleton for a number of years (Fosket, 1989). However, recent technological advances have provided the intrepid worker with the means to isolate, purify and characterize cytoskeletal proteins from plants successfully. For example, effective inhibitors of potent plant proteases are com-mercially available, rapidly growing tissue cultures provide vast quantities of plant material for protein purification, and micro-volume, analytical ultra-centrifuges facilitate rapid, quantitative polymer sedimentation. Indeed, plant biologists now have few stumbling blocks left in the path towards understanding the structure and function of microtubules. Information available on tubulin

The Cytoskeletal Basis of Plant Growth and Form   ISBN 0–12–453770–7

biochemistry, isolation and polymerization methods, and microtubule dynamics is summarized briefly in this section.

Structure and composition of microtubules
Microtubules are filamentous, protein polymers formed as hollow rods 24 nm in diameter and up to several micrometres in length. Tubulin, a dumbbell-shaped heterodimer, is arranged head-to-tail in 4–5 nm diameter linear proto-filaments, 13 of which interact laterally with a 10° helical pitch to compose the microtubule wall. Sequencing studies have shown that each tubulin subunit has a molecular weight of approximately 50 000, and that tubulins from plants and animals are 79–87% identical (Silflow et al., 1987). Although tubulin α- and β-subunits have been highly conserved over the course of eukaryotic evolution, plant tubulin is immunologically and pharmacologically different from tubulins of animals, fungi, and slime moulds (Morejohn and Fosket, 1986). The α-subunits of plants have particularly distinct peptide mapping patterns and electrophoretic mobilities (Morejohn and Fosket, 1982; Hussey and Gull, 1985). Both the α- and β-subunits of plant tubulin exhibit electrophoretic heterogeneity (isoforms) derived from combinations of polypeptides having slightly different sequences (isotypes), and presumably, post-translational modifications of these isotypes. Virtually no information has been reported on the post-translational modification of plant tubulin. Tubulin isotypes are encoded by small multi-gene families and are differentially expressed in plant organs (Fosket, 1989). Although tubulin dimer comprises most of the microtubule mass, 15–25% of the polymer is composed of assorted microtubule-associated proteins (MAPs), usually of high molecular weight, which provide microtubules with a wide variety of molecular functions.

Isolation and polymerization of tubulin
Purification of microtubule proteins from animal tissues has traditionally made use of the intrinsic properties of tubulin to polymerize spontaneously at warm temperatures and to depolymerize at cold temperatures. Typically, two or three cycles of GTP-dependent poly-merization and depolymerization, coupled with differential centrifugation, are necessary to purify microtubules. This method yields tubulin and various co-polymerizing MAPs. Because most animal cells possess an extensive microtubule cytoskeleton, supernatants contain tubulin levels exceeding the critical dimer concentration (concentration below which no microtubule polymerization occurs), and micro-tubule polymerization proceeds readily. Okamura (1984) used this method to isolate microtubule-like polymers from cultured cells of carrot, but the method has had limited value in plant systems because plant tissues generally are not rich in microtubule proteins, and supernatants can possess factors inhibitory to microtubule polymerization.

Typically, investigators isolate dimeric tubulin from supernatants by anion-exchange chromatography alone (Morejohn and Fosket, 1982; Morejohn et al., 1984), or in combination with herbicide-affinity chromatography (Mizuno et al., 1981). Isolated dimer can be polymerized into microtubules by addition of assembly-enhancing solution components such as glycerol (Morejohn and Fosket, 1982), dimethyl sulphoxide (Mizuno, 1985), or taxol (Morejohn and Fosket, 1984a). The micro-tubules obtained by these methods do not contain MAPs. Because taxol-induced assembly yields much more polymer than other means of polymerization, many more assembly experiments may be performed with the relatively small quantities of tubulin dimer obtained from plant materials (see below).

Dynamic properties of microtubules
Studies on animal systems have documented that microtubules undergo two different types of kinetic reaction, treadmilling and dynamic instability. When microtubules are polymerized in vitro (without nucleation by microtubule organizing centres (MTOCs)), they engage in an endwise mechanism of treadmilling (Margolis and Wilson, 1978). Treadmilling occurs when dimers preferentially are added to one end (plus end) of the microtubule and are lost from the other end (minus end), resulting in a GTP-driven net flux of dimers through the polymer from the plus to minus end (reviewed by

Margolis and Wilson, 1981). Transient tread-milling has been observed *in vivo* in metaphase kinetochore microtubules, with dimer incorporation at the kinetochore microtubule (plus) ends and dimer loss at the spindle pole microtubule (minus) ends (Mitchison *et al.*, 1986; Mitchison, 1989), although the function of treadmilling in the spindle is not understood (Mitchison and Swain, 1990).

When microtubules are nucleated *in vitro* from MTOCs, microtubules are anchored at their minus ends, and elongate at their plus ends. These microtubules exhibit dynamic instability, wherein a sub-population of microtubules growing slowly at their plus ends co-exists with another sub-population of microtubules shrinking rapidly from their plus ends (Mitchison and Kirschner, 1984a, b; Horio and Hotani, 1986). The mechanism regulating the transition between growing and shrinking phases is not known, but models have been proposed which include a fluctuating tubulin-GTP 'cap' (Chen and Hill, 1985), and a lateral tubulin-GTP 'cap' (Bayley *et al.*, 1989).

Although microtubules assembled from pure tubulin dimer undergo dynamic instability, it is suppressed and treadmilling predominates in microtubules having MAPs (Hotani and Horio, 1988). Job *et al.* (1985) reported that differential MAP distribution along microtubules may produce different stability 'subclasses' of microtubules, and that phosphorylation of MAPs reverses the MAP-dependent stabilization of microtubules. In cultured animal cells dynamic instability has been observed in interphase microtubules (Schulze and Kirschner, 1986; Cassimeris *et al.*, 1988). When growing microtubules presumably are stabilized either by interaction with MAPs, or by 'capture' of the plus ends, they remain polymerized at least transiently. Dynamic instability may provide the cell with continuous opportunities for cytoskeletal rearrangement, depending upon the need for microtubule-dependent activities at particular times during the cell cycle, or differentiation. Unfortunately, neither tread-milling, nor dynamic instability has been studied with plant microtubules *in vivo*, or *in vitro*.

Microtubule polymerization and stabilization by taxol

The only drug known to cause the polymerization and stabilization of microtubules is taxol, a complex taxane alkaloid obtained from plants within the genus *Taxus*. Taxol was identified as a antileukaemic agent by Wani *et al.* (1971), and has been under clinical investigation as a chemotherapeutic drug for the treatment of ovarian cancer. The structure of taxol is shown in Fig. 3.1.

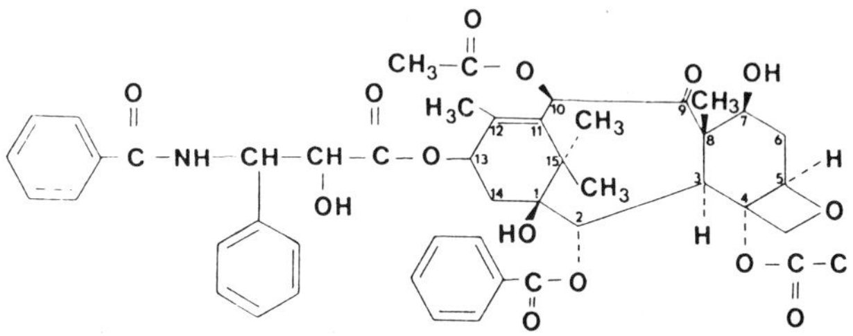

Fig. 3.1. Chemical structure of taxol.

Effects of taxol in cells

Treatment of cells with low micromolar concentrations of taxol results in the appearance of new microtubules, and the stabilization of previously existing microtubules at all stages of the cell cycle. These microtubule promotion effects have been observed in cells of phylogenetically diverse organisms including animals (Schiff and Horwitz, 1981; De Brabander *et al.*, 1981), higher plants (Bajer *et al.*, 1982), algae (Herth, 1983), protoctists (Baum *et al.*, 1981), and slime moulds (Wright *et al.*, 1982). The consistent microtubule stabilization effects of taxol among diverse species suggests that the taxol-binding site on tubulin has been highly conserved during evolution. Some noteworthy examples of the use of taxol in studies of microtubule function in plant cells are outlined below.

Taxol treatment of dividing endosperm cells of *Haemanthus* results in the formation of abundant new microtubules, and increases lateral associations of microtubules within the mitotic spindle (Bajer *et al.*, 1982; Molé-Bajer and Bajer, 1983). Taxol causes the spindle to become reorganized and mitosis is slowed.

Taxol promotes excessive polymerization of anaphase kinetochore microtubules, which stops poleward migration of chromosomes, and reverses their migration toward the equatorial plane. Taxol also stabilizes *Haemanthus* microtubules against depolymerization by cold treatment. The effects of taxol treatments are reversible in a time- and concentration-dependent manner (Molé-Bajer and Bajer, 1983).

Taxol may be used to polymerize microtubules in cells that have reduced numbers of microtubules. Hahne and Hoffmann (1984) found that nondividing *Hibiscus* protoplasts, which lose microtubules during cell wall digestion, contain microtubules after taxol treatment; the protoplasts then divide. Nondividing, colchicine-treated protoplasts of *Nicotiana* can be induced to divide after addition of taxol to the medium. Taxol treatment overcomes the colchicine inhibition, and induces the polymerization of microtubules.

The role of cortical microtubule reorientation in the control of xylem differentiation was studied in taxol-treated cultures of *Zinnia* (Falconer and Seagull, 1985). Stabilization of microtubules prevents the normal transition from longitudinal to transverse orientations of the cortical array. Secondary wall deposition occurs parallel to the stabilized microtubule array, demonstrating that microtubule reorientation is necessary for the normal transverse deposition of secondary wall during differentiation. Melan (1990) also found a strong correlation between taxol-induced microtubule stabilization and the orientation of cellulose microfibril deposition in regenerating primary wall of *Pisum* protoplasts.

Molecular interactions of taxol and microtubules

Most of our knowledge on the molecular action of taxol is based upon several studies using animal tubulin, and only a few studies with plant tubulin. Taxol promotes the *in vitro* polymerization of tubulins isolated from plants (Morejohn and Fosket, 1982) and animals (Schiff *et al.*, 1979). Taxol lowers the critical dimer concentration for tubulin polymerization, and promotes assembly in the absence of

microtubule-associated proteins (MAPs) and GTP. Animal microtubules polymerized with taxol are resistant to depolymerization by either anti-microtubule drugs, calcium, or cold, and taxol does not compete with MAPs in binding to microtubules (Schiff *et al.*, 1979; Kumar, 1981; Schiff and Horwitz, 1981; Parness and Horwitz, 1981). Although maximum polymerization of brain tubulin is obtained using equal concentrations of taxol and tubulin (Kumar, 1981), maximum polymerization of rose tubulin requires a 2:1 molar ratio of taxol to tubulin dimer (Morejohn and Fosket, 1984a). This slight difference in taxol-induced polymerization characteristics of tubulins from plants and animals supports the notion that taxol-binding sites on microtubules are highly conserved, and that taxol can be used effectively in plant microtubule biochemistry.

Kumar (1981) reported that treadmilling rates of taxol-stabilized microtubules are five-fold slower than microtubules stabilized by MAPs alone. Caplow and Zeeberg (1982) also found that taxol slows treadmilling, and proposed that it stabilizes the polymer by decreasing the dimer dissociation rate constants at both microtubule ends. Taxol probably suppresses dynamic instability as well. Collins and Vallee (1987) demonstrated that taxol-stabilized animal microtubules could be depolymerized by a combined treatment of cold and calcium, and that, if calcium is removed by chelation with EGTA, tubulin will repolymerize into microtubules upon warming.

Parness and Horwitz (1981) found that [^{3}H]-taxol binds to microtubules, but not to unpolymerized dimer, and reported a dissociation constant (K_d) of 8.7×10^{-7} M at 37°C. The estimated stoichiometry of binding is 0.6, indicating that most tubulin dimers in microtubules have an available taxol-binding site. Taxol binding to microtubules in animal cells demonstrated a single affinity-class of receptors and a similar intracellular dissociation constant ($K_d = 2 \times 10^{-7}$ M) (Manfredi *et al.*, 1982). These binding data identified microtubules as the only target of taxol binding in animal cells. The plant tubulin–taxol binding reaction and its specificity have not been examined. Because taxol is a plant secondary product, the possibility of metabolic inactivation of taxol by secondary pathway enzymes seems very real, depending upon the particular plant used. Clearly, more work will be necessary to characterize the plant microtubule/taxol interaction at the molecular level.

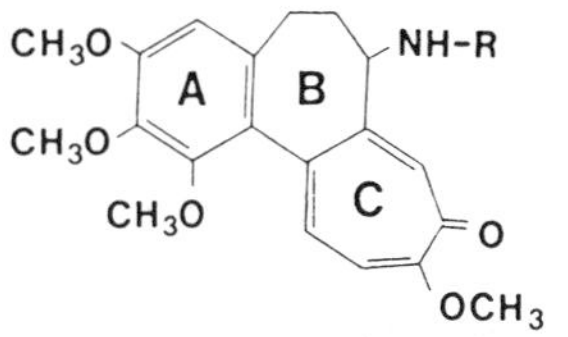

Microtubule 'depolymerizing' compounds

To date, the precise molecular mechanisms by which anti-microtubule drugs cause the disappearance of microtubules in cells are not understood, even in the case of the model interaction of animal tubulin and colchicine. Quite often it has been assumed that such drugs act simply and directly in cells to depolymerize existing microtubules. However, biochemical studies on purified animal microtubules and colchicine indicate that the *in vivo* situation is much more complicated. For example, treatment of mammalian cells with high nanomolar concentrations of colchicine results in the complete loss of microtubules within a few minutes, but purified microtubules are *not* depolymerized effectively *in vitro* by micromolar levels of colchicine (Lambeir and Engelborghs, 1980; Bergen and Borisy, 1983). Then how is it that microtubules in cells disappear following anti-microtubule drug treatment? Parts of the answer may be found by examining the currently available information on the molecular interaction of animal tubulin with colchicine.

Colchicine

Much of our understanding of anti-microtubule drug action has come from biochemical studies on the binding of colchicine to tubulin purified from mammalian cells. Colchicine is a highly soluble, plant alkaloid produced by the meadow saffron (*Colchicum autumnale* L.); the structures of colchicine and colcemid, one of its active analogues, are given in Fig. 3.2.

Molecular interactions of colchicine and animal tubulin
Colchicine binds non-covalently to a single high-affinity site on the unpolymerized animal tubulin dimer to form a tubulin–colchicine

Fig. 3.2. Chemical structure of colchicine and colcemid: R = $COCH_3$ for colchicine, R = CH_3 for colcemid.

complex. The binding proceeds via slow conformation changes in both the colchicine molecule and the tubulin dimer, a process that occurs more rapidly at elevated temperatures. Typically, affinity constants for the animal tubulin/colchicine interaction are in the range 1–40×10^6 M^{-1}, depending upon the method of analysis and the source of animal tubulin (e.g. Hains *et al.*, 1978). Another class of low-affinity colchicine-binding sites on the animal dimer has been observed as well (e.g. Ringel and Sternlicht, 1984; Ray *et al.*, 1984), but there may be little functional significance to this affinity-class at low colchicine concentrations. The high-affinity colchicine-binding site decays with first-order kinetics. Solution components such as GTP, dithiothreitol, organic acids, sucrose, and vinblastine will increase the half-life of the colchicine-binding site.

Colchicine does not bind to polymerized tubulin in microtubules (Wilson and Mesa, 1973). However, when colchicine is added at substoichiometric concentrations to soluble tubulin, complete inhibition of microtubule polymerization occurs. Colchicine binds first to unassembled dimeric tubulin to form a tubulin–drug complex, and this complex blocks the nucleation of new microtubules by unliganded tubulin. When a substoichiometric concentration of colchicine is added to a solution of steady state microtubules, the complex is proposed to bind to both the fast growing (plus) ends and slow growing (minus) ends of microtubules, which inhibits further elongation growth of microtubules at both ends (Bergen and Borisy, 1983). The mass of microtubules in such solutions rapidly decreases to a new steady state level, but for reasons that are not clear, many microtubules remain polymerized (Lambeir and Engelborghs, 1980). If disassembly events were to proceed unabated, a new steady state having no microtubules is predicted, but this does not happen. It has been proposed that the tubulin–colchicine complex co-polymerizes with unliganded tubulin to form co-polymers that have a lowered affinity for unliganded tubulin (Sternlicht *et al.*, 1983). Perhaps these co-polymers form because *in vivo* regulatory mechanisms controlling the kinetics of microtubule ends are absent.

The interpretation of many published biochemical experiments on colchicine and microtubules has become quite difficult within the context of microtubule treadmilling and dynamic instability. It is not known whether the tubulin–colchicine complex binds differentially to the plus end on growing or shrinking microtubules, or to either end on treadmilling microtubules within metaphase kinetochore bundles. These possibilities would be very difficult to test. The answer may come eventually from an understanding of the mechanisms controlling the transitions between growing and shrinking phases of dynamic instability, and the treadmilling of dimers through microtubules (reviewed by Bayley, 1990).

Molecular interactions of colchicine and plant tubulin

The effects of colchicine on plant microtubules vary greatly depending upon the type of cell, organ, or species examined. Usually millimolar concentrations of colchicine are required for the complete inhibition of microtubule-dependent processes in plants. The single most likely reason for the resistance of plant microtubules to colchicine is becoming apparent: the plant tubulin dimer has a low affinity for colchicine.

Plant tubulin has been isolated from a range of cultured plant cells, and used in colchicine-binding experiments. Colchicine-binding experiments with tubulins from cultured cells of *Rosa*, *Daucus*, and *Hibiscus* revealed the binding activity of each tubulin to be not only different from each other, but also much lower than that of purified bovine brain tubulin (Morejohn *et al.*, 1984). The effects of colchicine on the taxol-induced polymerization of micro-

tubules from rose and brain were compared (Morejohn and Fosket, 1984a; Morejohn *et al.*, 1987a). Inhibition of rose microtubule polymerization requires a large stoichiometric excess of colchicine, and the assembly reaction is approximately 160-fold more resistant to colchicine than brain microtubule assembly. Comparative [^{3}H]colchicine-binding experiments with these tubulins gave very different affinity constants, $K = 9.7 \times 10^3$ M^{-1} for rose tubulin and $K = 2.46 \times 10^6$ M^{-1} for brain tubulin at 24°C. Bound colchicine does not readily dissociate from the plant or animal dimer, and estimated maximum binding stoichiometries (r) are very similar for rose (0.47) and brain (0.45) tubulins. The only other colchicine-binding study on relatively pure plant tubulin is that of Kato *et al.* (1985). They found that tubulin from cultured cells of carrot binds colchicine with moderate affinity ($K = 5.7 \times 10^5$ M^{-1}), but only at a non-physiological temperature (37°C), and in the presence of a very high concentration (1.5 M) of sodium tartrate. The estimated binding stoichiometry is very low ($r = 0.18$).

Cumulatively, the results outlined above demonstrate that plant tubulins generally are low-affinity intracellular receptors for colchicine. Colchicine concentrations in the millimolar range can inhibit physiological processes that apparently are unrelated to microtubule function (Nooden, 1971; Sloan and Camper, 1981). Several workers have observed that plant cells may accumulate enormous tubulin aggregates, or paracrystals following colchicine treatment (Apostolakos *et al.*, 1990 and references therein). Neither the composition, nor the significance of such tubulin structures, is known. Because colchicine is a plant product, the possible metabolic inactivation of colchicine by secondary pathway enzymes in treated plants is another cause for concern. In general, colchicine does not seem to be a sensitive pharmacological probe of plant microtubule function, particularly in comparison with the potent anti-microtubule activities of the dinitroaniline and phosphoric amide herbicides described below.

Dinitroaniline herbicides

The dinitroaniline herbicides are agriculturally important chemicals commonly used to control the emergence of annual grasses and certain broadleaf weeds in a wide variety of crops. The dinitroaniline compounds most often used in plant microtubule research are trifluralin (α,α,α-trifluoro-2,6-dinitro-N,N-dipropyl-p-toluidine) and oryzalin (3,5-dinitro-N^4,N^4-dipropylsulphanilamide); the structures of trifluralin and oryzalin are shown in Fig. 3.3.

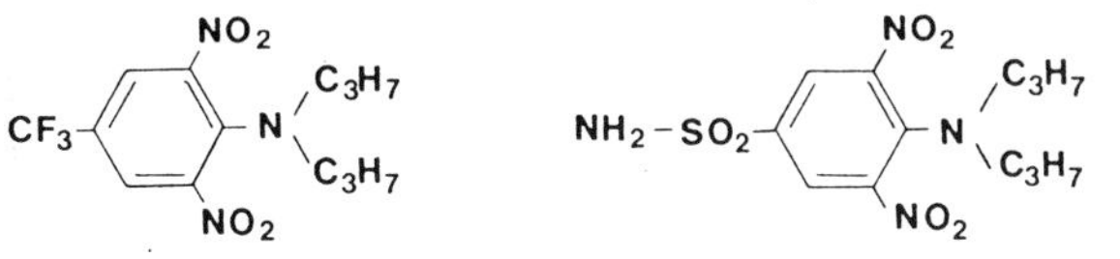

Fig. 3.3. Chemical structures of trifluralin (left) and oryzalin (right).

Effects of dinitroanilines in plants

Dinitroaniline herbicide treatment produces gross morphological abnormalities in plants, particularly in the root tip, a region containing high levels of meristematic activity. Following the inhibition of primary and lateral root growth, there is excessive radial expansion and swelling of the root tip. In general, dicotyledonous species are more tolerant to the effects of these compounds than are monocotyledonous plants, although the reasons for this are not clear (Upadhyaya and Nooden, 1987; Cleary and Hardham, 1988).

Cell division and mitosis are inhibited by brief dinitroaniline treatment, and affected cells do not form a cell plate and may contain restitution and polyploid nuclei. Interrupted xylem vessel differentiation is also a common feature of the dinitroanilines; the normally oriented secondary wall deposition becomes unordered. High nanomolar concentrations of dinitroanilines cause the disappearance of most of the microtubules in different arrays. The effects of brief dinitroaniline treatments at low concentrations can often be reversed by rinsing away the herbicide.

Molecular interactions of dinitroanilines and plant tubulin

To date, very few studies have been published on the molecular interaction of dinitroanilines with microtubule proteins. Early herbicide-binding studies showed that dintroanilines bind to algal tubulin. Hess and Bayer (1977), demonstrated that [^{14}C]trifluralin binds to flagellar tubulin from the protoctistan alga *Chlamydomonas*. However, trifluralin and most of the dinitroanilines are very insoluble in aqueous solutions (0.25 mg l^{-1}; 8.2×10^{-7} M) (Fedtke, 1982), and bind to a variety of surfaces, including glass and plastic (Strachen and Hess, 1982). Because this physical property makes difficult the maintenance of constant drug concentrations for careful biochemical studies, the significance of trifluralin-binding experiments was for a period uncertain.

More recently, oryzalin has been used as a model dinitroaniline because it is one of the more soluble dinitroanilines, having an estimated solubility constant of 2.5 mg l^{-1} (7.2×10^{-6} M) in water (Fedtke, 1982). Addition of 1–2% dimethyl sulphoxide, or ethanol to aqueous solutions will boost the solubility of oryzalin approximately two-fold (Strachen and Hess, 1982; Morejohn *et al.*, 1987b). Strachen and Hess (1983) reported that [^{14}C]oryzalin binds to *Chlamydomonas* tubulin independent of time and temperature. They obtained a binding constant (K) of 2.08×10^5 M^{-1}, and an estimated stoichiometry (r) of one mole oryzalin bound per mole tubulin at 25°C. These results pointed to tubulin as the primary subcellular receptor for dinitroaniline action.

Inhibition of plant microtubule polymerization *in vitro* by oryzalin has been demonstrated (Morejohn *et al.*, 1987b). The taxol-induced polymerization of tubulin from cultured cells of *Rosa* is inhibited by low micromolar concentrations of oryzalin. No inhibition of taxol-induced bovine brain microtubule polymerization is obtained. Morejohn *et al.* (1987b) used [^{14}C]oryzalin to examine the dinitroaniline-binding reaction with rose tubulin. The herbicide binds in a pH-dependent manner and with moderate affinity ($K_{app} = 1.2 \times 10^5$ M^{-1}) to the rose tubulin dimer at 24°C (pH 7.2). However, the estimated stoichiometry of binding in this preliminary study is very low ($r = 0.15$), indicating that the oryzalin-binding sites on tubulin may have decayed before and/or during the binding experiments, or that bound oryzalin may have dissociated from tubulin during the 'kinetic' ligand-binding method used. It is not known whether the dinitroanilines readily dissociate from tubulin dimers. An alternative possibility is that not all of the tubulin isotypes expressed in cultured rose cells possess an oryzalin-binding site. No specific binding of oryzalin to bovine brain tubulin or several other unrelated proteins was detected. This report not only confirmed the direct interaction of oryzalin with plant tubulin, but also demonstrated its direct inhibitory effect on microtubule polymerization.

Recent studies of oryzalin effects on cellular microtubules indicate that the drug causes the rapid loss of dimers from either the plus or minus ends of plant microtubules, depending upon the stability of the end (Bajer and Molé-Bajer, 1986; Morejohn *et al.* 1987b). High resolution immunogold-staining experiments with the giant microtubule cytoskeleton of *Haemanthus* endosperm cells have demonstrated that a 2-min treatment with 10^{-6} M oryzalin causes all interphase microtubules to be lost, with very short microtubule stubs remaining anchored stably to the nuclear envelope. In contrast, the same drug treatments cause the rapid loss of phragmoplast microtubules and kinetochore bundle microtubules from their minus ends. Only short microtubule stubs at plus ends of phragmoplast microtubules and kinetochore microtubules remain attached stably to the cell plate region and kinetochores, respectively. Electron microscopic observations on these structures reveal that the plus ends of microtubules in the phragmoplast and kinetochore are stabilized by extensive MAP cross-bridging (Hepler *et al.*, 1970). Interzonal microtubules having no extensive cross-bridging at either end are completely lost following the drug treatment. These data indicate that substoichiometric concentrations of the tubulin–oryzalin complex bind to either end of microtubules and cause their depolymerization, providing that the microtubule ends are not blocked by relatively stable molecular interactions. Also, these results suggest that particular microtubule ends in different arrays are kinetically distinct, i.e. they may participate differentially in a tubulin dimer/polymer equilibrium.

Phosphoric amide herbicides

The phosphoric amides comprise a small class of experimental organophosphorous pre-emergence herbicides used for the control of annual grasses and broadleaf weeds in a wide range of vegetable crops (Aya *et al.*, 1975). The most commonly used members of this class are amiprophos-methyl (APM; *N*-isopropyl *O*-methyl *O*-(2-nitro-*p*-tolyl) phosphorothioamidate) and butamiphos (cremart; *O*-ethyl-*O*-(3-methyl-6-nitrophenyl)-*N*-*sec*-butyl phosphorothioamidate). Phosphoric amides are much more soluble in aqueous solutions than are dinitroanilines; the water solubility constant for APM is 70 mg l^{-1} (2.3×10^{-4} M) (Fedtke, 1982). The structure of APM is illustrated in Fig. 3.4.

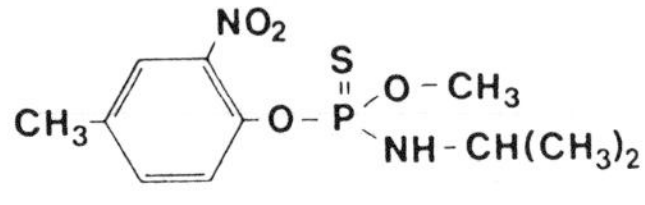

Fig. 3.4. Chemical structure of APM.

Effects of phosphoric amides in plants
The effects of phosphoric amides on plant growth and development are virtually indistinguishable from those described above for the dinitroaniline herbicides. At low micromolar concentrations these drugs inhibit root growth by arresting mitosis, and cells eventually become isodiametric in swollen root tips (Sumida and Ueda, 1976). Early studies on a wide range of microtubule-dependent processes in several plants demonstrated that low micromolar concentrations of either APM, or cremart cause the disappearance of microtubules. These organisms include the algae *Chlorella* (Sumida and Ueda, 1974), *Micrasterias* (Kiermayer and Fedtke, 1977), *Oocystis* (Robinson and Herzog, 1977), *Chlamydomonas* (Quader and Filner, 1980), and *Polytomella* (Sterns and Brown, 1981), and the higher plant *Allium* (Sumida and Ueda, 1976; Mita and Shibaoka, 1983). Following short periods of treatment with these herbicides, the effects on microtubules may be reversed by washing plant materials with drug-free medium.

Molecular interactions of phosphoric amides and plant tubulin

Convincing evidence that phosphoric amides interact directly with plant microtubules was reported by Morejohn and Fosket (1984b). They demonstrated that low micromolar concentrations of APM inhibit taxol-induced polymerization of *Rosa* microtubules *in vitro*. APM had no effect on the taxol-induced polymerization of bovine brain microtubules. Although no APM-binding experiments have been published, the inhibition of plant microtubule polymerization by APM strongly suggests that the phosphoric amides bind to the plant tubulin dimer. To date this is the only biochemical study on the effects of a phosphoric amide herbicide on plant microtubule polymerization.

Bajer and Molé-Bajer (1986) performed *in vivo* studies on *Haemanthus* endosperm cells which strongly support the idea that microtubules are the primary site of action of the phosphoric amides. During drug treatment, the rate of anaphase chromosome migration was followed with a video camera, and the observations were correlated with those obtained with indirect immunogold staining of microtubules in the same cells. Chromosome migration immediately slowed, but did not stop after perfusion with 10^{-8} M APM. Chromosome migration stops instantaneously after perfusion with 10^{-7} M APM. Indirect immunogold staining of these cells revealed that within 5 min of drug treatment, subsidiary ('fir tree') microtubules that normally emanate from kinetochore microtubule bundles are lost, polar microtubules are shorter, and interzonal microtubules are absent. These results with APM parallel closely the observations reported on the dinitroaniline herbicide oryzalin (Morejohn *et al.*, 1987b), and intimate that APM binds rapidly to plant tubulin to form tubulin–herbicide complexes which substoichiometrically poison microtubule dynamics at kinetic ends.

N-Phenyl carbamates and related herbicides

The *N*-phenyl carbamate herbicides are pre-emergence compounds having anti-mitotic activity in plants. Among the members of this herbicidal class isopropyl *N*-(3-chlorophenyl)-carbamate (CIPC) is relatively soluble in aqueous solutions (89 mg l^{-1}; 4.2×10^{-4} M) (Fedtke, 1982), and has been the most commonly used for studies on plant microtubules. The structurally similar herbicide pronamide (3,5-dichloro (*N*-1, 1-dimethyl-2-propynyl) benzamide), also known as propyzamide, has effects similar to those of CIPC. The structure of CIPC is given in Fig. 3.5.

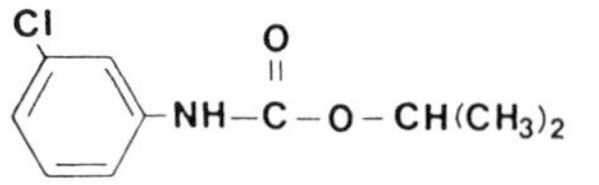

Fig. 3.5. Chemical structure of CIPC.

Effects of N-*phenyl carbamates in plants*

Treatment of plants with *N*-phenyl carbamate herbicides in the high micromolar concentration range produces root tip swelling, and affected cells are arrested at metaphase or anaphase, contain enlarged and polyploid nuclei, and become isodiametric. The effects of these compounds on microtubules vary greatly, depending upon the particular compound used and its concentration, and the type of cells and species of plant treated. In some cases microtubules remain long, apparently unaffected by herbicide treatment, and in other cases microtubules are short, disoriented, or lost completely, depending upon the stage in the cell cycle (Jackson, 1969; Bartels and Hilton, 1973; Clayton and Lloyd, 1984; Doonan *et al.*, 1985; Vaughn and Vaughn, 1987; Akashi *et al.*, 1988; Lehnen *et al.*, 1990). However puzzling the wide range of responses may seem, these compounds do have a particularly interesting and common characteristic: *N*-phenyl carbamate-treated algal and higher plant cells accumulate multiple foci of radiating microtubules. These multiple foci have been presumed to be supernumerary MTOCs (Hepler and Jackson, 1969).

Molecular interactions of N-*phenyl carbamates and plant tubulin*

Very little is understood of the molecular interaction of the *N*-phenyl carbamates with plant tubulin, or microtubules. Early ligand-binding experiments (Bartels and Hilton, 1973; Coss *et al.*, 1975) reported no binding of carbamates to chick brain tubulin, and no inhibition of brain microtubule polymerization. These results were taken to mean that the effects of carbamates were exerted on plant MTOCs, rather than plant microtubules, an interpretation that has been adopted subsequently by numerous investigators. However, Mizuno *et al.* (1981) published that EPC (ethyl *N*-(3-carboxyphenyl) carbamate), a derivative of pronamide, could be used in affinity chromatography to enrich tubulin from azuki bean proteins. In this methodology tubulin is one of several polypeptides binding to the EPC column; purification of tubulin requires subsequent DEAE-chromatography, and molecular sieve chromatography (Mizuno *et al.*, 1981). The same procedure has been used to purify tubulins from *Phaseolus*, *Pisum*, *Cucurbita*, *Cucumis*, and *Daucus* (Mizuno, 1985). It is not clear whether tubulin binds via a specific affinity interaction with the immobilized EPC moiety, because the proteins are eluted from the EPC columns with salt rather than with herbicide. Nevertheless, these experiments indicate that EPC binds to plant tubulin, and possibly several other plant proteins.

In a preliminary experiment Morejohn and Fosket (1986) found that 2×10^{-4} M CIPC inhibited taxol-induced polymerization of rose tubulin by 27% compared with the control at 24°C, and had no effect on the taxol-induced polymerization of brain tubulin. This effect of CIPC on plant microtubule assembly is much less than that of either oryzalin or APM, and about the same as that of colchicine. This polymerization assay identified tubulin as a subcellular target of a carbamate herbicide. Akashi *et al.* (1988) examined the effects of propyzamide on the GTP- and dimethyl sulphoxide-induced polymerization of tubulin purified from cultured cells of *Nicotiana*. They found that a propyzamide concentraton of 10^{-4} M inhibited the extent of polymer formation at 30°C by approximately 62% compared with the control polymer containing no drug. They found no effect of 10^{-4} M propyzamide on the polymerization of bovine brain microtubules at

30°C. Although these studies are the only published biochemical experiments on carbamate herbicides and plant microtubule polymerization *in vitro*, they provide compelling evidence that the carbamate herbicides do bind to plant tubulin and directly affect microtubule polymerization in cells.

Perhaps the binding of carbamate herbicides to plant tubulin in cells results in the formation of a tubulin–carbamate complex that not only is capable of limited polymerization, but also can nucleate from inappropriate sites within the cytosol. This working hypothesis suggests that the often observed multiple foci may not all be bona fide MTOCs, but rather a combination of real MTOCs and secondary, nonspecific nucleation sites. Alternatively, the faithful duplication and positioning of spindle poles may, in fact, be microtubule-dependent, and disrupted by carbamate binding to tubulin. Lastly, the binding of carbamates to tubulin may be only one of multiple modes of action for these herbicides, a possibility that should be investigated further. In any case, most of the previously published studies concluding that carbamates only affect MTOCs in plant cells will require some re-evaluation within the context of these findings and future biochemical studies on the *N*-phenyl carbamate herbicides.

Significance of future plant anti-microtubule drug research

Now that plant tubulins can be isolated and polymerized *in vitro*, the sites of anti-microtubule herbicide action in cells no longer need to be inferred from cytological observations. In most cases the mode of drug action and the plant response can be shown to have a realistic molecular basis. Since the appearance of the pharmacological results with plant tubulin and microtubules outlined above, numerous studies have successfully used dinitroanilines, phosphoric amides, and taxol to study a variety of microtubule-dependent processes in phylogenetically diverse species of plants. These organisms include the algae *Chlamydomonas* (Bolduc *et al.*, 1988; James *et al.*, 1988, 1989),

Ernodesmis and *Boergesenia* (LaClaire, 1987), and *Mougeotia* (Galway and Hardham, 1989), the mosses *Physcomitrella* (Doonan *et al.*, 1985) and *Funaria* (Wacker *et al.*, 1988), and the higher plants *Eleusine* (Vaughn, 1986), *Zinnia, Medicago, Daucus, Nicotiana* (Falconer and Seagull, 1987; Falconer *et al.*, 1988), *Potamogeton* and *Lolium* (Cleary and Hardham, 1988), *Tradescantia* (Salitz and Schmitz, 1989), *Nitella* (Wasteneys and Williamson, 1989a, b), and *Secale* (Kerr and Carter, 1990).

Recently Chan and Fong (1990) reported that trifluralin at micromolar levels inhibits the proliferation and differentiation of the parasitic protozoan *Leishmania*, and that [^{14}C]trifluralin binds to *Leishmania* tubulin. Does the binding of dinitroanilines to tubulins from *Leishmania* and the protistan alga *Chlamydomonas* (Strachen and Hess, 1983) indicate that they bind also to tubulins from diverse model protists such as *Euglena, Paramecium, Dunaliella, Naegleria,* and *Tetrahymena*? This possibility is predicted from the early work of Banerjee *et al.* (1975) which showed trifluralin to be the most potent agent for the arrest of mitosis and oral regeneration in the ciliated protoctist *Stentor*. Because protoctists are resistant to colchicine treatment, the anti-microtubule herbicides may become valuable probes for the analysis of microtubule function in these organisms as well, a possibility that can be tested directly.

Many intriguing questions remain on the molecular mechanisms of drug binding to plant tubulin. How many different drug-binding sites exist on the plant dimer? Do the different classes of herbicides compete for binding to the same site, or do they bind to distinct sites? Where are the herbicide-binding sites on α- and β-tubulin subunits? Do α- and β-tubulin isotypes bind herbicides differentially? How does the binding of herbicides to the dimer alter the ability of tubulin to polymerize? Do the drugs bind only to soluble tubulin, or can they bind to polymerized tubulin as well? Is the basis of the differential tolerance of dicots and monocots to herbicides encoded in their tubulins, or does this result from trivial differences in herbicide uptake, metabolism, or compartmentalization mechanisms? Although tubulin is the most likely primary receptor

for herbicide action, do other undocumented secondary receptors exist in plant cells? Are the herbicide-binding sites on plant tubulin actually allosteric sites for the binding of endogenous factors regulating microtubule dynamics *in vivo*? Do well chararacterized anti-microtubule drugs such as vinblastine, maytansine, rhizoxin, and griseofulvin bind to plant tubulin as they do to animal tubulin? Does taxol bind to the same number of sites, and with the same affinity on microtubules from plants and animals? These and numerous other questions remain to be addressed in future investigations on the molecular pharmacology of plant microtubules.

References

Akashi, T., Izumi, K., Nagano, E., Enomot, M., Mizuno, K. and Shibaoka, H. (1988). Effects of propyzamide on tobacco cell microtubules *in vivo* and *in vitro*. *Plant Cell Physiol.* **29**, 1053–1062.

Apostolakos, P., Galatis, B., Katsaros, C. and Schnepf, E. (1990). Tubulin conformation in microtubule-free cells of *Vigna sinensis*. An immunofluorescent and electron microscope study. *Protoplasma* **154**, 132–143.

Ashton, F.M. and Crafts, A.S. (1981). *Mode of Action of Herbicides*, pp. 180–223. John Wiley, New York.

Aya, M., Takase, I., Kishino, S. and Kurihara, K. (1975). Amiprophos-methyl, a new herbicide in upland crops. *Proc. 5th Asian Pac. Weed Sci. Soc. Conf.* 138–141.

Bajer, A.S. and Molé-Bajer, J. (1986). Drugs with colchicine-like effects that specifically disassemble plant but not animal microtubules. *Ann. N.Y. Acad. Sci.* **466**, 767–784.

Bajer, A.S., Cypher, C., Molé-Bajer, J. and Howard, H.M. (1982). Taxol action on mitosis. I. Pushing force by assembly of microtubules. *Proc. Natl Acad. Sci. U.S.A.* **79**, 6569–6573.

Banerjee, S., Kelleher, J.K. and Margulis, L. (1975). The herbicide trifluralin is active against microtubule-based oral morphogenesis in *Stentor coeruleus*. *Cytobios* **12**, 179–189.

Bartels, P.G. and Hilton, J.L. (1973). Comparison of trifluralin, oryzalin, pronamide, propham, and colchicine treatments on microtubules. *Pestic. Biochem. Physiol.* **3**, 462–472.

Baum, S.G., Wittner, M., Nadler, J.P., Horwitz, S.B., Dennis, J.E., Schiff, P.B. *et al.* (1981). Taxol, a microtubule stabilizing agent, blocks the replication of *Trypanosoma cruzii*. *Proc. Natl Acad. Sci. U.S.A.* **78**, 4571–4575.

Bayley, P.M. (1990). What makes microtubules dynamic? *J. Cell Sci.* **95**, 329–334.

Bayley, P., Schilstra, M. and Martin, S. (1989). A lateral cap model of microtubule dynamic instability. *FEBS Lett.* **259**, 181–184.

Bergen, L.G. and Borisy, G.G. (1983). Tubulin–colchicine complex inhibits microtubule elongation at both plus and minus ends. *J. Biol. Chem.* **258**, 4190–4194.

Bolduc, C., Lee, V.D. and Huang, B. (1988). β-Tubulin mutants of the unicellular green alga *Chlamydomonas reinhardtii*. *Proc. Natl Acad. Sci. U.S.A.* **85**, 131–135.

Caplow, M. and Zeeburg, B. (1982). Dynamic properties of microtubules at steady state in the presence of taxol. *Eur. J. Biochem.* **127**, 319–324.

Cassimeris, L., Pryer, N.K. and Salmon, E.D. (1988). Real-time observations of microtubule dynamic instability in living cells. *J. Cell Biol.* **107**, 2223–2231.

Chan, M.M.-Y. and Fong, D. (1990). Inhibition of leishmanias but not host macrophages by the antitubulin herbicide trifluralin. *Science* **249**, 924–926.

Chen, Y. and Hill, T.L. (1985). Monte Carlo study of the GTP cap in a five-start helix model of a microtubule. *Proc. Natl Acad. Sci. U.S.A.* **82**, 1131–1135.

Clayton, L. and Lloyd, C.W. (1984). The relationship between the division plane and spindle geometry in *Allium* cells treated with CIPC and griseofulvin: an antitubulin study. *Eur. J. Cell Biol.* **34**, 248–253.

Cleary, A.L. and Hardham, A.R. (1988). Depolymerization of microtubule arrays in root tip cells by oryzalin and their recovery with modified nucleation patterns. *Can. J. Bot.* **66**, 2353–2366.

Collins and Vallee (1987). Temperature-dependent reversible assembly of taxol-treated microtubules. *J. Cell Biol.* **105**, 2847–2854.

Coss, R.A., Bloodgood, R.A., Brower, D.L., Pickett-Heaps, J.D. and McIntosh, J.R. (1975). Studies on the mechanism of action of isopropyl N-phenyl-carbamate. *Exp. Cell Res.* **92**, 394–398.

Dawson, P.J. and Lloyd, C.W. (1987). Comparative biochemistry of plant and animal tubulins. In *The Biochemistry of Plants*, vol. 12 (ed. D.D. Davies), pp. 3–47. Academic Press, San Diego.

De Brabander, M., Geuens, G., Nuydens, R., Willebrords, R. and De Mey, J. (1981). Microtubule assembly in living cells after release from nocodazole block: the effects of metabolic inhibitors, taxol and pH. *Cell Biol. Int. Rep.* **5**, 913–920.

Doonan, J.H., Cove, D.J. and Lloyd, C.W. (1985). Immunofluorescence microscopy of microtubules in intact cell lineages of the moss, *Physcomitrella patens*. I. Normal and CIPC-treated tip cells. *J. Cell Sci.* **75**, 131–147.

Falconer, M.M. and Seagull, R.W. (1985). Xylogenesis in tissue culture: taxol effects on microtubule reorientation and lateral association in differentiating cells. *Protoplasma* **128**, 157–166.

Falconer, M.M. and Seagull, R.W. (1987). Amiprophos-methyl (APM): a rapid, reversible, anti-microtubule agent for plant cell cultures. *Protoplasma* **136**, 118–124.

Falconer, M.M., Donaldson, G. and Seagull, R.W. (1988). MTOCs in higher plant cells: an immunofluorescent study of microtubule assembly sites following depolymerization by APM. *Protoplasma*, **144**, 46–55.

Fedtke, C. (1982). Microtubules. In *Biochemistry and Physiology of Herbicide Action*, pp. 123–141. Springer-Verlag, Berlin.

Fosket, D.E. (1989). Cytoskeletal proteins and their genes in higher plants. In *The Biochemistry of Plants*, vol. 15 (eds P.K. Stumpf and E.E. Conn), pp. 393–454. Academic Press, San Diego.

Galway, M.E. and Hardham, A.R. (1989). Oryzalin-induced microtubule disassembly and recovery in regenerating protoplasts of the alga *Mougeotia*. *J. Plant Physiol.* **135**, 337–345.

Hahne, G. and Hoffmann, F. (1984). Dimethyl sulfoxide can initiate cell divisions of arrested callus protoplasts by promoting cortical microtubule assembly. *Proc. Natl Acad. Sci U.S.A.* **81**, 5449–5453.

Haines, F.O., Dickerson, R.M., Wilson, L. and Owellen, R.J. (1978). Differences in the binding properties of vinca alkaloids and colchicine to tubulin by varying protein sources and methodology. *Biochem. Pharmacol.* **27**, 71–76.

Hepler, P.K. and Jackson, W.T. (1969). Isopropyl N-phenylcarbamate affects spindle microtubule orientation in dividing endosperm cells of *Haemanthus katherinae* Baker. *J. Cell Sci.* **5**, 727–743.

Hepler, P.K., McIntosh, J.R. and Cleland, S. (1970). Intermicrotubule bridges in mitotic spindle apparatus. *J. Cell Biol.* **45** 438–444.

Herth, W. (1983). Taxol affects cytoskeletal microtubules, flagella and spindle structure of the chrysoflagellate alga *Poterioochromonas*. *Protoplasma* **115**, 228–239.

Hess, D. and Bayer, D.E. (1977). Binding of the herbicide trifluralin to *Chlamydomonas* flagella tubulin. *J. Cell Sci.* **24**, 351–360.

Horio, T. and Hotani, H. (1986). Visualization of the dynamic instability of individual microtubules by darkfield microscopy. *Nature (Lond.)* **321**, 605–607.

Hotani, H. and Horio, T. (1988). Dynamics of microtubules visualized by darkfield microscopy: treadmilling and dynamic instability. *Cell Motil. Cytoskel.* **10**, 229–236.

Hussey, P.J. and Gull, K. (1985). Multiple isotypes of α- and β-tubulin in the plant *Phaseolus vulgaris*. *FEBS Lett.* **181**, 113–118.

Jackson, W.T. (1969). Regulation of mitosis. II. Interaction of isopropyl *N*-phenylcarbamate and melatonin. *J. Cell Sci.* **5**, 745–755.

James, S.W., Ranum, L.P.W., Silflow, C.D. and Lefebvre, P.A. (1988). Mutants resistant to anti-microtubule herbicides map to a locus on the *uni* linkage group in *Chlamydomonas reinhardtii*. *Genetics* **188**, 141–147.

James, S.W., Silflow, C.D., Thompson, M.D., Ranum, L.P.W. and Lefebvre, P.A. (1989). Extragenic suppression and synthetic lethality among *Chlamydomonas reinhardtii* mutants resistant to anti-microtubule drugs. *Genetics* **122**, 567–577.

Job, D., Pabion, M, and Margolis, R.L. (1985). Generation of microtubule stability subclasses by microtubule-associated proteins: implications for the microtubule 'dynamic instability' model. *J. Cell Biol.* **101**, 1680–1689.

Kato, T., Kakiuchi, M. and Okamura, S. (1985). Properties of purified colchicine-binding protein from a cultured carrot cell extract. *J. Biochem. (Tokyo)* **98**, 1138–1143.

Kerr, G.P. and Carter, J.V. (1990). Relationship between freezing tolerance of root-tip cells and cold stability of microtubules in rye (*Secale cereale* L. cv Puma). *Plant Physiol.* **93**, 77–82.

Kiermayer, O. and Fedtke, C. (1977). Strong anti-microtubule action of amiprophos-methyl (APM) in *Micrasterias*. *Protoplasma* **92**, 163–166.

Kumar, N. (1981). Taxol-induced polmerization of purified tubulin. *J. Biol. Chem.* **256**, 10435–10441.

LaClaire, J.W. (1987). Microtubule cytoskeleton in intact and wounded coenocytic green algae. *Planta* **171**, 30–42.

Lambeir, A. and Engelborghs, Y. (1980). A quantitative analysis of tubulin–colchicine binding to microtubules. *Eur. J. Biochem.* **109**, 619–624.

Lehnen, L.P., Vaughn, M.A. and Vaughn, K.C. (1990). Terbutol affects spindle microtubule organizing centres. *J. Exp. Bot.* **41**, 537–547.

Manfredi, J.J., Parness, J. and Horwitz, S.B. (1982). Taxol binds to cellular microtubules. *J. Cell Biol.* **94**, 688–696.

Margolis, R.L. and Wilson, L. (1978). Addition of colchicine–tubulin complex to microtubule ends: the mechanism of substoichiometric colchicine poisoning. *Cell* **13**, 1–8.

Margolis, R.L. and Wilson, L. (1981). Microtubule treadmills: possible molecular machinery. *Nature (Lond.)* **293**, 705–711.

Melan, M.A. (1990). Taxol maintains organized microtubule patterns in protoplasts which lead to the resynthesis of organized cell wall microfibrils. *Protoplasma* **153**, 169–177.

Mita, T. and Shibaoka, H. (1983). Changes in microtubules in onion leaf sheath cells during bulb development. *Plant Cell Physiol.* **24**, 109–117.

Mitchison, T. (1989). Polewards microtubule flux in the mitotic spindle: evidence from photoactivation of fluorescence. *J. Cell Biol.* **109**, 637–652.

Mitchison, T. and Kirschner, M. (1984a). Microtubule assembly nucleated by isolated centrosomes. *Nature* (Lond.) **123**, 232–237.

Mitchison, T. and Kirschner, M. (1984b). Dynamic instability of microtubule growth. *Nature* (Lond.) **312**, 237–242.

Mitchison, T.J. and Sawin, K.E. (1990). Tubulin flux in the mitotic spindle: where does it come from, where is it going? *Cell Motil. Cytoskel.* **16**, 93–98.

Mitchison, T.J., Evans, L., Schulze, E. and Kirschner, M. (1986). Sites of microtubule assembly and disassembly in the mitotic spindle. *Cell* **45**, 515–527.

Mizuno, K. (1985). *In vitro* assembly of microtubules from tubulins of several higher plants. *Cell Biol. Int. Rep.* **9**, 13–21.

Mizuno, K., Koyama, M. and Shibaoka, H. (1981). Isolation of tubulin from Azuki bean epicotyls by ethyl *N*-phenylcarbamate-Sepharose affinity chromatography. *J. Biochem. (Tokyo)* **89**, 329–332.

Molé-Bajer, J. and Bajer, A.S. (1983). Action of taxol on mitosis: modification of microtubule arrangements and function of the mitotic spindle in *Haemanthus* endosperm. *J. Cell Biol.* **96**, 527–540.

Morejohn, L.C. and Fosket, D.E. (1982). Higher plant tubulin identified by self-assembly into microtubules *in vitro*. *Nature* (Lond.) **297**, 426–428.

Morejohn, L.C. and Fosket, D.E. (1984a). Taxol-induced rose microtubule polymerization *in vitro* and its inhibition by colchicine. *J. Cell Biol.* **99**, 141–147.

Morejohn, L.C. and Fosket, D.E. (1984b). Inhibition of plant microtubule polymerization *in vitro* by the phosphoric amide herbicide amiprophos-methyl. *Science* **224**, 874–876.

Morejohn, L.C. and Fosket, D.E. (1986). Tubulins from plants, animals and protists. In *Cell and Molecular Biology of the Cytoskeleton* (ed. J.W.

Sharp), pp. 257–329. Plenum Press, New York.

Morejohn, L.C., Bureau, T.E., Tocchi, L.P. and Fosket, D.E. (1984). Tubulins from different higher plant species are immunologically nonidentical and bind colchicine differentially. *Proc. Natl Acad. Sci. U.S.A.* **81**, 1440–1444.

Morejohn, L.C., Bureau, T.E., Tocchi, L.P. and Fosket, D.E. (1987a). Resistance of *Rosa* microtubule polymerization to colchicine results from a low-affinity interaction of colchicine and tubulin. *Planta* **170**, 230–241.

Morejohn, L.C., Bureau, T.E., Molé-Bajer, J., Bajer, A.S. and Fosket, D.E. (1987b). Oryzalin, a dinitroaniline herbicide, binds to plant tubulin and inhibits microtubule polymerization *in vitro*. *Planta* **172**, 252–264.

Nooden, L.D. (1971). Physiological and developmental effects of colchicine. *Plant Cell Physiol.* **12**, 759–770.

Okamura, S. (1984). Straight and helical filaments formed during microtubule-assembly cycling of cultured carrot cell extract. *J. Electron Microsc.* **33**, 182–185.

Parka, S.J. and Soper, O.F. (1977). The physiology and mode of action of the dinitroaniline herbicides. *Weed Sci.* **25**, 77–87.

Parness, J. and Horwitz, S.B. (1981). Taxol binds to polymerized tubulin *in vitro*. *J. Cell Biol.* **91**, 479–487.

Quader, H. and Filner, P. (1980). The action of antimitotic herbicides on flagellar regeneration in *Chlamydomonas reinhardtii*: a comparison with the action of colchicine. *Eur J. Cell Biol.* **21**, 301–304.

Ray, K., Bhattacharyya, B. and Biswas, B.B. (1984). Anion-induced increases in the affinity of colchicine binding to tubulin. *Eur. J. Biochem.* **142**, 577–581.

Ringel, I. and Sternlicht, H. (1984). Carbon-13 nuclear magnetic resonance study of microtubule protein: evidence for a second colchicine site involved in the inhibition of microtubule assembly. *Biochemistry* **23**, 5644–5653.

Robinson, D.G. and Herzog, W. (1977). Structure, synthesis and orientation of microfibrils. III. A survey of the actions of microtubule inhibitors on microtubules and microfibril orientation in *Oocystis solitaria*. *Cytobiologie* **15**, 467–474.

Salitz, A. and Schmitz, K. (1989). Influence of microfilament and microtubule inhibitors applied by immersion and microinjection on circulation streaming in the staminal hairs of *Tradescantia blossfeldiana*. *Protoplasma* **153**, 37–45.

Schiff, P.B. and Horwitz, S.B. (1981). Taxol assembles tubulin in the absence of exogenous guanosine 5'-triphosphate or microtubule associated proteins. *Biochemistry* **20**, 3247–3251.

Schiff, P.B., Fant, F. and Horwitz, S.B. (1979). Promotion of microtubule assembly *in vitro* by taxol. *Nature* (Lond.) **277**, 665–667.

Schulze, E. and Kirschner, M. (1986). Dynamic and stable populations of microtubules in cells. *J. Cell Biol.* **104**, 277–288.

Silflow, C.D., Oppenheimer, D.G., Kopczak, S.D., Ploense, S.E., Ludwig, S.R. and Snustad, D.P. (1987). Plant tubulin genes: structure and differential expression during development. *Develop. Genet.* **8**, 435–460.

Sloan, M.E. and Camper, N.D. (1981). Effects of colchicine on carrot callus – growth and energy status. *Plant Cell Tiss. Org. Cult.* **1**, 69–75.

Sternlicht, H., Ringel, I. and Szasz, J. (1983). Theory for modeling the co-polymerization of tubulin and tubulin–colchicine complex. *Biophys. J.* **42**, 255–267.

Sterns, M.E. and Brown, D.L. (1981). Microtubule organizing centers (MTOCs) of the alga *Polytomella* exert spatial control over microtubule initiation *in vivo* and *in vitro*. *J. Ultrastruct. Res.* **77**, 366–378.

Strachen, S.D. and Hess, D. (1982). Dinitroaniline herbicides adsorb to glass. *J. Agric. Food Chem.* **30**, 389–391.

Strachen, S.D. and Hess, D. (1983). The biochemical mechanism of the dinitroaniline herbicide oryzalin. *Pestic. Biochem. Physiol.* **20**, 141–150.

Sumida, S. and Ueda, M. (1974). In *Mechanism of Pesticide Action* (ed. G.K. Kohn), pp. 156–168. American Chemical Society, Washington, DC.

Sumida, S. and Ueda, M. (1976). Effect of *O*-ethyl *O*-(3-methyl-6-nitrophenyl) *N-sec*-butyl-phosphoroamidate (S-2846), an experimental herbicide, on mitosis in *Allium cepa*. *Plant Cell Physiol.* **17**, 1351–1354.

Upadhyaya, M.K. and Nooden, L.D. (1987). Comparison of [^{14}C]oryzalin uptake in root segments of a sensitive and resistant species. *Ann. Bot.* **59**, 483–485.

Vaughn, M.A. and Vaughn, K.C. (1987). Pronamide disrupts mitosis in a unique manner. *Pestic. Biochem. Physiol.* **28**, 182–193.

Vaughn, K. (1986). Cytological studies of dinitroaniline-resistant *Eleusine*. *Pestic. Biochem. Physiol.* **26**, 66–74.

Wacker, I., Quader, H. and Schnepf, E. (1988). Influence of the herbicide oryzalin on cytoskeleton and growth of *Funaria hygrometrica* protonemata. *Protoplasma* **142**, 55–67.

Wani, M.C., Taylor, H.L., Wall, M.E., Coggon, P. and McPhail, A.T. (1971). Plant antitumor agents. VI. The isolation and structure of taxol, a novel antileukemic and antitumor agent from *Taxus brevifolia. J. Am. Chem. Soc.* **93**, 2325–2327.

Wasteneys, G.O. and Williamson, R.E. (1989a). Reassembly of microtubules in *Nitella tasmanica*: assembly of cortical microtubules in branched clusters and its relevance to steady-state micro-tubule assembly. *J. Cell Sci.* **93**, 705–714.

Wasteneys, G.O. and Williamson, R.E. (1989b). Reassembly of microtubules in *Nitella tasmanica*: quantitative analysis of assembly and orientation.

Eur. J. Cell Biol. **50**, 76–83.

Wilson, L. and Mesa, I. (1973). The mechanism of action of colchicine. Colchicine binding properties of sea urchin sperm tail outer doublet tubulin. *J. Cell Biol.* **58**, 709–719.

Wright, M., Moisand, A. and Oustrain, M.L. (1982). Stabilization of monasters by taxol in the amoebae of *Physarum polycephalum* (Myxomycetes). *Proto-plasma* **113**, 44–56.

4. POTENTIAL SIGNIFICANCE OF MICROTUBULE REARRANGEMENT, TRANSLOCATION AND REUTILIZATION IN PLANT CELLS

Barry A. Palevitz

*Department of Botany, University of Georgia,
Athens, GA 30602, USA*

Microtubules (MTs) in the cells of higher plants are generally thought to comprise four major cytoskeletal arrays: the interphase cortical system, preprophase band (PPB), mitotic spindle and phragmoplast. MTs are also found in other locations in various cells, most notably emanating from the nuclear envelope (NE). The four major arrays play clear, decisive roles in cell development; the importance of NE-linked MTs is less obvious (see below).

One of the principal questions that has long puzzled 'green' cell biologists concerns the transition between various MT arrays during the cell cycle and differentiation. That is, do transitions involve the depolymerization and reassembly of new elements, or does re-arrangement and reutilization of MTs from one array to the other also occur? The debate is made more acute by two veins of research that have dominated recent work on MT behaviour: the role of dynamic instability in conjunction with differential spatial stabilization (a stimulating and relatively recent advance in our understanding of assembly behaviour) in generating new MT distributions (Kirschner and Mitchison, 1986), and the participation of motor proteins (kinesin, dynein) in the movement and positioning of various organelles in the cell (see Warner and McIntosh, 1989 for relevant reviews). The latter effort has been spurred by bioassays based on the demonstration that intact MTs can translocate over a substratum with the addition of soluble motor proteins and cofactors. Thus, it is reasonable to wonder whether MTs move in a similar manner *in vivo*.

Obviously, MTs are nucleated at specific sites in the cell, and evidence for the operation of such sites in plants can be readily amassed with even a cursory inspection of the literature. Clearly, the most striking example is found in the spindle apparatus. MTs assembled at the poles in dividing cells interact with kineto-chores and each other to form mitotic and meiotic spindles. The polar sites are initially associated with the NE (e.g. Wick and Duniec, 1983; Mineyuki and Palevitz, 1990; Zhang *et al.*, 1990). Other examples can be found in interphase. For example, the radial MT array in the cortex of *Allium* guard cells (GCs) is generated from a nucleating zone along the central part of the common ventral wall (Marc *et al.*, 1989; Marc and Palevitz, 1990).

Unfortunately, our understanding of MT dynamics in higher plants is complicated by the lack of discrete centrosomes. Centrosomes serve as the nucleation sites for the spindle and interphase cytoplasmic arrays in animal cells (Mazia, 1987). However, not only are the spindle poles in higher plant somatic cells less focused, but cortical arrays in interphase and preprophase do not emanate from a common point in most cases (see further discussion below). The purpose of this chapter is to examine evidence for rearrangement, translocation and reutilization of intact MTs between arrays. The author believes such evidence has become more compelling in the past few years. Thus, rearrangments and reutilization, in conjunction with controlled assembly and selective stabilization, can contribute to the generation of MT arrays.

Formation of the PPB

While the role of the preprophase band (PPB) in division plane determination has engendered an enormous amount of experimentation and discussion, its formation and relationship to antecedent MTs has not been resolved. Evidence indicates that in at least some cells the PPB forms by rearrangement of the preceding interphase MT array (e.g. Pickett-Heaps,

1969; Wick and Duniec, 1983). For example, in cells in which the PPB is parallel to the interphase cortical MTs, it seems reasonable to suppose that the PPB forms by progressively restricting the latter to a circumferential band. However, useful insights could be obtained by examining those cells in which the PPB does not assume the same orientation as prior cortical MTs. The guard mother cells (GMCs) of graminean species provide such material. The interphase MTs of these cells are arranged in a transverse band known as the IMB (Fig. 4.1a) (Busby and Gunning, 1980; Galatis, 1982; Cho and Wick, 1989; Cleary and Hardham, 1989; Mullinax and Palevitz, 1989). The IMB is replaced by a longitudinal PPB (Fig. 4.1f) before division, so a 90° change in orientation distinguishes the two structures. By taking advantage of the distinctive anatomical features of grass leaves, namely the arrangement of epidermal cells in longitudinal files and the gradient in stomatal development from the basal intercalary meristem to the leaf tip, it is possible to follow MT changes in succes-

(a) (b) (c)

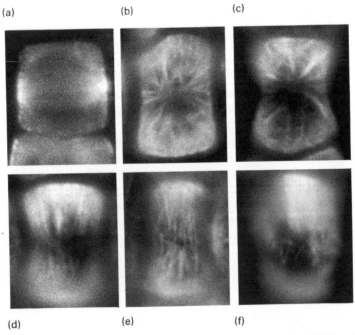

(d) (e) (f)

Fig. 4.1. Series of antitubulin images showing the transition between the IMB and PPB in *Avena sativa* GMCs. The IMB is seen in midplane in (a). In (b) the IMB appears to be splaying apart. In (c), MTs are now arranged in a radial pattern. Tuft-like MT bundles are also evident. In (d), the MTs and bundles are almost longitudinally aligned. A longitudinal orientation predominates in (e), while in (f) the PPB has consolidated into a narrow, bright band. × 1800. From Mullinax and Palevitz, 1989, with permission.

sive GMCs. Recent advances in antitubulin immunocytochemistry using large tissue segments markedly enhance this effort (e.g. Marc and Hackett, 1989). The images obtained with *Avena* in our laboratory indicate that IMB MTs are progressively rearranged to form the PPB (Mullinax and Palevitz, 1989). While MTs seem to diverge from the IMB even in very young cells, the band seems to progressively splay apart later on, yielding a radial MT arrangement (Fig. 4.1b, c). Moreover, the MTs also aggregate into tuft-like bundles (Fig. 4.1c). The MTs, including the bundles, gradually assume a more longitudinal orientation as successive cells in a file are examined, first as a broad PPB which then can narrow into a bright band (Fig. 4.1d–f). Similar images have been obtained by others (Cho and Wick, 1989; Cleary and Hardham, 1989), although different interpretations were attached to them (also see Busby and Gunning, 1980). Specifically, the IMB was seen as being replaced by new MTs. While all interpretations are based on the analysis of static, albeit successive, images of different cells, the gradual shift in the alignment of the tuft-like bundles (Fig. 4.1c, d) makes rearrangement seem more likely in our eyes.

MT rearrangements in differentiating stomatal cells

A surprising discovery made with recent immunocytochemical efforts has been the presence of fan-like radial arrays in graminean subsidiary cells (SCs) (Cleary and Hardham, 1989; Palevitz and Mullinax, 1989). These arrays, previously thought to be characteristic of GCs, emanate from a discrete spot or zone in the SC cortex bordering the GMC (Fig. 4.2a, b). After GMC division and the formation of opposing radial arrays in the sister GCs, the resulting four-celled complexes contain two sets of paired opposing arrays, with the middle of the ventral wall separating the GCs serving as the epicentre (Palevitz and Mullinax, 1989).

By following successive complexes distally along a file, changes in the arrays become evident. At first, the SC arrays become more spread out along the cortex and do not focus at

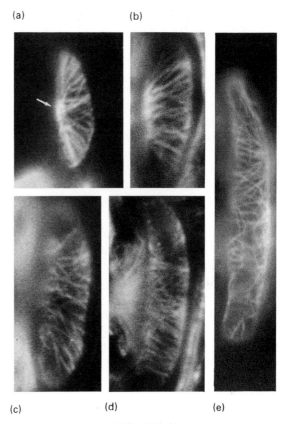

Fig. 4.2. Series of images showing the transition between radial and net-like MT patterns in the cortex of *Avena* SCs. In (a), MTs focus at a discrete spot in the cortex of a young SC. In (b), MTs focus at a zone more spread out than that seen in (a). In the older cell in (c), the array is more spread out still, but retains some of its radial characteristics. Net-like to transverse MTs are seen in the cells in (d) and (e). The cortex of an adjacent GC with radial MTs are seen to the left in (d). a–d, × 2000; e, × 1560. From Palevitz and Mullinax, 1989, with permission.

a restricted spot (Fig. 4.2c–e). At the same time, their orientation switches from a fan arrangement to a net or reticulate to transverse pattern, and the shape of the SC shifts from lenticular to sausage-like. The changes in MT arrangement are gradual enough and show intermediate patterns of such a nature as to suggest that actual rearrangements are occurring, although the assembly of MTs in new directions is certainly possible. Indeed, interpolation of new MTs occurs in the radial cortical arrays of *Allium* GCs, although overall orientation remains the same (Marc *et al.*, 1989). The graminean complexes then undergo a characteristic osmotic transient in which swelling of the GCs is associated with the initial opening of the stomatal pore (Palevitz, 1982). Following this transient, the GC MT arrays change orientation in a manner echoing the preceding shift in the SCs (Palevitz and Mullinax, 1989). Thus, once again all four cells have similar MT patterns, i.e. net-like to transverse. Eventually, MTs in both cell types assume longitudinal to helical orientations (Palevitz and Mullinax, 1989). Besides providing evidence for MT arrangements, the complexes also indicate that a remarkable degree of co-ordination in cytoskeletal arrangement is exerted across the four adjacent cells.

Generative cell division in *Tradescantia*

Perhaps one of the most striking examples of rearrangement and reutilization of MTs is found in generative cells. The generative cell plays a crucial role in angiosperm reproduction, since its division creates the two sperm cells that participate in double fertilization (Maheshwari, 1950). Thus, generative cell division has been the subject of considerable attention over the years, which has led to conflicting views on the mechanism of chromosome separation (see Palevitz and Cresti, 1989 and Palevitz, 1990 for review). In the generative cells of *Tradescantia virginiana*, kinetochores (visualized with CREST serum) remain distributed along the length and depth of the cell prior to anaphase (Fig. 4.3a), with chromosome arms entangled in complex, braided arrangements (Palevitz and Cresti, 1989; Palevitz, 1990). Curiously, kinetochore fibres are highly heterogeneous in length; very long fibres co-exist with much shorter ones, and all are linked to each other and to peripheral MT bundles present before chromosome condensation begins (Fig. 4.3a). Examination of cells early in division indicates that the kinetochore fibres form by capture through interaction of some of these MTs (Liu and Palevitz, 1991). While kinetochore capture of MTs generated from the spindle poles is now reasonably well accepted, the picture in

generative cells is unusual because the arrangement of 'interphase' MTs in these cells (quotation marks are used because details of the cell cycle in this system are scarce) is unlike that found in typical spindles; the MTs form flexuous bundles and display a high degree of cross-bridging (Lancelle *et al.*, 1987).

During anaphase, chromosome separation starts at various points along the cell, and filial kinetochores pass each other going in opposite directions (Fig. 4.3b) (Palevitz and Cresti, 1989; Palevitz, 1990). The kinetochore fibres of chromosomes destined for the same end of the cell (as well as surrounding MTs) join in large superbundles (Fig. 4.3b) (Liu and Palevitz, submitted). The appearance is one of two thick fishing lines to which are attached kinetochores (the fish) going in opposite directions. Elongating MT branches trail back from the superbundles and kinetochore fibres as interzonal elements (Fig. 4.3b). Finally, the resulting sperm cells separate by a cleavage process that constricts the trailing MTs. Each sperm cell inherits as its initial cytoskeleton a superbundle plus attached kinetochores and MT branches (Fig. 4.3b, inset). Thus, direct continuity seems to exist between interphase MT bundles, mitotic MTs and the sperm cytoskeleton. That is, there appears to be rearrangement, translocation and subsequent reutilization of MTs between successive arrays and cell generations.

Precedents

The idea that plant MTs exhibit such behaviour is not novel. For example, the PPB changes from a rather broad band in its early stages to a very narrow, compact array later on (Wick and Duniec, 1983; see Wick, Chapter 17, this volume), as if its component MTs become more tightly associated or gyred. Pickett-Heaps (1969) proposed long ago that PPB MTs are incorporated into the spindle. In early stages of its formation, the phragmoplast appears to utilize MTs left over from the karyokinetic apparatus (for recent results, see Zhang *et al.*, 1990). It has been proposed that changes in the orientation of new wall microfibrils are caused by spring or slinky-like alterations in the

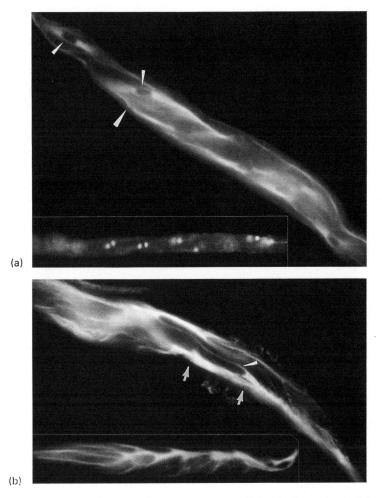

Fig. 4.3. (a) Antitubulin staining of a metaphase generative cell in *Tradescantia virginiana*. Kinetochores (at dark patches) are distributed along the length and depth of the cell. Only some are in focus. Kinetochore fibres are linked to each other and to surrounding MTs (e.g. large arrowhead). Thin MT branches (small arrowheads) link opposing kinetochore fibres and invest the kinetochores. The inset shows another metaphase cell specifically stained for kinetochores using CREST serum. The kinetochores appear as pairs of bright spots bordering the dark patches seen after antitubulin staining. In (b), an anaphase cell contains two superbundles to which are attached the now single kinetochores. One of the superbundles (the lower one) is in sharp focus. Kinetochore fibres consist of short stubs (arrows). Trailing MT branches (arrowhead) link separating kinetochore fibres. The inset shows a superbundle and trailing MTs in a newly formed sperm cell. CREST staining of similar cells shows that single kinetochores are attached at the notches along the superbundle. a, × 1400; inset, × 1100; b, × 1400; inset, × 1600. From Palevitz and Cresti, 1989, with permission; insets from Palevitz, 1990, with permission.

pitch of interphase cortical MTs mediated by bridging moieties (Lloyd and Seagull, 1985). Falconer and Seagull (1985) presented evidence that MTs shift in alignment from axial to transverse during the formation of second-ary wall bands in differentiating tracheary elements. MT realignments may also play a

role in the regeneration of cortical arrays and cell walls following protoplasting (Galway and Hardham, 1986). Precedent for translocation of cytoskeletal polymers also can be found in recent work on the cleavage furrow in animal cells, which shows that the contractile ring is formed by the recruitment of microfilaments

(MFs) from other parts of the cell (Cao and Wang, 1990).

Control of MT rearrangements

If MT rearrangements do occur, we must then examine the possible mechanisms and controls of such movements.

Microtubule-associated motor proteins

That a long, fibrous MT can be targeted and moved to a new location through a viscous cytoplasm filled with organelles and other fibres may seem unlikely at first glance. However, the concept is given credence by the fact that MTs can translocate *in vitro* over a substratum coated with motor proteins such as kinesin and dynein. Since cortical MTs are cross-bridged to the plasmalemma (e.g. Fig. 15 in Palevitz, 1982), it may be proper to ask whether they can also move on the surface of the membrane to targeted sites and thus achieve rearrangement.

While much has been learned about motor proteins in animal cells, their role in controlling organelle transport and MT behaviour in plants is just beginning to be explored. However, some progress in plant microtubule-associated proteins has been achieved. A common characteristic of plant MTs is their tendency to associate with each other in the form of planar aggregates or bundles. An obvious example is the PPB, but cross-bridged bundles are also found elsewhere, including generative cells (Lancelle *et al.*, 1987). Studies of interphase cortical arrays reveal that lateral associations of two to several MTs are common here as well (Hardham and Gunning, 1978). Moreover, the cortical arrays reorganize into bundles of significant dimensions in developing tracheary elements (Falconer and Seagull, 1985; Fukuda and Kobayashi, 1989) and certain other cells (e.g. Jung and Wernicke, 1990). The nature of the proteins that determine these interactions is obviously important. Recently, a protein named dynamin has been linked to MT–MT sliding in animal cells (Shpetner and Vallee, 1989). In plants, polypeptides have been obtained from carrot cells

which promote the bundling of neuronal and carrot MTs *in vitro* (Cyr and Palevitz, 1989; see Cyr, Chapter 5, this volume). The polypeptides confer cold stability on the MTs, and the MT–MT spacings in the bundles are similar to those seen *in situ*. Similar efforts should also be directed at the identification of proteins that link MTs with MFs and membranes, since interactions between these structures are important (see below). All these proteins may be subject to transcriptional and/or post-translational regulation (e.g. via phosphorylation), thereby linking developmental, cell cycle, and physiological signals to MT organization and arrangement.

Actin-mediated systems

MT–MF interactions have been the object of increased interest in recent years. Such interactions may function in specific motility phenomena, such as organelle transport, or they could be necessary for changes in organization of one cytoskeletal component or the other. Interactions with MFs in forming tracheary elements seem to be involved in the rearrangement of cortical MTs necessary for the deposition of secondary wall bands (Fukuda and Kobayashi, 1989). Our own studies on PPB formation during epidermal cell divisions in *Allium* seedlings may shed further light on this subject (Mineyuki *et al.*, 1989; Mineyuki and Palevitz, 1990). Our results support the idea that the PPB passes through two developmental phases in its development. The PPB first appears as a rather broad assemblage of MTs and MFs whose orientation matches that of the future division plane. Thus, division plane *orientation* is determined quite early. Next, the MT band narrows in a process that seems to determine the exact *position* of the future cell wall. The actin component of band remains broad, however (Mineyuki and Palevitz, 1990). Application of cytochalasin D (CD) to this system produces interesting results (Mineyuki and Palevitz, 1990). Namely, CD does not interfere with the first phase, but does inhibit band narrowing (Table 4.1 and Fig. 4.4). As a result, the width of PPBs in prophase, when the band should be at its narrowest, increases with increasing CD concentration,

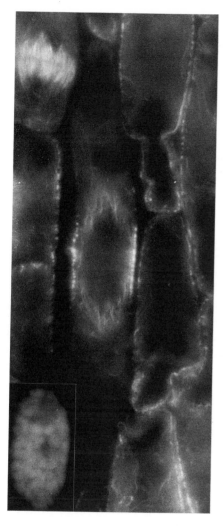

Fig. 4.4. Antitubulin immunofluorescence of an abnormally broad PPB in a symmetrically dividing cotyledon cell of *Allium cepa*. Initial elements of the spindle also invest the nucleus. A phragmoplast and cell plate are seen in another cell at the upper left. Inset: Hoechst fluorescence of the first cell's nucleus showing chromosome condensation indicative of prophase. × 1850. From Mineyuki and Palevitz, 1990, with permission.

both in terms of absolute dimensions and when normalized to nuclear length (Table 4.1 and Fig. 4.4). Since the effect is saturated in as little as 30 min, the drug may not only prevent broad bands from narrowing, but may also lead to rebroadening of previously narrowed PPBs. The effect on the PPB is also accompanied by abnormal displacements of prophase nuclei,

spindles and new cell plates in those cells destined to divide asymmetrically.

These experiments indicate that actin (and associated proteins) operates in the maturation of the PPB, perhaps by confining the broad band to a progressively more limited zone in the cortex. As already noted, an analogous process may operate during the formation of secondary wall bands in tracheary elements (Fukuda and Kobayashi, 1989). However, actin apparently does not function in other aspects of MT organization. For example, the generation and operation of the radial MT array in *Allium* is insensitive to cytochalasin treatment (Marc et al., 1989).

Environmental influences

Shifts in MT organization appear to be subject to endogenous influences as well as controls exerted by surrounding cells and other environmental factors. Generative cells are interesting in this regard, because typical somatic-like divisions as well as the more unorthodox arrangements described above have been reported, sometimes within the same genus and species (see Palevitz and Cresti, 1989 and Palevitz, 1990 for review). Such behaviour could be explained if the organization of the division apparatus is subject to interplay between space constraints imposed by the narrow confines of the pollen tube and the size/volume of the chromosomes (Sax and O'Mara, 1941). In other words, the spindle may be squeezed into seemingly bizarre configurations; nevertheless, it remains fully functional. Another example of such behaviour, albeit less drastic, may be evident in the oblique spindles seen in diverse plant cells (see Fig. 17 in Palevitz and Cresti, 1989). Regardless of the overriding constraints, the mitotic apparatus displays a remarkable degree of *plasticity*, with divergent arrangements still resulting in successful division.

One of the most exciting findings to emerge in the past few years has been the co-ordination in the arrangement of MTs across cell boundaries. However, hints of this possibility were also seen in earlier ultrastructural studies. It seemed reasonable, for example, that co-ordination in some form between sister GCs

Table 4.1. Effect of CD on cell length, nuclear length and PPB width in asymmetrically dividing prophase cells in the epidermis of *Allium* cotyledons

Treatment	Cell length (μm)	Nuclear length (μm)	PPB width (μm)	PPB width/ nuclear length	n
H₂O	38.2 ± 1.9	19.0 ± 0.7	3.9 ± 0.2	0.21 ± 0.01	15
DMSO (0.5%)	40.1 ± 2.6	18.9 ± 0.7	3.9 ± 0.3	0.21 ± 0.02	21
CD (2 μM)	36.7 ± 3.1	19.8 ± 0.9	8.3 ± 0.6	0.43 ± 0.05	6
CD (20 μM)	35.5 ± 2.4	18.2 ± 0.5	12.2 ± 2.0	0.68 ± 0.12	13

The length of cells and nuclei and the width of PPBs in the epidermis of the cotyledons of 3-day-old *Allium* seedlings incubated in 2 or 20 μM CD for 1 day were compared with values from distilled water and DMSO controls. All values represent the mean ± the standard error of the mean. Data from Mineyuki and Palevitz, 1990.

could account for the mirror image configurations of their cortical arrays (e.g. Palevitz and Hepler, 1976; Galatis, 1980). Later, Wick (1985) published a striking immunofluorescence micrograph in which continuities in cortical MT alignment across adjacent root cells were obvious. As discussed above, evidence has been found of cytoskeletal co-ordination across the four cells of the *Avena* stomatal complex (Palevitz and Mullinax, 1989). Radial arrays are generated in the GCs *and* SCs of this and other grass species, which then undergo similar shifts in arrangement to an intermediate network and a final longitudinal pattern. It seems possible that these patterns are keyed to longitudinal and transverse polarity vectors implicit in other aspects of the development of these complexes. For example, as already mentioned, the interphase MT array in grass GMCs consists of a transverse IMB. It may not be coincidental that during the asymmetric divisions that cut off SCs from lateral mother cells, the nucleus migrates towards the GMC (Pickett-Heaps, 1969). Moreover, one pole of the spindle apparatus during this division is firmly pressed against the wall adjacent to the IMB (Fig. 4.5) and cannot be displaced by centrifugation (Pickett-Heaps, 1969). This same region also contains a focal array of MFs (see Cho and Wick, 1990 for results and discussion of developmental consequences). After the SC forms, its radial MT array focuses at this point as well (Cleary and Hardham, 1989; Palevitz and Mullinax, 1989). Thus, the IMB is a manifestation of polarity influencing the cytoskeleton in all cells of the stomatal complex.

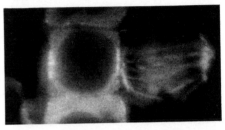

Fig. 4.5. SC formation next to a GMC in an *Avena* leaf. One pole of the mitotic figure is located at a position in the cortex adjacent to the IMB of the GMC. × 1600. From Mullinax and Palevitz, 1989.

Of course, the nature of any signals controlling events in these cells is obscure. However, chemical and environmental influences are known to affect the plant cytoskeleton. In grass stomatal cells, the shift in order from radial to net-like and longitudinal occurs around the time of a transient osmotic event responsible for the initial opening of the pore (Palevitz, 1982). Perturbation of the complexes at this time blocks the shift. In stomatal complexes of *Cyperus esculentus*, which share many similarities with those of grass species, similar shifts are also altered by environmental perturbations (e.g. water stress; see Palevitz, 1982 for review). Such perturbations could be transmitted via hormones such as ethylene (Lang *et al.*, 1982; Roberts *et al.*, 1985).

The NE and MT rearrangements

Having spent a good deal of time reviewing evidence in favour of MT rearrangements, it is also appropriate to address a situation where this process has been proposed but its occurrence is less certain. Considerable recent attention has been directed at the role of the NE in generating the interphase cortical array. A key part of the proposal involves rearrangement and translocation of MTs. The hypothesis is based on the appearance of MTs emanating from the new NE in telophase, and the binding of serum 5051, which marks centrosomal material in other eukaryotic cells, to the perinuclear region and spindle poles in root tips (Clayton et al., 1985; Wick, 1985, 1988; Lloyd, 1987). According to this hypothesis, the new perinuclear MTs are deployed to the plasmalemma, where they assume positions parallel to the membrane and to each other. However, while there can be little doubt that the NE is a site for MT formation during cytokinesis (and during S and prophase), its relationship to the cortex is still under investigation. In BY-2 tobacco suspensions, which can be synchronized with aphidicolin to obtain large populations of cells at specific points in the cell cycle, the appearance of NE-associated MTs bears no obvious relationship to the formation of cortical arrays (Hasezawa et al., 1990). When attention is specifically directed at the cytokinesis–G_1 interface, almost no overlap is seen between NE MTs and ordered elements in the cortex. The rarity of such overlap has been noted elsewhere as well (Wick, 1985; Simmonds et al., 1989). In the GCs of Allium, perinuclear MTs are seen for a brief time (c. 20 min) after cytokinesis but are then replaced by the radial MT array characteristic of these cells (Marc et al., 1989). The radial array is generated in the cortex along the middle part of the ventral wall. As in BY-2, no temporal overlap is seen between the perinuclear elements and the new cortical array. In GCs in which the array is depolymerized with colchicine and the nucleus then centrifuged out of position, MTs reappear in the expected region of the cortex following UV illumination, and bear no relation to the nucleus.

MT reassembly in the cortex following drug treatment and protoplasting has been seen in other cells as well (e.g. Galway and Hardham, 1989; Wasteneys and Williamson, 1989). MTs

reappear around the nucleus after the cortical array reforms (Wasteneys and Williamson, 1989). During normal cell growth, additional MTs are added to the cortical array by interpolation (e.g. Hardham and Gunning, 1979; Marc et al., 1989), and no sign of a nuclear origin is evident. Finally, the validity of serum 5051 as a probe for MT nucleation sites in higher plants is now in question (Harper et al., 1989; Marc et al., 1989).

These results indicate that the cortex can initiate and align its complement of MTs without intervention by the NE. On the other hand, that the NE serves as a MT nucleation site cannot be dismissed; MTs do form there late in division, as well as in S and prophase. However, explanations for the transition between NE-associated MTs and the cortical array in early G_1, other than a direct translocation between the two, are possible. These include the presence of mobile or flexible nucleation sites as well as differential activation of separate nucleating sites located at the NE and in the cortex.

Refrain and concluding remarks

An emerging body of evidence indicates that MTs can be translocated and reutilized within and between various arrays. Such reutilization undoubtedly complements specific depolymerization, assembly and differential stabilization processes in establishing and augmenting MT arrays. It must be realized, however, that almost all observations to date consist of static images obtained from fixed or permeabilized cells. In order to ascertain the actual contribution of assembly and rearrangement mechanisms in vivo, an accurate assessment of MT dynamics is needed, which will necessitate visualizing MTs in cells treated with exogenous or derivatized tubulin. Pioneering efforts in this direction are already underway (Vantard et al., 1990; Zhang et al., 1990; Asada et al., 1991).

Acknowledgements
I thank Mauro Cresti, Bo Liu, Jan Marc, Yoshinobu Mineyuki and J. Bennett Mullinax for their contributions to this work. Bill R.

Brinkley kindly provided the CREST serum. Supported by NSF grant DCB-8703292.

References

Asada, T., Sonobe, S. and Shibaoka, H. (1991). Microtubule translocation in the cytokinetic apparatus of cultured tobacco cells. *Nature* **350**, 238–241.

Busby, C.H. and Gunning, B.E.S. (1980). Observations on pre-prophase bands of microtubules in uniseriate hairs, stomatal complexes of sugarcane, and *Cyperus* root meristems. *Eur. J. Cell Biol.* **21**, 214–223.

Cao, L.-g. and Wang, Y.-l. (1990). Mechanism of the formation of contractile ring in dividing cultured animal cells. I. Recruitment of preexisting actin filaments into the cleavage furrow. *J. Cell Biol.* **110**, 1089–1096.

Cho, S.-O. and Wick, S.M. (1989). Microtubule orientation during stomatal differentiation in grasses. *J. Cell Sci.* **92**, 581–594.

Cho, S.-O. and Wick, S.M. (1990). Distribution and function of actin in developing stomatal complex of winter rye (*Secale cereale* cv. Puma). *Protoplasma* **157**, 154–164.

Clayton, L., Black, C.M. and Lloyd, C.W. (1985). Microtubule nucleating sites in higher plant cells identified by an auto-antibody against pericentriolar material. *J. Cell Biol.* **101**, 319–324.

Cleary, A.L. and Hardham, A.R. (1989). Microtubule organization during development of stomatal complexes of *Lolium rigidum*. *Protoplasma* **149**, 67–81.

Cyr, R.J. and Palevitz, B.A. (1989). Microtubule-binding proteins from carrot. *Planta* **177**, 245–260.

Falconer, M.M. and Seagull, R.W. (1985). Xylogenesis in tissue culture: taxol effect on microtubule reorientation and lateral association in differentiating cells. *Protoplasma* **128**, 157–166.

Fukuda, H. and Kobayashi, H. (1989). Dynamic organization of the cytoskeleton during tracheary-element differentiation. *Dev. Growth Differ.* **31**, 9–16.

Galatis, B. (1980). Microtubules and guard cell morphogenesis in *Zea mays* L. *J. Cell Sci.* **45**, 211–244.

Galatis, B. (1982). The organization of microtubules in guard cell mother cells of *Zea mays*. *Can. J. Bot.* **60**, 1148–1166.

Galway, M. and Hardham, A.R. (1986). Microtubule reorganization, cell wall synthesis and establishment of the axis of elongation in regenerating protoplasts of the alga *Mougeotia*. *Protoplasma* **135**, 130–143.

Galway, M. and Hardham, A.R. (1989). Oryzalin-induced microtubule disassembly and recovery in regenerating protoplasts of the alga *Mougeotia*. *J. Plant Physiol.* **135**, 337–345.

Hardham, A.R. and Gunning, B.E.S. (1978). Structure of cortical microtubule arrays in plant cells. *J. Cell Biol.* **77**, 14–34.

Hardham, A.R. and Gunning, B.E.S. (1979). Interpolation of microtubules into cortical arrays during cell elongation and differentiation in roots of *Azolla pinnata*. *J. Cell Sci.* **37**, 411–442.

Harper, J.D.I., Mitchison, J.M., Williamson, R.E. and John, P.C.L. (1989). Does the autoimmune serum 5051 specifically recognise microtubule organising centres in plant cells? *Cell Biol. Int. Rep.* **13**, 471–483.

Hasezawa, S., Marc, J. and Palevitz, B.A. (1990). Microtubule reorganization during the cell cycle in synchronized BY-2 tobacco suspensions. *Cell Motil. Cytoskel.*, **18**, 94–106.

Jung, G. and Wernicke, W. (1990). Cell shaping and microtubules in developing mesophyll of wheat (*Triticum aestivum* L.). *Protoplasma* **153**, 141–148.

Kirschner, M. and Mitchison, T. (1986). Beyond self-assembly: from microtubules to morphogenesis. *Cell* **45**, 329–342.

Lancelle, S.A., Cresti, M. and Hepler, P.K. (1987). Ultrastructure of the cytoskeleton in freeze-substituted pollen tubes of *Nicotiana alata*. *Protoplasma* **140**, 141–150.

Lang, J.M., Eisinger, W.R. and Green, P.B. (1982). Effects of ethylene on the orientation of microtubules and cellulose microfibrils of pea eipcotyl cells with polylamellate cell walls. *Protoplasma* **110**, 5–14.

Liu, B. and Palevitz, B.A. (1991). Kinetochore fiber formation in dividing generative cells of *Tradescantia*. Kinetochore reorientation associated with the transition between lateral microtubule interactions and end-on kinetochore fibers. *J. Cell Sci.* **98**, 475–482.

Lloyd, C.W. (1987). The plant cytoskeleton: the impact of fluorescence microscopy. *Ann. Rev. Plant Physiol.* **38**, 119–139.

Lloyd, C.W. and Seagull, R.W. (1985). A new spring for plant cell biology: microtubules as dynamic helices. *Trends Biochem. Sci.* **10**, 476–478.

Maheshwari, P. (1950). *An Introduction to the Embryology of Angiosperms*. McGraw-Hill, New York.

Marc, J. and Hackett, W.P. (1989). A new method for immunofluorescent localization of microtubules in surface cell layers; application to the shoot apical meristem of *Hedera*. *Protoplasma* **148**, 70–79.

Marc, J., Mineyuki, Y. and Palevitz, B.A. (1989). The generation and consolidation of a radial array of cortical microtubules in developing guard cells of *Allium cepa* L. *Planta* **179**, 516–529, 530–540.

Marc, J. and Palevitz, B.A. (1990). Regulation of the spatial order of cortical microtubules in developing guard cells of *Allium*. *Planta* **182**, 626–634.

Mazia, D. (1987). The chromosome cycle and the centrosome cycle in the mitotic cycle. *Int. Rev. Cytol.* **100**, 49–92.

Mineyuki, Y., Marc, J. and Palevitz, B.A. (1989). Development of the preprophase band from random cytoplasmic microtubules in guard mother cells of *Allium cepa*. *Planta* **178**, 291–296.

Mineyuki, Y. and Palevitz, B.A. (1990). Relationship between preprophase band organization, F-actin and the division site in *Allium*. Fluorescence and morphometric studies on cytochalasin-treated cells. *J. Cell Sci.* **97**, 283–295.

Mullinax, J.B. and Palevitz, B.A. (1989). Microtubule reorganization accompanying preprophase band formation in guard mother cells of *Avena sativa* L. *Protoplasma* **149**, 89–94.

Palevitz, B.A. (1982). The stomatal complex as a model of cytoskeletal participation in cell differentiation. In *The Cytoskeleton in Plant Growth and Development* (ed. C.W. Lloyd), pp. 345–376. Academic Press, London.

Palevitz, B.A. (1990). Kinetochore behavior during generative cell division in *Tradescantia virginiana*. *Protoplasma*, **157**, 120–127.

Palevitz, B.A. and Cresti, M. (1989). Cytoskeletal changes during generative cell division and sperm formation in *Tradescantia virginiana*. *Protoplasma* **150**, 54–71.

Palevitz, B.A. and Hepler, P.K. (1976). Cellulose microfibril orientation and cell shaping in developing guard cells of *Allium*: the role of microtubules and ion accumulation. *Planta* **132**, 71–93.

Palevitz, B.A. and Mullinax, J.B. (1989). Developmental changes in the arrangement of cortical microtubules in stomatal cells of oat (*Avena sativa* L.). *Cell Motil. Cytoskel.* **13**, 170–180.

Pickett-Heaps, J.D. (1969). Preprophase microtubules and stomatal differentiation; some effects of centrifugation on symmetrical and asymmetrical cell division. *J. Ultrastruct. Res.* **27**, 24–44.

Roberts, I.N., Lloyd, C.W. and Roberts, K. (1985). Ethylene-induced microtubule reorientations: mediation by helical arrays. *Planta* **164**, 439–447.

Sax, K. and O'Mara, J.G. (1941). Mechanism of mitosis in pollen tubes. *Bot. Gaz.* **102**, 629–636.

Simmonds, D.H., Conibear, E. and Setterfield, G. (1989). Microtubule organization during the cell cycle of cultured *Vicia hajastana* Grossh. *Biochem. Cell Biol.* **67**, 545–552.

Shpetner, H.S. and Vallee, R.B. (1989). Identification of dynamin, a novel mechanochemical enzyme that mediates interactions between microtubules. *Cell* **59**, 421–432.

Vantard, M., Levilleurs, N., Hill, A.-M., Adoutte, A. and Lambert, A.-M. (1990). Incorporation of *Paramecium* axonemal tubulin into higher plant cells reveals functional sites of microtubule assembly. *Proc. Natl Acad. Sci. U.S.A.* **87**, 8825–8829.

Warner, F.D. and McIntosh, J.R. (1989). *Cell Movement, vol. 2. Kinesin, Dynein and Microtubule Dynamics*. Alan R. Liss, New York.

Wasteneys, G.O. and Williamson, R.E. (1989). Reassembly of microtubules in *Nitella tasmanica*: assembly of cortical microtubules in branching clusters and its relevance to steady-state microtubule assembly. *J. Cell Sci.* **93**, 707–714.

Wick, S.M. (1985). Immunofluorescence microscopy of tubulin and microtubule arrays in plant cells. III. Transition between mitotic/cytokinetic and interphase microtubule arrays. *Cell Biol. Int. Rep.* **9**, 357–371.

Wick, S.M. (1988). Immunolocalization of tubulin and calmodulin in meristematic plant cells. In *Calcium-binding Proteins*, vol. 2 (ed. M.P. Thompson), pp. 21–45. CRC Press, Boca Raton, FL.

Wick, S.M. and Duniec, J. (1983). Immunofluorescence microscopy of tubulin and microtubule arrays in plant cells. I. Preprophase band development and concomitant appearance of nuclear-associated tubulin. *J. Cell Biol.* **97**, 235–243.

Zhang, D., Wadsworth, P. and Hepler, P.K. (1990). Microtubule dynamics in living dividing plant cells. Confocal imaging of microinjected fluorescent brain tubulin. *Proc. Natl Acad. Sci. U.S.A.*, **87**, 8820–8824.

5. MICROTUBULE-ASSOCIATED PROTEINS IN HIGHER PLANTS

Richard J. Cyr

Department of Biology, Pennsylvania State University,
208 Mueller Laboratory, University Park, PA 16802, USA

The accompanying chapters provide compelling evidence for the participation of MTs in many processes impacting upon the plant cell during growth and development. It is now appropriate to begin the biochemical dissection of plant MTs in order to discern, at the molecular level, how they carry out their various activities. Logically then, one of the first tasks is to discover the composition of MTs in order to learn how various MT proteins interact to affect MT activity.

As discussed in Chapter 3, we know the major protein constituent of MTs to be tubulin, which upon self-assembly forms the basic 25-nm MT structure. Typically, highly purified tubulin is capable of forming MTs *in vitro* with only GTP and Mg^{2+} as requisite cofactors (Dustin, 1984). However, the MTs formed in the presence of tubulin alone show only a limited behaviour; they either assemble or disassemble. Although assembly and disassembly is an important spatial aspect of MT activity, MTs must be able to do more in order to carry out their various functions. Biochemically, this is accomplished using auxiliary, or associated, proteins.

Microtubule associated proteins (MAPs) describe a class of proteins which upon binding to a MT affect its function and/or behaviour (Vallee and Bloom, 1984; Olmsted, 1986; Diaz-Nido *et al.*, 1990). In a broad sense, MAPs function to extend the biochemical repertoire of tubulin; all eukaryotes studied to date possess them. However, unlike tubulin, which is highly conserved between organisms and even kingdoms, there seems to be considerable diversity among MAPs. Evolution appears to have preserved tubulin whilst selecting for novel MAPs to create functional heterogeneity, yet maintain structural constancy. Evidence for this evolutionary trend is three-fold; biochemical, genetic, and immunological.

Biochemically, tubulin from plant and animal sources is remarkably similar (Fosket, 1989). Although there are some important differences in how they interact with drugs (see Morejohn, Chapter 3, this volume), they have similar weights, isoelectric points, cofactor requirements, assembly-temperature optima, and peptide maps. Genetically, the tubulin genes sequenced to date confirm the evolutionary conservation of these proteins (see Hussey *et al.*, Chapter 2, this volume). Immunologically, antibodies raised against tubulin derived from species in one kingdom are often found to cross-react with species in another kingdom, in spite of the fact that immune systems tend to recognize divergent epitopes.

However, this trend towards evolutionary conservation does not appear to hold for MAPs. First, MAPs comprise a relatively diverse group of polypeptides and proteins with different molecular weights and biochemical behaviour. Although we have limited published data for plant MAPs, what is known indicates they are different from their animal counterparts. First, the molecular weights of plant MAPs differ from those of animals. In our work with cultured cells of carrot we find prominent polypeptides having apparent molecular weights of 114 000, 111 000, 102 000, 98 000, 76 000, 58 000 and 39 000 to bind MTs (Cyr and Palevitz, 1989; Fig. 5.1). In animals, the MAPs are typically classified into high molecular weight MAPs (HMW-MAPs; >200 kD) and low molecular weight MAPs (<200

The Cytoskeletal Basis of Plant Growth and Form   ISBN 0–12–453770–7

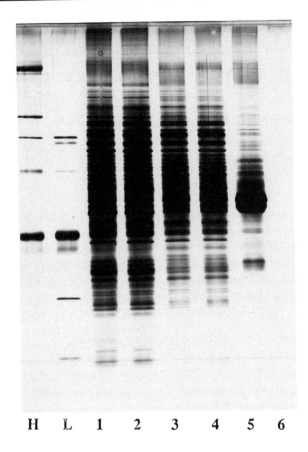

Fig. 5.1. SDS-PAGE analysis of carrot proteins binding to taxol-stabilized neuronal MTs. To the high-speed homogenate (lane 1) was added taxol-stabilized MTs. After 30 min at 4°C, the MTs were removed by centrifugation; the supernatant after the removal of MTs is shown in lane 2. The microtubules were taken up in PM + 0.5 M NaCl + 10 μM taxol and the microtubules collected by centrifugation. The supernatant (lane 3) was desalted and taxol-stabilized MTs were added once again. After 30 min at 4°C, the MTs were removed by centrifugation in an Airfuge. The supernatant from this second binding cycle is shown in lane 4 and the pellet is shown in lane 5. Lane 6 shows a pellet after the same reaction represented in lane 5, except that MTs were omitted. Lanes H and L contain high and low ranges, respectively, of molecular weight standards. M_r values for H = 200 000; 116 250; 97 400; 66 200; 42 699. M_r values for L = 97 400; 66 200; 42 699; 31 000; 21 500; 14 400. (Cyr, R.J. and Palevitz, B.A. (1989) *Planta* **177**, 245–260, used with permission.)

kD); the values as listed above for carrot cells would therefore fall into the latter category. The apparent lack of HMW-MAPs is perhaps not surprising in view of the observation that within the animal kingdom there seems to be a preponderance of lower weight MAPs (at least in non-neuronal tissues).

Additional evidence for diversity of MAPs within the animal kingdom is provided by immunological studies which have shown that many antibodies raised against MAPs derived from nervous tissue fail to cross-react in an interpretable fashion with animal cells of non-neuronal origin. Naturally then it is not surprising that antibodies raised against both high and low molecular weight MAPs derived from animal sources have, as of yet, failed to cross-react in an interpretable fashion to plant polypeptides.

The limited cross-reactivity of antibodies against MAPs is not surprising, as the most extensively studied MAPs are derived from neuronal material which is probably the most specialized of all animal tissues. In the hundreds of millions of years since the plant and animal kingdoms diverged, there have been unarguable differences in selective pressures affecting MT behaviour in nervous tissues compared with those found in any plant cell type.

In summary, evolution appears to have worked to preserve the basic structure of MTs via conservation of tubulins. Functional divergence of MTs seems to have been achieved via selection for a relatively diverse group of MAPs which, upon binding to MTs, dramatically extend the functional activity of the basic MT to suit the unique physiological requirement of different organisms at various points in their development. Prudently, a discussion of MAPs should therefore begin by addressing how they influence MT function.

What roles might MAPs play in the functioning of MTs?

In discussing the role which MAPs play in the functioning of MTs, it is convenient to address two fundamental microtubular questions, i.e. how do MTs come to be spatially organized in the cell, and once positioned, how do they carry out their various activities? In discussing spatial organization of MTs there are three possible ways MAPs may affect the placement of a MT.

Tubulin is known to assemble into MTs by itself. However, the critical concentration (Cc) of pure tubulin necessary for assembly *in vitro* is relatively high, in the range of 40–80 μM (Dustin, 1984). On the other hand, typical cellular concentrations of tubulin are lower than this (1–20 μM; Fulton and Simpson, 1979); how then do MTs form *in vivo*? Insight into this problem came during early biochemical studies. It was noted that upon purification of neuronal MTs by cyclic warm/cold, assembly/disassembly purification some non-tubulin proteins co-purified with tubulin. These 'contaminants' of tubulin preparations were later found to affect the assembly characteristics of tubulin (by lowering the Cc) and the term 'MAPs' was coined with reference to these proteins. Thus,

one of the first activities attributed to MAPs was the dramatic alteration of the Cc for tubulin assembly. In a typical *in vitro* assembly reaction containing MAPs, the Cc is typically reduced to 1–2 μM, a level approximating that found within many cells. The MT-assembly promoting activity has also been observed in MAP preparations obtained from cultured plant cells (Cyr and Palevitz, 1989).

More direct evidence for a role of MAPs in affecting MT polymerization *in vivo* was obtained by studying the developmental expression of MAPs in differentiating cells. In neuronal cells, where this has been most thoroughly studied, the appearance of MAPs (both high and low molecular weight) correlates well with the state of MT polymerization occurring in developing neurites (Matus, 1988). Furthermore, by using immunolocalization techniques on differentiating cells it has been observed that MAPs presage the appearance of MTs within growing dendritic extensions (Bernhardt and Matus, 1982). This suggests that in order for MTs to form at a particular site within the cell, there must first be MAPs present at the location. This type of evidence leads one to consider the molecular composition of the microtubule organizing centres (MTOC) which have been described in structural studies (Gunning and Hardham, 1982). Perhaps an MTOC is a complex which has two molecular functions: one to allow correct cellular targeting and the second, as a MAP which lowers the Cc for tubulin in the region it occupies.

In considering how a MAP influences the Cc of tubulin, it is relevant to recognize that the Cc has two components: nucleation and stabilization. Small aggregates or oligomers of tubulin are thought to form the 'seed' for a MT. If the addition of subsequent tubulin dimers is stable, a MT then forms. However, seeds themselves may be unstable. Thus, a MAP which stabilizes or promotes seed assembly may decrease the Cc; the neuronal MAPs 1 and 2 appear to function in this fashion (Olmsted, 1986). A MAP with such activity can function at substoichiometric concentrations to tubulin.

Besides stabilizing MT seeds, a MAP may exert a stabilizing influence along the length of a MT. Once a MT is seeded there is a net addition to one end of the polar MT (the plus end, with the seed at the minus end). In the 'dynamic instability' model of MT behaviour (Mitchison and Kirschner, 1984) the growing MT may be stable and persist, or unstable and depolymerize. It is unclear how many factors influence the stability of a growing MT, however, the τ subclass of MAPs is known to stabilize MTs by binding along their length (Olmsted, 1986). This type of MAP-induced MT stabilizing effect requires a higher stoichiometric relationship than one which acts on seed stability/formation. In addition to stabilization by binding along the length of the MT, one type of MAP of neuronal origin, termed Stable Tubule Only Protein (STOP), induces MT stability by forming substoichiometric 'caps' which apparently block depolymerization past their point of association on the MT (Job *et al.*, 1982).

We know that cultured plant cells contain proteins which stabilize MTs (at low tubulin concentrations and at low temperatures; Cyr and Palevitz, 1989), although we do not know if these stabilizing MAPs affect seed stability, bind along the entire MT length, form stable caps or a combination of all three. Owing to the fact that many plant MTs (especially those in the cortical arrays) are often cold stable, it will be particularly important to investigate further the role which MAPs play in this stabilizing process.

Not only can MAPs affect MT nucleation and stability, they can also affect the rate at which tubulin is added to the MT. For example, a MAP has been isolated from *Xenopus* oocytes which accelerates the tubulin dimer addition rate 10-fold (Gard and Kirschner, 1987).

Besides affecting the appearance of a MT, a role for MAPs in MT disassembly must also be entertained. This may involve simply the dissociation of a MAP from a MT resulting in its destabilization. This dissociation may be brought about by specific regulatory factors; for example, the binding of several MAPs to MTs is known to be affected by the phosphorylation state of the MAP. It is believed that several kinases, by their interactions with MAPs, are pivotal components of transduction pathways serving to communicate spatial information by altering the polymerization characteristics of MTs (Matus, 1988; Diaz-Nido *et al.*, 1990; Verde *et al.*, 1990).

In considering the intracellular cues to which MTs may be receptive we must consider a putative role for calcium. Of particular relevance is the Ca^{2+}/calmodulin complex which has been reported in animal cells to depolymerize MTs via some form of interaction with MAPs (Allan and Hepler, 1989). It is currently not well understood how the Ca^{2+}/calmodulin complex depolymerizes MTs; however, because of the requirement for MAPs it is intriguing to consider that MAPs may serve as an intracellular receptor for this important regulatory ligand. Our laboratory has been using a lysed cell model (carrot protoplasts) to study this phenomenon in plants cells. We find Ca^{2+}/calmodulin to be a potent MT-destabilizing agent in this system and furthermore the addition of exogenous plant MAPs dramatically amplifies the effect. We are currently examining the possibility that MAPs are important components of a transduction chain in higher plants permitting MTs to receive information from second messengers such as calcium.

The spatial arrangement of MTs is not governed solely by polymerization phenomena. In plants there is compelling evidence that cortical MTs can be moved *in toto* within the cell (Lloyd and Seagull, 1985; Palevitz, Chapter 4, this volume). A natural question then is how this is accomplished. In a growing plant cell, the movement may simply be passively coupled to cell extension with water potential ultimately providing the motive force. However, in a non-growing cell other mechanisms must be considered. One class of MAP yet to be described in plant cells is the so-called 'MT-motors' which include dynein (and its cytoplasmic relatives) and kinesin (Vale, 1987). These MAPs use the energy of ATP hydrolysis to move MTs physically, or objects along MTs. It would be important to know, for plant cells, whether these MAPs can power the sliding of MTs past one another. Because many MTs are bundled in plant cells in parallel groups (especially in the cortex) it is intriguing to contemplate a role for this type of MAP in the spatial rearrangement of MTs. That is, they may function to move bundled MTs past one another in a telescopic fashion to extend the array. Thus, a plant cell which has undergone elongation may use a motor to control MT-pitch angle by extending the effective length of the cortical array; in the spatially constricted cortex the force generated by MT-extension could only be relieved via a lateral moment. Precedence for this basic type of MT-sliding motion is found within the bundled MT arrays in neurites where MT sliding is thought to be partly responsible for the MT-based morpho-genesis of neuronal extension (Diaz-Nido *et al.*, 1990).

Besides providing the motive force for MT–MT sliding, these motor proteins are also capable of moving subcellular structures along MTs. In higher plants, the most likely role for such a MAP would be chromosomal movement during karyokinesis and vesicular transport in the phragmoplast during cytokinesis. In cells where it has been closely studied it appears that a MT-motor is located at the kinetochore (Pfarr *et al.*, 1990; Steuer *et al.*, 1990). The hydrolysis of ATP is converted into mechanical energy with the kinetochore MTs serving as vectors for the movement of chromosomes; the kinetochore fibres progressively depolymerize as the chromosomes move towards the poles. There is no direct evidence that higher plants use a similar mechanism for moving chromo-somes. However, given that chromosomal segregation is a fundamental feature of all eukaryotes, it is unlikely that the proteins involved in the process evolved independently. Indeed, from an evolutionary standpoint the movement of chromosomes is probably the most important role for MTs and therefore the proteins involved in the process would be the most conserved of all MAPs. Thus, plant homologues probably exist.

Phragmoplast MTs function in the convey-ance of secretory vesicles to the site of cell plate formation. It is likely that a MT-motor of some sort is involved in powering the movement of these vesicles along the MTs. Whether these motors are homologous or analogous to motor proteins found in animal cells is moot and we will have to await the appearance of data obtained from higher plants. Besides simply

transporting vesicles in the phragmoplast, MT-motors might also be involved in the posi-tioning of the coalescing cell plate. Careful observations of cytokinetic plant cells reveal the growing plate to be capable of rotation just prior to fusing with the plasma membrane (Palevitz, 1986). This rotation appears to require ATP and therefore it is possible that a MT-motor in the phragmoplast may interact with cortical determinants to align the cell plate in its final location.

MAPs may also serve as linkers between MTs and other cytoskeletal components or organelles. Although there is no biochemical evidence for these linker-MAPs in higher plants, there are at least three locations where they would be predicted to occur.

Actin microfilaments and MTs are often seen to co-localize in plant cells (see Chapters 1, 14, 17, 18 and 21, this volume). Moreover, the organization of microfilaments has been shown in some cases to be dependent upon the organization of MTs (but not vice versa). Because pure actin and tubulin do not them-selves associate it is likely that a MAP may serve to interface these two cytoskeletal systems. Cytoskeletal preparations from animal tissues indicate that some MAPs indeed fulfil this function (Griffith and Pollard, 1978) but we have no information in higher plants as to whether a similar activity is present in plant MAPs.

There are several reports which describe the distribution of intermediate filament (IF) antigens in the cells of higher plants (see Shaw *et al.*, Chapter 6, this volume). Interestingly some of the distribution patterns are similar to those of MTs. If indeed this reflects the distribution of intermediate filaments then, as with actin, some MAPs may be involved in the linkage of IFs with MTs.

Besides interfacing MTs with other com-ponents of the cytoskeleton, some MAPs may link MTs to organelles, particularly the plasma membrane (PM). Due to the role of MTs in somehow influencing the orientation of cellulose microfibrils (see Giddings and Staehelin, Chapter 7, this volume) it will be essential to investigate the presence of such a linker-MAP(s). The existence of MT/PM bridging MAPs is indicated by electron microscopic observations showing electron-dense material (presumably proteins) between cortical MTs and the PM. Additionally, when protoplasts are anchored onto poly-L-lysine surfaces and then lysed, the MTs remain attached to the inner surface of the membrane.

Little information is available about MAPs which may fulfil this function in higher plants and it is an area which clearly deserves extensive study. At this early date it is difficult to know the best avenue of pursuit for the study and identification of this class of plant MAPs. The putative MT/PM linker-MAP would be expected to have characteristics of both a MAP as well as a peripheral/integral membrane protein. It will therefore undoubtedly be important to apply appropriate membrane as well as cytoskeletal techniques to study these proteins.

How are MAPs studied?

Over the years a variety of approaches has been developed to isolate MAPs. As previously mentioned, this class of proteins was first observed as contaminants during the cyclic purification of tubulin. Neuronal tissue has a high percentage of tubulin (up to 42% of soluble protein) and it is relatively easy to assemble MTs within crude extracts. The assembled MTs can subsequently be sedimented under moderate g forces ($\times 48\,000$). Many of these MTs are cold labile and can be resolubilized upon cold treatment ($< 10°C$). Upon recentrifugation at low temperature (to remove cold-stable MTs and aggregates), the tubulin-rich supernatant can be rewarmed thereby reassembling the tubulin. After three or four rounds of such cyclic assembly/disassembly, tubulin is the major protein ($> 90\%$), with both high and low molecular weight MAPs present in varying proportions depending upon the exact procedure used. Different buffers, as well as the presence of DMSO and/or glycerol, can affect the final MAP composition. Significantly, further cycling reveals a constant stoichiometric association between tubulin and MAPs and this provided the first clue that

these proteins are not simple contaminants of the preparation. It was originally believed that constant stoichiometric association of a protein to tubulin during cyclic purification was the litmus test to define a MAP. This is no longer justified for two reasons; first, poly-L-lysine (which is not a MAP) cycles with constant stoichiometry (Timasheff, 1978). Secondly, many MAPs (as ascertained by other criteria) do not cycle well (if at all) with tubulin. The ambiguity probably arises because of differences in affinities of various MAPs for tubulin. MAPs having high affinity to tubulin/MTs probably outcompete low-affinity MAPs for binding, and furthermore, MAPs having lower affinities may not stoichiometrically maintain their associations during the extensive dilutions accompanying warm/cold cyclic purification.

There are no authenticated reports of warm/cold cyclic purification being applicable to studies in higher plants. We have tried this approach (with and without glycerol, and/or DMSO) and found that a good deal of non-MT material sediments in the first assembly pellet. Although there is an enrichment for tubulin (as assayed by Western blots and ELISA) in the first assembly pellet, it is very difficult to resolubilize. Until better methods for reversibly solubilizing plant MTs in bulk biochemical preparations are developed, this approach will not be useful in plant studies.

Collins and Vallee (1987) have developed an elegant variation of the warm/cold cyclic purification method which utilizes taxol, a MT-stabilizing drug. MTs treated with taxol are usually very resistant to depolymerization but can be disassembled using a combination of cold and Ca^{2+}. The advantage of this approach is that tubulin in crude homogenates is more efficiently polymerized. We have used this approach with cultured carrot cells and found MT yields to be very low after two rounds of assembly/disassembly. Furthermore, we have observed aberrant forms of tubulin polymer (as well as authentic MTs) in crude extracts in the presence of taxol. This approach may prove useful in future studies; however, it will be important to establish buffer conditions which favour the assembly of authentic MTs and furthermore, to define conditions where MTs are efficiently resolubilized after assembly.

The problem of MT resolubilization may lie in the unique nature of plant MTs. We know there to be factors in plants which hyperstabilize MTs *in vitro* (Cyr and Palevitz, 1989). Perhaps if we know more about these factors we will be in a better position to understand how they can be manipulated to give efficient recovery of MTs in cyclic purification studies.

Another method of MAP isolation, as pioneered by Bloom *et al.* (1985) utilizes taxol to pre-assemble MTs prior to their addition to crude homogenates. This procedure is therefore not dependent upon the ability to form authentic MTs in total protein extracts. An additional advantage of this technique is that in the presence of taxol, MTs can form without nucleotides; this can aid in the isolation of MAPs (mostly MT-motors) which are dissociated from MTs by nucleotides. After the addition of MTs to crude homogenates the MTs can then be collected by centrifugation. As MAPs bind to MTs via ionic bonds they can subsequently be removed by salt treatment. We have found this method to work well when applied to preparations from cultured carrot cells. Additionally, following dialysis the proteins can be rebound once more to taxol-stabilized MTs, thus 'salt-cycling' the proteins. It is important to note that, although plant and animal MAPs may be different, the tubulin binding sites are apparently conserved, as the same plant proteins associate with taxol-stabilized neuronal MTs as with MTs derived from plant sources. Additionally, the bundling of MTs is a prominent feature of MAP preparations from carrot cells (Fig. 5.2), but upon the addition of neuronal MAPs the bundles dissociate, thus providing additional evidence for the conservation of MAP–tubulin binding sites (Cyr and Palevitz, 1989).

All approaches to the isolation of MAPs have their difficulties and it is impossible to predict which method, or combination thereof, will prove to be the best in the study of MAPs from higher plants. Undoubtedly, we will find that some methods work best when applied to specific tissues and species. Regardless of the methodology applied in their isolation, the challenge in MAP research will be to demon-strate the role they play in the functioning of MTs within the cell.

Minimally, there are three areas which should be addressed in the study of a MAP. First, it is necessary to show *in vitro* that the putative MAP binds to MTs. Secondly, it is important to show the putative MAP has some effect upon a relevant MT function. Lastly, it is important to demonstrate that the protein has some relationship with MTs *in situ*.

In vitro assays for MAP activity

The ability to affect MT assembly is one function attributed to MAPs and many, but not all, lower the critical concentration of MT assembly. Quantitatively this is most often studied using light scattering in a spectrophotometer. However, because tubulin can form aberrant structures, it is important to confirm that authentic MTs are associated with any change in light scattering observed in the assay. Furthermore, the spectrophotometric analysis of MT formation should only be used with relatively pure material, as crude homogenates tend to form amorphous light-scattering aggregates upon standing.

Obviously, in order for a MAP to associate it must bind to the MT. This binding can be visualized using the electron microscope; MTs free of MAPs have smooth side walls, whereas those possessing MAPs have decorated side walls. We have noted that MAP preparations from carrot cells are capable of forming these types of decorations on MTs of neuronal origin (Cyr and Palevitz, 1989). Depending upon the stoichiometry of the interaction, this may or may not appear as a periodic association.

The bundling of MTs can be studied directly by microscopy (either dark-field light microscopy or electron microscopy) or quantitatively in purified fractions by light scattering in a spectrophotometer. In a light-scattering assay, bundling is typically noticed as a very high change in optical density (OD) (rarely does a typical MT assembly reaction vary more than 0.1 OD units). The reason bundling gives such a high OD change is the formation of lateral MT associations, the width of which approaches

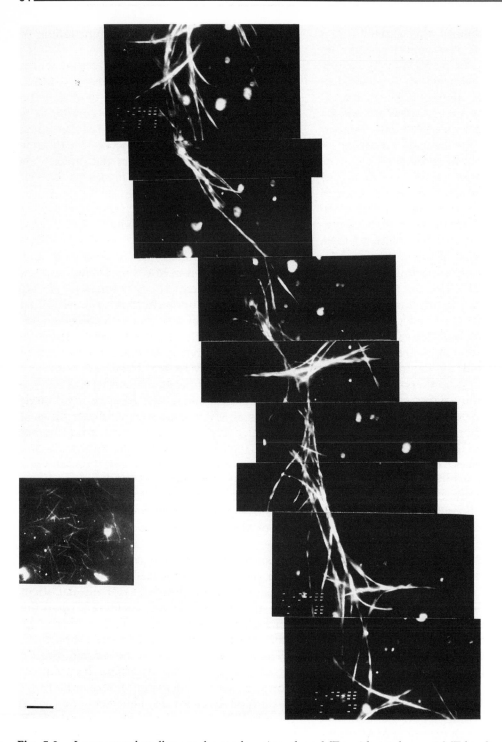

Fig. 5.2. Long superbundles can be made using short MTs with total carrot MT-binding proteins. Taxol-stabilized neuronal MTs were added to MT-binding proteins and visualized using dark-field microscopy. Long arrays are frequently seen. Shown here is a photomontage of a portion of one such superbundle which spanned several high-magnification fields. Magnification = ×806. Scale bar =10 μm. (Ibid., Fig. 5.1., used with permission.)

or exceeds the wavelength of visible light. In a pure MT assembly reaction, light is scattered only because the length of a MT is greater than the wavelength of visible light.

MAPs which function as MT motors can be studied *in vitro* with various motility assays (Vale, 1987). Typically these are performed as gliding assays in which the MT-motor is adsorbed to a glass surface and taxol-stabilized MTs are added along with ATP. Motility is judged by directly examining the gliding of MTs along the surface. The MTs can be visualized with dark-field or computer-enhanced DIC microscopy. Another method used to assess MT motility is to immobilize the MT-motor onto a bead and then visualize bead movement along an immobilized MT.

Study of MAP function *in situ*

Immunolocalization studies

Obviously, biochemical studies performed *in vitro* can be very informative, especially with respect to the quantitative behaviour of MT–MAP interactions. However, it is essential that direct relationships be established with *in situ* studies as well. The most common method for demonstrating these relationships is to use specific antibodies in immunolocalization studies.

The ability to use antibodies against MAPs to show co-localization with MTs in cells is the most straightforward approach to bridge biochemical and cell biological studies. We have used this approach to verify that plant polypeptides identified *in vitro* as binding to MTs also bind to MTs *in situ* in lysed protoplasts. However, the inability to demonstrate co-localization should not, by itself, be used as evidence for non-association of a putative MAP with MTs *in vivo*. Numerous reports have indicated that the fixation of MAPs is not always as easy as that of tubulin. Also MAPs such as kinesin, do not strictly associate with MTs, resulting in punctate staining in some instances (Pfister *et al.*, 1989). The presence of detergents is another source of false negative staining with anti-MAP antibodies as some MAPs apparently do not stay bound, or fix

well after detergent treatment. An additional source of concern in immunolocalization studies, especially those which employ detergents, is the redistribution of proteins during extraction (Allan and Hepler, 1989). Unfortunately, it is frequently impossible to achieve good localization of cytoskeletal epitopes without extraction, so the use of detergent remains problematic. Perhaps we need to work on sectioning methods which avoid detergent use.

Presently, the most rigorous method of performing immunolocalization is at the electron microscopic level. Although time consuming, the advent of new freeze-substitution fixation techniques gives excellent fixation of cytoskeletal elements and permits the use of gold-conjugated antibodies (Tang *et al.*, 1989). These methods convincingly demonstrate co-localization with a cytoskeletal structure directly observable at high resolution, without the need for detergent treatment.

It is important to keep in mind that immunolocalization studies, by themselves, rarely provide insight into functional relationships. In order to deduce these relationships other *in vivo* experimental methods must be used.

In vivo depolymerization/repolymerization studies

One powerful approach which has been used to examine spatial aspects of MT organization is to depolymerize the MTs *in situ* and follow the subsequent repolymerization patterns. This approach should lend itself well to the study of MAPs which function *in vivo* to affect the site of MT nucleation. In higher plants these studies have shown that cortical MTs can repolymerize at discrete locations (Marc *et al.*, 1989) as well as with a diffuse distribution (Falconer *et al.*, 1988), depending upon the cell type examined. As previously noted, MAPs are likely participants in this repolymerization process, serving to stabilize small seeds or nascent MTs. MAPs which serve to stabilize seeds would be expected to predict the future location (at least initially) of new MTs during experimentally induced repolymerization *in situ*.

The development of relevant assays for MAP

function will be pivotal in understanding how this class of proteins interacts with MTs. The following is a brief, and assuredly incomplete, list of possible approaches which may prove beneficial in understanding these relationships.

Electroporation has proven to be a useful tool for the introduction of a variety of molecules into protoplasts. We have found this approach to be satisfactory for the introduction of antibodies, and other proteins should work as well. It should therefore be feasible to introduce purified MAPs into appropriate plant cells with this technique. Where a MAP shows a particular *in vitro* function, one could introduce the MAP into a cell type which either does not express the function, or expresses it at a low level. If the MAP functions similarly *in vivo*, then the effect should be amplified with the introduction of additional protein, thereby serving as an effective link between *in vitro* assay and *in vivo* function.

The electroporation of relevant antibodies into living cells may also prove to be an important tool for dissecting functional aspects of MAP behaviour. The success of this approach relies on the production of antibodies against MAPs which recognize an essential epitope on the molecule. The use of such 'blocking antibodies' has proven to be an important general tool in studying functional aspects of cellular function. Although this approach has yet to be reported in protoplasts, it has been shown to be effective in animal cells (Chakrabarti *et al.*, 1989).

Summary and prospectus

One key to the understanding of MT function will be to learn more about specific MAPs which contribute to the behaviour of MTs in the various arrays within a developing plant. In pursuit of a greater understanding of MAP function, it will be important to keep in mind that the evolutionary process has probably used MAPs to 'tinker' with MT function (see Jacob, 1983 for a general discussion of 'molecular tinkering'). It may therefore prove unwise to concentrate solely upon homologies with known MAPs described in animal systems

where MTs may be put to very different uses. For example, the MAPs which undoubtedly participate in the MT-based ordering of cellulose microfibrils are probably very different from those found in the interphase arrays of most animal cells.

We also need to consider that the cytoskeleton is but one component of the cell. It is likely that MAPs are biochemical integrators between MTs and general cellular physiology. Therefore, we need to understand not only how MAPs interact with MTs, but how the MAPs themselves interface with other cellular components and/or signalling mechanisms.

Acknowledgements

Thanks to Linda Maxson (Pennsylvania State University) and Barry Palevitz (University of Georgia) for helpful discussions during the preparation of this chapter.

References

Allan, E. and Hepler, P. (1989). Calmodulin and calcium-binding proteins. In *The Biochemistry of Plants* (ed. P.K. Stumpf), vol. 15, pp. 455–484. Academic Press, New York.

Bernhardt, R. and Matus, A. (1982). Initial phase of dendrite growth: Evidence for the involvement of high molecular weight microtubule-associated proteins (HMWPs) before the appearance of tubulin. *J. Cell Biol.* **92**, 589–593.

Bloom, G.S., Luca, F.C. and Vallee, R.B. (1985). Identification of high molecular weight microtubule-associated proteins in anterior pituitary tissue and cells using taxol-dependent purification combined with microtubule-associated protein specific antibodies. *Biochemistry* **24**, 4185–4191.

Chakrabarti, R., Wylie, D.E. and Schuster, S.M. (1989). Transfer of monoclonal antibodies into mammalian cells by electroporation. *J. Biol. Chem.* **264**, 15494–15500.

Collins, C.A. and Vallee, R.B. (1987). Temperature-dependent reversible assembly of taxol-treated microtubules. *J. Cell Biol.* **105**, 2847–2854.

Cyr, R.J. and Palevitz, B.A. (1989). Microtubule-binding proteins from carrot. I. Initial characterization and microtubule bundling. *Planta* **177**, 245–260.

Diaz-Nido, J., Hernandez, M.A. and Avila, J. (1990). Microtubule proteins in neuronal cells. In *Micro-

tubule Proteins (ed. J. Avila), pp. 193–258. CRC Press, Boca Raton, FL.

Dustin, P. (1984). *Microtubules*, 2nd edn. Springer-Verlag, New York.

Falconer, M.M., Donaldson, G. and Seagull, R.W. (1988). MTOCs in higher plant cells: An immuno-fluorescent study of microtubule assembly sites following depolymerization by APM. *Protoplasma* **144**, 46–55.

Fosket, D.E. (1989). Cytoskeletal proteins and their genes in higher plants. In *The Biochemistry of Plants* (ed. P.K. Stumpf), vol. 15, pp. 392–454. Academic Press, New York.

Fulton, C. and Simpson, P.A. (1979). Tubulin pools, synthesis and utilization. In *Microtubules* (eds K. Roberts and J.S. Hyams), pp. 118–149. Academic Press, New York.

Gard, D.L. and Kirschner, M.W. (1987). A micro-tubule-associated protein from *Xenopus* eggs that specifically promotes assembly at the plus-end. *J. Cell Biol.* **105**, 2203–2215.

Griffith, L.M. and Pollard, T.D. (1978). Evidence for actin filament microtubule interaction mediated by microtubule-associated proteins. *J. Cell Biol.* **78**, 958–968.

Gunning, B.E.S. and Hardham, A.R. (1982). Micro-tubules. *Ann. Rev. Plant Physiol.* **33**, 651–698.

Jacob, F. (1983). Molecular tinkering in evolution. In *Evolution from Molecules to Man* (ed. D.S. Bendell) pp. 131–134. Cambridge University Press, Cambridge, UK.

Job, D., Rauch, C.T., Fischer, E.H. and Margolis, R.L. (1982). Recycling of cold-stable microtubules: Evidence that cold stability is due to substoichio-metric polymer blocks. *Biochemistry* **21**, 509–515.

Lloyd, C.W. and Seagull, R.W. (1985). A new spring for plant cell biology: microtubules as dynamic helices. *Trends Biochem. Sci.* **10**, 476–478.

Marc, J., Mineyuki, Y. and Palevitz, B.A. (1989). A planar microtubule-organizing zone in guard cells of *Allium*: Experimental depolymerization and reassembly of microtubules. *Planta* **179**, 530–540.

Margolis, R.L., Rauch, C.T. and Job, D. (1986). Purification and assay of a 145 kDa protein (STOP$_{145}$) with microtubule-stabilizing and motility behavior. *Proc. Natl Acad. Sci. U.S.A.* **83**, 639–643.

Matus, A. (1988). Microtubule-associated proteins: Their potential role in determining neuronal morphology. *Ann. Rev. Neurosci.* **11**, 29–44.

Mitchison, T.J. and Kirschner, M. (1984). Dynamic instability of microtubule growth. *Nature* (Lond.) **312**, 232–237.

Olmsted, J. (1986). Microtubule-associated proteins. *Annu. Rev. Cell Biol.* **2**, 421–457.

Palevitz, B.A. (1986). Division plane determination in guard mother cells of *Allium*: Video time-lapse analysis of nuclear movements and phragmoplast rotation in the cortex. *Dev. Biol.* **117**, 644–654.

Pfarr, C.M., Coure, M., Grissom, P.M., Hays, T.S., Porter, M. and McIntosh, J.R. (1990). Cytoplasmic dynein is localized to kinetochores during mitosis. *Nature* **345**, 263–265.

Pfister, K.K., Wagner, M.C., Stenoien, D.L., Brady, S.T. and Bloom, G.S. (1989). Monoclonal anti-bodies to kinesin heavy and light chains stain vesicle-like structures, but not microtubules, in cultured cells. *J. Cell Biol.* **108**, 1453–1463.

Steuer, E.R., Wordeman, L., Schroer, T.A. and Sheetz, M.P. (1990). Localization of cytoplasmic dynein to mitotic spindles and kinetochores. *Nature* **345**, 266–268.

Tang, X., Lancelle, S.A. and Hepler, P.K. (1989). Fluorescence microscopic localization of actin in pollen tubes: Comparison of actin antibody and phalloidin staining. *Cell Motil. Cytoskel.* **12**, 216–224.

Timasheff, S.N. (1978). Thermodynamic examination of the self-association of brain tubulin to micro-tubules and other structures. In *Physical Aspects of Protein Interactions* (ed. N. Catsimpoolas), pp. 219–290. Elsevier, New York.

Vale, R. (1987). Intracellular transport using micro-tubule-based motors. *Ann. Rev. Cell Biol.* **3**, 347–348.

Vallee, R.B. and Bloom, G.S. (1984). High molecular weight microtubule-associated proteins (MAPs). *Mod. Cell Biol.* **3**, 21–75.

Verde, F., Labbe, J.-C., Doree, M. and Karsenti, E. (1990). Regulation of microtubule dynamics by cdc2 protein kinase in cell-free extracts of *Xenopus* eggs. *Nature* **343**, 233–238.

6. CYTOPLASMIC AND NUCLEAR INTERMEDIATE FILAMENT ANTIGENS IN HIGHER PLANT CELLS

Peter J. Shaw, David J. Fairbairn and Clive W. Lloyd

*Department of Cell Biology, John Innes Institute,
Colney Lane, Norwich NR4 7UH, UK*

Intermediate filaments (IFs) in plants seem only to have a provisional listing in the cytoskeletal directory. This, quite reasonably, can be attributed to the fact that no more than a handful of papers has been published on the subject. Another reason may be the failure of electronmicroscopists to detect 10-nm filaments in plant cytoplasm. However, a third reason, their lack of obvious function, has been no barrier to animal studies where an enormous amount of work has failed to turn up a universally agreed role. In this chapter we will review the evidence that plant cells do indeed possess nuclear and cytoplasmic intermediate-filament antigens.

General background

IFs, together with microtubules (MTs) and actin filaments, form the major structural elements of the cytoskeleton. They are so called because they are intermediate in diameter (*c.* 10 nm) between thick and thin muscle filaments (for recent review, see Robson, 1989). They occur widely in a range of vertebrate cells and are known to occur in invertebrates also (Bartnik and Weber, 1989). An earlier operational definition of the cytoskeleton considered it to be the insoluble residue of the cell that remains after extraction with detergent. On this basis, IFs are the most stable because they remain in animal cells after MTs and actin have been selectively extracted by salt washes and Ca^{2+}, etc. However, there is concern that cell extraction methods and EM sectioning of resin-embedded material fail to preserve or illustrate cytoskeletal material that may be more fine and pervasive than the major structural elements (see Schliwa, 1986). Probably because of their chemical stability (concentrated solutions of urea are required for depolymerization) IFs are regarded as functionally stable – the still centre around which all else revolves. They have been pictured as the stable matrix which maintains integrity whilst actin and MTs disassemble and assemble to form different conformations. However, it now appears from microinjection studies with labelled polypeptides that IFs are not necessarily static and may exchange their subunits (Vikstrom *et al.*, 1989). Certainly, during mitosis, IFs depolymerize, probably in a phosphorylation-dependent manner (Chou *et al.*, 1990).

IFs have been placed in five groups according to protein primary sequence comparisons (Osborn and Weber, 1986). Although sequence homology may not necessarily be high between the groups, they none the less form virtually indistinguishable filaments. This can be attributed to the central α-helical rod domain which is conserved amongst the groups. The C-terminal head and N-terminal tail are hypervariable in length and sequence but the rod domain of over 300 residues is constructed by common sequence principles whereby hydrophobic amino acids occur at positions 1 and 4 within a heptadic repeat. The helices are able to form coiled-coils and one characteristic of IFs is their propensity for filament formation when renatured from strong solutions of urea. Apart

The Cytoskeletal Basis of Plant Growth and Form   ISBN 0–12–453770–7

from common sequence principles, probably all IFs (including lamins) contain a common epitope detected at the C-terminal end of the helical rod domain (Geisler *et al.*, 1983). This epitope is recognized by a monoclonal antibody designated anti-IFA (Pruss *et al.*, 1981). Another monoclonal antibody, ME101 (Escurat *et al.*, 1989) has been reported to cross-react with most classes of IFs, but – in distinction to anti-IFA – at the amino terminal half of coil 1a.

Despite the uncertainty about *in vivo* function, IFs are turning out to be useful diagnostic markers for different developmental or pathological states. Keratins, for instance, occur as acidic (type I) and neutral-basic (type II) chains that are obliged to co-polymerize in order to form filaments. There are 20 or so different keratins, occurring in different combinations, which are markers for epithelial differentiation (although there are notable exceptions). Type III is comprised of vimentin, desmin, glial fibrillary acidic protein and peripherin found, respectively, in mesenchymal cells, muscle, glial cells/astrocytes and neurones. Type IV neurofilament proteins and α-internexin occur in neurones and type V is composed of the lamins which constitute the nuclear lamina.

Fibrillar bundles (FBs) in plants contain intermediate filament antigens

In 1982, Powell *et al.* prepared cytoskeletons from carrot protoplasts by extracting them with detergent. In addition to cortical MTs, thick bundles of 7-nm fibrils in negatively stained preparations were seen to be associated with the nucleus at one end and, occasionally, at the other end to fray into filaments amongst the cortical MTs. In terms of their optical diffraction pattern, molecular mass and lack of rhodamine–phalloidin staining, the FBs were shown not to be composed of F-actin.

In that study, FBs were also detected in oil palm, Italian ryegrass and maize cell suspensions. Olesen and Jensen (1984) also reported the occurrence of FBs in the cytoplasm and within the nucleus of another maize suspension line. Chris Hawes (personal communication) has also detected FBs in an embryogenic carrot suspension culture (see Fig. 6.1). Such bundles

of fibrils may not, however, be restricted to suspension cells since the authors have recently identified negatively stained bundles in burst protoplasts derived from pea leaf epidermis (Catherine Duckett and Lloyd, unpublished). That is, FBs are not peculiarities of cells maintained under the artificial conditions of suspension culture. To characterize the FBs further, Dawson *et al.* (1985) fractionated carrot suspension cells, purifying the bundles. On immunoblots of 1-D gels, the 'universal' monoclonal antibody, anti-IFA, recognized bands at 50 and 68 kD (and intermediate bands) in whole-cell homogenates of carrot suspension cells and of onion root-tip cells. The 50-kD species was enriched in a purified fraction of FBs, solubilized in 9.5 M urea. Anti-IFA was unsuccessful (see below) for the immunofluorescent staining of FBs. However, in onion root-tip cells this antibody stained the cortical array, the pre-prophase band (PPB), the spindle and the phragmoplast in a fuzzy, punctate manner that roughly reflected normal MT patterns. To address the possibility that anti-IFA was recognizing plant tubulin, carrot tubulin was isolated using taxol, separated by 2-D gel electrophoresis and its immunoblotting characteristics compared with that of FB polypeptides separated on 2-D gels. Anti-IFA does not recognize tubulin nor does anti-tubulin recognize FB proteins; the polypeptides being clearly separable by 2-D gel electrophoresis.

Parke *et al.* (1987) also used anti-IFA on onion root tips as well as a panel of monoclonal antibodies (mAbs) raised against a high salt, detergent-insoluble fraction from *Chlamydomonas reinhardtii* cells. The mAbs were shown to stain IFs in animal tissue culture cells. With plants, however, only the spindle and phragmoplast stained, not the PPB nor the cortical array. The linear-punctate staining pattern of the cortical array reported by Dawson *et al.* (1985) is – as we will see – difficult to obtain but the PPB staining is stable and it is difficult to account for the discrepancy in the results. In addition to staining onion cells, anti-IFA has subsequently been shown to produce the MT-like staining pattern in carrot and black Mexican sweetcorn (BMS) suspension cells and *Datura stramonium* root-tip cells. In immunoblots of plant, whole-cell extracts anti-IFA consistently

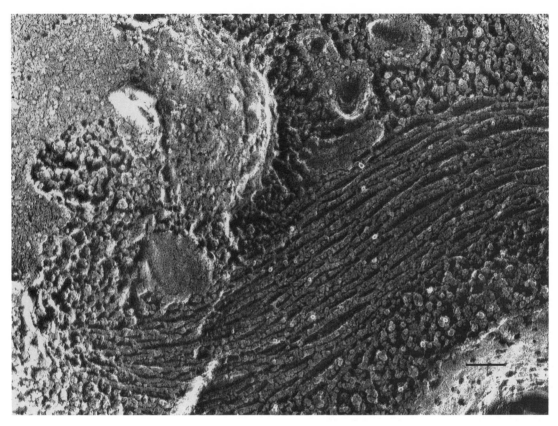

Fig. 6.1. A deep-etch replica of a carrot suspension cell, showing a fibrillar bundle in the cytoplasm, apparently associated with the nucleus. Scale bar = 100 nm. Provided by Dr Chris Hawes and Barry Martin, Oxford Polytechnic.

recognizes two bands between 50 and 70 kD (Table 6.1). The anti-IFA antigen appears to be widely distributed in higher plants.

Escurat *et al.* (1989) raised a mAb to human peripherin which recognized the amino terminal end of several vertebrate IF proteins. This antibody also cross-reacts with plants (Table 6.1). The ability of two pan-specific anti-IFAs, which recognize either ends of the conserved α-helical rod region in vertebrates, to immunoblot the same polypeptide bands of carrot protoplasts (Table 6.1), 62 and 58 kD, indicates that the α-helix-rich rod region is likely to be conserved in plants.

In 1989, Hargreaves *et al.* raised a mAb (anti-fibrillar bundle antibody (AFB)) against carrot FBs. This antibody recognizes polypeptides at 58–62 kD on Western blots of FBs. Again, it does not recognize tubulin but does stain all four MT arrays in onion root-tip cells. The cortical staining pattern is the most difficult to

reproduce but provided the cells are not damaged, patches of staining can be resolved which – by comparison with the anti-dog brain tubulin staining pattern produced by double immunofluorescence – lines up with cortical MTs. The AFB recognizes desmin, vimentin and glial fibrillary acidic protein which compose the type III group of IFs in animal cells. This again demonstrates a broad cross-reactivity between plant and animal IF antigens. In carrot suspension cells, AFB immunostains the thick FBs but not the MT-like arrays. This was the converse of results previously obtained with anti-IFA (Dawson *et al.*, 1985) and the contrariwise labelling of FBs and MT-like arrays was the focus of a further study (Goodbody *et al.*, 1989). Using a slightly different method for staining protoplasts, both AFB and anti-IFA could be shown to label the cortical MTs of protoplasts derived from a BMS suspension. The key difference between this and the

Table 6.1. Summary of labelling of plant tissues by anti-IF antibodies

Antibody	Source of antigen	Plant tissue	Immunoblotted proteins (kD)	Immunofluorescence
Anti-IFA	Human GFAP	Carrot protoplasts[1]	58, 62	IMTA, PPB, S, PHR
		BMS protoplasts[1]	58, 62	IMTA, PPB, S, PHR
		Pea leaves[2]	55, 63	–
		Onion root tips[3]	50, 68	IMTA, PPB, S, PHR
		Onion root tips[5]	50, 60, 68, 90, 120, 190	–
		Datura root tips[2]	50, 60	IMTA, PPB, S, PHR
MAC322	Carrot cytoskeletons	Wheat root tips[4]	–	IMTA, PPB, S, PHR
		Datura root tips[4]	–	IMTA, PPB, S, PHR
		Carrot cells[6]	–	FB
		Carrot protoplasts[6]	42, 50, 55, (92[2])	IMTA
		BMS protoplasts[6]	–	IMTA
		Pea leaves[2]	92, 125	–
LE41	Cytokeratin 8	Carrot protoplasts[6]	42, 50, 55	–
34g5	*Chlamydomonas* basal bodies	Onion root tips[1]	–	IMTA, PPB, S, PHR
		Carrot protoplasts[1]	–	IMTA, PPB, S, PHR
		BMS protoplasts[1]	–	IMTA, PPB, S, PHR
371	Flagella C tektin	Carrot cells[1]	–	FB
ME101	Human peripherin	Carrot protoplasts[2]	42, 58, 62, 77, 105, 145, 150	IMTA
		Tobacco protoplasts[4]	–	IMTA
		Datura root tips[4]	–	IMTA, PPB, S, PHR
C3	*Chlamydomonas* cytoskeletons	Onion root tips[5]	50, 60, 90, 120, 190	–
C22		Onion root tips	50	–
C26		Onion root tips	50, 90, 120, 190	–
C31		Onion root tips	50, 60, 90	–
AFB	Carrot fibrillar bundles	Onion root tips[7]	–	IMTA, PPB, S, PHR
		Carrot cytoskeletons[7]	58, 62	FB
		Carrot protoplasts[1]	–	IMTA, PPB, S, PHR
		BMS protoplasts[1]	–	IMTA, PPB, S, PHR
		Datura root tips[4]	–	IMTA, PPB, S, PHR

References: [1] Goodbody *et al.*, 1989; [2] Fairbairn and Lloyd, unpublished; [3] Dawson *et al.*, 1985; [4] Goodbody and Lloyd, unpublished; [5] Parke *et al.*, 1987; [6] Ross *et al.*, 1991; [7] Hargreaves *et al.*, 1989a.
Abbreviations: IMTA – interphase microtubule array; PPB – preprophase band; S – spindle; PHR – phragmoplast; FB – fibrillar bundles.

method reported in Hargreaves *et al.* (1989a) was that the latter used methanol-dried cells for screening large numbers of monoclonal supernatants. The use of methanol is critical for unmasking epitopes on the carrot FBs, enabling them to be stained with AFB. However, this method does not permit the MT-like arrays to be stained; they are only stained when methanol is omitted. The MT-like arrays are not destroyed by methanol treatment since anti-IFA is fully able to stain the meshwork form treated in this way. Bundling of MTs with taxol causes the AFB antigens to be similarly bundled around the nucleus, and depolymerization of MTs disperses the AFB antigens. Cytochalasin D treatment perturbs the actin meshwork but not the AFB staining patterns. Therefore, the cytoplasmic distribution of the AFB and anti-IFA antigens seems to depend entirely upon MTs.

Apart from underlining the breadth of cross-reactivity between species, these studies establish that IF antigens exist in plants in two different forms: the paracrystalline FBs and the MT-associated form. The latter is particularly interesting because of the apparently close relationship between two classes of cytoskeletal element.

A striking example of a close association between IFs and MTs is provided by filaments found in the wall of the A-tubule of the doublet MTs in sea urchin sperm flagella. These detergent-resistant 2–6-nm filaments are composed of a group of polypeptides (tektins) and resemble IFs in several respects, including immunological cross-reactivity (Steffen and Linck, 1988). mAbs against flagellar tektins stain vimentin and desmin in vertebrate tissue culture cells (Chang and Piperno, 1987). To test whether, as in flagella, a tektin-like moleule was incorporated into the wall of plant MTs, cortical MTs on footprints were extracted with a mixture of urea and the detergent sarkosyl. This did not leave resistant filaments to be stained with anti-tektins although too much emphasis cannot be placed on the failure of a particular protocol. However, a mAb against C tektin (Chang and Piperno, 1987) does immunostain carrot FBs (Goodbody et al., 1989), increasing the range of IF antigens contained in the FBs.

Other studies reinforce the conserved nature of IF epitopes and their existence in higher plants. For instance, in addition to anti-IFA and AFB, Goodbody et al. (1989) reported that an anti-vimentin (34g5) raised against Chlamydomonas basal bodies (Klymkowsky, 1988) also stains the MT-associated form in BMS cells.

AFB immunoblots the desmin, vimentin and glial fibrillary acidic protein, indicating a relationship between plant FBs and a core group of type III IFs. However, more recent work, showing that plants and animals share a keratin epitope, implies an even wider familial relationship. A mAb (MAC 322) raised against a carrot cytoskeleton preparation has been found to stain the cytokeratin network of PtK$_2$ cells, not the vimentin system stained by AFB (Ross et al., 1991). By immunoblotting cells containing a range of the major type I and type II keratins, the antibody binds only to K8. Using wheat root-tip cells, MAC 322 stains the MT arrays in the linear-punctate manner found for all other anti-IF antibodies (Fig. 6.2). It stains the cortical array particularly well (as determined by confocal laser microscopy) in addition to a fine, wispy meshwork which radiates from the nucleus of BMS cells. From the phylogenetic point of view the recognition of K8 by MAC 322 is interesting since this cytokeratin has been claimed to have archetypal features (Franz and Franke, 1986). It is often the only type II cytokeratin in many amphibian and mammalian epithelia, is found in simple epithelia, and even occurs in non-epithelial cell types. K8 is expressed in the Xenopus oocyte and is therefore not a hallmark polypeptide for epithelial differentiation. Rungger-Brandle et al. (1989) have concluded that its expression correlates with what has been called an embryonal proliferative state.

All indications are that the epitopes so far detected in plants are generic, and being conserved in phylogenetically distant species, are likely to be retained for functional purposes. The conspicuous feature of IF primary sequence is the α-helix rich central portion that forms coiled-coils. Although primary sequence may vary, the existence of heptatic repeats with hydrophobic residues at positions 1 and 4, indicates that sequence principles are probably phylogenetically constant since such motifs are found in invertebrate IFs (Weber et al., 1989). The immunocytochemical data establish that plants possess the C-terminal (anti-IFA) and the N-terminal (ME101) ends of the helical region necessary for filament formation. Indeed, there is evidence that carrot IF antigens are capable of forming filaments.

Reconstitution studies

One characteristic of IFs is their strong tendency to reconstitute filaments upon removal of denaturing agents. Hargreaves et al. (1989b)

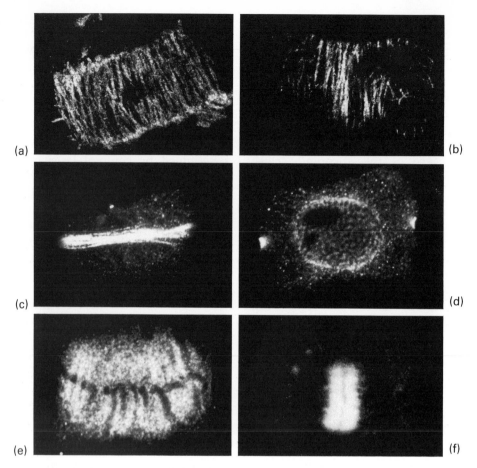

Fig. 6.2. Confocal laser microscopy of wheat root-tip cells immunostained with MAC322. This mAb was raised against carrot cytoskeletons and recognizes keratin 8 in mammalian cells (Ross *et al.*, 1991). In (a) and (b), the cortical MTs can be seen to be labelled in a coarser fashion than is seen with anti-tubulin. The PPB is labelled in (c) and at a deeper focal level within the same cell, MAC322 also stains the nuclear periphery (d). Kinetochore bundles (e) are coarsely stained with granular material. The phragmoplast, too, stains in a fuzzy (but specific) manner (f).

used conditions employed for dissociating animal IFs in order to depolymerize carrot FBs to see whether they could be reconstituted. Purified carrot FBs solubilized in 9 M urea were centrifuged at high speed, and the supernatant then subjected to a two-step dialysis, finally against 150 mM NaCl. Negative staining confirmed that tangles of 10-nm filaments, and bundles of filaments, were reconstituted in this way. The major polypeptides of this material (58 and 62 kD) both positively immunoblotted with anti-IFA and AFB, thereby confirming that the filament-forming fraction contained these epitopes. In a similar manner, the detergent-resistant, high salt-resistant fraction

from BMS suspension cells was shown to be capable of reconstituting filaments upon dialysis from 9 M urea. The major polypeptides, like those of carrot, were at 58 and 62 kD and reacted on Western blots with anti-IFA and AFB (Goodbody *et al.*, 1989).

This shows that plant cells possess IF antigen-containing polypeptides that are intrinsically capable of forming 10-nm filaments. The FB fraction is composed of filaments, can reconstitute filaments, and contains the anti-IFA epitope known to be essential for forming a normal IF network (Albers and Fuchs, 1987). However, this reconstituted fraction represents only a part of the plant cell's total IF antigen.

Mutation studies of keratin cDNAs (Albers and Fuchs, 1989) have shown that carboxy-terminal mutants give rise to aggregates of keratin in the cytoplasm of transfected animal cells, whereas amino-terminal mutants produce aggregates at the nuclear surface. The fact that plants contain IF antigens is therefore no guarantee that they all contain sufficient sequence to enable them to constitute a network which connects the nuclear envelope to the plasma membrane. Furthermore, not all IF antigens need necessarily form filaments. Franke *et al.* (1982) have described cytokeratin and vimentin filaments unravelling into 2–4-nm protofilaments during mitosis and forming 'alternative, non-fila-mentous structures'. This issue of whether plant IF antigens are filamentous or non-filamentous is particularly pertinent to the MT-associated form. By bursting protoplasts on EM grids, adherent discs of plasma membrane are formed, with MTs exposed on the cytoplasmic face. These footprints stain with AFB in a linear-punctate manner by immunofluorescence (Goodbody *et al.*, 1989). Using AFB and col-loidal gold-tagged second antibody, electron-dense patches could be seen at the EM level to be stained by AFB. These patches occur along and between the negatively-stained cortical MTs.

This, then, is the major problem in discussing IFs in plants: FBs contain filaments but do not appear to constitute a cytoskeletal meshwork, whereas the 'meshwork' which co-distributes with MTs does not appear to be composed of filaments. Either the meshwork form is non-filamentous (but gains order from association with MTs) or is a finer filamentous form which we have yet to learn how to preserve. Such a fine meshwork – resembling a cytomatrix – has been seen by investigators using cleaving (Traas, 1984), high voltage EM (Hawes, 1985) and freeze-slamming methods (Hawes and Martin, 1986). All of these methods avoided resin embedding on the grounds that very fine elements might be obscured in sectioned plastic. Even though the nature of a cytomatrix in eukaryotic cytoplasm is still under debate (see Schliwa, 1986), it seems unreasonably optimistic to assume that electron microscopy of sectioned, embedded material preserves and

reveals the finest detail of the cytoskeleton. Unfortunately, the fine cortical reticulum seen in glutaraldehyde-fixed, critical-point dried, cleaved cells is not preserved by formaldehyde (Traas and Kengen, 1986) – the fixative which allows positive AFB staining (Goodbody *et al.*, 1989).

Sequencing studies will undoubtedly throw light on the extent of sequence similarity between plant and animal IF antigens, and on the likelihood of plants forming cytoplasmic filaments. However, it is clear from studies so far that plants do not conform to any simple model derived from animal tissue-culture cells and that further advances in ultrastructural preservation may be required to understand fine distribution.

Nuclear lamins – animal studies

The nuclear lamins form the best characterized group of nuclear structural proteins in animal cells (for reviews, see Krohne and Benavente, 1986; Franke, 1987; Gerace and Burke, 1988). They were initially identified as a prominent group of polypeptides in SDS-PAGE gels of the detergent/salt insoluble fraction of a number of animal cell types (i.e. the residue from similar extraction conditions to those for cytoplasmic IF proteins). Immunological studies showed that these proteins were antigenically related to the cytoplasmic IF family of proteins. Immuno-fluorescence and immunogold electron micro-scopy showed that the proteins were located at the nuclear lamina – a thin layer of stain-dense material seen just within the inner nuclear membrane in certain types of cell – and thus they were called lamins. More recent sequence analysis has allowed a detailed comparison with the IF proteins and confirmed that they are closely related, sharing a central α-helical core domain, flanked by more globular N- and C-terminal regions (McKeon *et al.*, 1986). The lamins are now often classified as a further class of IF proteins.

Electron microscopy of an oocyte lamina has shown that it comprises filaments of similar dimensions to cytoplasmic IFs arranged in an orthogonal network. Under certain conditions the lamin subunits can be reconstituted into 10-nm filaments showing the characteristic approximately 24-nm IF repeat. In some cells the lamina may be only a single molecule in thickness; in other animal cells the lamina can be many layers thick. It is suggested that the organization is in all cases similar to the orthogonal network seen in oocyte laminas, but there is little direct evidence for this. Immunofluorescence staining with lamin antibodies often shows a smooth nuclear rim, however, Paddy *et al.* (1990) have recently shown a discontinuous network at the nuclear periphery by anti-lamin staining and 3-D optical microscopy.

In rat liver nuclei – one of the most intensively studies sources – there are three major lamin components, denoted A, B and C. Lamins A and C are most closely related, whereas the more acidic lamin B is more distantly related. Lamin B is suggested to be the nuclear membrane binding component, and a lamin B receptor has been identified in the nuclear membrane of turkey erythrocytes (Worman *et al.*, 1988). There are different spectra of lamin isotypes in different species and in different cell types within a species. For example, avian erythrocytes have two major lamins – an A type and a B type. Differential lamin expression has been studied in most detail in the amphibian, *Xenopus laevis*. Four major lamin components, denoted L_I to L_{IV}, have been identified and have been shown to be expressed in a cell type-dependent manner. The oocyte, fertilized egg and early blastula express L_{III}. L_I subsequently begins to appear, followed by L_{II}. L_{III} is then lost by the tadpole stage, and most somatic cells contain L_I and L_{II}. Some somatic cell types express L_I, L_{II}, and L_{III}, and the sperm cells express only L_{IV}. It is not clear what is the significance of this differential expression. Apart from these major lamins, there may also be other minor lamin components present.

Function of nuclear lamins

As with cytoplasmic IFs the function of the nuclear lamina is still not clear. Interactions have been shown, both by *in vivo* binding studies, and by structural studies between cytoplasmic IF networks (vimentin) and the nuclear lamina. This had led to suggestions that this type of interaction is important in positioning the nucleus. It is probable that the lamina is also involved in mediating interactions between chromatin and the nuclear envelope. The finding of different lamin isotypes during development in amphibians and in other vertebrates has led to suggestions of a role for lamins in determining cell-type specific chromatin structure, which, in turn, might be a mechanism for differential gene expression.

The lamins are phosphorylated. During cell division they become hyperphosphorylated and this causes the subunits to become soluble. The resulting breakdown of the lamina is part of the cascade of events leading to nuclear envelope breakdown in most somatic cell division (Gerace and Blobel, 1980). However, in meiosis of chickens and amphibians, the nuclear lamins are solubilized, but the nuclear envelope retains intact. Nuclear envelope breakdown, lamina dissolution and chromatin condensation can also be decoupled using cell-free extracts from *Xenopus* eggs to induce mitotic changes in isolated rat liver nuclei (Newport and Spann, 1987). Thus these three key events in cell division are not related in a simple causal sequence.

The functions suggested for the nuclear lamina would imply that its presence is an absolute requirement for cell viability. All animal cells so far examined seem to possess at least a B-type lamin, although certain cells may lack any A/C lamins. However, outside the animal kingdom the evidence is extremely sketchy. Antigenic cross-reactivity and peptide mapping have been used to demonstrate the presence of A-type and B-type lamins in yeast as well as a homologue to the lamin B receptor (Georgatos *et al.*, 1989). However there has, as yet, been no structural demonstration that these lamin-like components do in fact form a nuclear lamina, although the lamin B analogue was shown to partition with the nuclear envelope fraction during purification. The lamin A analogue was distributed both in the cytoplasmic and nuclear envelope fractions.

Neither of the yeast lamin analogues has been cloned and sequenced as yet.

Nuclear lamina/IF proteins in plants

In plants, the available data are also still inconclusive. Electron microscopy has never shown a lamina in any plant nucleus. However this cannot be taken to prove that no lamina exists; many animal cells possess only a very thin lamina which is often not revealed by thin section electron microscopy but which is detectable by immunofluorescence staining with lamin antibodies. Worman *et al.* (1989) have given a preliminary report of a lamin B analogue, again by demonstrating cross-reactivity with antibodies to turkey lamin B on immunoblots of purified pea nuclei. They showed that a 65-kD protein was recognized by the anti-avian lamin B, and also by human auto-antibodies against lamin B. Immuno-fluorescence with anti-lamin B antibodies on isolated nuclei showed nuclear rim staining. Galcheva-Gargova *et al.* (1988) have reported the isolation of a nuclear 'shell' from *Zea mays* and *Phaseolus vulgaris*. In this preparation they found a rather simple profile of polypeptides in the range 45–65 kD and showed by immuno-blotting a possible relation between these proteins and animal IF/lamina proteins.

Beven *et al.* (1991) have recently purified a nuclear matrix fraction from carrot suspension culture cells. This preparation contained a complex spectrum of many polypeptides none of which predominated on SDS-PAGE gels. This contrasts with similar preparations from such sources as rat liver nuclei, where the lamins are the most prominent bands. Extraction with various concentrations of urea gave a supernatant enriched in several proteins – in particular a group at approximate molecular weight 92 kD and one at 60 kD. Two mono-clonal antibodies – JIM63 and JIM62 – raised against total carrot nuclear matrices labelled the 92-kD group on 1-D and 2-D blots, both of nuclear matrix preparations and of total nuclear proteins. The 92-kD proteins were also labelled on blots by ME 101 (Escurat *et al.*, 1989) – an IF antibody, by AFB (see above), and by human

auto-immune anti-lamin A/C (see Fig. 6.3). The 60-kD band was labelled by anti-IFA, another general monoclonal antibody to IFs and lamins. Immunofluorescence staining of carrot culture cell nuclei with both JIM63 and some of the other IF antibodies showed a nuclear rim staining (which was greatly increased by a methanol dip treatment after fixation – see above) and also intranuclear fibre bundles and networks (see Fig. 6.4a, b). JIM63 is quite a widely reactive antibody; for example, similar fibre bundles have been labelled within tobacco suspension culture cell nuclei. Figure 6.5 shows confocal sections from a thick tissue section of pea root treated *in situ* as for nuclear matrix preparation and then labelled with JIM63. Prominent nuclear rims are seen, often also with staining which extends into the body of the nuclei. JIM63 also cross-reacts with antigens in animal cells. In rat liver nuclei, staining was observed which was virtually identical to that obtained with anti-lamins (see Fig. 6.4c, d). In whole PtK2 cells, cytoplasmic staining apparently similar to the vimentin network was observed, together with nuclear rim staining, and also specific staining inside the nucleus – in particular, the nucleolus was labelled, together with several other smaller bodies within the nucleus.

The fibre bundles labelled were very remi-niscent of cytoplasmic FBs, and it might be suggested that they are, in fact, cytoplasmic bundles which had become incorporated into the nucleus. This seemed unlikely: firstly the proteins that comprise the cytoplasmic FBs are very different in molecular weight from the nuclear matrix components recognized by JIM63 and the other antibodies, with no components as high as 92 kD in molecular weight. Secondly the fibre bundles are often observed in a rather complex network, which may enmesh the nucleolus completely. This suggests the fibre bundle networks are formed at the same time as or after the nucleolus – i.e. after reformation of the nuclear envelope. Although Beven *et al.* (1991) only observed the nuclear fibre bundles in cultured cells, nuclear inclusions have been previously observed by electron microscopy in intact tissue from various plant sources which bear a striking

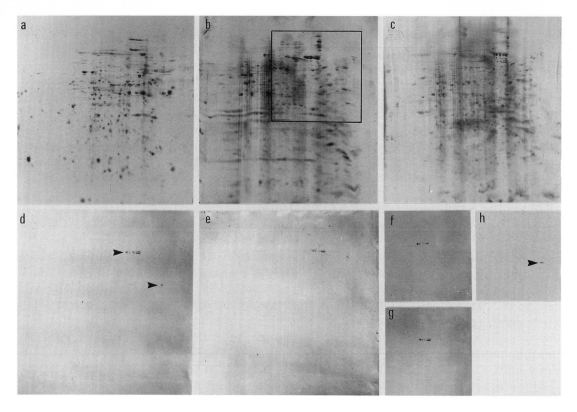

Fig. 6.3. 2-D gels and immunoblots of carrot suspension cell extracts. (a) Whole carrot cell extract 2-D gel, blotted onto nitrocellulose and labelled with biotin/avidin horseradish peroxidase for total protein; (b) whole nuclei, labelled for total protein as in (a); (c) nuclear matrix, labelled for total protein as in (a); (d) total nuclear proteins (as (b)) immunolabelled with JIM63; (e) total nuclear proteins immunolabelled with JIM62; (f) total nuclear proteins immunolabelled with AFB; (g) total nuclear proteins immunolabelled with ME101; (h) total nuclear proteins immunolabelled with IFA. In (f), (g) and (h) only the area of the immunoblot which includes the proteins reorganized by the antibodies (equivalent to the box in (b)) is shown. The 92-kD group of proteins is arrowed in (d). A lower molecular weight, more acidic protein is also recognized by JIM63 (arrow), and appears to be the 60-kD protein recognized by IFA (arrow in (h)).

resemblance to the cytoplasmic FBs on purely morphological criteria. The function of these inclusions is unknown, but it is possible that they are produced as a response to heat shock or other stress.

Thus it seems fairly certain that plant nuclei contain IF-related proteins. What is less certain is their exact relation to animal lamins and cytoplasmic IFs. The polypeptide described by Worman *et al.* (1989) seems a good candidate for a plant lamin B. It is possible that this is the same as the 60-kD nuclear matrix band labelled by anti-IFA in the study of Beven *et al.* (1991). The 92-kD proteins identified by Beven *et al.* may be other lamin-like components – possibly

more analogous to A/C lamins. However the location of these antigens does not appear to be exclusively in a peripheral lamina. Further progress will require sequence information about the putative plant lamin and IF proteins. Structural studies in plants are currently hampered by the lack of monospecific antisera; all the antibodies which have been used so far, whether monoclonal or polyclonal, label more than one antigen. This is presumably because, as with animal cells, there is an extensive family of related IF proteins which share many antigenic determinants. Nevertheless, specific antibodies have been developed in animal studies, and progress is being made in

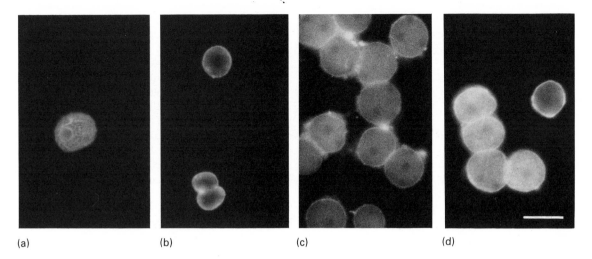

(a)   (b)   (c)   (d)

Fig. 6.4. Immunofluorescence staining of carrot and rat liver nuclei with JIM63. (a) and (b) carrot suspension culture nuclei stained with JIM63 show both an internal matrix and nuclear rim labelling. The nuclear rim labelling is enhanced by treatment with cold methanol (b). (c) rat liver nuclei show staining of the nuclear rim with JIM63. (d) for comparison, rat liver nuclei are stained with antibody to lamin A/C. Scale bar = 10 μm.

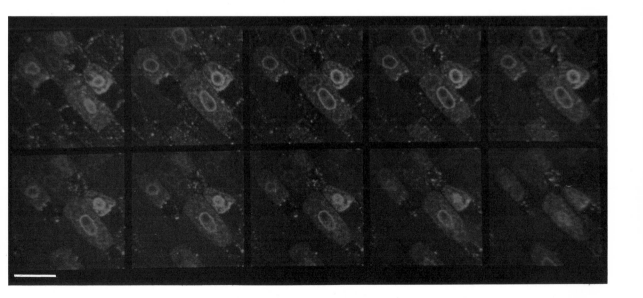

Fig. 6.5. A series of confocal optical sections through a vibratome thick section of pea root tissue labelled with JIM63. Sections approximately 40 μm in thickness were sequentially extracted in an isotonic buffer with 1% Triton X100, DNase I, and 1 M NaCl then fixed in 3.7% formaldehyde before immunolabelling. Successive confocal sections separated by 2 μm are shown. All nuclei show clear nuclear rim staining; in some nuclei there is also internal labelling. Scale bar = 20 μm.

unravelling the complex patterns of expression of the various IF proteins. It is essential that similarly specific antibodies are produced to elucidate this important area of plant studies.

References

Albers, K. and Fuchs, E. (1987). The expression of mutant epidermal keratin cDNAs transfected in simple epithelial and squamous cell carcinoma lines. *J. Cell Biol.* **105**, 791–806.

Albers, K. and Fuchs, E. (1989). Expression of mutant keratin cDNAs in epithelial cells reveals possible mechanism for initiation and assembly of intermediate filaments. *J. Cell Biol.* **108**, 1477–1493.

Bartnik, E. and Weber, K. (1989). Widespread occurrence of intermediate filaments in invertebrates; common principles and aspects of diversion. *Eur. J. Cell Biol.* **50**, 17–33.

Beven, A., Guan, Y., Peart, J., Cooper, C. and Shaw, P. (1991). Intermediate filament-related components of plant nuclear matrix revealed by monoclonal antibodies. *J. Cell Sci.* **98**, 293–302.

Chang, X.-J. and Piperno, G. (1987). Cross-reactivity of antibodies specific for flagellar tektin and intermediate filament subunits. *J. Cell Biol.* **104**, 1563–1568.

Chou, Y.-H., Bischoff, T.R., Beach, D. and Goldman, R.D. (1990). Intermediate filament reorganizing during mitosis is mediated by p34 cdc$_2$ phosphorylation of vimentin. *Cell* **62**, 1063–1071.

Dawson, P.J., Hulme, J.S. and Lloyd, C.W. (1985). Monoclonal antibody to intermediate filament antigen cross-reacts with higher plant cells. *J. Cell Biol.* **100**, 1793–1798.

Escurat, M., Phamgia, H., Huc, C., Pouplard-Barthelaix, A., Boitard, C., Bach, J.-F. *et al.* (1989). A new monoclonal antibody recognizing the amino-terminal consensus sequence of vertebrate intermediate filament proteins. *FEBS Letters* **253**, 157–162.

Franke, W.W. (1987). Nuclear lamins and cytoplasmic intermediate filament proteins: a growing multigene family. *Cell* **48**, 3–4.

Franke, W.W., Schmid, E., Grund, C. and Geiger, B. (1982). Intermediate filament proteins in non-filamentous structures: Transient disintegration and inclusion of subunit proteins in granular aggregates. *Cell* **30**, 103–113.

Franz, J.K. and Franke, W.W. (1986). Cloning of cDNA and amino acid sequence of a cytokeratin expressed in oocytes of *Xenopus laevis*. *Proc. Natl Acad. Sci. U.S.A.* **83**, 6475–6479.

Galcheva-Gargova, Z.I., Marinova, E.I. and Koleva, S.T. (1988). Isolation of nuclear shells from plant cells. *Plant, Cell and Environment* **11**, 819–825.

Geisler, N., Kaufmann, E., Fischer, S., Plessman, U. and Weber, K. (1983). Neurofilament architecture combines structural principles of intermediate filaments with carboxyterminal extensions increasing in size between triplet proteins. *EMBO J.* **2**, 129–130.

Georgatos, S.D., Maroulakou, I. and Blobel, G. (1989). Lamin A, lamin B, and lamin B receptor analogues in yeast. *J. Cell Biol.* **108**, 2069–2082.

Gerace, L. and Blobel, G. (1980). The nuclear envelope is reversibly depolymerized during mitosis. *Cell* **19**, 277–287.

Gerace, L. and Burke, B. (1988). Functional organization of the nuclear envelope. *Annu. Rev. Cell Biol.* **4**, 335–374.

Goodbody, K.C., Hargreaves, A.J. and Lloyd, C.W. (1989). On the distribution of microtubule-associated intermediate filament antigens in plant suspension cells. *J. Cell Sci.* **93**, 427–438.

Hargreaves, A.J., Dawson, P.J., Butcher, G.W., Larkins, A., Goodbody, K.C. and Lloyd, C.W. (1989a). A monoclonal antibody raised against cytoplasmic fibrillar bundles from carrot cells, and its cross-reaction with animal intermediate filaments. *J. Cell Sci.* **92**, 371–378.

Hargreaves, A.J., Goodbody, K.C. and Lloyd, C.W. (1989b). Reconstitution of intermediate filaments from a higher plant. *Biochem J.* **261**, 679–682.

Hawes, C. (1985). Conventional and high voltage electron microscopy of the cytoskeleton and cytoplasmic matrix of carrot (*Daucus carota* L.) cells grown in suspension culture. *Eur. J. Cell Biol.* **38**, 201–210.

Hawes, C. and Martin, B. (1986). Deep etching of plant cells: cytoskeleton and coated pits. *Cell Biol. Int. Rep.* **10**, 985–992.

Klymkowsky, M.W. (1988). Metabolic inhibitors and intermediate filament organization in human fibroblasts. *Exp. Cell Res.* **174**, 282–290.

Krohne, G. and Benavente, R. (1986). The nuclear lamins: a multigene family of proteins in evolution and differentiation. *Exp. Cell Res.* **162**, 1–10.

McKeon, F.D., Kirschner, M.W. and Caput, D. (1986). Homologies in both primary and secondary structure between nuclear envelope and intermediate filament proteins. *Nature* **319**, 463–467.

Newport, J. and Spann, T. (1987). Disassembly of the nucleus in mitotic extracts: membrane vesicularization, lamin disassembly, and chromosome condensation are independent processes. *Cell* **48**, 219–230.

Olesen, P. and Jensen, C.J. (1984). Ultrastructure of

intermediate filament-type bundles in cells and protoplasts from suspension-cultured maize (*Zea mays* L.). *J. Ultrastruct. Res.* **88**, Abstract 305.

Osborn, M. and Weber, K. (1986). Intermediate filament proteins: a multigene family distinguishing major cell lineages. *Trends Biochem. Sci.* **11** (11), 469–472.

Paddy, M.R., Belmont, A.S., Saumweber, H., Agard, D.A. and Sedat, J.W. (1990). Interphase nuclear envelope lamins form a discontinuous network that interacts with only a fraction of the chromatin in the nuclear periphery. *Cell* **62**, 89–106.

Parke, J.M., Miller, C.C.J., Cowell, I., Dobson, A., Dowding, A., Downes, M. *et al.* (1987). Monoclonal antibodies against plant proteins recognize animal intermediate filaments. *Cell Motil. Cytoskel.* **8**, 312–323.

Powell, A.J., Peace, G.W., Slabas, A.R. and Lloyd, C.W. (1982). The detergent-resistant cytoskeleton of higher plant protoplasts contains nucleus-associated fibrillar bundles in addition to microtubules. *J. Cell Sci.* **56**, 319–335.

Pruss, R.M., Mirsky, R., Raff, M.C., Thorpe, R., Dowding, A. and Anderton, B.H. (1981). All classes of intermediate filaments share a common antigenic determinant defined by a monoclonal antibody. *Cell* **27**, 419–428.

Robson, R.M. (1989). Intermediate filaments. *Current Opinion in Cell Biol.* **1**, 36–43.

Ross, J.H.E., Hutchings, A., Butcher, G.W., Lane, E.B. and Lloyd, C.W. (1991). The intermediate filament-related system of higher plant cells shares an epitope with cytokeratin 8. *J. Cell Sci.* **99**, 91–98.

Rungger-Brändle, E., Achstatter, T. and Franke, W.W. (1989). An epithelium-type cytoskeleton in a glial cell; astrocytes of amphibian optic nerves contain cytokeratin filaments and are connected by desmosomes. *J. Cell Biol.* **109**, 705–716.

Schliwa, M. (1986). *The Cytoskeleton*. Springer-Verlag, Wien.

Steffen, W. and Linck, R.W. (1988). Evidence for tektins in centrioles and axonemal microtubules. *Proc. Natl Acad. Sci. U.S.A.* **85**, 2643–2647.

Traas, J.A. (1984). Visualisation of the membrane bound cytoskeleton and coated pits of plant cells by means of dry cleaving. *Protoplasma* **119**, 212–218.

Traas, J.A. and Kengen, H.M.P. (1986). Gold labelling of microtubules in cleaved whole mounts of cortical root cells. *J. Histochem. Cytochem.* **34**, 1501–1504.

Vikstrom, K.L., Borisy, G. and Goldman, R.D. (1989). Dynamic arrays of intermediate filament networks in BHK-21 cells. *Proc. Natl Acad. Sci. U.S.A.* **86**, 549–553.

Weber, K., Plessmann, U. and Ulrich, W. (1989). Cytoplasmic intermediate filament proteins of invertebrates are closer to nuclear lamins than are vertebrate intermediate filament proteins, sequence characterization of two muscle proteins of a nematode. *EMBO J.* **8**, 3221–3227.

Worman, H.J., Yuan, J., Blobel, G. and Georgatos, S.D. (1988). A lamin B receptor in the nuclear envelope. *Proc. Natl Acad. Sci. U.S.A.* **85**, 8531–8534.

Worman, H.J., Henriquez, D.J., Schnell, D.J., Newman, S.D., Georgatos, S.D. and Blobel, G. (1989). Identification of a lamin B-like protein in plant cell nuclei. *J. Cell Biol.* **109**, 133a (Abstract from 29th meeting of ASCB).

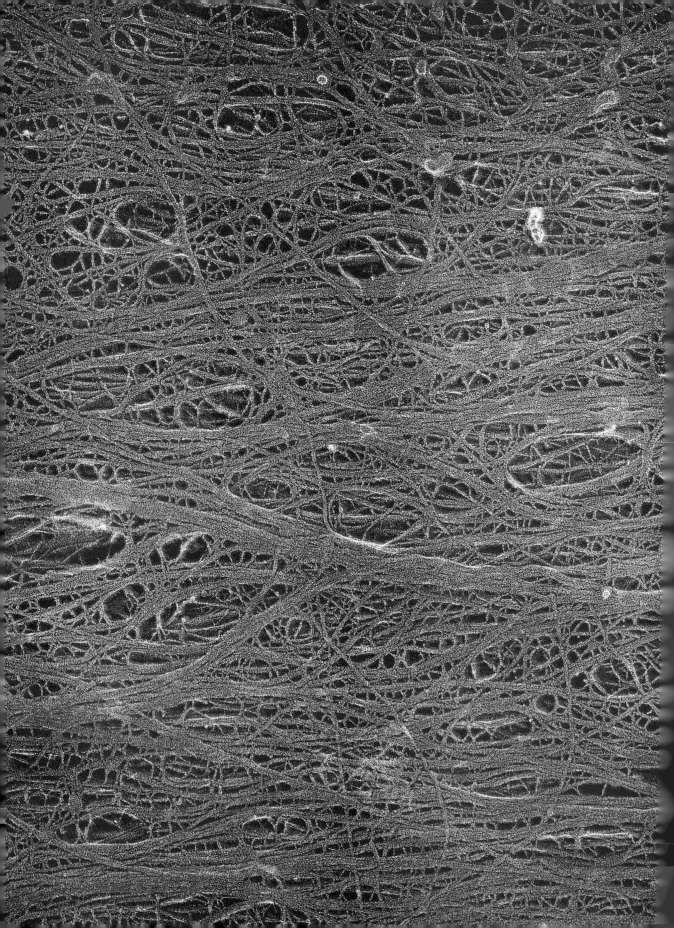

THE CELL WALL

The protoplast without a cell wall is powerless to develop an asymmetric shape and it cannot divide properly until the wall is regenerated. By itself, the cytoskeleton is unable to make these processes come about. This brings to mind the dissociated animal cell which is unable to spread or be stimulated into division until proper contact with an extracellular matrix has been restored and the cytoskeleton allowed, once more, to develop. Even though the transmembrane links between the extracellular matrix and the cytoskeleton have not been identified in plants the circumstantial evidence overwhelmingly points towards an equivalent kind of interaction. It is difficult, and probably quite incorrect therefore, to draw artificial boundaries around 'the cell' which do not recognize the integral nature and essential contribution of the wall. Cellulose microfibrils constitute the major fibrillar components of walls and it is now almost 30 years since Ledbetter and Porter first observed these tough, inelastic microfibrils to run parallel to cortical microtubules. The relationship between microfibrils and microtubules has been under scrutiny ever since. There are numerous examples of parallelism as well as a significant minority of cases where the two elements do not co-align, leading to suggestions of wall self-assembly without cytoplasmic control. In Chapter 7, Giddings and Staehelin re-evaluate the role of microtubules in microfibril alignment.

There seems little doubt that cellulose is polymerized at the cell surface although recreating this process, *in vitro*, beyond the confines of normal cellular controls and organization, has been a difficult process. Recent progress, particularly with *Saprolegnia* and *Dictyostelium* is summarized by Delmer in Chapter 8.

Perceptions of the wall change as more becomes known about it. It is clear that it is a dynamic metabolic compartment comprised of a variety of shapely polysaccharides as well as families of proteins. It is just as clear that there is no one model wall since there are different wall domains within a cell. Furthermore, the composition of the extracellular matrix varies according to its location (i.e. whether it is adjoined by another cell, an air space or the environment) as well as to its stage of development. Ideas on how the different wall components may interact within the matrix, and their regional specialization, are explored by McCann and Roberts in Chapter 9.

In Chapter 10, the last chapter in this section, Hayashi underlines the importance of non-cellulosic wall components to cellular responses. Xyloglucans interact with cellulose microfibrils but this interaction is sensitive to auxins which can break the bonds holding a cell in a particular shape. This provides a link between physiological control and cell elongation which is explored further in the succeeding section on directional cell expansion.

7. MICROTUBULE-MEDIATED CONTROL OF MICROFIBRIL DEPOSITION: A RE-EXAMINATION OF THE HYPOTHESIS

Thomas H. Giddings, Jr and L. Andrew Staehelin

Department of Molecular, Cellular and Developmental Biology, University of Colorado, Boulder, CO 80309–0347, USA

Introduction

The final shape of a growing plant cell is determined by the controlled expansion of its cell walls. Foremost among the factors that control this turgor-driven expansion is the direction in which the innermost cellulose fibrils are laid down in the growing cell wall (Green, 1980). Thus, understanding the mechanism(s) responsible for the directed deposition of cellulose fibrils has been of central interest to researchers studying plant cytomorphogenesis. Ever since Ledbetter and Porter (1963) first described transverse cortical microtubules (MTs) in elongating root-tip cells and noted that cell wall microfibrils exhibited a similar orientation, cortical MTs have been postulated to control the orientation in which cellulose fibrils are deposited (reviewed by Gunning and Hardham, 1982; Robinson and Quader, 1982; Staehelin and Giddings, 1982; Lloyd, 1984, 1987).

A number of themes appear to be emerging from recent research on MT control of microfibrillar orientation. First, the evidence that MTs are involved in orienting newly deposited microfibrils in a wide variety of plant cells is more compelling than ever. The numerous reports of exceptions to the MT–microfibril parallelism rule are increasingly viewed not as evidence against a role for MTs but as an indication that other mechanisms, as well as the one based on MTs, may operate to a variable extent in different cells. Second, progress has been made toward understanding at a molecular level how MTs and the plasma membrane may interact to regulate the pattern of microfibril deposition but the results have not as yet confirmed any one of several conflicting models. Finally, recent advances in visualizing the distribution of selected molecules in whole cells have fostered a more global view of how the entire network of MTs (Lloyd, 1984, 1987), possibly interacting with other cytoskeletal networks, may function to bring about patterned microfibril deposition. Ultimately, of course, the molecular and global approaches will have to arrive at compatible models.

The current theoretical framework for most research on the role of MTs in oriented cellulose deposition has its origin in the 'unified hypothesis' proposed by Heath (1974). The hypothesis was 'unified' in the sense that it incorporated several key concepts that had been proposed separately, including the suggestion by Roelofsen (1958) and Preston (1964, 1974) that an enzyme complex associated with the ends of nascent microfibrils may polymerize multiple β-1,4 glucan polymers in such a way that they assembled into crystalline microfibrils; the concept of a fluid bilayer membrane that permits the terminal complexes to move within the plane of the membrane; an array of cortical MTs that predicts the orientation of the cellulose microfibrils as

The Cytoskeletal Basis of Plant Growth and Form   ISBN 0–12–453770–7

initially reported by Ledbetter and Porter (1963); and a mechanism by which the MTs could direct the movement of the complexes. Heath (1974) proposed that the complexes could have a component that contacts the MTs and generates force via a dynein-like ATPase mechanism to move the complex. This chapter will focus on the role of both the plasma membrane and cortical MTs in orientating cellulose microfibril deposition.

MT–microfibril parallelism

The observation that microfibrils are deposited parallel to cortical MTs has been documented for a wide variety of plant cells, but in some cell types microfibril deposition has been shown to occur in the absence of cortical MTs or in different orientations (reviewed in Gunning and Hardham, 1982; Robinson and Quader, 1982; Lloyd, 1984). When the morphological observation of parallelism is made, a functional correlation between MT and microfibril orientations has typically been demonstrated by using colchicine or other inhibitors to destabilize the MTs which in turn disrupts the pattern of microfibril deposition. In the 'classic' case, new microfibrils are synthesized and deposited in a transverse fashion along the entire length of elongating cylindrical cells, and the cortical array of transverse MTs is likewise found along the full length of the cell. One elegant demonstration of this type of parallelism was obtained in a study of elongating semicells of the unicellular green alga, *Closterium* (Hogetsu and Shibaoka, 1978). Comparing colchicine-treated cells to controls revealed that MTs were absent from treated cells, the microfibrils of the primary cell wall were deposited randomly rather than transversely, and the new semicell became spherical rather than the normal tapered cylinder.

Another highly suggestive example of MT–microfibril parallelism comes from cells where the deposition of microfibrils is more localized. Groups of transverse MTs were shown to be grouped over developing secondary cell wall thickenings in differentiating xylem cells, and the microfibrils in the cell wall thickenings also exhibited a transverse orientation (Hepler and Newcomb, 1964; Cronshaw and Bouch, 1965;

Brower and Hepler, 1976). The clustering of the MTs into groups precedes the appearance of the thickenings. In response to MT destabilizing agents, secondary cell wall material was found to be deposited homogeneously along the surface rather than in rings (Hepler and Fosket, 1971; Brower and Hepler, 1976). Recently, a similar phenomenon has been described in differentiating protophloem sieve elements (Eleftheriou, 1987). In this case, the secondary cell wall thickenings are relatively broad and are separated by narrow strips where the wall remains thin and in which many plasmodesmata are found. MTs are clustered over the thickenings during deposition and microfibrils visualized in tangential thin sections appear to run parallel to the MTs. Cell wall thickenings reminiscent of those seen in xylem have also been observed by fluorescence microscopy of calcofluor-stained wheat mesophyll cells (Jung and Wernicke, 1990). Double staining of the same cells with antibodies to tubulin revealed a remarkably precise coincidence of the distribution of the MTs and the thickenings.

While supporting a role for MTs in controlling microfibril orientation, these studies indicate that MTs are not required for cellulose synthesis or the formation of microfibrils. This distinction was clearly illustrated in a study of the alga *Oocystis solitaria* (Robinson and Quader, 1981; Lloyd, 1984). The cell wall is laid down in layers of parallel microfibrils, with each consecutive layer deposited at an angle to the previous one. In the presence of colchicine, deposition of parallel microfibrils continues, but the normal shift to a new orientation does not. If the colchicine is washed out, the orientation of subsequent layers resumes shifting normally.

Melan (1990) has recently demonstrated that MT–microfibril parallelism is maintained even when MT alignment is artificially manipulated. In control protoplasts obtained from elongating pea epicotyls, anti-tubulin immunofluorescence demonstrated that MTs became randomized during protoplast formation. When such protoplasts regenerated a cell wall, the microfibrils also exhibited a random orientation. In contrast, when 40 μM taxol was applied, ordered arrays of MTs were preserved and the microfibrils in regenerating cell walls became parallel to the orientation of the MTs. Perhaps the most interesting observation was that the arrays of parallel MTs in taxol-treated protoplasts were sometimes shifted away from the transverse orientation found in intact cells. The pattern of microfibril deposition nevertheless followed the orientation of the MTs, not the orientation that would have been formed in the intact cell. These results provide compelling evidence that MT arrays are involved in the orientation of cellulose microfibrils.

In summary, the studies with MT-disrupting drugs and the observation that many plant cells synthesize microfibrils in the absence of MTs indicate that the role of cortical MTs is to influence the pattern (location and orientation) of microfibril deposition. They also suggest that cellulose synthetases can move in the plane of the plasma membrane without direct, dynamic interactions with MTs (e.g. via a dynein-like motor), although the possibility that such interactions occur in some cases can not be ruled out.

Exceptions to the MT–microfibril parallelism rule

A role for MTs in orienting the deposition of cellulose microfibrils has been disputed based on the observation of a number of cells in which the MTs and most recently synthesized (innermost) microfibrils are not parallel. As reviewed previously (Lloyd, 1984; Seagull, 1989), some of these 'exceptions' can be explained by dynamic cortical MT arrays shifting to their new orientation just prior to fixation, and being fixed before they had a chance to reorient the microfibrils of the next cell wall layer. However, there are numerous other examples where MTs are clearly not involved in controlling the ultimate microfibril orientation. The best characterized exceptions to the MT–microfibril parallelism rule are cells that exhibit tip growth, such as root hairs and pollen tubes, and/or form helicoidal walls. The orientation of the microfibrils within each lamella of a helicoidal wall is helical and the pitch of the helices varies by some angle from one layer to the next (Levy, 1987). Emons and

co-workers have reported that in the root hairs of several aquatic and terrestrial plants, individual microfibrils generally exhibit a helical architecture whereas MTs exhibit a net axial orientation (Emons, 1982; Emons and Wolters-Arts, 1983; Emons and van Maaren, 1987). Direct visualization of microfibrils deposited at an angle to adjacent MTs in the same root hair was reported by Traas and Derksen (1989) in dry-cleaved preparations of *Equisetum*. In root hairs of *Allium*, however, the deposition of crossed-helical ($\pm 45°$) microfibrils has been postulated to be controlled by a shifting helical MT network which also forms 45° helices (Lloyd, 1983; Lloyd and Wells, 1985). It has also been suggested that the organization of cellulose microfibrils into helicoidal structures occurs post-synthetically in the cell wall via a self-assembly process involving other cellulose-associated polysaccharides such as hemicelluloses (Roland et al., 1977; Neville, 1985; Neville and Levy, 1985). The mechanistic aspects of this process are still not understood but may be dependent on a certain amount of cell expansion, the action of specific enzymes, and a specific pH and ionic environment (Vian et al., 1986; Neville, 1988).

Other tip-growing cells appear to exhibit a more classical MT-mediated deposition of transversely oriented microfibrils The protonemata of *Adiantum* exhibit uniform diameter when grown under red light and contain a band of transverse cortical MTs just below the apex (Murata et al., 1987). The microfibrils in the innermost layer of the cell wall in the subapical region were also orientated transversely (Murata and Wada, 1989). Upon exposure to blue light, the orientation of cortical MTs in the subapical region shifted from transverse to more random, perhaps somewhat longitudinal. The innermost layer of microfibrils was then deposited in a more random looking mesh, and the diameter of the apex began to swell (Murata and Wada, 1989). These findings appear to support a role for cortical MTs in regulating microfibril deposition in a tip-growing cell.

In summary, whereas some cells appear to employ a MT-based system for orienting microfibril deposition, others apparently do not.

Furthermore, not all MTs in the cortical cyto- plasm are necessarily involved in regulating microfibril deposition. The axial MTs that occur in the cortical cytoplasm of tip-growing cells, for example, could be engaged in some other function, possibly interacting with micro- filaments to guide new cell wall material and various organelles to the growing tip (Doonan *et al.*, 1988). In *Closterium*, MTs that are apparently involved in orienting the deposition of microfibrils in the primary cell wall could still be detected adjacent to the plasma mem- brane during onset of secondary cell wall formation, but the new secondary wall micro- fibrils were not deposited parallel to the MTs (Giddings and Staehelin, 1988). Secondary cell wall microfibrils were also deposited at sharp angles to cortical MTs in *Valonia* protoplasts (Itoh and Brown, 1984). Thus, it appears that something more than the mere presence of cortical MTs is required to bring about control of microfibril deposition.

Involvement of plasma membrane 'terminal complexes' in the synthesis of cellulose microfibrils

There is a general consensus that cellulose microfibrils are synthesized by 'terminal' enzyme complexes that are localized in the plasma membrane, and that these complexes determine to a large extent the organization of the cellulosic component of the cell wall (Brown, 1985). The fact that cellulose is secreted in the form of crystalline microfibrils with sufficient tensile strength to act as a cellular exoskeleton must be due to the ability of terminal complexes not only to synthesize β-1,4 glucan polymers, but to do so in such a way that the individual polymers coalesce into microfibrils of a defined size. As pointed out by Delmer (1987), one may hypothesize that the diameter of the microfibril depends on the number of polymerases located in a single terminal complex. Higher orders of structure, consisting of grouped microfibrils, may then be generated if multiple terminal complexes are held together in a larger array.

Although relatively little is known at present concerning the biochemistry of the cellulose synthetase complex of plants (Delmer, 1987;

see also Chapter 8, this volume), there is considerable evidence that certain complexes visualized in the plasma membranes of plant cells by freeze-fracture electron microscopy are the structures that synthesize cellulose and secrete microfibrils.

Two broad classes of terminal complexes have been identified to date (summarized in Brown, 1985); those exhibiting linear arrays of particles, and those in which particles are clustered in rings called rosettes. Brown and Montezinos (1976) provided the first evidence of a distinct linear array of plasma membrane particles associated with the growing tips of microfibrils in the alga *Oocystis apiculata*. Linear complexes have been visualized in a number of other algae (Brown, 1985; Delmer, 1987). The evidence correlating them with a role in microfibril assembly (summarized by Delmer, 1987) includes the observation that the com- plexes change morphology and orientation when microfibril assembly is disrupted by dyes that hydrogen-bond to the glucan polymers, and resume a normal configuration when the dye is washed out (Robinson and Quader, 1981; Itoh *et al.*, 1984). Delmer (1987) also draws attention to the observation that larger linear arrays of particles in the terminal com- plexes appear to give rise to cellulose fibrils of larger diameter. As discussed below, a similar concept has been proposed to explain how cells with a rosette type of terminal complex can generate larger fibrils.

Many other plants, from algae to angio- sperms, have been shown to possess terminal complexes consisting of a ring of about six individual particles on the P-face of the plasma membrane (summarized in Brown, 1985). In *Micrasterias*, hexagonal arrays of rosettes were visualized at the ends of bands of parallel fibrils (Giddings *et al.*, 1980). Due to the high degree of order in the arrangement of the rosettes and the resulting pattern of cellulose fibrils in the secondary cell walls of *Micrasterias*, detailed correlations could be drawn that have implications for the function of the rosette terminal complexes. The centre-to-centre spacing of the fibrils within a band was found to be the same as that of parallel rows of rosettes within a hexagonal array (Fig. 7.1).

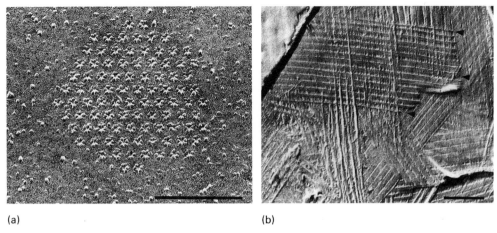

(a)   (b)

Fig. 7.1. Freeze-fracture electron micrographs of *Micrasterias denticulata* semicells engaged in secondary cell wall deposition. (a) Rows of rosettes within this typical hexagonal array exhibit a constant centre-to-centre spacing of 28 nm. (b) A band of parallel cellulose fibrils in the secondary cell wall. Although the width of the fibrils varies from wide in the centre to narrow at the edges (arrowheads), their centre-to-centre spacing remains constant at about 28 nm. Scale bar = 0.2 μm. From Giddings *et al.* (1980) with permission.

Furthermore, larger fibrils of cellulose occur in the centre of the band and progressively smaller ones are found towards the edge. This correlates with a greater number of rosettes in the central rows of the hexagonal array. As illustrated diagrammatically in Fig. 7.2, these results suggest that each rosette forms a single 'elementary' microfibril and when rosettes occur in rows, these microfibrils associate laterally to form larger fibrils.

The clustering of rosettes into hexagonal arrays appears to be unique to certain groups of algae (Zygnematales) but the morphology of the rosettes themselves appears to be the same in algae, ferns (Wada and Staehelin, 1981) and a wide range of higher plants (Mueller and Brown, 1980; Brown, 1985). Rosettes are most reliably visualized in intact plant tissues or cells which have been ultrarapidly frozen without prior exposure to chemical fixatives or cryoprotectants. They appear to be even more labile in suspension culture cells (Herth and Weber, 1984; Chapman and Staehelin, 1985) and are not generally found in protoplasts. Whether this lability correlates with that of the biochemical activity of cellulose biosynthesis (Delmer, 1987) remains to be seen. Rosette-type terminal complexes are probably delivered to the plasma membrane in vesicles derived

from the Golgi apparatus. During secondary cell wall formation in *Micrasterias*, characteristic flattened vesicles containing arrays of rosettes are formed by the Golgi apparatus and can be seen to fuse with the plasma membrane (Dobberstein and Kiermayer, 1972; Giddings *et al.*, 1980). Individual rosettes were visualized in *trans* Golgi cisternae and in vesicles budding off the *trans* Golgi in isolated *Zinnia* cells (Haigler and Brown, 1986). Using monensin to inhibit Golgi-mediated secretion, Rudolph and Schnepf (1988) were able to deplete the plasma membrane of moss protonemata of rosettes, and to estimate their lifetime to be about 20 min.

Mechanism of MT control of microfibril order

Visualization of the putative cellulose synthesizing complexes has made it possible to achieve some progress in understanding how MTs may regulate the pattern of microfibril deposition. As discussed in the previous section, the terminal complexes, either rosettes or linear particle arrays, occur at the ends of microfibrils in growing cell walls, and their distribution is invariably correlated with the observed pattern of microfibril deposition. The most likely explanation for this is that the

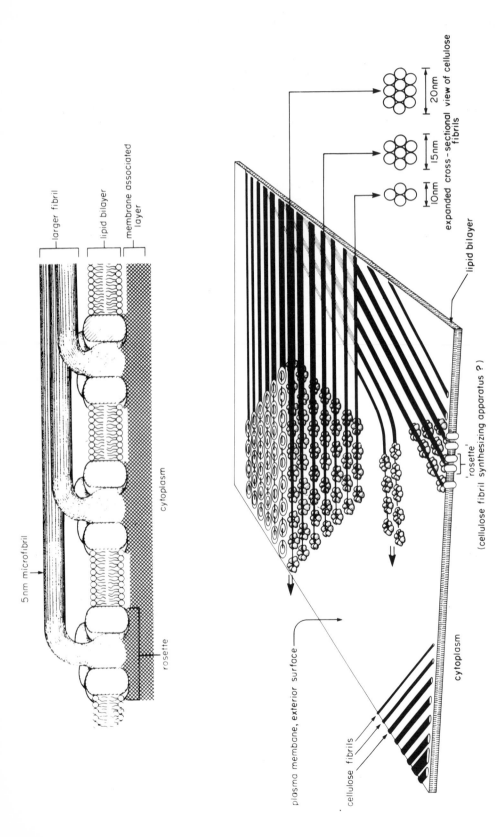

Fig. 7.2. Model of cellulose fibril deposition during secondary wall formation in *Micrasterias*. Each rosette is believed to form one 5-nm microfibril. A row of rosettes forms a set of 5-nm microfibrils, which aggregate laterally to form the larger fibrils of the secondary cell wall. Above: side view. Below: surface view with expanded cross-sectional view of cellulose fibrils. From Giddings *et al.* (1980) with permission.

glucan polymerases responsible for cellulose synthesis comprise these complexes. However, even if the actual polymerases are located on the cytoplasmic surface of the plasma membrane, it seems almost certain that the microfibril exits the plasma membrane via the terminal complex. Therefore, the terminal complexes can be considered, at a minimum, to represent markers of the site of microfibril secretion from the cell. It follows that a mechanism capable of regulating the pattern of microfibril deposition may do so by controlling the distribution and/or direction of travel of the terminal complexes. Quader (1986) provided direct support for this hypothesis in a study of cell wall formation in *Oocystis* in which the orientation of the terminal complexes in the plasma membrane was randomized by a combination of MT-disrupting drugs and dyes that disrupt the coalescence of cellulose polymers into microfibrils. When these inhibitors were washed out, microfibrils were deposited in random orientations, suggesting that MTs exert their influence over microfibril deposition by controlling the orientation of the terminal complexes.

The original hypothesis of Heath (1974) states that the terminal complexes may be directly attached to MTs by a dynein-like linkage that propels the complex within the plane of the membrane. Other models (reviewed in Lloyd, 1984; Seagull, 1989) include a permanent but non-mechanical linkage between the terminal complex and a short MT with a cytoplasmic mechanism moving the MT (e.g. via interaction with actin filaments or sliding against another MT). As indicated previously however, the observation that cellulose microfibrils can elongate in the absence of MTs would seem to argue against the necessity of invoking any mechanism that relies on MTs to elongate the microfibril. The alternative 'membrane channel' hypothesis proposed by Herth (1980) to account for chitin deposition in *Poterioochromonas* and by Staehelin and Giddings (1982) for cellulose deposition suggests that forces resulting from the polymerization and crystallization of microfibrils are sufficient to propel the terminal complexes in the plane of the plasma membrane.

Simultaneous visualization of terminal complexes and MTs

One test of these conflicting models of how MTs could 'steer' terminal complexes is to demonstrate whether or not the two structures are in direct contact. This question was addressed in a study of *Closterium* (Giddings and Staehelin, 1988), a unicellular alga in which both MTs and microfibrils are oriented circumferentially, and in which colchicine disruption of MTs results in randomly deposited microfibrils in the primary cell wall (Hogetsu and Shibaoka, 1978). Freeze-fracture replicas of cells forming primary walls revealed short rows of two to five rosettes in the plasma membrane, oriented parallel to the microfibrils and perpendicular to the long axis of the cell (Fig. 7.3). The position of the underlying cortical MTs was revealed by etching the sample which caused much of the P-face of the plasma membrane to collapse except where supported by underlying MTs. Thus, as shown in Fig. 7.4, the rosettes were found not directly on top of MTs but rather between them or sometimes adjacent to them. The distance between the rosettes and MTs argues against a mechanical linkage between them. Instead, these results support the hypothesis that terminal complexes travel in membrane channels delineated by cortical MTs, illustrated schematically in Fig. 7.5. The hypothesis implies that MTs interact with the plasma membrane in such a way that terminal complexes can not cross the MT 'guide rails' as they are propelled forward by forces generated from the polymerization and crystallization of cellulose microfibrils.

In the case of secondary wall thickenings in differentiating xylem, Herth and co-workers (Herth, 1985; Schneider and Herth, 1986) found that rosette terminal complexes are concentrated in the regions of the plasma membrane over the forming thickenings but are not arranged in straight lines as one would expect if they were all in direct contact with the cortical MTs. These authors therefore also favoured the membrane channel hypothesis.

MT interaction with the plasma membrane
The nature of the interaction between cortical MTs and the plasma membrane remains

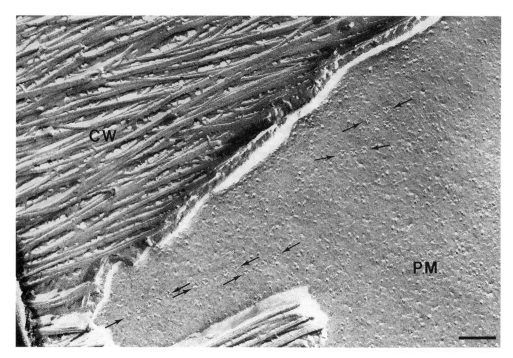

Fig. 7.3. Rows composed of three to five rosettes (bracketed by arrows) on the P-face of the plasma membrane (PM) of an elongating *Closterium* semicell. The rows are roughly parallel to the cellulose fibrils of the primary cell wall (CW). Scale bar = 0.1 μm. From Giddings and Staehelin (1988) with permission.

Fig. 7.4. Rows of rosettes (bracketed by arrows) in the plasma membrane *between* cortical MTs in *Closterium*. The position of the underlying MTs (open arrowheads) was revealed by freeze-etching which caused the P-face of the plasma membrane to collapse toward the cytoplasm. Scale bar = 0.2 μm. From Giddings and Staehelin (1988) with permission.

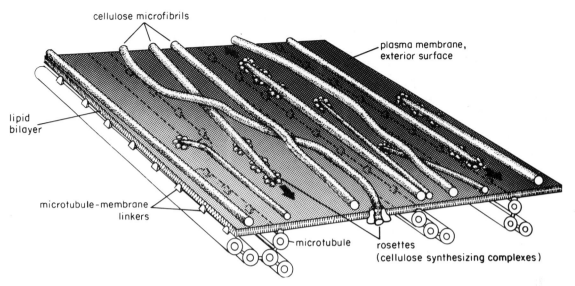

Fig. 7.5. Model depicting MT-mediated control of microfibril deposition during primary cell wall formation in *Closterium*. The rosette complexes are believed to be propelled forward by forces resulting from the polymerization and crystallization of cellulose microfibrils. MTs interact with the plasma membrane via a putative bridging protein to form barriers which constrain the path of the rosettes. From Staehelin and Giddings (1982) with permission.

incompletely defined in molecular terms. Structures that bridge the gap between cortical MTs and the plasma membrane have often been visualized in electron micrographs (Hardham and Gunning, 1978; Gunning and Hardham, 1982). Improved ultrastructural preservation, especially as a result of cryo-fixation and freeze-substitution, has provided more reproducible evidence of periodic cross-bridge structures, some of the clearest examples including those visualized in stamen hairs (Lancelle *et al.*, 1986) and pollen tubes (Lancelle *et al.*, 1987). The cross-bridges were observed between MTs, between MTs and the plasma membrane, and between MTs and adjacent, parallel microfilaments. The periodicity of the cross-bridges, approximately 22 nm, appeared to be roughly constant regardless of whether they were attached to a MT, microfilament or the plasma membrane. These studies demonstrate that the extensive MT–plasma membrane cross-bridging predicted by some models can occur.

Similar structures, with a periodicity of about 30 nm, were observed on cortical MTs in freeze-etch preparations of elongating *Clos-*

terium semicells (Fig. 7.6), a case in which the MTs apparently do play a role in orienting the deposition of transverse cellulose microfibrils (Giddings and Staehelin, 1988). Whereas terminal complexes are readily observed in freeze-fracture electron micrographs of the plasma membrane of plant cells engaged in cell wall formation, no rows of particles that could represent intramembrane domains of the hypothesized cross-linkers have been detected to date. Clearly, better visualization and more precise description of the contacts between cortical MTs and the plasma membrane are needed.

The cross-bridges that have been visualized by electron microscopy have not yet been identified in biochemical terms. However, an integral membrane protein derived from a plasma membrane preparation was shown to associate with plant microtubules (Cyr *et al.*, 1986). Cyr and Palevitz (1989) also identified a set of proteins in carrot suspension cells that were capable of binding to MTs. These putative plant microtubule-associated proteins (MAPs) induced MT bundling and the assembly of tubulin into MTs, and also decorated the

Fig. 7.6. Freeze-etch micrograph of cortical MTs in *Closterium*. Structures perpendicular to the MTs (arrows) have a periodicity of 30 nm and may contact the plasma membrane (PM) or other MTs. Scale bar = 0.2 μm. From Giddings and Staehelin (1988) with permission.

surface of isolated MTs with a periodicity of about 100 nm, suggesting that they bind to specific sites on MTs. An affinity purified antibody to a 76-kD protein from this set of putative MAPs co-localized with cortical MTs in taxol-stabilized, lysed protoplasts in immuno-fluorescence micrographs. These experiments suggest that it may soon be possible to equate the periodic structures visualized on cortical MTs with specific proteins.

Still, a crucial gap remains in our under-standing of the nature of the interaction between cortical MTs and the plasma membrane. Even if the MAPs do not interact directly with the terminal complexes, the question remains whether they play a passive or active role in cellulose fibril orientation. Passive MT–plasma membrane cross-linkers that could form stationary barriers in the membrane which restrict the travel of mobile terminal complexes have been discussed above. Alternatively, it has been suggested that active, dynein-like molecules could cause bulk flow of the plasma membrane (Hepler and Fosket, 1971; Hepler and Palevitz, 1974; Brower and Hepler, 1976; Mueller and Brown, 1982; Brown, 1985). A flowing membrane might either orient microfibrils by shearing forces or might sweep along synthesizing complexes. The observation that cell wall lamellae com-posed of straight microfibrils can be deposited in the presence of MT-destabilizing agents, as observed in *Oocystis* (Robinson and Quader,

1981) for example, would seem to argue against the need for such a mechanism.

Dynamic MT arrays and changes in microfibril orientation

Microfibril orientation and deposition patterns often change during cell growth and differen-tiation, implying that if MTs govern these patterns the distribution of the MTs must also change. This phenomenon was observed in the polylamellate cell wall of *Oocystis*, in which each successive layer of microfibrils has a different orientation from the preceding one. As mentioned previously, each shift in micro-fibril orientation is preceded by a prior shift in MT distribution (Robinson and Quader, 1981). Changes in microfibril-deposition patterns also play an important part in photo- and gravitropic responses (see Sievers *et al.*, Chapter 13, this volume), as well as other hormone-regulated morphogenetic processes by favouring asym-metrical elongation rates of cell walls on different sides of growing shoots and roots. If MTs are responsible for these changes, the entire cortical MT array within certain cells must be altered in order to bring about the reorientation of the microfibrils.

Evidence of a hormone-induced growth response mediated by MT reorientation and consequent change in microfibril deposition has been described in growing maize coleoptiles (Bergfeld *et al.*, 1988; Nick *et al.*, 1990). Segments of maize coleoptiles from which the apices had been removed stopped elongating, apparently due to the absence of an endogenous source of auxin. Application of exogenous auxin to the segments restored elongation, an effect that is believed to be regulated primarily by the outer epidermal cells (Kutschera et al., 1987). In the absence of auxin, cortical MTs and the most recently synthesized cell wall microfibrils were more longitudinally oriented than in intact, growing coleoptiles. In response to applied auxin, the entire MT network of the epidermal cells underwent a reorientation from longitudinal to transverse and the microfibrils were also deposited transversely, a response which presumably accounts for the resumption of cell elongation. A similar reorientation was observed on the rapidly growing sides of the coleoptiles during photo- or gravitropically induced curvature.

As shown by Roberts et al. (1985), ethylene applied to epicotyls and hypocotyls can induce similar MT rearrangements in the outer cell layers. Based on the visualization of MT arrays in whole mounts by means of anti-tubulin immunofluorescence, they describe the transversely oriented MTs of control cells as forming a very tight multistart helical array. In response to ethylene, the entire array becomes shifted toward a more longitudinal orientation which appears to consist of very steep helices. Thus, the authors contend that the switch between these two configurations could be accomplished by varying the torque on the helical arrays, as in winding or unwinding a coiled spring. The concomitant cessation of elongation and increase in lateral expansion of ethylene-treated tissue segments appears to arise from a change in orientation of the cellulose microfibrils from predominantly transverse to more longitudinal (reviewed in Roberts et al., 1985). As was the case in the previous example, none of the treated cells was found to lack a cortical MT array, arguing against the hypothesis that the change in the orientation of the MT array was achieved by means of depolymerization followed by nucleation of a new array. The mechanistic basis of this reorientation of whole cortical MT arrays, however, remains an enigma, although actin filaments have been implicated in these changes (Kobayashi et al., 1988; see next section).

Other cytoskeletal elements

In addition to MTs, actin microfilaments and intermediate filaments are significant elements of the plant cytoskeleton (see McCurdy and Williamson; Shaw et al., Chapters 1 and 6 respectively, this volume). Interpretation of the role of actin has been slowed by problems in faithfully preserving its in situ distribution for immunofluorescence microscopy, the lack of effective and specific inhibitors, and the difficulty of visualizing microfilaments (relative to MTs) by electron microscopy (Staiger and Schliwa, 1987). Nevertheless, the involvement of actin in cytoplasmic streaming and transport of organelles is well established and there is evidence that actin plays a role in transport of cell wall material in cells that exhibit localized tip-growth. Active roles for microfilaments in orienting microfibril deposition also have been proposed, such as moving rosettes, moving MTs that are linked to rosettes, or causing the whole membrane to flow, thereby sweeping terminal complexes along (Brown, 1985; Seagull, 1989). However, no direct involvement in orienting cellulose microfibrils has been demonstrated to date.

Improved fixation methods have recently led to more reliable visualization of actin in plant cells by electron microscopy (Parthasarathy et al., 1985; Lichtscheidl et al., 1990). In samples prepared by freeze-substitution, microfilaments often run parallel to cortical MTs and are seen to be cross-linked to them (Lancelle et al., 1987). It is likely that these cortical microfilaments are composed of actin, since similar looking filaments that occur in larger cytoplasmic bundles have been labelled with anti-actin probes (Lancelle and Hepler, 1989). The function of these microfilaments has not yet been elucidated.

One intriguing report has appeared which suggests that actin may play an indirect role in regulating the pattern of microfibril deposition (Kobayashi et al., 1988). Double fluorescent staining of MTs and actin was employed to visualize the distribution of these cytoskeletal elements during the differentiation of cultured Zinnia mesophyll cells into tracheary elements, a process accompanied by the deposition of generally transverse, localized secondary cell wall thickenings. At the onset of secondary wall deposition, MTs reoriented from longitudinal to transverse and became clustered at sites where the thickenings were to be deposited. Actin labelling was detected between the bundles of MTs. Exposure of the cells to cytochalasin B prevented the reorientation of MTs and resulted in the deposition of longitudinal thickenings. The authors suggest that the MTs are responsible for localizing cell wall deposition in the thickenings, and that the role of actin would be the reorientation of the MTs. It is premature to generalize based on this study, but the possibility has been raised that actin could play a significant role in reorienting MT arrays, especially where the distribution of the latter has been shown to undergo dynamic changes.

Future directions

The proteins that bridge cortical MTs to the plasma membrane in cells engaged in cellulose microfibril deposition can be considered the 'missing links' in the MT/microfibril parallelism story. Once these are identified, a number of questions at the core of this problem can be addressed. Do the bridging molecules have ATPase activity, and if so, what do they act on? Do they insert directly into the bilayer membrane or are they connected to other membrane proteins? What distinguishes cortical MTs that affect the orientation in which cellulose fibrils are deposited from those that appear to have no such effect? Detailed immuno-localization studies at both the cellular and ultrastructural levels could provide significant insights into the function of such a cross-linking molecule, and could help define more precisely the role of MTs in cell wall assembly.

With an increasing number of systems identified in which oriented microfibril deposition occurs in the absence of a demonstrable effect of MTs, attention may now turn to identifying the mechanisms that do operate in these cases. Improved visualization techniques may help draw more attention to previously overlooked cytoskeletal elements; but the extent to which microfibrils can be oriented without cytoskeletal involvement (e.g. by self-assembly processes) also has to be studied in greater detail.

Based on the extensive ultrastructural literature describing putative cellulose synthesizing complexes, a consensus seems to be emerging that the mechanism that orients newly deposited cellulose microfibrils does so by acting on the direction taken by mobile terminal synthesizing complexes. An equally significant question concerns rearrangements in the organization of macromolecules, including cellulose microfibrils, *after* deposition in the wall. There are proposals that 'self-assembly' processes involving certain components of the plant's extracellular matrix may play a more significant role than the initial deposition of cellulose in generating some of the observed cell wall textures (Roland *et al.*, 1977, 1989; Neville and Levy, 1985; Vian *et al.*, 1986; Neville, 1988); and ultimately it is the properties of the cell wall that regulate morphogenesis.

If one accepts that MTs somehow determine the pattern of microfibril deposition, the question naturally arises as to what controls the orientation of the MTs. Microtubule organizing centres (MTOCs) presumably play a role (Gunning and Hardham, 1982), and it is possible that these dictate the structure of the cortical MT array to the same extent that centrioles or MTOCs do in mitotic spindles. However, recent results, such as the evidence that actin plays a role in reorienting MT networks (Kobayashi *et al.*, 1988), demonstrate the need for a more comprehensive model. An integrated cytoskeletal network including MTs, microfilaments, intermediate filaments and possibly other elements, could be assembled by various organizing centres, but then be subject to reorientations controlled by more global sensing and transducing systems. Plant growth hormones, for example, could trigger a cascade of events including calcium/calmodulin control of the cytoskeleton.

Acknowledgements

This work was supported by NSF grant DCB-8615763 and NIH grant GM 18639 to LAS. We are grateful to Dr Sam Levy for helpful discussions and critical reading of the manuscript.

References

Bergfeld, R., Speth, V. and Schopfer, P. (1988). Reorientation of microfibrils and microtubules at the outer epidermal wall of maize coleoptiles during auxin-mediated growth. *Bot. Acta* **101**, 57–67.

Brower, D.L. and Hepler, P.K. (1976). Microtubules and secondary wall deposition in xylem: The effects of isopropyl N-phenylcarbamate. *Protoplasma* **87**, 91–111.

Brown, R.M. (1985). Cellulose microfibril assembly and orientation: Recent developments. *J. Cell Sci. Suppl.* **2**, 13–32.

Brown, R.M. and Montezinos, D. (1976). Cellulose microfibrils: Visualization of biosynthetic and orienting complexes in association with the plasma membrane. *Proc. Natl Acad. Sci. U.S.A.* **73**, 143–147.

Chapman, R.L. and Staehelin, L.A. (1985). Plasma membrane 'rosettes' in carrot and sycamore suspension culture cells. *J. Ultrastruct. Res.* **93**, 87–91.

Cronshaw, J. and Bouck, G.B. (1965). The fine structure of differentiating xylem elements. *J. Cell Biol.* **24**, 415–431.

Cyr, R.J. and Palevitz, B.A. (1989). Microtubule-binding proteins from carrot. *Planta* **177**, 245–260.

Cyr, R.J., Bustos, M., Guiltinan, M.J. and Fosket, D.E. (1986). Identification of a 74 kD integral membrane protein that associates with microtubules in carrot protoplasts. *J. Cell Biol.* **103**, 395a.

Delmer, D.P. (1987). Cellulose synthesis. *Ann. Rev. Plant Physiol.* **38**, 259–290.

Dobberstein, B. and Kiermayer, O. (1972). Das Auftreten eines besonderen Typs von Golgivesikeln wahrend der Sekundarwandbildung von *Micrasterias denticulata* Breb. *Protoplasma* **75**, 185–194.

Doonan, J.H., Cove, D.J. and Lloyd, C.W. (1988). Microtubules and microfilaments in tip growth: evidence that microtubules impose polarity on protonemal growth in *Physcomitrella patens*. *J. Cell Sci.* **89**, 533–540.

Eleftheriou, E.P. (1987). Microtubules and cell wall development in differentiating protophloem sieve elements of *Triticum aestivum* L. *J. Cell Sci.* **87**, 595–607.

Emons, A.M.C. (1982). Microtubules do not control microfibril orientation in a helicoidal cell wall. *Protoplasma*, **113**, 85–87.

Emons, A.M.C. and van Maaren, N. (1987). Helicoidal cell-wall texture in root hairs. *Planta* **170**, 145–151.

Emons, A.M.C. and Wolters-Arts, A.M.C. (1983). Cortical microtubules and microfibril deposition in the cell wall of root hairs of *Equisetum hyemale*. *Protoplasma* **117**, 68–81.

Giddings, T.H. and Staehelin, L.A. (1988). Spatial relationship between microtubules and plasma-membrane rosettes during the deposition of primary wall microfibrils in *Closterium* sp. *Planta* **173**, 22–30.

Giddings, T.H., Brower D.L. and Staehelin, L.A. (1980). Visualization of particle complexes in the plasma membrane of *Micrasterias denticulata* associated with the formation of cellulose fibrils in primary and secondary cell walls. *J. Cell Biol.* **84**, 327–339.

Green, P.B. (1980). Organogenesis – a biophysical view. *Ann. Rev. Plant Physiol.* **31**, 51–82.

Gunning, B.E.S. and Hardham, A.R. (1982). Microtubules. *Ann. Rev. Plant Physiol.* **33**, 651–698.

Haigler, C.H. and Brown, R.M. (1986). Transport of rosettes from the Golgi apparatus to the plasma membrane in isolated mesophyll cells of *Zinnia elegans* during differentiation to tracheary elements in suspension culture. *Protoplasma* **134**, 111–120.

Hardham, A.R. and Gunning, B.E.S. (1978). Structure of cortical microtubule arrays in plant cells. *J. Cell Biol.* **77**, 14–34.

Heath, I.B. (1974). A unified hypothesis for the role of membrane bound enzyme complexes and microtubules in plant cell wall synthesis. *J. Theor. Biol.* **48**, 445–449.

Hepler, P.K. and Fosket, D.E. (1971). The role of microtubules in vessel member differentiation in *Coleus*. *Protoplasma* **72**, 213–236.

Hepler, P.K. and Newcomb, E.H. (1964). Microtubules and fibrils in the cytoplasm of *Coleus* cells undergoing secondary wall deposition. *J. Cell Biol.* **20**, 529–533.

Hepler, P.K. and Palevitz, B.A. (1974). Microtubules and microfilaments. *Ann. Rev. Plant Physiol.* **25**, 309–362.

Herth, W. (1980). Calcofluor white and congo red inhibit chitin microfibril assembly of *Poteriochromonas*: Evidence for a gap between polymerization and microfibril formation. *J. Cell Biol.* **87**, 442–450.

Herth, W. (1985). Plasma membrane rosettes involved in localized wall thickening during xylem vessel formation of *Lepidum sativum* L. *Planta* **164**, 12–21.

Herth, W. and Weber, G. (1984). Occurrence of the putative cellulose-synthesizing 'rosettes' in the plasma membrane of *Glycine max* suspension culture cells. *Naturwissenschaften* **71**, 153–154.

Hogetsu, T. and Shibaoka, H. (1978). Effects of colchicine on cell shape and on microfibril arrangement in the cell wall of *Closterium acerosum*. *Planta* **140**, 15–18.

Itoh, T. and Brown, R.M. (1984). The assembly of cellulose microfibrils in *Valonia macrophysa* Kutz. *Planta* **160**, 372–381.

Itoh, T., O'Neill, R.M. and Brown, R.M. (1984). Interference of cell wall regeneration of *Boergesenia forbesii* protoplasts by Tinopal LPW, a fluorescent brightening agent. *Protoplasma* **123**, 174–183.

Jung, G. and Wernicke, W. (1990). Cell shaping and microtubules in developing mesophyll of wheat (*Triticum aestivum* L.). *Protoplasma* **153**, 141–148.

Kobayashi, H., Fukuda, H. and Shibaoka, H. (1988). Interrelation between the spatial disposition of actin filaments and microtubules during the differentiation of tracheary elements in cultured *Zinnia* cells. *Protoplasma* **143**, 29–37.

Kutschera, U., Bergfeld, R. and Schopfer, P. (1987). Cooperation of epidermis and inner tissues in auxin-mediated growth of maize coleoptiles. *Planta* **170**, 168–180.

Lancelle, S.A. and Hepler, P.K. (1989). Immungold labelling of actin on sections of freeze-substituted plant cells. *Protoplasma* **150**, 72–74.

Lancelle, S.A., Callaham, D.A. and Hepler, P.K. (1986). A method for rapid freeze-fixation of plant cells. *Protoplasma* **131**, 153–165.

Lancelle, S.A., Cresti, M. and Hepler, P.K. (1987). Ultrastructure of the cytoskeleton in freeze-substituted pollen tubes of *Nicotiana alata*. *Protoplasma* **140**, 141–150.

Ledbetter, M.C. and Porter, K.R. (1963). A 'microtubule' in plant cell fine structure. *J. Cell Biol.* **19**, 239–250.

Levy, S. (1987). A 3-D computer representation of helicoidal superstructures in biological materials, exemplified by the *Nitella* cell wall. *Eur. J. Cell Biol.* **44**, 27–33.

Lichtscheidl, I.K., Lancelle, S.A. and Hepler, P.K. (1990). Actin–endoplasmic reticulum complexes in *Drosera*. Their structure relationship with the plasmalemma, nucleus, and organelles in cells prepared by high pressure freezing. *Protoplasma* **155**, 116–126.

Lloyd, C.W. (1983). Helical microtubular arrays in onion root hairs. *Nature* **305**, 311–313.

Lloyd, C.W. (1984). Toward a dynamic helical model for the influence of microtubules on wall patterns in plants. *Int. Rev. Cytol.* **86**, 1–51.

Lloyd, C.W. (1987). The plant cytoskeleton: the impact of fluorescence microscopy. *Ann. Rev. Plant Physiol.* **38**, 119–139.

Lloyd, C.W. and Wells, B. (1985). Microtubules are at the tips of root hairs and form helical patterns corresponding to inner wall fibrils. *J. Cell Sci.* **75**, 225–238.

Melan, M.A. (1990). Taxol maintains organized microtubule patterns in protoplasts which lead to the resynthesis of organized cell wall microfibrils. *Protoplasma* **153**, 169–177.

Mueller, S.C. and Brown, R.M. (1980). Evidence for an intramembrane component associated with a cellulose microfibril synthesizing complex in higher plants. *J. Cell Biol.* **84**, 315–326.

Mueller, S.C. and Brown, R.M. (1982). The control of cellulose microfibril deposition in the cell wall of higher plants. I. Can directed membrane flow orient cellulose microfibrils? Evidence from freeze-fractured plasma membrane of maize and pine seedlings. *Planta* **154**, 489–500.

Murata, T. and Wada, M. (1989). Organization of cortical microtubules and microfibril deposition in response to blue-light-induced apical swelling in a tip-growing *Adiantum* protonema cell. *Planta* **178**, 334–341.

Murata, T., Kadota, A., Hogetsu, T. and Wada, M. (1987). Circular arrangement of cortical microtubules around the subapical part of a tip-growing fern protonema. *Protoplasma* **141**, 135–138.

Neville, A. (1985). Molecular and mechanical aspects of helicoid development in plant cell walls. *Bioessays* **3**, 4–8.

Neville, A. (1988). A pipe-cleaner molecular model for morphogenesis of helicoidal plant cell walls based on hemicellulose complexity. *J. Theor. Biol.* **131**, 243–254.

Neville, A. and Levy, S. (1985). The helicoidal concept in plant cell wall ultrastructure and morphogenesis. In *Biochemistry of Plant Cell Walls* (eds C. Brett and J. Hillman), *Soc. Exp. Biol. Sem.* **28**, 99–124. Cambridge University Press, Cambridge, UK.

Nick, P., Bergfeld, R., Schafer, E. and Schopfer, P. (1990). Unilateral reorientation of microtubules at the outer epidermal wall during photo- and gravitropic curvature of maize coleoptiles and sunflower hypocotyls. *Planta* **181**, 162–168.

Parthasarathy, M.V., Perdue, T.D., Witztum, A. and Alvernaz, J. (1985). Actin network as a normal component of the cytoskeleton in many vascular plant cells. *Am. J. Bot.* **72**, 1318–1323.

Preston, R.D. (1964). Structural and mechanical aspects of plant cell walls with particular reference to synthesis and growth. In *Formation of Wood in Forest Trees* (ed. M.H. Zimmermann), pp. 169–188.

Academic Press, New York.

Preston, R.D. (1974). *The Physical Biology of Plant Cell Walls*. Chapman and Hall, London.

Quader, H. (1986). Cellulose microfibril orientation in *Oocystis solitaria*: Proof that microtubules control the alignment of the terminal complexes. *J. Cell Sci.* **83**, 223–234.

Roberts, I.N., Lloyd, C.W. and Roberts, K. (1985). Ethylene-induced microtubule reorientations: mediation by helical arrays. *Planta* **164**, 439–447.

Robinson, D.G. and Quader, H. (1981). Structure, synthesis, and orientation of microfibrils. IX. A freeze-fracture investigation of the *Oocystis* plasma membrane after inhibitor treatments. *Eur. J. Cell Biol.* **25**, 278–288.

Robinson, D.G. and Quader, H. (1982). The microtubule–microfibril syndrome. In *The Cytoskeleton in Plant Growth and Development* (ed. C.W. Lloyd), pp. 109–126. Academic Press, London.

Roland, J., Vian, B. and Reis, D. (1977). Further observations on cell wall morphogenesis and polysaccharide arrangement during plant growth. *Protoplasma* **91**, 125–141.

Roland, J.C., Reis, D., Vian, B. and Roy, S. (1989). The helicoidal plant cell wall as a performing cellulose-based composite. *Biology of the Cell* **67**, 209–220.

Roelofsen, A. (1958). Cell wall structure as related to surface growth. *Acta Bot. Neerl.* **7**, 77–89.

Rudolph, U. and Schnepf, E. (1988). Investigations of the turnover of the putative cellulose-synthesizing particle 'rosettes' within the plasma membrane of *Funaria hygrometrica* protonema cells. I. Effects of monensin and cytochalasin B. *Protoplasma* **143**, 63–73.

Schneider, B. and Herth, W. (1986). Distribution of plasma membrane rosettes and kinetics of cellulose formation in xylem development of higher plants. *Protoplasma* **131**, 142–152.

Seagull, R.W. (1989). The plant cytoskeleton. *Crit. Rev. Plant Sci.* **8**, 131–167.

Staehelin, L.A. and Giddings, T.H. (1982). Membrane-mediated control of cell wall microfibrillar order. In *Developmental Order: Its Origins and Regulation* (eds S. Subtelny and P.B. Green), pp. 133–147. Liss, New York.

Staiger, C.J. and Schliwa, M. (1987). Actin localization and function in higher plants. *Protoplasma* **141**, 1–12.

Traas, J.A. and Derksen, J. (1989). Microtubules and cellulose microfibrils in plant cells; simultaneous demonstration in dry cleave preparations. *Eur. J. Cell Biol.* **48**, 159–164.

Vian, B., Reis, D., Mosiniak, M. and Roland, J. (1986). The glucuronoxylans and the helicoidal shift in cellulose microfibrils in linden wood: cytochemistry *in muro* and on isolated molecules. *Protoplasma* **131**, 185–199.

Wada, M. and Staehelin, L.A. (1981). Freeze-fracture observations on the plasma membrane, the cell wall and the cuticle of growing protonemata of *Adiantum capillus-veneris* L. *Planta* **151**, 462–468.

8. THE BIOCHEMISTRY OF CELLULOSE SYNTHESIS

Deborah P. Delmer

Department of Botany, Institute of Life Sciences, The Hebrew University,
Jerusalem 91904, Israel

Preface

The process of cellulose biosynthesis was reviewed extensively a few years ago (Delmer, 1987), and more recently with emphasis on its relationship to the plasma membrane (Delmer, 1990) and on the process in bacteria (Ross *et al.*, 1991). The main purpose of this chapter is to provide a concise summary of our state of knowledge of the biochemistry of the process with emphasis on recent developments, which may be used as a guideline for those interested in the cytoskeleton and its relationship to cellulose biosynthesis. A description of the morphology of putative cellulose synthase complexes as viewed by freeze-fracture is found elsewhere in this volume (see Giddings and Staehelin, Chapter 7, this volume) as is an extensive discussion of the role of the cytoskeleton in orientation of microfibril deposition in chapters in the next section.

There are some features which are probably common to all organisms which synthesize cellulose. The formal name given to the enzyme which polymerizes glucose residues into cellulose is UDP-glucose: $(1{\rightarrow}4)$-β-D-glucan glucosyltransferase, but is referred to more generally as cellulose synthase. Direct evidence is available that UDP-glucose is the substrate for the enzyme in the bacterium *Acetobacter xylinum* (Aloni *et al.*, 1983) and $(1{\rightarrow}4)$-β-glucan can also be synthesized *in vitro* from UDP-glucose in preparations from lower eukaryotes (Girard and Fevre, 1984; Blanton and Northcote, 1990). Indirect evidence indicates it is the substrate in higher plants as well (see Delmer, 1987). The evidence is only indirect, since the major enigma in this field is that no one has succeeded in demonstrating activity for cellulose synthase *in vitro* using cell-free preparations derived from higher plants. Instead, high levels of $(1{\rightarrow}3)$-β-glucan (callose) are synthesized *in vitro* from UDP-glucose via another plasma membrane-localized glucan synthase, and a study of the possible relationship between the two synthases represents part of the current efforts with higher plants.

The sections below outline recent advances in the biochemistry of cellulose biosynthesis in bacteria, lower forms of eukaryotes, and higher plants.

Recent advances for the bacterium *Acetobacter xylinum*

This system currently stands as the most well characterized in terms of biochemical knowledge. Low levels of *in vitro* cellulose synthase activity were demonstrated with membrane preparations as early as 1958 (Glaser, 1958), but a major breakthrough came with the discovery that the enzyme is specifically activated by a unique cyclic di-nucleotide: bis-$(3'\text{-}5')$-cyclic diguanylic acid (Ross *et al.*, 1987, 1991). Inclusion of this compound in assays results in achievement of *in vitro* rates of synthesis from UDP-glucose which are similar to those observed *in vivo*. Benziman's group has also identified the enzyme which synthesizes the activator and another pathway which is responsible for its degradation (Ross *et al.*, 1987, 1991).

The question of the precise nature of the difference between the crystal structures of native cellulose (cellulose I) and mercerized cellulose (cellulose II) has still not been adequately resolved (see Delmer, 1987). It is interesting that Bureau and Brown (1987) have recently determined that cellulose synthesized *in vitro* using isolated membrane preparations from *A. xylinum* was shown to have the non-native cellulose II structure. Therefore, as more *in vitro* studies are pursued, it may be that the field of biosynthesis may have insights to contribute to this still-unresolved question, particularly with respect to the question of how precise organization of the complex affects the size of microfibrils and the interactions of the glucan chains within these structures.

Progress has also been made in purification of the enzyme and characterization of its subunit composition. Most success has come from use of a technique called 'product entrapment' developed by Kang *et al.* (1984) for the purification of yeast chitin synthase. When detergent-solubilized enzyme is incubated with substrate and appropriate effectors, it apparently remains associated with its glycan product and can be recovered with the product by low-speed centrifugation. Whether the association is covalent or simply some entangleglement or non-covalent association is not clear, but with some systems it can give a very effective purification. In the experience of Benziman's group, the *A. xylinum* cellulose synthase is so tightly associated with its product that the only effective means of release is treatment with cellulase or in sample buffers for SDS-PAGE (M. Benziman, personal communication). This probably explains the success in purification by his group (Mayer *et al.*, 1991) and that of Lin *et al.* (1990) since the pellets can be easily washed free of contaminating proteins.

The most purified preparation of Lin *et al.* (1990) was highly enriched for polypeptides of 90 and 83 kD; using the photo-affinity substrate analogue azido-^{32}P-UDP-glucose, they observed labelling only of the 83-kD polypeptide, and they have concluded that this is the catalytic subunit of the enzyme. Using a different strain of *A. xylinum*, Benziman's group (Mayer *et al.*, 1991) found polypeptides of 91, 67, 54 kD highly enriched in their most purified preparations. The 67-kD polypeptide could be directly

photo-labelled with ^{32}P-c-di-GMP and its N-terminal region was shown to share sequence homology with an internal sequence of the 91-kD polypeptide. The 54-kD polypeptide could be photo-labelled with ^{32}P-UDP-glucose. Using a polyclonal antibody directed against a fusion protein of the cloned gene coding for the 91-kD polypeptide (see below), they demonstrated that all three polypeptides reacted on Western blots. Thus, they have concluded that both the 67- and 54-kD species may be cleavage products of the 91-kD polypeptide. Why the large species does not label with either probe has not yet been clarified, nor do we yet have a clear explanation for the differences in subunit sizes obtained by the two different groups. However, a comparison of polypeptide and gene sequences has allowed this author to suggest a model which can reconcile the differing results (see below).

The most exciting new discovery in this field concerns two recent reports claiming cloning of the cellulose synthase genes in *A. xylinum*. Using oligonucleotide probes based on the N-terminal sequence of the 83-kD polypeptide proposed to be the catalytic subunit, Saxena *et al.* (1990) cloned a corresponding gene from *A. xylinum*. Wong *et al.* (1990) used a different technique of cloning involving complementation of a cellulose synthase negative mutant. In this case, the clone studied contained an operon of four genes; of these, complementation with the second (*bcsB*) gene alone restored synthase activity, providing evidence that this gene codes for a subunit of the cellulose synthase. This conclusion is also supported by the finding that amino acid sequences in both the 91- and 67-kD polypeptides found in purified synthase preparations from this strain indicate these are derived from the *bcsB* gene (Mayer *et al.*, 1991). Based only upon antibody reaction (see above), these authors also concluded that the 54-kD UDP-glc-binding fragment is also derived from this gene. However, the first gene of the operon (*bcsA*) shows striking sequence homology to the gene cloned by Saxena *et al.* (1990) based upon the sequence of their 83-kD UDP-glc-binding polypeptide. At present, the most likely conclusion is that the synthase contains at least two types of nonidentical subunits: one coded by the *bcsA* gene represents the catalytic subunit (of about 83 kD, perhaps cleaving to 54 kD in the preparations of Mayer *et al.*, 1991), and that coded for by the *bcsB* gene which codes for a regulatory subunit (of 91 kD which cleaves to a 67-kD fragment binding c-di-GMP). The function of the two other gene products produced by the operon identified by Wong *et al.* (1990) are not known, but their existence suggests that other polypeptides are also required for cellulose synthesis *in vivo*.

Recent advances with lower forms of eukaryotes

Dictyostelium discoideum
Some years ago, the author suggested that *Dictyostelium discoideum* should be a very attractive system for studying cellulose biosynthesis (Delmer, 1983). The process is highly regulated, being expressed only at certain stages of development, and there is extensive background on developmental biology and genetics available with this organism; there was even a preliminary report that low levels of synthase activity could be detected *in vitro*. Therefore, the author was delighted at the appearance in the literature (Blanton and Northcote, 1990) of a report concerning a very promising initial characterization of a cellulose synthase activity. The enzyme uses UDP-glucose, requires Mg^{2+} (and perhaps also low levels of Ca^{2+}), and clearly synthesizes small rodlets of $(1\rightarrow4)$-β-glucan. Of special interest is the fact that appearance of activity correlates well with the timing of deposition of cellulose *in vivo*. Like higher plants, these preparations also synthesize some $(1\rightarrow3)$-β-glucan, but, unlike plants, this activity is much reduced compared with cellulose synthase. This system, therefore, offers good promise for future work in the field.

Saprolegnia
This fungal system has been very nicely studied by Fevre's group in Lyon. It has the advantage of depositing both $(1\rightarrow3)$-β-glucan and $(1\rightarrow4)$-β-glucan during growth, and that

both synthetic activities can be detected *in vitro* (Girard and Fevre, 1984; Bulone and Fevre, 1990). Synthesis of the former glucan is favoured when high concentrations of substrate are used and cations omitted; the latter enzyme has a lower K_m for UDP-glucose and requires Mg^{2+}. The relationship between the two enzymes is not yet clear. Monoclonal antibodies, selected for ability to immunoprecipitate solubilized enzyme activity, can immobilize both activities, implying some common antigenic deteminants (Nodet *et al.*, 1988). However, during glycerol gradient centrifugation, some separation of the two activities can be achieved (Bulone and Fevre, 1990). Nevertheless, conditions which allow synthesis of essentially only $(1\rightarrow3)$-β-glucan during product entrapment, lead to recovery of both activities in the product-entrapped pellets. When these pellets are washed, a newer, smaller form of the $(1\rightarrow3)$-β-glucan synthase is solubilized, leaving the residue enriched for $(1\rightarrow4)$-β-glucan synthase activity. Both of these most purified fractions are enriched for several polypeptides, but the residue also contains an additional 60-kD polypeptide which is suggested to be unique to the $(1\rightarrow4)$-β-glucan synthase (Bulone and Fevre, 1990). Thus, we can expect in the near future a clarification of the subunit compositions and relationships between these two enzymes, a subject which is very relevant to the process in higher plants (see below).

The search for cellulose synthase in higher plants

Enzymology
In one sense, there is no progress in this area, since no one has yet detected higher plant cellulose synthase *in vitro*. The reasons for this obviously remain obscure but may relate to the interesting regulation of cellulose and callose synthesis *in vivo*. When plants are perturbed they often respond by deposition of callose (Kauss, 1987; Delmer, 1987). The callose synthase is now known to require micromolar levels of Ca^{2+} for activity, and such elevation is presumed to occur during the perturbed state and be one factor responsible for activation of

this normally latent enzyme (Kauss, 1987; Delmer, 1987). It seems to this author most likely that elevation of Ca^{2+} may also be a key factor responsible for the corresponding loss of cellulose synthase activity. It is exceedingly difficult to isolate membranes from plant cells without them being exposed to the high levels of Ca^{2+} released by cell walls and vacuoles during cell breakage. Acting as a rapid second messenger, Ca^{2+} may stimulate a specific protease, kinase, or phosphatase which modifies and inactivates the cellulose synthase. While this may all be part of the plant's regulatory strategy *in vivo*, it can make the life of the biochemist very difficult. Another possibility is that a specific activator of the enzyme, such as c-di-GMP, is required as discussed in more detail below.

Some years ago, the author suggested that the callose synthase and the cellulose synthase may be different forms of the same enzyme (Delmer, 1977). This still remains an open question today. However, in the intervening years, considerable progress has been made on characterization of the callose synthase, and the question should be able to be addressed more seriously in the near future. Affinity-labelling studies using substrate or substrate analogues by Frost *et al.* (1989) with red beet, and by Delmer *et al.* (1991a) with cotton fibres, have identified 57-kD (beet) and 52-kD (cotton) polypeptides as the likely candidate for catalytic polypeptide. Dhugga and Ray (1991) have indicated that a 55-kD polypeptide in pea is the most likely candidate, and reported that an antibody to this polypeptide can immuno-precipitate enzyme activity. The author has also identified a doublet at 60–62 kD in cotton fibres and mung beans which is recognized by a monoclonal antibody specifically able to immobilize callose synthase activity (Delmer, 1989); these do not bind UDP-glucose (Delmer *et al.*, 1991a), and appear to associate with the enzyme in a Mg^{2+}-dependent manner. Fink *et al.* (1990) have identified a 32-kD polypeptide which is particularly enriched in their most purified preparations from soybean; we have also seen enrichment for a similar 34-kD species under certain conditions (Delmer *et al.*, 1990) as has Fevre's group in *Saprolegnia*

(Bulone and Fevre, 1990). If all these identifications are valid, then this implies that the enzyme, known to be of very high molecular weight when solubilized, is comprised of a number of non-identical subunits. However, rigorous purification will be necessary to further validate these subunit identifications.

Recent studies of ours (Delmer *et al.*, 1990) indicate that the digitonin-solubilized cotton fibre callose synthase can assume a very high molecular weight form when incubated with certain combinations of Ca^{2+} and Mg^{2+} and the question of whether polypeptides such as the 32–34-kD and 60-kD species are true components of the complex or represent an artefactual cation-mediated association with the lipid-detergent-micelles of the enzyme must be considered. However, it is interesting that similar molecular weights are being singled out in a variety of different systems. Nevertheless, identification of these opens the way for cloning of the genes coding for them, and eventually to modifying their expression and testing effects on expression of both callose and cellulose synthesis in transgenic plants.

Returning to the subject of cellulose synthesis, both our group and that of Benziman's have attempted to elicit cellulose synthase activity *in vitro* in plant systems by addition to assays of the activator of the bacterial synthase, c-di-GMP. To date, these experiments have all yielded negative results. As a result, the author had tentatively concluded that this activator may have no role in higher plants (Delmer, 1987). However, in a recent collaboration between our group and that of M. Benziman (Delmer *et al.*, 1991b and unpublished information), we have made the interesting discovery of two cotton fibre membrane polypeptides of 83 and 48 kD which bind c-di-GMP with high affinity and specificity. Both show striking developmental regulation consistent with their playing a role in cellulose synthesis; this conclusion is further supported by the fact that both show antigenic relatedness to the product of the *bcsB* gene of *A. xylinum*. Analogous to the situation in *A. xylinum*, it appears that the larger 83-kD polypeptide is cleaved to form the 48-kD species. We have also identified a distinct 84-kD polypeptide which photolabels with ^{32}P-UDP-glucose with high affinity and specificity which we consider a likely candidate for the catalytic subunit (Delmer et al., 1991a). Thus, evidence is beginning to accumulate which suggests that higher plants, like bacteria, contain a c-di-GMP-dependent cellulose synthase which may be composed of distinct catalytic and regulatory subunits.

Thus, the direction for the future would seem to be in terms of concentrating on purifying and characterizing specific polypeptides already suggested to be involved in β-glucan synthesis. This should allow development of probes for future efforts in gene cloning. A very exciting possibility is that plant genes involved in cellulose synthesis may soon be able to be cloned based upon sequence homology between these genes and those already cloned in A. xylinum. The development of the polymerase chain reaction (Erlich, 1989) makes this increasingly possible even when minimal homology is conserved. With respect to enzyme activity, it would seem that efforts could usefully be directed at renewed attempts to find activity in the presence of c-di-GMP, and to assess more rigorously the possibility that proteolytic or other modifying events occur upon cell breakage and to find ways to minimize these during enzyme isolation and assay.

Inhibitors of cellulose synthesis

It has long been known that the herbicide 2,6-dichlorobenzonitrile (DCB) is an effective and specific inhibitor of cellulose synthesis in algae and higher plants, although apparently not in bacteria (see references in Delmer, 1987). Using a radioactive photoaffinity analogue of DCB, 2,6-dichlorophenylazide, we have identified an 18-kD polypeptide as a specific DCB-binding protein in cotton fibres (Delmer et al., 1987). The specificity of labelling and the fact that ability to detect the labelled protein increases in parallel with the large increase in rate of cellulose synthesis in vivo which occurs during fibre development, lead us to believe that this polypeptide plays some role in the process. The author speculated that it might be some regulatory protein (Delmer, 1987), but no precise role has yet been established. We have

recently succeeded in obtaining specific antibodies against this protein (unpublished), and should be in a position to analyse it in more detail in the near future.

Rather surprising results were obtained by Shedletzky et al. (1990) in analysing a cell line of suspension-cultured tomato cells adapted to growth on DCB. The process of cellulose synthesis remains sensitive to DCB in these cells, and the major mechanism of adaptation rests on the fact that the cells have 'learned' to grow with cell walls which virtually lack cellulose. These unusual walls also lack xyloglucan; this polymer is synthesized normally, but lacking cellulose with which to associate, it is all secreted into the medium. The walls appear to consist largely of pectic components held together by ionic and covalent linkages. The surprising finding that plant cells can survive without cellulose has suggested to us that it may be possible under certain special conditions to select for mutants which lack or have reduced levels of cellulose synthase, and we have embarked on such a project using Arabidopsis. Also of interest is a recent report (Kokubo et al., 1989) that single gene mutations leading to the brittle culm phenotype in barley, also result in markedly reduced levels of cellulose, providing another indication that such mutants can be found. Although such mutants may have very altered phenotypes, at least the mutations need not be lethal events.

Heim et al. (1990b) have recently made the interesting finding that another herbicide, Isoxaben, has a mode of action which very closely resembles that of DCB, suggesting that it also inhibits cellulose synthesis in vivo. However, it is interesting that, on a molar basis, it is at least 100 times more effective than DCB; this, and the fact that Isoxaben-resistant mutants have been selected and are being characterized (Heim et al., 1989, 1990a) offers hope that future studies with this herbicide may shed more light on the biochemistry of cellulose biosynthesis.

Finally, of special interest to those concerned with the plant cytoskeleton, is a recent report implicating the existence of an integrin-like protein in plants (Schindler et al., 1989). Thus, the presence of a protein with antigenic cross-reactivity to human vitronectin receptor was reported in soybean. Such integral membrane proteins in animal cells serve to provide a connection between the extracellular matrix and the intracellular cytoskeleton (Hynes, 1987). This makes such proteins intriguing candidates for involvement in cell wall biosynthesis in plants, and the fact that Schindler *et al.* reported gross alterations in wall structure in cells grown on a peptide which inhibits integrin function in animals, is certainly intriguing and suggests that this class of proteins merits further examination in plants.

References

Aloni, Y., Cohen, R., Benziman, M. and Delmer, D. (1983). Solubilization of the UDP-glucose: 1,4-β-D-glucosyl-transferase (cellulose synthase) from *Acetobacter xylinum*. *J. Biol. Chem.* **258**, 4419–4423.

Blanton, R.L. and Northcote, D.H. (1990). A 1,4-β-D-glucan synthase system from *Dictyostelium discoideum*. *Planta* **180**, 324–332.

Bulone, V. and Fevre, M. (1990). Separation and partial purification of β-1,3- and β-1,4-glucan synthases from *Saprolegnia*. *Plant Physiol.* **94**, 1748–1755.

Bureau, T.E. and Brown, R.M. Jr (1987). *In vitro* synthesis of cellulose II from a cytoplasmic membrane fraction of *Acetobacter xylinum*. *Proc. Natl Acad. Sci. U.S.A.* **84**, 6985–6989.

Delmer, D.P. (1977). Biosynthesis of cellulose and other plant cell wall polysaccharides. *Recent Adv. Phytochem.* **11**, 45–77.

Delmer, D.P. (1983). Biosynthesis of cellulose. *Adv. Carbohyd. Chem. and Biochem.* **41**, 105–153.

Delmer, D.P. (1987). Cellulose biosynthesis. *Ann. Rev. Plant Physiol.* **38**, 259–290.

Delmer, D.P. (1989). The relationship between the synthesis of cellulose and callose in higher plants. In *Cellulose and Wood* (ed. C. Schuerch), pp. 749–764. John Wiley, New York.

Delmer, D.P. (1990). The role of the plasma membrane in cellulose biosynthesis. In *The Plant Plasma Membrane* (eds C. Larsson and I.M. Moller), pp. 256–268. Springer-Verlag, Berlin, Heidelberg.

Delmer, D.P., Cooper, G. and Read, S.M. (1987). Identification of a receptor protein in cotton fibers for the herbicide 2,6-dichlorobenzonitrile. *Plant Physiol.* **84**, 415–420.

Delmer, D.P., Solomon, M., Andrawis, A. and Amor, Y. (1990). Cation-induced formation of a macroglucan synthase complex. *Plant Physiol.* **93** (1), Abstract 327.

Delmer, D.P., Solomon, M. and Read, S.M. (1991a). Direct photolabeling with [^{32}P]UDP-glucose for identification of a subunit of cotton fiber callose synthase. *Plant Physiol.* **95**, 556–563.

Delmer, D.P., Amor, Y., Solomon, M., Shedletzky, E., Shmuel, M., Mayer, R. and Benziman, M. (1991b). Identification and characterization of genes and gene products involved in β-glucan synthesis. *J. Cell. Biochem. Suppl.* **15A**, 13.

Dhugga, K.S. and Ray, P.M. (1991). A 55 kDa plasma membrane-associated polypeptide is involved in β-1,3-glucan synthase activity in pea tissue. *Plant Physiol.* **278**, 283–286.

Erlich, H.A. (ed.) (1989). *PCR Technology: Principles and Applications for DNA Amplification*. Stockton Press, New York.

Fink, J., Jeblick, W. and Kauss, H. (1990). Partial purification and immunological characterization of 1,3-β-glucan synthase from suspension cells of *Glycine max*. *Planta* **181**, 343–348.

Frost, D.J., Read, S.M., Drake, R.R., Haley, B.E. and Wasserman, B.P. (1989). Identification of the UDPG-binding polypeptide of (1,3)-β-glucan synthase from a higher plant by photoaffinity labeling with 5-azido-UDP-glucose. *J. Biol. Chem.* **265**, 2162–2167.

Girard, V. and Fevre, M. (1984). β-1,4- and β-1,3-glucan synthases are associated with the plasma membrane of the fungus *Saprolegnia*. *Planta* **160**, 400–406.

Glaser, L. (1958). The synthesis of cellulose in cell-free extracts of *Acetobacter xylinum*. *J. Biol. Chem.* **232**, 627–636.

Hayashi, T., Read, S.M., Bussell, J., Thelen, M.T., Lin, F.C., Brown, R.M. *et al.* (1987). UDP-glucose: (1→3)-β-glucan synthases from mung bean and cotton. Differential effects of Ca^{2+} and Mg^{2+} on enzyme properties and on macromolecular structure of the glucan product. *Plant Physiol.* **83**, 1054–1062.

Heim, D.R., Roberts, J.L., Pike, P.D. and Larrinua, I.M. (1989). Mutation of a locus of *Arabidopsis thaliana* confers resistance to the herbicide Isoxaben. *Plant Physiol.* **90**, 146–150.

Heim, D.R., Roberts, J.L., Pike, P.D. and Larrinua, I.M. (1990a). A second locus 1xnB1 in *Arabidopsis thaliana* that confers resistance to the herbicide Isoxaben. *Plant Physiol.* **92**, 858–861.

Heim, D.R., Skomp, J.R., Tschabold, E.E. and Larinua, I.M. (1990b). Isoxaben inhibits the synthesis of acid soluble wall materials in *Arabidopsis thaliana*. *Plant Physiol.* **93**, 695–700.

Hynes, R.O. (1987). Integrins: a family of cell surface receptors. *Cell* **48**, 549–554.

Kang, M.S., Elango, N., Mattie, E., Au-Young, J., Robbins, P. and Cabib, E. (1984). Isolation of chitin synthetase from *Saccharomyces cerevisiae*. Purification of an enzyme by entrapment in the reaction product. *J. Biol. Chem.* **259**, 14966–14972.

Kauss, H. (1987). Some aspects of calcium-dependent regulation in plant metabolism. *Ann. Rev. Plant Physiol.* **38**, 47–72.

Kokubo, A., Kuraishi, S. and Sakurai, N. (1989). Culm strength of barley. Correlation amount maximum bending stress, cell wall dimensions, and cellulose content. *Plant Physiol.* **91**, 876–882.

Lin, F.C., Brown, R.M. Jr, Drake, R.R. Jr and Haley, B.E. (1990). Identification of the uridine-5'-diphosphoglucose (UDP-glc) binding subunit of cellulose synthase in *Acetobacter xylinum* using the photoaffinity probe 5-azido-UDP-glc. *J. Biol. Chem.* **265**, 4782–4784.

Mayer, R., Ross, P., Weinhouse, H., Amikam, D., Volman, G., Ohana, P. *et al.* (1989). The polypeptide substructure of bacterial cellulose synthase and its occurrence in higher plants. *Proceedings of the Fifth Cell Wall Meeting* (eds S.C. Fry, C.T. Brett and J.S.G. Reid), Abstract 38, University of Edinburgh.

Mayer, R., Ross, P., Weinhouse, H., Amikam, D., Volman, G., Ohana, P. *et al.* (1991). Polypeptide composition of bacterial c-di-GMP-dependent cellulose synthase and the occurrence of immunologically cross-reacting proteins in higher plants. *Proc. Natl Acad. Sci. U.S.A.* (in press).

Nodet, P., Grange, J. and Fevre, M. (1988). Dot-blot assays and their use as a direct antigen-binding method to screen monoclonal antibodies to 1,4-β- and 1,3-β-glucan synthases. *Anal. Biochem.* **174**, 662–665.

Ross, P., Weinhouse, H., Aloni, Y., Michaeli, D., Weinberger-Ohana, P., Mayer, R. *et al.* (1987). Regulation of cellulose synthesis in *Acetobacter xylinum* by cyclic diguanylic acid. *Nature* **325**, 279–281.

Ross, P., Mayer, R. and Benziman, M. (1991). Cellulose biosynthesis and function in bacteria. *Microbiol. Rev.* **55**, 35–58.

Saxena, I.M., Lin, F.C. and Brown, R.M. Jr (1990). Cloning and sequencing of the cellulose synthase catalytic subunit gene of *Acetobacter xylinum*. *Plant Mol. Biol.* **15**, 673–683.

Schindler, M., Meiners, S. and Cheresh, D.A. (1989). RGD-dependent linkage between plant cell wall and plasma membrane: consequences for growth. *J. Cell Biol.* **108**, 1955–1965.

Shedletzky, E., Shmuel, M., Delmer, D.P. and Lamport, D.T.A. (1990). Adaptation and growth of tomato cells on the herbicide 2,6-dichlorobenzo-nitrile leads to production of unique cell walls virtually lacking a cellulose-xyloglucan network. *Plant Physiol.* **94**, 980–987.

Wong, H.C., Fear, A.L., Calhoon, R.D., Eichinger, G.H., Mayer, R., Amikam, D. *et al.* (1990). Genetic organization of the cellulose synthase operon in *Acetobacter xylinum*. *Proc. Natl Acad. Sci. U.S.A.* **87**, 8130–8134.

9. ARCHITECTURE OF THE PRIMARY CELL WALL

Maureen C. McCann and Keith Roberts

Department of Cell Biology, John Innes Institute,
Colney Lane, Norwich NR4 7UH, UK

Introduction

The wall forms a single continuous extracellular matrix through the body of the plant. This continuity has implications for cell signalling and communications, the rate and direction of cell growth, and the organization of cellular events at the tissue level. The plant extracellular matrix is also a site of protein and carbohydrate metabolism and there must be mechanisms to regulate and compartmentalize this metabolic machinery. The matrix provides the framework for generating the complex spatial organization of cells. Only by analysing the molecular architecture of the primary cell wall can we approach an understanding of the processes of growth, development and decay within the wall and the influence of the wall on these processes within their enclosed cells.

Cell walls are a composite of skeletal cellulose microfibrils, which form the scaffolding framework of the wall, and so-called matrix polymers, which include hemicelluloses, pectins and proteins.

Cellulose, an unbranched β-1,4-glucan in linear chain form, is the major component of the primary cell wall, 20–30% of the wall's dry weight and 10–15% of its volume. The chains associate into microfibrils, crystalline or para-crystalline arrays virtually free of water, of diameter 5–15 nm and of elliptical cross-section. Cellulose is a component of all higher plant cell walls but a schism exists with respect to matrix polymers between the graminaceous monocots and the dicots and non-graminaceous monocot cell walls: the former are characterized by very low levels of pectin, xyloglucan and protein, and very high levels of arabinoxylans and

β-1-3,1-4-glucans. This chapter will consider the major components and architecture of dicot and non-graminaceous monocot cell walls.

Hemicelluloses are non-cellulosic wall polysaccharides other than pectins. Although xylans are the major hemicellulose components of graminaceous monocots (15–20% dry weight), cell walls of dicots and parenchymatous tissues are virtually devoid of xylans (2%) (Darvill *et al.*, 1980); the bulk of the xylose being present in xyloglucans. The backbone of xyloglucan is identical to cellulose, but attached to 70–80% of the glucose residues are side-chains, mostly of xylose residues α-linked to C-6 of glucose. Purified xyloglucans associate strongly *in vitro* with purified microfibrils (Hayashi *et al.*, 1987). *In muro*, it seems likely that they coat microfibrils by tight hydrogen-bonding between the cellulosic backbone and cellulose molecules themselves.

Pectins are a heterogeneous, branched and highly hydrated group of polysaccharides rich in D-galacturonic acid, the side-chains of such polysaccharides and chemically similar polysaccharides. Whilst the definitions of 'pectin' and 'hemicellulose' given here may seem vague, they do reflect the current state of carbohydrate chemistry with respect to wall components. Pectins are described as 'block' polymers since they contain 'smooth' blocks (homogalacturonans) consisting mainly of continuous unbranched α-D-galacturonic acid residues punctuated by the occasional rhamnose, with the intervening homoGalA block sometimes being fully methylesterified and sometimes not esterified at all (Jarvis, 1984) and 'hairy' blocks containing numerous other sugars. Pectinase-resistant 'hairy' blocks

The Cytoskeletal Basis of Plant Growth and Form   ISBN 0–12–453770–7

include Rhamnogalacturonan I and II (RGI, RGII) best characterized in walls of cultured sycamore cells (Darvill *et al.*, 1980).

Homogalacturonans, which together with RGI and II comprise the bulk of dicot pectins, are known to form rigid insoluble gels in the presence of Ca^{2+}; cross-linking requires about 12 consecutive unesterified galacturonic acid residues, so block-wise distribution of ester groups enables local Ca^{2+} bridging (Yamaoka and Chiba, 1983). Pectins cement adjacent cells together since the use of pectinolytic enzymes and chelating agents causes rapid release of single cells from many soft plant tissues.

Many enzymes and enzyme inhibitors are also present in the wall reinforcing the notion of the wall as a metabolic compartment.

The primary cell wall is a focus of particular interest because it controls plant growth and morphogenesis. But to stray from observed physiology to proposed biochemical mechanisms is to become lost in the micro-forest of cell wall architecture. The problem is that, although wall components have been chemically characterized, there is no single cohesive model of wall architecture at the molecular level. Without this fundamental information, absent for a lack of structural methods and suitable probes, molecular mechanisms of growth and development cannot be elucidated.

Previous models which consider the wall as one covalently linked macromolecule (Keegstra *et al.*, 1973), or as a protein network coupled to a cellulose network (Lamport, 1986), or as a cellulose–xyloglucan network (Fry, 1989a, b; Hayashi, 1989), are fragmentary in the sense that they all ignore major wall components. We propose new components of a model of cell wall architecture that are based on direct visualization of the cell wall by the fast-freeze, deep-etch, rotary-shadowed (FDR) replica technique (Heuser, 1981), and the use of monoclonal antibody probes to specific wall components. The basic wall parameters we have derived by direct methods (McCann *et al.*, 1990) impose constraints on possible cell wall models.

Previous models

Albersheim's model

Albersheim's original model, derived from enzymic analysis of an *Acer pseudoplatanus* (sycamore) in cell suspension culture, represented the first attempt to describe wall ultra-structure at the molecular level. It was proposed that each cellulose fibre is coated in a layer of xyloglucan one molecule thick, the glucose backbone of which lies parallel to the fibre axis on one side only – bonding being impeded on the other side by the protruding fucose and galactose residues (Valent and Albersheim, 1974). Each xyloglucan binds to a single arabinogalactan chain, which in turn binds to a single rhamnogalacturonan chain. Each rhamnogalacturonan chain can receive several arabinogalactan molecules, radiating from different cellulose fibres and similarly each cellulose fibre can be connected to several rhamnogalacturonan chains. As a result of this extensive cross-linking, the fibres are immobilized in a seemingly rigid matrix (Keegstra *et al.*, 1973).

In this model, each polysaccharide is joined by the reducing end of the main chain to become the side branch of another polymer. But bridge molecules, having both the connected molecules or parts of these molecules still attached to the bridge, have never been isolated. The method used to identify them has been to try to establish the linkages from the reducing end of the polymer – identified by gas liquid chromatography (GLC) analysis in which each constituent was assumed to be fully methylated. Evidence for covalent attachment of xyloglucan to other polysaccharides is based on co-elution of xyloglucan with other polysaccharides on ion-exchange, gel-permeation and hydroxyapatite chromatography columns. Unfortunately, owing to its polydispersity, xyloglucan gives broad peaks upon chromatography; the observed co-elution could, therefore, be coincidental and does not prove covalent linkage. Later work (Monro *et al.*, 1976; Darvill *et al.*, 1980) has failed to verify the reported covalent linkages, nor have the covalent linkages been demonstrated in cell walls taken from intact plant tissues.

Protein-based models

In 1960, Lamport and Northcote presented the first evidence for an integral primary cell wall protein containing virtually all the cell's hydroxyproline, and it was later proposed that the hydroxyproline-rich glycoprotein (HRGP) must play a structural role in the wall and, that since the protein component of secondary walls is negligible, it therefore must be involved in extension – the 'extensin' hypothesis (Lamport, 1965).

Although a direct role in extension growth was never demonstrated, indirect evidence supported a structural role because of the difficulty of extraction without using degradative treatments. Even after enzymic attack by hemicellulase/cellulase/protease mixtures much of the extensin remained as an insoluble wall-shaped residue (Lamport, 1965). Mort and Lamport (1977) used anhydrous hydrofluoric acid specifically to dissolve wall polysaccharides whilst leaving peptide bonds intact. An insoluble fraction, accounting for about 10% of the wall, consisted of equal amounts of wall protein and an 'unknown (phenolic?) compound'. As the residue was insoluble in protein solvents such as sodium dodecyl-sulphate and urea, the presence of cross-links between the proteins was inferred. Later, pulse-chase and peptide-mapping experiments demonstrated that soluble carrot extensin is slowly made insoluble in the cell wall (Cooper and Varner, 1983). Extensin monomers eluted from tomato cell (*Lycopersicon esculentum*) suspension cultures can be cross-linked to form oligomers by a peroxidase eluted from intact cells of the same culture and extensin dimers, trimers and oligomers so formed can be visualized in the electron microscope (Heckman *et al.*, 1988). Extensins are basic proteins with a high lysine content, a generally-assumed poly-proline II helical structure, and appear as flexuous rods by electron microscopy. Many other 'extensin-like' proteins have been identified, and as they are all HRGPs, this latter term seems more satisfactory.

No linkages to cell wall polysaccharides or between extensin monomers have been identified. Isodityrosine (IdT), an unusual amino acid found *in muro* (Fry, 1982a), has been postulated to cross-link an extensin precursor weft around a cellulose microfibrillar warp, thus forming a mechanically coupled extensin–cellulose network by peroxidase-catalysed formation of isodityrosyl bridges (Lamport, 1986). Isodityrosine is present in carrot root slices and is formed in carrot cell walls during the insolubilization of extensin, both *in vivo* and *in vitro* (Cooper and Varner, 1983). Carrot extensin can also be cross-linked *in vitro* by peroxidase and hydrogen peroxidase, but dityrosine and not isodityrosine is the phenolic cross-link formed. Isodityrosine has so far only been found as the intramolecular cross-link in the sequence $^1/_2$ IDT-Lys-$^1/_2$ IDT (Epstein and Lamport, 1984). There is, at present, no evidence for IdT as an intermolecular cross-link. The extensin network model is the only cell wall model to stress the importance of a structural protein component in the primary cell wall.

Hemicellulose-based models

Xyloglucan has been shown to hydrogen-bond tightly to cellulose fibrils and probably helps to keep cellulose fibrils anchored into the cell wall matrix (Valent and Albersheim, 1974). Fry has suggested that xyloglucans are hydrogen-bonded onto microfibrils only as discrete segments with intervening lengths crossing between microfibrils. It is further proposed that oxidative coupling of other matrix polymers (acidic polysaccharides and/or basic glyco-proteins), via their phenolic side-chains, 'straps' xyloglucans to their microfibrils so that the existing architecture can be rendered more nearly permanent (Fry, 1989a). With such cross-linking, further growth is impossible unless the inter-microfibrillar segments of xylo-glucan are hydrolysed by an endo-β-(1→4)-D-glucanase (a well-established plant cell wall enzyme) or the xyloglucan–cellulose hydrogen bonds are physically pulled apart.

For 'strapping' to be a real possibility, phenolics must be attached at specific sites on the 'strap' molecules. In acidic polysaccharides, Driselase digestion shows that much of the ferulic acid from cultured spinach cells is linked very specifically, particularly notable in the case of Fer-Ara$_2$ since the feruloylated arabinose residues are in the rare pyranose ring-form (Fry, 1982b). Ferulic and *p*-coumaric residues occur in dicot pectic arabinogalactans and monocot hemicellulosic arabinoxylans. Specific examples of phenolic coupling products that have been found in plant cell wall polymers incude isodityrosine, which is a biphenyl ether derived from the tyrosine residues of extensin, and diferulic acid, which is a biphenyl derived from ferulic acid. Monocot feruloyl esters are subject to peroxidase-catalysed coupling to yield diferuloyl groups, thereby cross-linking the xylan molecules and perhaps affecting the physical properties of the wall, its ability to grow and to resist enzymic digestion. Spinach tissue culture cell wall pectins contain feruloyl and *p*-coumaroyl groups ester-linked through their carboxyl groups to specific pectic sugar residues, thereby proving the opportunity for phenolic coupling of pectins to each other and to other components.

Direct visualization of the cell wall

Using a relatively mild chemical extraction procedure (Redgwell and Selvendran, 1986), the authors have sequentially extracted wall components from isolated cell walls and have visualized both these and the remaining structures at each stage by the FDR replica technique. They have concentrated initially upon onion parenchyma cell wall, a chemically simple system low in protein and phenolics, and an abundant and homogeneous source of primary cell walls. The principal advantage of 'native' wall-imaging is the preservation of three-dimensional architecture at high resolution. No chemical fixatives or dehydrants are used, the wall is as close to the *in vivo* state as possible.

FDR replica technique

At each extraction step, a pellet of cell wall material (CWM) is frozen by dropping under gravity on to a highly polished copper block cooled to −180°C with liquid nitrogen (metal-mirror freezing). Impact on the block produces a thin – less than 10 μm thick – layer of extremely good freezing in which ice crystals

are too small to obscure overall macromolecular organization. In this layer, vitreous ice forms, at the same volume as water, and there is therefore no damage to ultrastructure from the formation of ice crystals. The slammed samples are transferred under liquid nitrogen to a Balzer's freeze-fracture unit where the temperature is raised to −115°C and its surface scraped clean of hoar frost acquired during transfer and then to −100°C, at which temperature ice sublimes away from the scraped surface leaving the molecular architecture exposed. The etched surface is replicated by rotary-shadowing with a film of platinum and then strengthened with a film of carbon. The replica is removed from the Balzer's unit, floated on to the surface of sulphuric acid and the tissue underneath dissolved away. After careful washing, pieces of replica are lifted on to grids and viewed in the transmission electron microscope. Micrographs are printed on to reverse contrast paper in which objects receiving platinum are white and shadows appear black.

Extraction

Treatment of isolated onion cell walls with cyclohexanediaminetetraacetic acid (CDTA) preferentially solubilizes pectic substances from the middle lamella by complexing Ca^{2+} from the pectin gel. Sodium carbonate at 1°C de-esterifies pectins and thus minimizes trans-eliminative degradation. Subsequent extraction at 20°C solubilizes pectic substances from the primary cell wall presumably by breaking ester cross-links. Hemicelluloses are extracted by aqueous solutions of alkali; strong alkali solubilizes hemicelluloses that are strongly associated with cellulose by hydrogen bonding, such as xyloglucans. This extraction leaves an α-cellulosic residue which can be further extracted with acidified chlorite to remove some of the remaining 15% neutral pectic and hemicellulosic material associated with the residue.

Polymers are operationally defined since each extraction step is not completely efficient, but sugar analysis (Redgwell and Selvendran, 1986), Fourier transform infra-red spectroscopy (FTIR) (McCann *et al.*, unpublished results) and the reproducibility of wall and polymer images

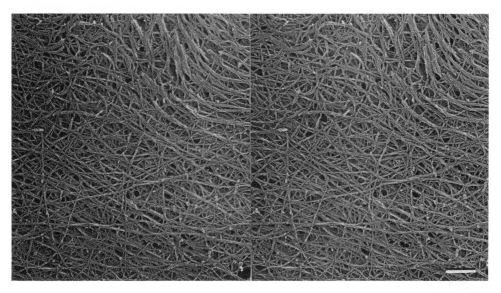

Fig. 9.1. Stereo pair of reverse contrast micrographs of depectinated (CDTA and Na$_2$CO$_3$-extracted) onion cell wall material. Fibres at the same visual depth run in roughly the same direction. Scale bar = 200 nm.

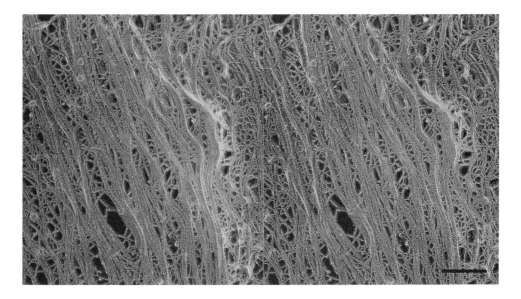

Fig. 9.2. Stereo pair of reverse contrast micrographs of onion cell wall material after extraction with CDTA, Na$_2$CO$_3$ and 1 M KOH to remove pectins and some xyloglucans. Cellulose microfibrils have laterally associated. Scale bar = 200 nm.

shows a high degree of specificity of extraction.

By visualizing the removed components and the remaining wall structure at each step (McCann *et al.*, 1990), an accurate model of wall construction has been assembled.

Wall components

Role of cellulose

Although our understanding of the wall as an ordered macromolecular assembly is primitive, a common assumption is that the typical wall is polylamellate in construction (Roelofson, 1965). At all stages of extraction, microfibrils in the same plane run roughly parallel (Figs 9.1 and 9.2). Lamellae are respected – there is no detectable weaving of microfibrils between lamellae. Areas of local order occur in the wall where microfibrils appear to have a limited number of preferential orientations but in other areas the microfibrils appear to be more randomly disposed. Isodiametric parenchyma cells expand uniformly and so it is not surprising that microfibrils are not rigorously ordered, as they are in rapidly elongating cells. The lamellate structure of the wall is highly resistant to extraction protocols. Recognizable wall fragments are still visible in the light microscope after full extraction or even after deglycosylation with anhydrous HF for 1 h at 0°C.

Cellulosic microfibrils, width 5–12 nm, are the thickest filaments seen in the FDR replica images and their identity is confirmed by cellulase digestion. The distance between adjacent microfibrils varies between 20 and 40 nm. Microfibril length is indeterminate since ends are only seen at the broken edges of a cell wall fragment. Their diameter swells during alkali treatment (to 16–20 nm) (Fig. 9.3) and this may be due to disruption of hydrogen

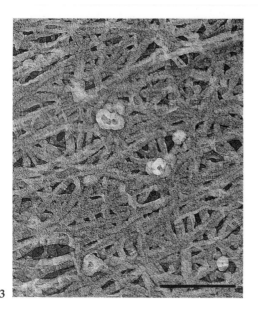

9.3   9.4

Figs 9.3 and **9.4** are images at the same magnification and of the same onion cell wall material prepared by the FDR replica technique and thin section electron microscopy. Scale bars = 200 nm.

Fig. 9.3. FDR replica of acidified chlorite-extracted α-cellulose residue.

Fig. 9.4. Acidified chlorite-extracted α-cellulosic residue imaged by conventional thin section electron microscopy, stained with uranyl acetate and lead citrate. None of the fibres is thick enough to represent the swollen microfibrils in Fig. 9.3. The spacing between fibres suggests that a polymer laterally associated with the microfibril is being imaged, perhaps a neutral galactan.

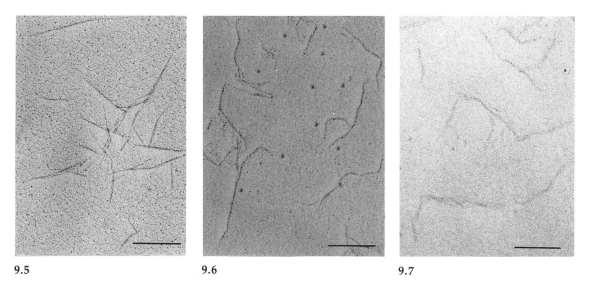

9.5   9.6   9.7

Figs 9.5, 9.6 and **9.7.** Extracted onion cell wall pectins were imaged by spraying polymers in glycerol on to a clean mica surface, drying off the glycerol *in vacuo*, and then rotary-shadowing with platinum then carbon. The replica was floated off the mica surface and picked up on copper grids. Scale bars = 200 nm.

Fig 9.5 Pectins removed from the middle lamella by the first CDTA extraction step are rigid rod-like molecules unlike those removed by subsequent extractions.

Fig. 9.6. Pectins extracted by a second CDTA extraction step. The small round stipples are an artefact of the mica surface.

Fig. 9.7. Large molecular aggregates are present in fractions extracted with Na_2CO_3.

bonding in the amorphous regions around the crystalline core of the microfibril.

The minimum width of unextracted CWM is 200 nm measured on thin sections, of which 50 nm is middle lamella, leaving *c.* 75 nm of primary cell wall between middle lamella and plasma membrane. If cross-links only occur in the plane of the plasma membrane, a maximum of 10 layers of cellulose microfibrils can be accommodated. However, microfibrils are not imaged with less than 20 nm spacing between them, and planes of fibrils collapse together as extraction proceeds, so it is likely that lamellae are spaced by matrix components. In this case, the wall may house as few as three or four layers of microfibrils (Roberts, 1989). Lamellae must be free to move with respect to each other during elongation, a process which must involve matrix components. A cellobiohydrolase probe and other matrix probes showed that cellulose fibres are discontinuously

coated with matrix polymers, including xyloglucans (Reis, personal communication, Edinburgh 5th Cell Wall Meeting).

The concept of the wall as a set of skeletal microfibrils embedded in an amorphous matrix is clearly problematic. The matrix polymers do not form 'jelly' – only extended fibrous components are seen in images and there is no reason to suppose the matrix phase is any less organized than the microfibrils themselves.

Role of pectins

Limiting porosity
Direct measurements of pore size in unextracted walls, i.e. the diameter of 'holes' created by the meshwork, show most apparent pores to be less than 10 nm diameter, with a rare maximum of 20 nm. This is in good agreement with previous indirect determinations (Carpita *et al.*, 1979) of 5 nm *in vivo*. Removal of pectins

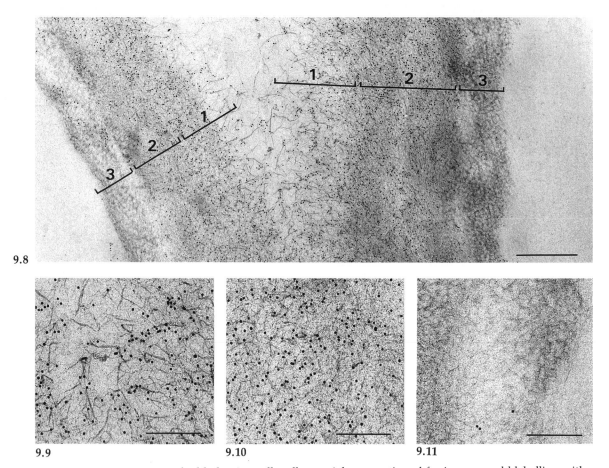

9.8

9.9   **9.10**   **9.11**

Fig. 9.8.   Low temperature embedded onion cell wall material was sectioned for immunogold labelling with a monoclonal antibody (JIM5) that recognizes an unesterified pectic epitope. Cell wall material extracted with CDTA to split apart the middle lamella shows at least three domains within the wall on immunogold labelling; a fuzzy outermost region (3) which does not label; a fuzzy inner region (2) which does label; and an area of electron-dense rigid rods in the middle lamella (1) which also labels. Gold colloids used are 5 nm diameter. Scale bar = 500 nm.

Figs 9.9, 9.10 and 9.11.   Details from regions labelled 1, 2 and 3 respectively in Fig. 9.8. Scale bars = 200 nm.

increases pore size to 20–40 nm with the wall having a more ordered fibrous appearance. Removal of hemicelluloses causes collapse of microfibrils so that, whilst xyloglucans must maintain cellulose spacing, pectins are a limiting factor in wall porosity. FRAP (fluorescence recovery after photobleaching) has been used to examine macromolecular transport of fluorescently-labelled dextrans and proteins of graded sizes across soybean cell walls (Baron-

Epel *et al.*, 1988). Treatment with pectinase, but not cellulysin or protease, apparently enlarged putative trans-wall channels from which it was concluded that the organization of pectic substances is a major controlling element in defining the sieving properties of the wall. The functional range of diameters for putative trans-wall channels was determined to be 6.6–8.6 nm (Baron-Epel *et al.*, 1988). Free diffusion of molecules across the wall is not

possible for molecules with a Stokes radius
>4.6 nm, but the estimated widths of extracted
wall polymers of 1–2 nm explains in part how
large linear polysaccharides can still be secreted
through the cell wall.

Middle lamella

Extraction with CDTA causes cells to separate
and isolated walls to shear down the middle
lamella by chelating Ca^{2+} from the wall. A first
CDTA extraction removes a set of polymers of
rigid appearance when visualized by spraying
on mica (Fig. 9.5) and which strongly label
with a monoclonal antibody probe for de-
esterified pectin. A second CDTA extraction is
known to solubilize a distinct population of
polysaccharides (Redgwell and Selvendran,
1986) but how their linkage differs from the
initially extracted material is unclear and rather
puzzling. These polymers are not rigid in
appearance (Fig. 9.6) and have increased
methyl-esterification. Whilst the rigidity in the
first case may be an artefact of the mica, it is
more likely to reflect true chemical or conformational differences. Occasional glancing
sections of the wall reveal a distinct layer of
rigid polymers in the middle lamella of highly
electron-dense material which splits apart on
extraction with CDTA (Figs 9.8–9.12).

Since a typical primary cell wall is 0.2 μm
thick, most pectin molecules are more than
long enough to span the distance between two
cells (250–400 nm). Middle lamella pectins may
couple adjacent cells together by cross-linking
with primary cell wall pectins on either side,
possibly mediating mechanisms that control
synchronized elongation.

After removal of middle lamella pectins, it is
no longer possible to guess which region of the
wall is being imaged in FDR replicas. The wall
may be too thin for significant differences in
appearance to be visible once the middle
lamella has been removed.

An independent co-extensive network

The length of cross-links in wall material that
has been depectinated by extraction with
CDTA and Na_2CO_3 is the same as those in
unextracted wall material. However, porosity
increases, suggesting that removal of pectins

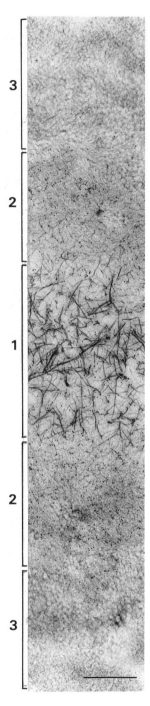

Fig. 9.12. A glancing section through an onion cell
wall shows the presence of rigid polymers in the
middle lamella. This section was labelled with an
antibody that recognizes an unesterified pectic
epitope which helps define three layers across the
wall: a fuzzy outermost layer (3), a fuzzy inner layer
which labels (2) and a layer of rigid electron-dense
rods also labelled (1). Scale bar = 500 nm.

does not affect the integrity of the cellulose/hemicellulose network, and that pectin may form an independent co-extensive network. Immunogold labelling with monoclonal antibody probes shows the presence of pectic epitopes throughout the wall, and its independence from the cellulose network is implied by the relative ease of its extraction with chemical agents which break Ca^{2+} and ester-linked bonds. Tomato cell walls adapted to growth on 2,6-dichlorobenzonitrile, a cellulose synthesis inhibitor, make a functional wall of over 99% pectin (Shedletzky et al., 1990). The formation of one network (cellulose/hemicellulose) has been inhibited, but the other (pectin network) can still function as a wall.

In a careful study of cell walls regenerated on carrot protoplasts made with highly purified enzymes, the presence of callose could not be detected, and synthesis of pectins and hemicelluloses was initiated as quickly as in normal cells with acid-resistant cellulose forming more slowly (Shea et al., 1989). It was proposed that de-esterification of the carboxyl groups of pectin uronic acid residues permits the formation of a Ca^{2+} cross-linked network that envelopes the protoplast, and the rigid cellulose–hemicellulose framework forms along with this gel matrix (Shea et al., 1989). Wall structure is lost in 6-day-old protoplasts when pectins and hemicelluloses are removed (Hayashi et al., 1986).

Na_2CO_3-extractable polymers are extremely long (300–500 nm) (Fig. 9.7) and this will limit their possible orientations in the wall. FTIR spectroscopy has demonstrated that pectins take up a specific conformation in muro (McCann et al., in preparation), although the computer models of Walkinshaw and Arnott (1981) have shown that pectate chains are sterically capable of packing in quite a number of ways. Polarized infra-red spectra showed that pectins have an oriented structure in pea epidermal cell walls with their molecular chains oriented preferentially parallel to the direction of cell elongation (Morikawa et al., 1978).

The pectin network may be established by self-assembly mechanisms as occurs in the formation of in vitro gels. Certain polysaccharides that form gels by a mechanism that is known, or suspected to involve the cross-linking of molecular chains in multiple helix formation, have been subjected to selective chain cleavage by Smith degradation at helix-terminating residues (Dea et al., 1972). The products retain the ability to form helices but do not now form a network and their gel properties are thus abolished. Addition of certain galacturonans to such fragmented products from agarose and α-carrageenan can evidently re-establish the network because characteristic bulk properties reappear (Dea et al., 1972). However, any gel system used to model the specific conformations (and orientation?) of pectin in the primary cell wall must introduce the pressure/volume constraints of the wall itself since the volume of most gels is much greater than the available volume of the wall from which the pectin is extracted. Conformation, modifications, and sequence specificity may permit the establishment of wall domains by self-assembly; polysaccharide structure becoming a functional code for final topographical location. Fine control of local wall architecture can occur by control of pectin concentration, pectin methylesterase activity, calcium concentration, and pH. Nari et al. (1986) have proposed that the change in polyanionic character caused by pectin methylesterase controls the enzymes that allow the wall loosening necessary for cell growth. Soybean pectin methylesterase generates fixed negative charges in the wall, resulting in proton attraction in the wall. The response of the multi-enzyme plant cell wall system to slight changes of proton concentration relies on the opposite pH sensitivities of the two types of enzyme involved in the growth process: pectin methylesterase, maximally active at pH 8 and inhibited by protons, and wall-loosening enzymes, active around pH 5.

Elongation depends on wall loosening, and all networks must yield. Images of isolated pectins suggest a tendency of the polymers to aggregate, particularly Na_2CO_3-extractable pectins (Fig. 9.7). On high performance size exclusion chromatography (HPSEC), tomato cell wall pectin behaves 'as if it were an aggregate mosaic held together at least partially through non-covalent interactions' (Fishman *et al.*, 1989). Changes in pH may directly modify such interactions, possibly providing non-enzymic mechanisms of wall loosening. Growth responses of cucumber seedling hypocotyls to light are completely and rapidly reversible (Cosgrove, 1981). It has been suggested that the rapidly imposed inhibitory effect of blue light on the growth of stems is mediated by peroxidase-catalysed cross-linking of wall polymer-bound phenolics (Shinkle and Jones, 1988); but since the initial growth rate is rapidly restored when the blue light is switched off, the hypothesis only stands if the mechanism of growth inhibition is reversible; the suggested effects of peroxidase are unidirectional. Gradient growth of dark-grown hypocotyls changes to an even distribution of growth along the stem when transferred to dim red light, and the phenomenon is reversible (Shinkle, personal communication, Edinburgh 5th Cell Wall Meeting). The rapidity of these responses (and the normal presence of a gradient) again suggests physical disaggregation rather than an enzymic process, perhaps mediated by the local concentration of protons.

Pectin heterogeneity

In many different tissues the composition of the wall pectins changes during development and differentiation of the constituent cells. Onion cell wall pectins are a highly heterogeneous group of molecules by sugar analysis and anion exchange chromatography (Redgwell and Selvendran, 1986), antibody labelling, FTIR spectra, length measurements and appearance on mica.

Na$_2$CO$_3$-extractable polymers can be distinguished from CDTA-extractable pectins in a variety of ways. They have a much higher ratio of rhamnose to galacturonic acid (1:10 rather than 1:35) and so may be rhamnogalacturonans rather than homogalacturonans. They have a much greater length (or aggregate to a greater extent) at 300–500 nm, and they behave differently when run on a high performance liquid chromatography (HPLC) system or with other fractionation techniques. The second Na$_2$CO$_3$-extractable fraction is more highly branched than the first (Redgwell and Selvendran, 1986).

A neutral galactan is associated with the α-cellulosic residue throughout the extraction procedure and can be detected in this residue by FTIR spectroscopy. Galactose is also a major constituent of the residue left after treatment of the onion cell wall with anhydrous HF. There is chemical evidence for a neutral pectic fraction still strongly associated with cellulose (Chambat *et al.*, 1981) and solubilized by cellulase, and a highly branched pectic polysaccharide is tightly associated with α-cellulose in potato. Acid hydrolysis releases galactose from cellulose. This neutral pectin must laterally associate with the microfibrils since no cross-links are seen in the electron microscope after treatment with 4 M KOH and acid chlorite (see Fig. 9.3). A very fine meshwork can be seen in thin section images of acid chlorite-treated CWM which may show preferential staining of this pectin – the spacing of the network is about the right size for imaging a molecule lying along the cellulose backbone (Fig. 9.4).

Role of xyloglucans

It has been suggested that xyloglucan cross-links exist between microfibrils (Fry, 1989b; Hayashi, 1989) and this is consistent with the observations: (i) that some wall polymers seem to cross several microfibrils (Fig. 9.1), (ii) that microfibrils collapse together after the removal of alkali-extractable polymers (Fig. 9.2), (iii) that cross-links in FDR replica images finally disappear with the removal of a fraction rich in glucose and xylose (Fig. 9.3).

After removal of pectins, some wall polymers can be seen to underlie several microfibrils; this is particularly clear with stereoscopic viewing. Xyloglucan is thought to form a tightly bound molecular monolayer on the surface of cellulose but there is more xyloglucan in the primary cell wall than would be required for an uninterrupted monolayer and so a substantial proportion of the xyloglucan cannot be in contact with the microfibrils (Hayashi *et al.*, 1987). It seems plausible that the ends of an individual xyloglucan molecule could be hydrogen-bonded on to two different microfibrils, and there is some evidence for this arrangement in the architecture of pea stem cell walls (Hayashi and Maclachlan, 1984). Lengths of extracted xyloglucans range up to 400 nm, very much longer than the cross-link distance between microfibrils of 20–40 nm. Proton magnetic resonance results also support a model in which the hemicellulose forms an immobile layer surrounding a core of cellulose with more freely intervening pectin and hemicellulose components (Mackay *et al.*, 1982).

Sugar analysis shows removal of a xylose and glucose-rich fraction (Redgwell and Selvendran, 1986) and FTIR spectra identify a xyloglucan component, at the stage when FDR replica images show cellulose microfibrils laterally associating to form bundles of from two to more than 20 fibres (Fig. 9.2). Some fibres anastamose, and the spaces that result contain apparently branched bridging molecules, each bridge between 10 and 50 nm long. Either cross-linking polymers can have more complex forms than just simple connecting rods, that is, all bonds are glycosidic and a single molecule is being imaged, or more than one polymer can be involved in forming the cross-link. As (i) these complex cross-links are mostly seen in the 1 M KOH-extracted cell walls where remaining cross-links may be under tension from the bundling of microfibrils, (ii) all polymers imaged are simple rods, (iii) polymers are extremely long and show a tendency to aggregate, we suggest that these cross-links form as a result of lateral interaction between polymers and that more than one xyloglucan molecule can form the cross-link between microfibrils. Then, as microfibrils bundle leaving large holes bridged by cross-links, the tension in the cross-link disrupts lateral interactions between the xyloglucans, pulling them apart.

Removal of hemicelluloses causes collapse of microfibrils so these polymers must play a role in maintaining microfibril spacing, important for wall porosity and assembly. Xyloglucan may be woven into amorphous regions of the microfibril which would explain why concentrated alkali, which causes microfibrils to swell, is required for the extraction of xyloglucan (Hayashi, 1989). Monocot xylans are extracted by mild alkali (4% KOH) and may bind only at the surface of microfibrils.

Cross-links disappear completely with removal of a 4 M KOH-extractable xyloglucan fraction. Phenol ester/ether linkages which strap the xyloglucans to microfibrils (Fry, 1989a) have been proposed, but the absence of ferulic acid in onion pectin makes this mechanism unlikely in onion tissue.

Short-range order (isotropic and anisotropic phases) is typical of systems including cellulose microcrystals that display liquid crystallinity (Marchessault *et al.*, 1959). Such systems commonly contain asymmetric rigid or semirigid macromolecules. The formation of PTFE (polytetrafluoroethylene) whiskers is dominated by the use of a surfactant. Currently it is thought that tetrafluoroethylene in aqueous emulsion tends to polymerize and simultaneously crystallize into extended chain whiskers. Polymerization occurs at temperatures far below the melting point of the final polymers provided the growing particle is protected by surfactant against the very high polymer/water interfacial forces (Folda *et al.*, 1988). It is possible that the assembly of cellulose molecules into microfibrils is directly influenced by the availability of xyloglucans at the assembly sites (Hayashi *et al.*, 1987). The surfactant also serves to prevent flocculation of the growing rods. The analogy with cellulose and xyloglucan is further strengthened by evidence that xyloglucan is present in the cell plate before extrusion of microfibrils (Moore and Staehelin, 1988) and by FDR replica images which show that xyloglucans are necessary to keep microfibrils from laterally associating. Exogenous xyloglucan fails to complex with newly formed cellulose microfibrils in protoplast walls (Hayashi *et al.*, 1986) which suggests that the final macromolecular organization is intimately involved with the secretion process at the plasma membrane. A threshold ratio of cellulose to hemicellulose is required to generate a helicoidal pattern in the periplasmic space in maize coleoptiles and mung bean hypocotyls (Satiat-Jeunemaitre, 1987).

Xyloglucan may act as a morphogenetic key in these and other polylamellate walls. Cellulose microfibrils do not spontaneously realize a helicoidal assembly. If cellulose synthesis is inhibited with 2,6-dichlorobenzonitrile (DCB), hemicellulose deposited in the periplasmic space is also without preferential direction. The structural quality of the pattern depends on the quantity of cellulose synthesized (a graduation in the perturbations of the helicoidal pattern occurs according to the concentration of DCB) – a threshold value of hemicellulose/cellulose ratio can be a determinant in the realization of an ordered wall pattern (Satiat-Jeunemaitre, 1987). During the elongation of pea stems, cellulose content of the walls increases 12-fold while xyloglucan content increases only two-fold (Hayashi et al., 1984). Such compositional changes may be necessary to key architectural rearrangements during cell elongation.

McDougall and Fry (1989) have proposed that the growth-promoting oligosaccharins act by stimulating endogenous enzymes to loosen the wall by cleavage of load-bearing bonds. The breaking and remaking of specific linkages may allow microfibrils to slip past one another (Fry, 1989a; Hayashi, 1989), but ends of xyloglucans may be left attached to microfibrils, maintaining specific sites for bonding. The interconnection between cellulose and xyloglucan is stable at low pH (less than 6) and unstable at alkaline pH (Hayashi et al., 1987).

As the length of cross-links measured is rarely less than 20 nm, then elements of long range order must be present in either the xyloglucan or the microfibril, since there must be specific sites at which the hydrogen bonding between the xyloglucan molecule and cellulose is disrupted to allow 'lift off'. There are two possibilities: either the cellulose structure or some associated molecule (such as a neutral galactan) has a periodicity along the microfibril and binding is interrupted, or xyloglucan itself has some long range order with regions which reverse the polarity of the side-chains with respect to their backbone. Either is feasible but we have made a further interesting observation with respect to xyloglucans; when extracted, their measured lengths have a regular length distribution with polymers being preferentially a multiple of 30 nm long. Given that cross-links in depectinated wall measure between 20 and 40 nm, this result may imply some long range order in xyloglucans of about 60 backbone sugar residues.

Role of proteins

The warp–weft model (Lamport, 1986), which gives a central role to protein cross-links, is inappropriate to the onion cell wall given the small percentage of (hydroxyproline-poor) protein in the wall (Lamport, personal communication). However, extensin can account for 20% of some cell walls and in these may play a more significant role. Extensin levels vary between different cells by as much as 100-fold, and by a factor of 20 in the walls of different tissues of the same plant, making a structural role unlikely to be general. Soybean seed coat extensin is regulated in a developmental and tissue-specific manner; it can first be detected at 16–18 days after anthesis, increasing to high levels at 24 days with marked deposition in the walls of palisade epidermal cells and hour-glass cells (Cassab and Varner, 1987).

HRGPs often increase in direct response to pathogens or at the site of strictly mechanical wounding such as the cut ends of epicotyls. Hydroxyproline levels increase in ageing sections of sweet pepper fruits and on wounding in bean leaves. It has been suggested that production of HRGPs in stress situations may be controlled by ethylene, as large amounts of ethylene are released by plants on wounding, ageing and infection but different carrot extensin mRNA transcripts have been found to accumulate in response to wounding and ethylene treatment. Extensin may provide a flexible and rapid response in situations requiring added architectural strength and impenetrability. Formation of an extensin network from in muro soluble extensin or from extensin currently being synthesized in response to wounding, infection or other stress could reinforce a damaged polysaccharide network; it may make the wall indigestible being highly resistant to proteases or, as a polycation, agglutinate bacteria and thereby prevent their spread.

Protein levels vary dramatically in different walls. Soybean proline-rich proteins (PRPs) are organ-specific and developmentally regulated in expression. Glycine-rich proteins (GRPs) display localized expression, occurring only in the protoxylem cells of the vascular system of bean hypocotyl and the five GRP genes in Arabidopsis show clear differences in patterns of expression throughout the plant and in addition one of the genes is up-regulated in response to stress. Little is known of the cell wall proteins of monocots though a threonine-rich HRGP has been identified in maize and immuno-localized to the cell walls of maize root tips. All classes of cell wall protein also show variation as a function of stress. In carrot, the PRP protein accumulates in the cell wall after wounding and is insolubilized. Soybean seedling repetitive proline-rich protein (RPRP) also gradually becomes insoluble in the cell wall.

Proteins may have more subtle architectural roles: chaperoning polymers, forming nucleation sites for wall assembly, or directly binding polymers together like the clamps which interlock scaffolding poles.

Wall–cytoskeletal interactions

One of the most important aspects of extracellular matrix assembly and its subsequent behaviour is the set of interactions it has with the cell that gave rise to it. In animal cells this is well recognized and there is now solid evidence that matrix molecules not only perform a structural role, but also act to regulate cell division, gene expression, cell differentiation, cell motility and cell adhesion. In plants it seems likely that many similar responses to matrix molecules occur, but the evidence in this case is still largely circumstantial. It is certainly true that protoplasts, plant cells without their walls, cannot proliferate without first regenerating a new cell wall. This is partly because there is no stable wall for the phragmoplast to join up with but it is probably more complex than that. It has been shown that adding exogenous HRGP to protoplasts can help stabilize the cortical microtubule array, for example against cold-induced disassembly (Akashi et al., 1990). As in animal cells it would seem likely that there are complex sets of molecules that connect cytoskeletal elements within the cell (both actin cables and microtubules) to transmembrane linker proteins that then interact with matrix molecules, either directly, or more likely indirectly, through extracellular matrix attachment proteins. It is of interest in this connection that hyaluronic acid in animals could easily be viewed as a highly

homologous molecule to pectin in plants. Both are high molecular weight, highly anionic polysaccharides that are found at the cell surface. Both are capable of forming continuous networks at low concentration, are viscoelastic, and form a water-retaining compression-resistant matrix material. For hyaluronans, however, membrane-bound receptors are known, structural binding proteins have been described, and their effects on cell behaviour are well characterized (Toole, 1990). We urgently need such information for the pectin family. Polygalacturonic acid is certainly found in a zone close to the plasma membrane (see Figs 9.13, 9.14 and 9.15), but many other molecules such as arabinogalactan proteins also occupy this zone.

Several observations demand that not only transmembrane links must be present in plant cells but also intercellular links across the apoplast. For example, the remarkable alignment of actin-cable-containing cytoplasmic strands across groups of epidermal cells prior to wound-induced cell divisions, demands not only indirect connections between actin cables and matrix but in turn between matrix and the neighbouring cell (Goodbody and Lloyd, 1990). The parallel development of wall thickenings in adjacent xylem vessel elements makes similar demands for transcellular microtubule–matrix–microtubule connectivity. At present we have no idea what matrix molecules are involved in such transcellular communication although all those looked at by electron microscopy (McCann et al., 1990) are plenty long enough to span the wall from cell to cell. A subtle manifestation of matrix–cell interactions is the remarkable capacity plant cells have of regulating the exact thickness of any matrix that they lay down, even when the cells are isolated from each other in suspension culture. There is indirect evidence that this mechanism may be mediated by pectin, as cells adapted to DCB, that have no cellulose or hemicellulose in their wall but only a pectin network, are still capable of regulating their wall thickness (Wells, Shedletzky and Delmer, personal communication, and see Delmer, Chapter 8, this volume).

The most complex, and most controversial aspect of cell–matrix interactions has been the

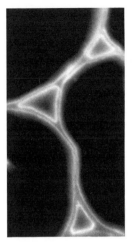 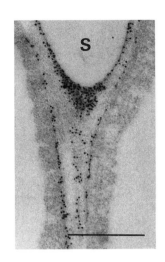 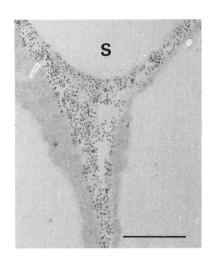

9.13   9.14   9.15

Fig. 9.13. Frozen thin section across the cortex of a pea root, stained by indirect immunofluorescence with JIM5, an antibody recognizing an unesterified pectic epitope. Unesterified pectin is localized in two zones, one adjacent to the plasma membrane of each cell and one lining the intercellular spaces that form by the splitting of the middle lamella (photograph courtesy of Dr Paul Knox).

Fig. 9.14. Thin section of carrot root tip cells stained by immunogold electron microscopy with JIM5. Unesterified pectin is found in a thin (20 nm) zone just outside the plasma membrane, in the middle lamella and in a region surrounding the intercellular space(s) (photograph courtesy of Paul Linstead). Scale bar = 500 nm.

Fig. 9.15. In a similar section to Fig. 9.14 but treated with dilute sodium carbonate to remove methyl ester groups before reacting with JIM5, the staining pattern is now found throughout the wall, suggesting that esterified and unesterified pectin are localized in different regions of the wall (photograph courtesy of Paul Linstead). Scale bar = 500 nm.

relationship between cortical microtubule alignment inside the cell, and cellulose microfibril alignment outside the cell. The problem is reviewed in detail elsewhere in this volume (Giddings and Staehelin, Chapter 7) but some general points can be made here. Again, animal models offer us clear examples in which the cytoskeleton can influence the orientation of structural polymers in the extracellular matrix, for example, collagen and fibronectin. Equally there are examples where structured matrices in turn influence cytoskeletal behaviour and orientation, the essence being that there is a complex interplay, with responsive feedback mechanisms, between cellular architecture and matrix architecture. We see no reason to believe that plant cells will turn out to be any different, and such a view shifts the emphasis away from the current plant paradigm (that

microtubules determine cellulose orientation) towards a more complex, dynamic and interactive relationship. Indeed it may turn out that the ordering of other matrix molecules, in particular pectin, is just as important as cellulose ordering and that both may be able to influence cytoskeletal dynamics. At present we are rich in ideas and structural observations, but very poor in molecules and mechanisms. It is only a matter of time before that changes.

Preliminary model of the onion cell wall

From the images obtained by the FDR replica technique, the following parameters have been derived, putting constraints on current cell wall models:

(1) Microfibrils (diameter 5–12 nm) are cross-linked together in the primary cell wall of onion.

(2) These cross-links, 20–40 nm in length, are hemicellulosic.

(3) The primary cell wall can accommodate only about four layers of parallel-running microfibrils and no weaving is seen between the layers.

(4) Xyloglucans maintain the spacing of microfibrils, preventing lateral association of cellulose.

(5) Middle lamella pectins are Ca^{2+}-cross-linked.

(6) Ester-linked primary cell wall pectins form a network that is co-extensive with, and independent of, the cellulose/hemicellulose network, and which can be removed without affecting the structural integrity of the cellulose/hemicellulose network.

(7) Pectins adopt a precise conformation in the cell wall and may be oriented in specific directions.

(8) Pectins limit wall porosity.

(9) A neutral galactan is associated with microfibrils throughout extraction.

(10) There is no evidence for the existence of a cross-linked extensin network in this tissue but this network may be present in other, extensin rich tissues, or be formed as a stress response to wounding or invasion, in order to reinforce existing networks.

However, the diversity of wall components and even the ability of oligosaccharins to elicit a variety of responses from the cell (Albersheim et al., 1986) indicate that a wide range of architectural modifications must occur across a wall, between tissues, and in specialized wall domains. Even in onion parenchyma tissue, three distinct layers can be delimited across the cell wall: an outer zone containing methyl-esterified pectins, an inner zone which labels with a monoclonal antibody that recognizes unesterified pectin and a middle lamella zone which also labels with the antibody, but that contains very densely stained rod-like polymers (Figs 9.8–9.12). Thus, the original idea of a wall being polylamellate may be elaborated, and we

should perhaps be encouraged to view each lamella, not as identical to the next, but subtly modified in composition and properties as we move across the wall from the plasma membrane to the middle lamella. Localization must occur even at the level of a single layer of microfibrils. In sycamore suspension cells, xyloglucans are found in both wall and middle lamella, whilst rhamnogalacturonan I (RGI) is localized to the middle lamella only (Moore et al., 1986). However, organized tissue has xyloglucan, and carrot storage roots have extensin-I in the wall only (Stafstrom and Staehelin, 1988).

Pectin expression is also controlled at a cell and tissue level. Unesterified and methylesterified pectic epitopes are spatially regulated at the earliest stages of development. In oat coleoptile and leaf, unesterified pectin is present in all walls, while methylesterified pectin is present in none, but in oat root apex, unesterified pectin is in the walls of cortical cells while methylesterified pectin is in the cortex and stele (Knox et al., 1990).

A further important addition to our emerging idea of cell complexity and microheterogeneity is the notion of cell wall domains (Roberts, 1990). Apart from the heterogeneity of cell wall layers, the wall is also locally modified in structure and composition at different regions of the cell surface (Figs 9.14 and 19.15). Such wall domains may allow 'recognition' of one wall by another, for example, where cross walls meet expanding walls so that thickening is restricted. Further evidence of specific domains within walls comes from electron micrographs of osmiophilic structures in the middle lamella of storage parenchyma of pea cotyledons which limit separation of contiguous walls during the formation of intercellular spaces (Kollofel and Linssen, 1984). A clear example of differing wall domains around a single cell can be seen in Figs 9.16–9.18 where a carrot root cap cell shows polarized secretion of an unesterified pectic epitope on different faces of its wall. Wall thickness, composition, and distribution of the pectic epitope differs between the wall facing the root surface to that facing the epidermal cell. Also noteworthy is

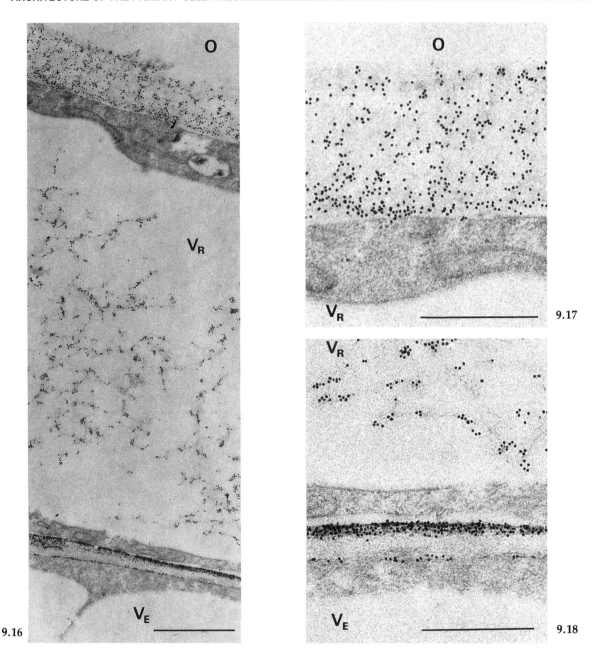

Fig. 9.16. Thin section of a carrot root tip labelled with JIM5. Two points can be made about the distribution of unesterified pectin. The first is that domains are present. The wall of the root cap cell facing the outside (o) is thicker and contains more evenly distributed pectin than the inner wall (shown in more detail in Figs 9.17 and 9.18 respectively). The second is that endocytosis of pectin is occurring in the root cap cell and pectin is ending up in the vacuole (v_R) whereas no pectin is found in the vacuole of the epidermal cell (v_E) or indeed in the cortical cells (photographs for Figs 9.16 to 9.18 courtesy of Paul Linstead). Scale bar = 1 μm.

Fig. 9.17. Detail of the outside wall shown in Fig. 9.16. Scale bar = 500 nm.

Fig. 9.18. Detail of the inner wall shown in Fig. 9.16. Scale bar = 500 nm.

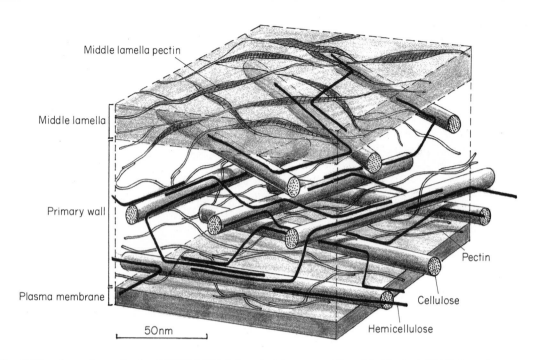

Middle lamella pectin

Middle lamella

Primary wall

Plasma membrane

Pectin

Cellulose

Hemicellulose

50nm

Fig. 9.19. An extremely simplified and schematic representation of how three very broad classes of wall polymer might be spatially arranged in the onion cell wall. Although simplistic, the sizes and spacings of the polymers are based on direct measurements of native walls (McCann *et al.*, 1990) and are drawn to scale. There are two main interpenetrating networks, one of cellulose and hemicellulose and one of pectin. We have not speculated about the nature of cross-links in either network beyond suggesting that hemicellulose may hydrogen bond to two or more cellulose microfibrils. In the onion wall there is room for perhaps three lamellae of cellulose microfibrils between the plasma membrane and the middle lamella. No protein is shown as onion has very little protein in its cell wall. For the sake of clarity, we have not shown all the hemicellulose molecules that presumably coat the cellulose microfibrils. We have also not shown any matrix attachment molecules or arabinogalactan proteins (AGPs) associated with the membrane although these are undoubtedly present.

the distinction between the root cap cell wall and its adjoining epidermal cell wall (Fig. 9.18) where differences in cell type are reflected in differences in pectic composition of the two walls. Many of the measurements we have made on replicas of the onion cell wall have been incorporated in a very schematic and provisional scale model of this primary wall (Fig. 9.19).

Conclusion

Onion parenchyma has proven to be a useful model system by virtue of its relatively simple chemical composition, abundance and the homogeneity of preparations of primary cell walls. However, other model systems active in development and morphogenesis must now be considered and different architectures compared: ordered versus disordered, graminaceous monocot versus dicot and non-graminaceous monocot, primary versus secondary, and between tissues, if some unifying cell wall architectural principles are to be defined.

Cell wall architecture must also be considered in the context of cells, in whole tissue rather than isolated wall fragments. How does the architecture change as a function of tissue type? What happens at tissue boundaries where two architectures meet? How does the architecture of wall domains vary – the middle lamella, intercellular spaces, the interface of the wall with the plasma membrane?

References

Akashi, T., Kawasaki, S. and Shibaoka, H. (1990). Stabilization of cortical microtubules by the cell wall in cultured tobacco cells: Effects of extensin on the cold-stability of cortical microtubules. *Planta* **182**, 363–369.

Albersheim, P., Darvill, A.G., Davis, K.R., Doares, S.H., Gollin, D.J., O'Neill, R. *et al.* (1986). Oligosaccharins: Regulatory molecules in plants. *Hort. Sci.* **21**, 842–845.

Baron-Epel, O., Gharyal, P.K. and Schindler, M. (1988). Pectins as mediators of wall porosity in soybean cells. *Planta* **175**, 389–395.

Carpita, N., Sabularse, D., Montezinos, D. and Delmer, D.P. (1979). Determination of the pore size of cell walls of living plant cells. *Science* **205**, 1144–1147.

Cassab, G.I. and Varner, J.E. (1987). Immunocyto-localisation of extensin in developing soybean seed coats by immunogold-silver staining and by tissue printing on nitrocellulose paper. *J. Cell Biol.* **105** (6/1), 2581–2588.

Chambat, G., Joseleau, J.P. and Barnoud, F. (1981). The carbohydrate constituents of the cell wall of suspension cultures of *Rosa glauca*. *Phytochemistry* **20**, 241–244.

Cooper, J.B. and Varner, J.E. (1983). Insolubilization of hydroxyproline-rich cell wall glycoprotein in aerated carrot root slices. *Biochem. Biophys. Res. Comms* **112** (1), 161–167.

Cosgrove, D.J. (1981). Rapid suppression of growth by blue light: occurrence, time course and general characteristics. *Plant Physiol.* **67**, 584–590.

Darvill, A., McNeill, M., Albersheim, P. and Delmer, D.P. (1980). The primary cell walls of flowering plants. In *The Plant Cell* (ed. N.E. Tolbert), pp. 91–162. Academic Press, New York.

Dea, I.C.M., McKinnon, A.A. and Rees, D.A. (1972). Tertiary and quaternary structure in aqueous polysaccharide systems which model cell wall cohesion: reversible changes in conformation and association of agarose, carrageenan and galactomannans. *J. Mol. Biol.* **68**, 153–172.

Epstein, L. and Lamport, D.T.A. (1984). An intra-molecular linkage involving isodityrosine in extensin. *Phytochemistry* **23**, 1241–1246.

Fishman, M.L., Gross, K.C., Gillespie, D.T. and Sondey, S.M. (1989). Macromolecular components of tomato fruit pectin. *Arch. Biochem. Biophys.* **274**, 179–191.

Folda, T., Hoffman, H., Chanzy, H. and Smith, P. (1988). Liquid crystalline suspensions of poly (tetrafluoroethylene) 'whiskers'. *Nature* **333**, 55–56.

Fry, S.C. (1982a). Isodityrosine, a new cross-linking amino acid from plant cell-wall glycoprotein. *Biochem. J.* **204**, 449–455.

Fry, S.C. (1982b). Phenolic components of the cell wall. *Biochem. J.* **203**, 493–504.

Fry, S.C. (1989a). Analysis of cross-links in the growing cell walls of higher plants. In *Modern Methods in Plant Analysis*, New Series, vol. 10 (ed. H.S. Linskens and J.S. Jackson), pp. 1–42. Springer-Verlag, Berlin.

Fry, S.C. (1989b). The structure and functions of xyloglucan. *J. Exp. Bot.* **40** (210), 1–11.

Goodbody, K.C. and Lloyd, C.W. (1990). Actin filaments line up across *Tradescantia* epidermal cells, anticipating wound-induced division planes. *Protoplasma* **157**, 92–101.

Hayashi, T. (1989). Xyloglucans in the primary cell wall. *Ann. Rev. Pl. Physiol. Pl. Mol. Biol.* **40**, 139–168.

Hayashi, T. and Maclachlan, G. (1984). Pea xylo-glucan and cellulose. I. Macromolecular organisation. *Plant Physiol.* **75**, 596–604.

Hayashi, T., Wong, Y.S. and Maclachlan, G. (1984). Pea xyloglucan and cellulose. II. Hydrolysis by pea endo-1,4-β-glucanases. *Plant Physiol.* **75**, 605–610.

Hayashi, T., Polonenko, D.R., Camiraud, A. and MacLachlan, G. (1986). Pea xyloglucan and cellu-lose. IV. Assembly of β-glucans by pea protoplasts. *Plant Physiol.* **82**, 301–306.

Hayashi, T., Marsden, M.P.F. and Delmer, D.P. (1987). Pea xyloglucan and cellulose. V. Xyloglucan-cellulose interactions *in vitro* and *in vivo*. *Plant Physiol.* **83**, 384–389.

Heckman, J.W., Terhune, B.T. and Lamport, D.T.A. (1988). Characterization of native and modified extensin monomers and oligomers by electron microscopy and gel filtration. *Plant Physiol.* **86**, 848–856.

Heuser, J. (1981). Preparing biological samples for stereomicroscopy by the quick-freeze, deep-etch, rotary-replication technique. *Methods in Cell Biology* **22**, 97–122.

Jarvis, M.C. (1984). Structure and properties of pectin gels in plant cell walls. *Plant Cell Env.* **7**, 153–164.

Keegstra, K., Talmadge, K.W., Bauer, W.D. and Albersheim, P. (1973). The structure of plant cell walls. III. A model of the walls of suspension-cultured sycamore cells based on the interconnections of the macromolecular components. *Plant Physiol.* **51**, 188–198.

Knox, J.P., Linstead, P.J., King, J., Cooper, C. and Roberts, K. (1990). Pectin esterification is spatially regulated both within cell walls and between developing tissues of root apices. *Planta* **181**, 512–521.

Kollofel, C. and Linssen, P.W.T. (1984). The formation of intercellular spaces in the cotyledons of developing and germinating pea seeds. *Protoplasma* **120**, 12–19.

Lamport, D.T.A. (1965). The protein component of primary cell walls. In *Advances in Botanical Research*, vol. 2 (ed. R.D. Preston), pp. 152–218. Academic Press, New York.

Lamport, D.T.A. (1986). The primary cell wall: a new model. In *Cellulose: structure, modification and hydrolysis* (ed. R.A. Young and R.M. Rowell). John Wiley, New York.

Lamport, D.T.A. and Northcote, D.H. (1960). Hydroxyproline in the cell walls of higher plants. *Nature* **188**, 665–666.

McCann, M.C., Wells, B. and Roberts, K. (1990). Direct visualisation of cross-links in the primary cell wall. *J. Cell Sci.* **96**, 323–334.

McDougall, G.J. and Fry, S.C. (1989). Structure-activity relationships for xyloglucan oligosaccharides with anti-auxin activity. *Plant Physiol.* **89**, 883–887.

Mackay, A.L., Bloom, M., Tepfer, M. and Taylor, I.E.P. (1982). Broadline proton magnetic resonance study of cellulose, pectin and bean cell walls. *Biopolymers* **21**, 1521–1534.

Marchessault, R.H., Morehead, F.F. and Walters, N. M. (1959). Liquid crystal systems from fibrillar polysaccharides. *Nature* **184**, 632–633.

Monro, J.A., Penny, D. and Bailey, R.W. (1976). The organization and growth of primary cell walls of lupin hypocotyl. *Phytochemistry* **15**, 1193–1198.

Moore, P.J. and Staehelin, L.A. (1988) Immunogold localisation of the cell wall matrix polysaccharides rhamnogalacturonan I and xyloglucan during cell expansion and cytokinesis in *Trifolium pratense* L.: implications for secretory pathway. *Planta* **174**, 433–445.

Moore, P.J., Darvill, A.G., Albersheim, P. and Staehelin, L.A. (1986). Immunogold localisation of xyloglucan and rhamnogalacturonan I in the cell walls of suspension-cultured sycamore cells. *Plant Physiol.* **82**, 787–794.

Morikawa, H., Hayashi, R. and Senda, M. (1978). Infrared analysis of pea stem cell walls and oriented structure of matrix polysaccharides in them. *Plant and Cell Physiology*, **19**, 1151–1159.

Mort, A.J. and Lamport, D.T.A. (1977). Anhydrous hydrogen fluoride deglycosylates glycoproteins. *Anal. Biochem.* **82**, 289–309.

Nari, J., Noat, G., Diamantidis, G., Woudstra, M. and Ricard, J. (1986). Electrostatic effects and the dynamics of enzyme reactions at the surface of plant cells. 3. Interplay between limited cell-wall autolysis, pectin methyl esterase activity and electrostatic effects in soybean cell walls. *Eur. J. Biochem.* **155**, 199–202.

Redgwell, R.J. and Selvendran, R.R. (1986). Structural features of cell wall polysaccharides of onion *Allium cepa. Carbohyd. Res.* **157**, 183–199.

Roberts, K. (1989). The plant extracellular matrix. *Curr. Op. in Cell Biol.* **1** (5), 1020–1027.

Roberts, K. (1990). Structures at the plant cell surface. *Curr. Op. in Cell Biol.* **2**, 920–928.

Roelofson, P.A. (1965). Ultrastructure of the wall of growing cells and its relation to the direction of growth. *Adv. Bot. Res.* **2**, 69–149.

Satiat-Jeunemaitre, B. (1987). Inhibition of the helicoidal assembly of the cellulose-hemicellulose complex by 2,6-dichlorobenzonitrile (DCB). *Biology of the Cell* **59**, 89–96.

Shea, E.M., Gibeaut, D.M. and Carpita, N.C. (1989). Structural analysis of the cell walls regenerated by carrot protoplasts. *Planta* **179**, 293–308.

Shedletzky, E., Shmuel, M., Delmer, D.P. and Lamport, D.T.A. (1990). Adaptation and growth of tomato cells on the herbicide 2,6-dichlorobenzonitrile (DCB) leads to production of unique cell walls virtually lacking a cellulose-xyloglucan network. *Plant Physiol.* **94**, 980–987.

Shinkle, J.R. and Jones, R.L. (1988). Inhibition of stem elongation in *Cucumis* seedlings by blue light requires calcium. *Plant Physiol.* **86**, 960–966.

Stafstrom, J.P. and Staehelin, L.A. (1988). Antibody localisation of extensin in cell walls of carrot storage roots. *Planta* **174**, 321–332.

Toole, B.P. (1990). Hyaluronan and its binding proteins, the hyaladherins. *Curr. Op. in Cell Biol.* **2**, 839–844.

Valent, B. and Albersheim, P. (1974). Structure of plant cell walls. V. On the binding of xyloglucan to cellulose fibres. *Plant Physiol.* **54**, 105–109.

Walkinshaw, M.D. and Arnott, S. (1981). Conformations and interactions of pectins. I. X-ray diffraction analysis of sodium pectate in neutral and acidified forms. II. Models for junction zones in pectinic acid and calcium pectate gels. *J. Mol. Biol.* **153**, 1055–1085.

Yamaoka, T. and Chiba, N. (1983). Changes in the coagulating ability of pectin during the growth of soybean hypocotyls. *Plant Cell Physiol.* **24**, 1281–1290.

10. BIOCHEMISTRY OF XYLOGLUCANS IN REGULATING CELL ELONGATION AND EXPANSION

Takahisa Hayashi

Wood Research Institute, Kyoto University,
Gokasho, Uji, Kyoto 611, Japan

Introduction

Xyloglucans occur universally in the primary walls of higher plants, both dicotyledons and monocotyledons (Hayashi, 1989). The polysaccharides contain a $(1\rightarrow4)$-β-D-glucan backbone with $(1\rightarrow6)$-linked α-D-xylosyl residues along the backbone. There is considerable variation in the levels of xylosyl residues and distribution of additional branching galactosyl and fucosyl–galactosyl residues, depending on the plant species.

Xyloglucans interfere with the ribbon assembly of cellulose microfibrils produced by the bacterium *Acetobacter xylinum* (Hayashi *et al.*, 1987). It appears that xyloglucans associate with ribbon subunits and prevent fasciation into larger bundles. This may also stimulate the rate of cellulose synthesis, when calcofluor and carboxymethylcellulose are added to *A. xylinum* (Haigler and Benziman, 1982). This results not only in the spreading of a fibre network, but also gives the flexibility necessary for microfibrils to slide. This phenomenon probably occurs *in vivo* in the primary walls, where cellulose is synthesized at plasma membranes during the secretion of xyloglucan via secretory vesicles. Although the microfibril is more disordered in the primary wall, it is reasonable to consider that the plane of the backbone glucan of the xyloglucan molecule is parallel to the molecular plane of cellulose at the microfibril surface (Ogawa *et al.*, 1990). Xyloglucan and cellulose probably combine with each other at the edges of their planes by hydrogen bonds.

Xyloglucans are biodegraded rapidly after acid or auxin treatment. Such treatment activates and/or induces the development of endo-1,4-β-glucanase activity in peas (Verma *et al.*, 1975), leading to the hydrolysis of xyloglucan cross-links. The phenomenon probably contributes to auxin-induced loosening of cellulose microfibril networks to render the wall susceptible to turgor-driven expansion.

This chapter describes the biochemistry of xyloglucan in regulating cell elongation and expansion in auxin-stimulated growth.

Biochemistry of xyloglucans

Occurrence of xyloglucans

Xyloglucans in most dicotyledons possess a glucan backbone with three of every four consecutive residues substituted with 1,6-α-D-xylosyl residues and minor additional side chains containing α-L-fucosyl-$(1\rightarrow2)$-β-D-galactosyl residues attached to the 2 position of xylosyl residues. Figure 10.1 shows the chemical repeating units of pea xyloglucan (Hayashi and Maclachlan, 1984a). The terminal α-L-fucosyl-$(1\rightarrow2)$-β-D-galactosyl residues as side chains are associated with the serological activity of the human blood group substance H. Therefore, fucosylated xyloglucans complex with α-L-fucose-binding lectins from *Ulex europaeus* and *Lotus tetragonolobus*. Recently, an arabinosyl residue was found linked at C-2 of the β-glucosyl residue at the non-reducing end of the heptasaccharide unit in the sycamore

The Cytoskeletal Basis of Plant Growth and Form ISBN 0–12–453770–7

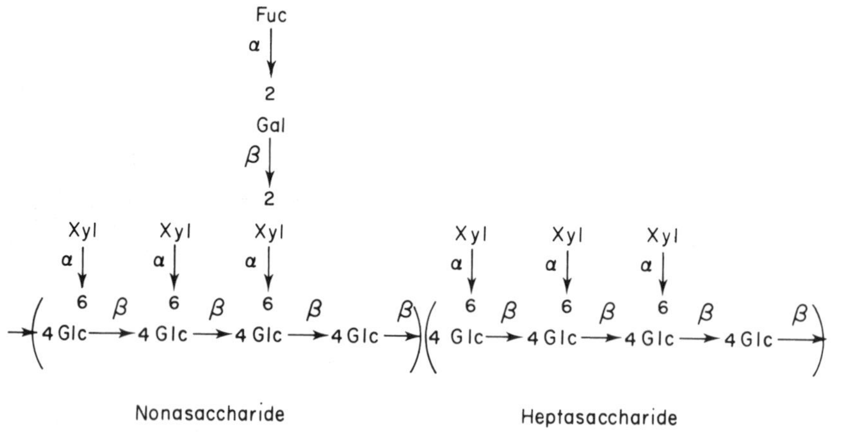

Fig. 10.1   Structure of pea xyloglucan repeating units.

extracellular xyloglucan (Kiefer *et al.*, 1990). Xyloglucans in monocotyledons are less substituted with xylosyl residues, and in some monocotyledonous xyloglucans there is further minor addition of β-D-galactose residues to the 2-position of the xylosyl residues (Kato and Matsuda, 1985). The glucans resemble cellulose with fewer substitutions. The polysaccharides are insoluble in water, although xyloglucans from dicotyledons are all soluble in water.

Localization of xyloglucan on thin sections was determined by Moore and Staehelin (1988) using a polyclonal antibody against sycamore xyloglucans with a protein carrier ovalbumin and also by Misaki *et al.* (1988) using antibody IgG against octasaccharide (glucose/xylose/galactose, 4:3:1) with a protein carrier BSA. Visualization using both antibody IgGs with protein A-gold or fluorescein isothiocyanate-labelled antiserum showed xyloglucan to be localized in the inner layer of the primary walls and not in the middle lamella. A cell wall ghost composed of xyloglucan and cellulose was examined by light microscopy with radioautography after labelling with [^{3}H]fucose, by fluorescence microscopy using a fluorescent fucose-binding lectin, and by electron microscopy after shadowing (Hayashi and Maclachlan, 1984a). Xyloglucans were found located both on and between cellulose microfibrils.

Xyloglucan levels per unit length or fresh weight of tissue steadily decline during development. As a result, the xyloglucan: cellulose weight ratio decreases from about 0.7 in growing regions to 0.11 in maturing regions of pea stems (Hayashi *et al.*, 1984).

Cross-linking of xyloglucan to cellulose microfibrils

Xyloglucans bind to cellulose microfibrils because they contain the cellulose-like 1,4-β-glucan backbone. The binding of xyloglucan to cellulose is inhibited above pH 6 (Fig. 10.2). Based on the binding capacity for cellulose microfibrils of differing surface areas, the capacity was dependent on the surface of the microfibrils (Hayashi *et al.*, 1987). The binding of xyloglucan to cellulose was very specific and not affected by the presence of a 10-fold excess of (1→2)-β-glucan, (1→3)-β-glucan, (1→6)-β-glucan, (1→3, 1→4)-β-glucan, arabinogalactan, or pectin. Thus, binding possibly occurs specifically with structures containing (1→4)-β-glucosyl linkages in polymers having a conformation complementary to cellulose.

The native xyloglucan: cellulose complex (cell ghosts) contains 14-fold higher levels of xyloglucan than those in the reconstituted complex. In reconstitution experiments, 100 μg of pea cellulose were saturated with 5 μg of xyloglucan, although 70 μg of xyloglucan were embedded in 100 μg of pea cellulose in the native complex. In the primary walls of the pea stem, xyloglucan probably not only binds to the surface of

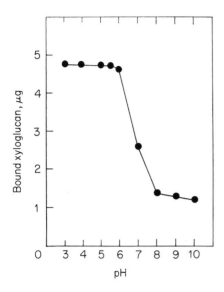

Fig. 10.2. Effects of pH on the binding of xyloglucan to cellulose. (From Hayashi *et al.*, 1987 with permission.)

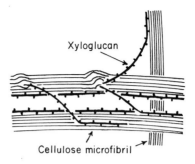

Fig. 10.3. Potential linkages betwen xyloglucan and cellulose. (From Hayashi, 1989 with permission.)

cellulose microfibrils but also weaves into their amorphous parts. This would perhaps explain why concentrated alkali, which causes microfibril swelling, is required for the extraction of xyloglucan; mild alkali which does not cause swelling does not dissociate the complex even though it prevents new association of xyloglucan and cellulose. Native association may occur close to or directly at the site of cellulose synthesis to enhance binding *in vivo*.

One potential role of xyloglucans that bind to cellulose could be to cover the cellulose microfibril, thereby preventing hydrogen bonding between microfibrils in the primary walls. This may help each microfibril to slide during cell enlargement, because xyloglycans have mutual affinity. Another possible role associated with the fact that xyloglucans weave into, as well as bind to, microfibrils, is the cross-linking of each cellulose microfibril network. This may provide a rigidity to cell walls by the binding of adjacent microfibrils and has actually been confirmed by the observation that cell ghosts composed of cellulose only appear fragile when xyloglucan is extracted with concentrated alkali from cell ghosts composed of xyloglucan and cellulose. Cross-linking between perpendicular

fibrils may function as a bracket, and that between parallel fibrils as a beam (Fig. 10.3). Microfibrils may thus be embedded in cement-like xyloglucan polysaccharides, and the resulting xyloglucan:microfibril complex may form reinforced concrete-like walls.

Biodegradation of xyloglucan

Growing pea stem cells generate two distinct endo-1,4-β-glucanase activities (Byrne *et al.*, 1975). These hydrolyse the internal 1,4-β-glucosyl linkages of xyloglucan, as well as various (1→4)-β-glucans, (1→3, 1→4)-β-glucans and cellodextrins (Hayashi *et al.*, 1984). The (Michaelis constant) Km of two pea endo-1,4-β-glucanases acting on pea xyloglucan, *Tamarindus indica* xyloglucan, cellohexaose and carboxymethylcellulose were found virtually the same (3.5 mg/ml). Endo-1,4-β-glucanases reduce the viscosity of solutions of carboxymethylcellulose and *Tamarindus* xyloglucan with a minor concurrent increase in the number of reducing groups generated, this being typical of endohydrolysis kinetics. The action patterns of pea endo-1,4-β-glucanases against pea xyloglucan are essentially the same as those of fungal cellulases, to hydrolyse internal linkages adjacent to unsubstituted glucose residues, thereby introducing a free reducing end-group at these points.

Bal *et al.* (1976) showed the localization of enzymes in ultrathin sections of pea tissues, using ferritin conjugate antibody and electron microscopy. One of the enzymes, with a molecular weight of 70 kD, is firmly associated

Table 10.1. Hydrolysis of native xyloglucan:cellulose complex by purified pea endo-1,4-β-glucanase (reproduced from Hayashi *et al.*, 1984). Tritiated xyloglucan and cellulose (accessible reducing end groups) were assayed and converted to reducing equivalents

Time (h)	Residual amount			Residual reducing ends	
	Xylo-glucan (mg)	Cellulose (mg)	Solubilized sugar (mg)	Xylo-glucan (ng Glc)	Cellulose (ng Glc)
0	2.9	3.9	0	0	0
3	2.8	3.9	0.18	393	10
6	2.4	3.8	0.56	600	11
9	2.1	3.8	1.12	1200	15
48	1.6	3.6	1.79	1155	19

with the inner surface of cell walls. The other enzyme, with a molecular weight of 20 kD, is also associated with rough endoplasmic reticulum. Although the relationship between the endo-1,4-β-glucanases is unknown, the former appears to be involved in the biodegradation of xyloglucan.

The capacity of wall-bound pea endo-1,4-β-glucanase to degrade native xyloglucan: cellulose complex (cell ghosts) *in vitro* was examined directly (Table 10.1) (Hayashi *et al.*, 1984). During the early stages of incubation with purified glucanase, the amount of cellulose-bound xyloglucan decreased and there was concomitant release of buffer-soluble saccharide. There was also a marked increase in the number of free reducing end groups (as determined by reduction with sodium borotritide) in the xyloglucan remaining bound to the complex, and a lesser but detectable increase in the chain ends of the cellulose component. Cellulose hydrolysis may thus only occur following the removal of xyloglucan, since all fibrils appear to be coated with the polysaccharide. Wall-bound endo-1,4-β-glucanase obviously acts *in vitro* as a xyloglucanase rather than cellulase.

Auxin action on the primary cell wall

Effects of auxin on plant cells

Cell divisions occur in both the apical meristems of shoots and/or roots for tissue elongation and in lateral meristems for tissue expansion during growth. A cell in the meristematic area divides into two daughter cells, one of which again divides while the other elongates and/or expands causing its volume to increase. This volume increase does not result from an increase in cytoplasmic volume but from the accumulation of water in expanding vacuoles while new wall materials are being supplied. Volume stability is dependent on both wall and turgor pressure. The latter is a function of the osmolality of cell solutes and the former on auxin action to mediate wall extension.

Auxin first mediates the permeability of plasma membranes and activates an H^+ pump in the membranes. The H^+ formed acidifies the inner surface of the cell walls, thus possibly weakening the cell wall network, particularly the xyloglucan–cellulose network, and causing wall extension. Auxin also has long-term effects for inducing the gene expression of glycan hydrolases and synthases required for polysaccharide deposition. Figure 10.4 shows auxin action on the primary wall with cell wall loosening and turgor-driven cell elongation. The orientation of cellulose microfibrils determines the direction of cell expansion.

Table 10.2 shows latent periods of the effects of auxin on enzyme activity and polysaccharide metabolism in cell walls, along with miscellaneous changes (Evans, 1974a). Several of the enzymes may mediate auxin-induced wall loosening. It would not be likely that β-galactosidase activity has the function of

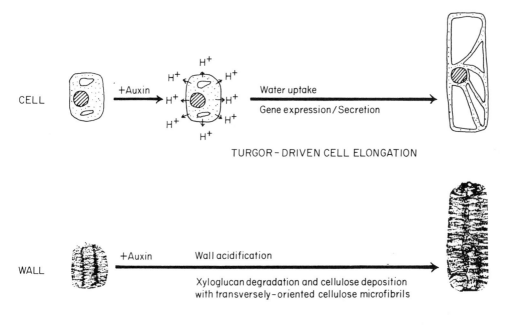

Fig. 10.4 Auxin action on the primary wall with cell wall loosening and turgor-driven cell elongation.

Table 10.2. Latent periods in the effects of auxin on cell wall metabolism

Phenomenon	Material	Latent period (min)	References
Enzyme activity			
β-Galactosidase	*Avena* coleoptile	60	Johnson *et al.*, 1974
1,3-β-Glucanase	*Avena* coleoptile	10	Masuda and Yamamoto, 1970
Sclerotan hydrolase	Barley coleoptile	180	Tanimoto and Masuda, 1968
	Pea stem	180	Tanimoto and Masuda, 1968
Endo-1,4-β-glucanase	Pea stem	360	Fan and Maclachlan, 1966
β-Glucan synthase	Pea stem	15	Ray, 1973
Polysaccharide			
Xyloglucan turnover	Pea stem	10–15	Labavitch and Ray, 1974a
Miscellaneous			
Permeability to water	Pea stem	1	Kang and Burg, 1971
Osmolality of cell sap	Pea stem	20	Yoda and Asida, 1961
Hydrogen ion leakage	*Avena* coleoptile	15–20	Cleland, 1973
Wall extensibility	*Avena* coleoptile	20	Cleland, 1967
	Pea stem	2–3	Burstrom *et al.*, 1970
	Sunflower hypocotyl	4	Uhrström, 1969
Ethylene biosynthesis	Pea stem	60	Burg and Burg, 1966

auxin-stimulated wall-loosening because the activity is inhibited by galactonolactone without blocking auxin- and/or acid-stimulated growth (Evans, 1974b). Since Sclerotan contains β-1,3- and β-1,6-glucoside linkages, the hydrolase of the polysaccharide is probably due to 1,3-β-glucanase (Tanimoto and Masuda, 1968). The glucanase is probably the enzyme responsible for loosening walls in monocotyledons by the hydrolysis of (1→3, 1→4)-β-glucan, although auxin inhibits 1,3-β-glucanase production at the level of mRNA in dicotyledons (Mohnen *et al.*, 1985).

Auxin induces the biosynthesis of endo-1,4-β-glucanases *in vivo* and the glucanase synthesized is secreted on to the inner surface of the wall, where xyloglucan is present. Theologis (1986) suggests that mRNAs rapidly induced by auxin mediate H$^+$ secretion and cell elongation, and that late mRNAs may code for endo-1,4-β-glucanases. In fact, glucanase activity increased after a lag of about 6–12 h with supraoptimal auxin, although mRNA for the glucanase developed linearly without a lag phase (Fig. 10.5) (Verma *et al.*, 1975).

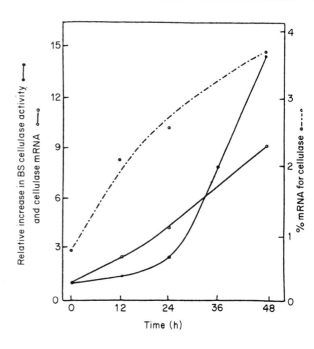

Fig. 10.5. Increase in endo-1,4-β-glucanase activity and mRNA following auxin treatment. (From Verma *et al.*, 1975 with permission.)

Wall loosening as a rapid response to auxin and/or acid

Pectins – consisting of galacturonic acid and minor arabinose, galactose, and rhamnose – constitute as much as 20–30% of the primary walls of dicotyledons and 2–3% of monocotyledons (Darvill *et al.*, 1980). The carboxy groups of the galacturonic residues are usually methyl-esterized in the pectins. Pectinesterase (EC 3.1.1.11) present in the cell walls potentially catalyses trans-esterification to form a gel made from galacturonic acid residues and Ca^{2+} (Komae and Misaki, 1989; Komae *et al.*, 1989, 1990). The activity of pectinesterase is optimum above pH 7 and inhibited by low pH (Fig. 10.6). Acidic pH may thus also inhibit the formation of Ca-bridge cross-links of cell walls and maintains wall extensibility.

The primary walls of dicotyledons are composed of 10% of the basic glycoprotein, extensin (Cassab and Varner, 1988). Extensin cross-linking *in vitro* is highly dependent on the pH of the cell wall because the tyrosine

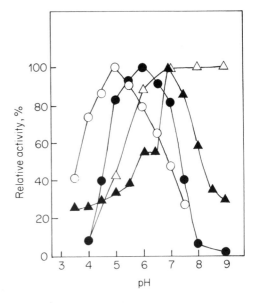

Fig. 10.6. Effects of pH on the activity of maize exo-β-glucanase (○—○), pea endo-1,4-β-glucanase (●—●), pectinesterase (△—△), and peroxidase (▲—▲).

residues of extensin are coupled by peroxidase (EC 1.11.1.7) to form isodityrosine cross-links (Fry, 1987). The level of insolubilization *in vivo* is probably inhibited by acid in the physiological pH range (Cooper and Varner, 1984). The primary walls contain diferulate which may function as a cross-link between arabinoxylans (Kato and Nevins, 1985; Harvey *et al.*, 1986; Kato *et al.*, 1987; Ishii and Hiroi, 1990), arabinogalactans (Fry, 1982) and xyloglucans (Ishii *et al.*, 1990). Peroxidase activity is also involved in coupling between two ferulate residues to form diferulate cross-links. The activity of peroxidase is optimum at pH 9 and inhibited at low pH (Fig. 10.6). Thus, low pH potentially inhibits the formation of isodityrosine and/or diferulate cross-links to maintain wall extensibility, although no direct evidence has been published to date that the cross-links occur between chains of extensins and between polysaccharides.

Labrador and Nevins (1989) found an exo-β-glucanase prepared from maize coleoptiles enhanced the elongation of maize coleoptile segments by 1.3-fold in both the presence and absence of exogenous indole-3-acetic acid. The enzyme activity was maximal at pH 5 (Fig. 10.6) with the substrate of laminarin and lichenan, the latter being similar to $(1\rightarrow3, 1\rightarrow4)$-β-glucan in the walls of monocotyledons (Kato and Nevins, 1984). The $(1\rightarrow3, 1\rightarrow4)$-β-glucan is hydrolysable not only by exo-β-glucanase and but also by 1,3-β-glucanase induced by auxin in monocotyledons (Table 10.2) (Masuda and Yamamoto, 1970). Although the localization of $(1\rightarrow3, 1\rightarrow4)$-β-glucan and relationship between β-glucan and microfibril network are unclear, the β-glucan is a probable contributor to cell wall loosening in monocotyledons.

Cell wall extension may occur by a hydrogen bond creep between xyloglucan and cellulose microfibrils (Keegstra *et al.*, 1973). However, interconnection between the two polymers is stable at low pH (less than 6) and unstable at alkaline pH as previously shown in Fig. 10.2 (Hayashi *et al.*, 1987). Therefore, creep between the two β-glucans does not occur at low pH.

Acid- and/or auxin-induced growth is always associated with the degradation of xyloglucan.

The activity of endo-1,4-β-glucanase located on the inner surface of the primary walls is higher at acidic pH (Fig. 10.6). Labavitch and Ray (1974a) showed, for the first time, that auxin promotes the turnover of xyloglucan. This promotion begins within 15 min following auxin treatment, and the effects of this are not only augmented with increased indole-3-acetic acid concentration, but also correspond to the elongation rate. Nishitani and Masuda (1982) observed that acid pH induces a decrease in the average molecular weight of xyloglucans in the walls of *Vigna angularis* hypocotyls. Jacobs and Ray (1975) found a water-soluble xyloglucan to form from cell walls on incubation with acidic pH when the elongation of pea stems was induced. Thus, xyloglucan is degraded at an early stage of auxin and/or acid growth without a lag time. This rapid response is probably due to xyloglucan degradation by activation of endo-1,4-β-glucanase activity. The degradation of xyloglucan may not be the only event involved in cell wall loosening since the loosening may also accompany the hydrolysis of $(1\rightarrow3, 1\rightarrow4)$-β-glucan under maintaining wall extensibility. Nevertheless, auxin- and/or acid-induced growth has indeed been shown to accompany xyloglucan degradation in *Pisum* (Labavitch and Ray, 1974a, b; Jacobs and Ray, 1975; Gilkes and Hall, 1977; Terry and Bonner, 1980; Terry *et al.*, 1981), *Vigna* (Nishitani and Masuda, 1983), *Pinus* (Lorences *et al.*, 1987a,b; Lorences and Zarra, 1987) and even monocotyledons, *Avena* (Inouhe *et al.*, 1984) and *Oryza* (Revilla and Zarra, 1987). Endo-1,4-β-glucanase activity responsible for xyloglucan degradation was also confirmed to be associated with auxin-induced cell expansion in pea (Hayashi *et al.*, 1984; Hayashi and Maclachlan, 1986), cell growth in mung bean (Kato and Matsuda, 1981) and, even in monocotyledons, with developing vessels of barley roots (Sassen, 1965) and developing pistils and pollen tubes of *Hemerocallis fulva* (Konar and Stanley, 1969).

Wall loosening as a response to supraoptimal auxin

Supraoptimal auxin treatment of pea epicotyls *in vivo* causes increase in the net deposition of xyloglucan and cellulose despite marked

Table 10.3. Amounts and molecular weights of xyloglucan and cellulose in a growing pea cell (reproduced from Hayashi *et al.*, 1984). The apical 5 mm of intact seedling (segment) elongated about five-fold in control tissue in 48 h while there was lateral expansion after treatment with supraoptimal auxin (1 mM 2,4-dichlorophenoxyacetic acid)

	Time (h)	Treatment	Amount (μg)	Molecular weight
Xyloglucan	0	Initial	28	330 000
	48	Control	59	230 000
	48	Auxin	137	50 000
Cellulose	0	Initial	40	1 300 000
	48	Control	515	1 700 000
	48	Auxin	615	1 500 000

induction of endo-1,4-β-glucanase activity (Verma *et al.*, 1975). As shown in Table 10.3, net bound xyloglucan levels per segment doubled in untreated elongating tissue and increased five-fold following auxin treatment. Soluble (buffer soluble) xyloglucan constituted less than 1% of the total and slightly increased following auxin treamtent. Cellulose levels increased 13-fold in control and 15-fold following auxin treatment.

Total extractable endo-1,4-β-glucanase activity per segment in this experiment, as assayed using carboxymethylcellulose as substrate, increased about three-fold in untreated segments, but 70-fold in auxin-treated segments. The amounts of cellulose deposited during growth with or without auxin treatment showed relatively high average molecular weights. However, the average molecular weight of the resulting xyloglucan (50 000) with supraoptimal auxin was much lower than that (230 000) of the controls, and decrease in molecular weight may possibly lead to the solubilization of the polysaccharide. The molecular weight of a small amount of soluble xyloglucan was less than 20 000. Since these events probably occur *in vivo* mainly on the inner surface of the wall, endohydrolysis of xyloglucan may weaken cross-linkages that restrain microfibril slip and creep. Thus, there is a correlation between xyloglucan deposition,

endo-1,4-β-glucanase activity and xyloglucan degradation following auxin treatment. In the primary walls of higher plants, xyloglucan is apparently more accessible to hydrolysis. Supraoptimal auxin-induced endo-1,4-β-glucanase undoubtedly functions *in vivo* primarily as a xyloglucan-degrading enzyme.

The average molecular weight (50 000) of xyloglucan after auxin treatment at 48 h was much lower than that at zero time (330 000) as shown in Table 10.3. Nevertheless, molecular weight continued to decrease from 330 000 in growing cells to 230 000 in elongated control cells (Hayashi *et al.*, 1984) as well as ethylene-treated swollen cells (Hayashi and Maclachlan, 1984b). This decrease was due possibly to the partial degradation of xyloglucan molecules by endogenous endo-1,4-β-glucanase activity. Nishitani and Masuda (1982) found a modest decrease (20%) in the average molecular weight of the polysaccharide in excised segments of Azuki bean stems after indole-3-acetic acid or acid treatment, although they did not show endo-1,4,-β-glucanase activity responsible for xyloglucan degradation. The small decrease in xyloglucan molecular weight at the early stage of the treatment may be sufficient to loosen cellulose microfibrils so as to expand or elongate whole cells during growth, if xyloglucan is not actively deposited on their inner surface. If there is high xyloglucan deposition following supraoptimal auxin treatment, the strain may necessitate induction of more endo-1,4-β-glucanase activity to maintain a constant rate of cell expansion, leading consequently to marked decrease in xyloglucan molecular weight. That excess strain may possibly trigger induction of new endoglucanase production is consistent with an observed lag period (6–12 h) for the induction of endoglucanase activity (Fig. 10.5 and Table 10.2), while there is continuous maintenance of high xyloglucan synthase activity after auxin treatment (Hayashi and Maclachlan, 1984b). Supraoptimal auxin may prevent a normal decline in the gene expression of the system for xyloglucan synthesis, and then, high tension of cell walls as a result of enhanced xyloglucan deposition induces endo-1,4-β-glucanase. Or, xyloglucan fragments derived from the hydrolysis of deposited xyloglucan with endogenous endoglucanases may induce new endo-1,4-β-glucanase activity, since fragments have recently been found to promote the elongation of pea stems probably by activation of endo-1,4-β-glucanase activity (McDougall and Fry, 1990). Thus, the effect of auxin on xyloglucan metabolism may be related to altered extensibility of cell walls, which is consistent with the concept that xyloglucan degradation is required for loosening of cellulose microfibrils necessary for cell expansion during growth.

Does xyloglucan metabolism affect cell shape?

The growth response of the apical 5 mm (elongating region) of intact pea epicotyl was examined after treatment with supraoptimal auxin (1 mM 2,4-dichlorophenoxyacetic acid) with or without the ethylene inhibitor aminoethoxyvinylglycine (AVG) during growth. The presence of AVG prevented to a great extent auxin-induced lateral expansion and brought about elongation almost equalling that of the controls. Elongation was dependent on the concentration of AVG (Fig. 10.7). Swelling levels (X-section) measured as fresh weight per unit length were markedly reduced.

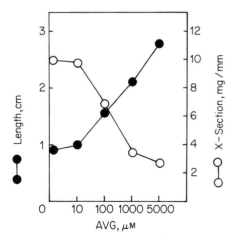

Fig. 10.7. Effects of the ethylene inhibitor AVG on the lateral expansion of pea epicotyls after treatment with supraoptimal auxin (1 mM 2,4-dichlorophenoxyacetic acid) in 48 h.

AVG did not alter xyloglucan levels due to auxin, even though AVG changed the growth from swelling to elongation. During growth, the synthesis of cellulose and xyloglucan occurred in proportion to each other at the same rates between auxin-treated tissue and auxin-plus AVG-treated tissue. Net bound xyloglucan and total extractable endo-1,4-β-glucanase activity increased and xyloglucan molecular weight decreased to 50 000 at 48 h after auxin treatment with or without AVG. Thus, the xyloglucan in auxin-treated tissue is metabolized in the same way (Hayashi et al., 1984) regardless of whether AVG is present or not.

Auxin-induced lateral expansion of pea growing regions was inhibited by AVG, and this supports the concept that expansion could be due to ethylene effect because AVG is a specific inhibitor of ethylene production (Yu and Yang, 1979). However, there is no difference in xyloglucan and cellulose metabolism in tissues swollen (auxin only) or elongated (auxin plus AVG). However, even though the mechanism for the deposition of these two polysaccharides is very similar, the morphogenetic effects of these two treatments clearly differ.

Microtubules may control the orientation of cellulose microfibrils; e.g. the effect of ethylene on pea cells is believed to be due to cellulose microfibril orientation as a result of microtubule reorganization (Steen and Chadwick, 1981). Roberts et al. (1984) found microtubule orientation to change mainly to a longitudinal direction in a short time after ethylene treatment and proposed that the mechanism for this is the realignment of transversely-oriented microtubules. This is consistent with our data, indicating microtubules to control the orientation of microfibrils.

Auxin-induced endo-1,4-β-glucanases have been proposed (Hayashi et al., 1984) as wall-loosening enzymes which weaken the cellulose–xyloglucan network, thus allowing expansion of the entire cell during growth. The levels of xyloglucan synthase declined during normal growth and disappeared after ethylene treatment, but were maintained in auxin (Hayashi and Maclachlan, 1984b). During elongation or expansion of pea epicotyl cells, supraoptimal auxin always enhanced by as much as four times the deposition of xyloglucan in cell walls, where cellulose microfibrils had been newly deposited on the inner surface. This may possibly cause the number of xyloglucan cross-links to the cellulose framework to increase. Following this, auxin-induced endo-1,4-β-glucanase activity is probably responsible for the degradation of such newly formed xyloglucan cross-links in order to loosen microfibrils. The present data indicate that the concurrent action of both xyloglucan degradation and cellulose synthesis in the cell wall is required for the auxin-induced enlargment of cells during growth, but that enhanced wall-loosening, as judged by increase in xyloglucan turnover, is not a factor in determining cell shape.

Feedback regulation of auxin-stimulated growth

Xyloglucan fragments, derived from the hydrolysis of xyloglucan with endo-1,4-β-glucanase, could block the action of auxin on elongation of pea stem segments (York et al., 1984). This effect was specifically observed for the α-L-fucosyl-(1→2)-β-D-galactosyl residue containing oligosaccharide fragments. Pentasaccharide (fucose/galactose/xylose/glucose, 1:1:1:2), nonasaccharide (fucose/galactose/xylose/glucose, 1:1:3:4) and undecasaccharide (fucose/galactose/xylose/glucose, 2:2:3:4) including human milk trisaccharide α-L-fucosyl-(1→2)-β-D-galactosyl-(1→4)-D-glucose are all effective as antiauxin (McDougall and Fry, 1989a, b). The oligosaccharides inhibited the auxin effect at 10^{-9}–10^{-8} M, although 2,4-dichlorophenoxyacetic acid was used at 10^{-6} M. The oligosaccharide fragment may thus serve as feedback inhibitors of auxin-stimulated growth, whereas an auxin may promote plant growth by activation and induction of endo-1,4-β-glucanase activity. The inhibitory effect was observed only in a certain range of oligosaccharide concentrations at 10^{-9}–10^{-8} M, and high concentrations of oligosaccharides (above 10^{-6} M) might be ineffective. Essentially the same was noted for the action of auxin since high concentrations of auxin inhibit cell elongation.

Farkas and Maclachlan (1988) reported that xyloglucan oligosaccharides at 5×10^{-6}–2×10^{-4} M stimulate pea endo-1,4-β-glucanase activity. Longer chain lengths are the most effective and α-L-fucosyl-(1→2)-β-D-galactosyl residue is not essential for stimulation. Cello-oligosaccharides at similar concentrations were inhibitory, e.g. activity at different concentrations of cellopentaose yielded Dixon plots with identical apparent Ki values (1.5 mM) for pea endo-1,4-β-glucanase (Hayashi et al., 1984). The stimulatory effect was substrate- and enzyme-specific, being observed only when xyloglucan was used as the substrate and pea endo-1,4-β-glucanase as the enzyme; e.g. it was not detected when carboxymethylcellulose was used as the substrate or fungal endo-1,4-β-glucanase (Trichoderma viride cellulase) was used with xyloglucan for the substrate. Xyloglucan fragments thus appear to act as feedback activators for plant endo-1,4-β-glucanase.

McDougall and Fry (1990) recently reported that xyloglucan oligosaccharides themselves promote the elongation of pea stem segments. The promotion seems due to the activation of wall-bound endo-1,4-β-glucanase activity with oligosaccharides and was noted at 10^{-7}–10^{-5} M of the oligosaccharides. At higher concentrations the promotion was less, although the activation of endo-1,4-β-glucanase in vitro was observed at higher concentrations (more than 5×10^{-6} M). Thus apparently, xyloglucan oligosaccharides function as an auxin since high concentrations of auxin inhibit cell elongation.

Figure 10.8 shows a scheme for auxin-stimulated growth with regulation of xyloglucan degradation. The action of auxin on cell enlargement, involving xyloglucan degradation, is also regulated by xyloglucan fragments, showing that new regulatory molecules should be involved in plant cell growth.

Future prospects

Xyloglucan is obviously the most probable component for loosening the cellulose micro-

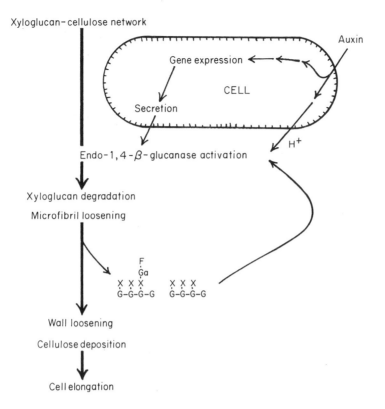

Fig. 10.8 Scheme for auxin-stimulated growth with regulation of xyloglucan degradation.

fibril network since it is always associated with cellulose microfibrils in plant cell walls. However, the regulation of cellulose synthase is not known, e.g. no one has so far succeeded in cellulose synthesis in a cell-free system (Delmer, 1987). The action of xyloglucan oligosaccharides which inhibit auxin-stimulated elongation or promote elongation at different concentrations of oligosaccharides has not been well characterized. However, we hope that the present study provides some clarification of the function of the primary wall for regulating cell elongation and expansion in higher plants.

Acknowledgements

I thank Drs D.P. Delmer, G. Maclachlan, A. Camirand and N. Sakurai for critically reading the manuscript. I also thank Ms C. Ohsumi for numerous useful discussions.

References

Bal, A.K., Verma, D.P.S., Byrne, H. and Maclachlan, G.A. (1976). Subcellular localization of cellulases in auxin-treated pea. *J. Cell Biol.* **69**, 97–105.

Burg, S.P. and Burg, E.A. (1966). The interaction between auxin and ethylene and its role in plant growth. *Proc. Natl Acad. Sci. U.S.A.* **55**, 262–269.

Burstom, H.G., Uhrstrom, I. and Olausson, B. (1970). Influence of auxin on Young's modulus in stems and roots of *Pisum* and the theory of changing the modulus in tissues. *Physiol. Plant.* **23**, 1223–1233.

Byrne, H., Christou, N.V., Verma, D.P.S. and Maclachlan, G.A. (1975). Purification and characterization of two cellulases from auxin-treated pea epicotyls. *J. Biol. Chem.* **250**, 1012–1018.

Cassab, G.I. and Varner, J.E. (1988). Cell wall proteins. *Annu. Rev. Plant Physiol. Plant Mol. Biol.* **39**, 321–353.

Cleland, R. (1967). Extensibility of isolated cell walls: measurement and changes during cell elongation. *Planta* **74**, 197–209.

Cleland, R. (1973). Auxin-induced hydrogen ion

excretion from *Avena* coleoptiles. *Proc. Natl Acad. Sci. U.S.A.* **70**, 3092–3093.

Cooper, J.B. and Varner, J.E. (1984). Cross-linking of soluble extensin in isolated cell walls. *Plant Physiol.* **76**, 414–417.

Darvill, A., McNeil, M., Albersheim, P. and Delmer, D.P. (1980). The primary cell walls of flowering plants. In *The Biochemistry of Plants* (ed. N.E. Tolbert), vol. 1, pp. 91–162. Academic Press, New York.

Delmer, D.P. (1987). Cellulose biosynthesis. *Annu. Rev. Plant Physiol.* **38**, 259–290.

Evans, M.L. (1974a). Rapid responses to plant hormones. *Annu. Rev. Plant Physiol.* **25**, 195–223.

Evans, M.L. (1974b). Evidence against the involvement of galactosidase or glucosidase in auxin- or acid-stimulated growth. *Plant Physiol.* **54**, 213–215.

Fan, D.F. and Maclachlan, G.A. (1966). Control of cellulase activity by indoleacetic acid. *Can. J. Bot.* **44**, 1025–1034.

Farkas, V. and Maclachlan, G. (1988). Stimulation of pea 1,4-β-glucanase activity by oligosaccharides derived from xyloglucan. *Carbohydr. Res.* **184**, 213–219.

Fry, S.C. (1982). Phenolic components of the primary cell wall: feruloylated disaccharides of D-galactose and L-arabinose from spinach polysaccharide. *Biochem J.* **203**, 493–504.

Fry, S.C. (1987). Formation of isodityrosine by peroxidase isozymes. *J. Exp. Bot.* **38**, 853–862.

Gilkes, N.R. and Hall, M.A. (1977). The hormonal control of cell wall turnover in *Pisum sativum* L. *New Phytol.* **78**, 1–15.

Haigler, C.H. and Benziman, M. (1982). Biogenesis of cellulose I microfibrils occurs by cell-directed self-assembly in *Acetobacter xylinum*. In *Cellulose and Other Natural Polymer Systems* (ed. R.M. Brown, Jr), pp. 273–297. Plenum, New York.

Harvey, I.M., Hartley, R.D., Harris, P.J. and Curzon, E.H. (1986). Linkage of *p*-coumaroyl and feruloyl groups to cell-wall polysaccharides of barley straw. *Carbohydr. Res.* **148**, 71–85.

Hayashi, T. (1989). Xyloglucans in the primary cell wall. *Annu. Rev. Plant Physiol. Plant Mol. Biol.* **40**, 139–168.

Hayashi, T. and Maclachlan, G. (1984a). Pea xyloglucan and cellulose. I. Macromolecular organization. *Plant Physiol.* **75**, 596–604.

Hayashi, T. and Maclachlan, G. (1984b). Pea xyloglucan and cellulose. III. Metabolism during lateral expansion of pea epicotyl cells. *Plant Physiol.* **76**, 739–742.

Hayashi, T. and Maclachlan, G. (1986). Pea cellulose and xyloglucan: biosynthesis and biodegradation.

In *Cellulose: Structure, Modification, and Hydrolysis* (eds R.A. Young and R.M. Rowell), pp. 67–76. Wiley, New York.

Hayashi, T., Wong, Y.-S. and Maclachlan, G. (1984). Pea xyloglucan and cellulose. II. Hydrolysis by pea endo-1,4-β-glucanases. *Plant Physiol.* **75**, 605–610.

Hayashi, T., Marsden, M.P.F. and Delmer, D.P. (1987). Pea xyloglucan and cellulose. V. Xyloglucan-cellulose interactions *in vitro* and *in vivo*. *Plant Physiol.* **83**, 384–389.

Inouhe, M., Yamamoto, R. and Masuda, Y. (1984). Auxin-induced changes in the molecular weight distribution of cell wall xyloglucans in *Avena* coleoptiles. *Plant Cell Physiol.* **25**, 1341–1351.

Ishii, T. and Hiroi, T. (1990). Isolation and characterization of feruloylated arabinoxylan oligosaccharides from bamboo shoot cell-walls. *Carbohydr. Res.* **196**, 175–183.

Ishii, T., Hiroi, T. and Thomas, J.R. (1990). Feruloylated xyloglucan and *p*-coumaroyl arabinoxylan oligosaccharides from bamboo shoot cell-walls. *Phytochemistry* **29**, 1999–2003.

Jacobs, M. and Ray, P.M. (1975). Promotion of xyloglucan metabolism by acid pH. *Plant Physiol.* **56**, 373–376.

Johnson, K.D., Daniels, D., Dowler, M. and Rayle, D.L. (1974). Activation of *Avena* coleoptile cell wall glycosidases by hydrogen ions and auxin. *Plant Physiol.* **53**, 224–228.

Kang, B.G. and Burg, S.P. (1971). Rapid change in water flux induced by auxins. *Proc. Natl Acad. Sci. U.S.A.* **68**, 1730–1733.

Kato, A., Azume, J. and Koshijima, T. (1987). Isolation and identification of a new feruloylated tetrasaccharide from bagasse lignin–carbohydrate complex containing phenolic acid. *Agric. Biol. Chem.* **51**, 1691–1693.

Kato, Y. and Matsuda, K. (1981). Occurrence of soluble and low molecular weight xyloglucan and its origin in etiolated mung bean hypocotyls. *Agric. Biol. Chem.* **45**, 1–8.

Kato, Y. and Matsuda, K. (1985). Xyloglucan in the cell walls of suspension-cultured rice cells. *Plant Cell Physiol.* **26**, 437–445.

Kato, Y. and Nevins, D.J. (1984). Enzymic dissociation of *Zea* shoot cell wall polysaccharides. II. Dissociation of (1→3),(1→4)-β-D-glucan by purified (1→3),(1→4)-β-D-glucan 4-glucanohydrolase from *Bacillus subtilis*. *Plant Physiol.* **75**, 745–752.

Kato, Y. and Nevins, D.J. (1985). Isolation and identification of *O*-(5-*O*-feruloyl-α-L-arabinofuranosyl)-(1→3)-*O*-β-D-xylopyranosyl-(1→4)-D-xylopyranose as a component of *Zea* shoot cell-walls. *Carbohydr. Res.* **137**, 139–150.

Keegstra, K., Talmadge, K.W., Bauer, W.D. and Albersheim, P. (1973). The structure of plant cell walls. III. A model of the walls of suspension-cultured sycamore cells based on the interconnections of the macromolecular components. *Plant Physiol.* **54**, 105–108.

Kiefer, L.L., York, W.S., Albersheim, P. and Darvill, A.G. (1990). Structural characterization of an arabinose-containing heptasaccharide enzymically isolated from sycamore extracellular xyloglucan. *Carbohydr. Res.* **197**, 139–158.

Komae, K. and Misaki, A. (1989). Isolation and characterization of the gel-forming polygalacturonide from seeds of *Ficus awkeotsang*. *Agric. Biol. Chem.* **53**, 1237–1245.

Komae, K., Sone, Y., Kakuta, M. and Misaki, A. (1989). Isolation of pectinesterase from *Ficus awkeotsang* seeds and its implication in gel-formation of the awkeotsang polygalacturonide. *Agric. Biol. Chem.* **53**, 1247–1254.

Komae, K., Sone, Y., Kakuta, M. and Misaki, A. (1990). Purification and characterization of pectinesterase from *Ficus awkeotsang*. *Agric. Biol. Chem.* **54**, 1469–1476.

Konar, R.N. and Stanley, R.G. (1969). Wall-softening enzymes in the gyneocium and pollen of *Hemerocallis fulva*. *Planta* **84**, 304–310.

Labavitch, J.M. and Ray, P.M. (1974a). Turnover of cell wall polysaccharides in elongating pea stem segments. *Plant Physiol.* **53**, 669–673.

Labavitch, J.M. and Ray, P.M. (1974b). Relationship between promotion of xyloglucan metabolism and induction of elongation by indoleacetic acid. *Plant Physiol.* **54**, 449–502.

Labrador, E. and Nevins, D.J. (1989). An exo-β-D-glucanase derived from *Zea* coleoptile walls with a capacity to elicit cell elongation. *Physiol. Plant.* **77**, 479–486.

Lorences, E.P. and Zarra, I. (1987). Auxin-induced growth in hypocotyl segments of *Pinus pinaster* Aiton. Changes in molecular-weight distribution of hemicellulosic polysaccharides. *J. Exp. Bot.* **38**, 960–967.

Lorences, E.P., Suárez, L. and Zarra, I. (1987a). Hypocotyl growth of *Pinus pinaster* seedlings. Changes in α-cellulose, and in pectic and hemicellulosic polysaccharides. *Physiol. Plant.* **69**, 461–465.

Lorences, E.P., Suárez, L. and Zarra, I. (1987b). Hypocotyl growth of *Pinus pinaster* seedlings. Changes in the molecular weight distribution of hemicellulosic polysaccharides. *Physiol. Plant.* **69**, 466–471.

Masuda, Y. and Yamamoto, R. (1970). Effect of auxin on β-1,3-glucanase activity in *Avena* coleoptile. *Develop. Growth & Differentiation* **11**, 287–296.

McDougall, G.J. and Fry, S.C. (1989a). Structure-activity relationship for xyloglucan oligosaccharides with antiauxin activity. *Plant Physiol.* **89**, 883–887.

McDougall, G.J. and Fry, S.C. (1989b). Anti-auxin activity of xyloglucan oligosaccharides: the role of groups other than the terminal α-L-fucose residue. *J. Exp. Bot.* **40**, 233–238.

McDougall, G.J. and Fry, S.C. (1990). Xyloglucan oligosaccharides promote growth and activate cellulase: evidence for a role of cellulase in cell expansion. *Plant Physiol.* **93**, 1042–1048.

Misaki, A., Sone, Y., Nagata, A., Komae, K. and Shibata, S. (1988). Immunochemical specificities and histochemical application of the xyloglucan-recognizing antibodies, derived from its oligosaccharides. In *Abstract for XIVth International Carbohydrate Symposium*, Stockholm, pp. 359. XIVth International Carbohydrate Symposium Committee.

Mohnen, D., Shinshi, H., Felix, G. and Meins, Jr F. (1985). Hormonal regulation of β1,3-glucanase messenger RNA level in cultured tobacco tissues. *EMBO Journal* **4**, 1631–1635.

Moore, P.J. and Staehelin, L.A. (1988). Immunogold localization of the cell-wall-matrix polysaccharides rhamnogalacturonan I and xyloglucan during cell expansion and cytokinesis in *Trifolium pratense* L.; implication for secretory pathways. *Planta* **174**, 433–445.

Nishitani, K. and Masuda, Y. (1982). Acid pH-induced structural changes in cell wall xyloglucans in *Vigna angularis* epicotyl segments. *Plant Sci. Lett.* **28**, 87–94.

Nishitani, K. and Masuda, Y. (1983). Auxin-induced changes in the cell wall xyloglucans: effects of auxin on the two different subfractions of xyloglucans in the epicotyl cell wall of *Vigna angularis*. *Plant Cell Physiol.* **24**, 345–355.

Ogawa, K., Hayashi, T. and Okamura, K. (1990). Conformational analysis of xyloglucans. *Int. J. Biol. Macromol.* **12**, 218–222.

Ray, P.M. (1973). Regulation of β-glucan synthethase activity by auxin in pea stem tissue. I. Kinetic aspects. *Plant Physiol.* **51**, 601–608.

Revilla, G. and Zarra, I. (1987). Changes in the molecular weight distribution of the hemicellulosic polysaccharides from rice coleoptiles growing under different conditions. *J. Exp. Bot.* **38**, 1818–1825.

Roberts, I.N., Lloyd, C.W. and Roberts, K. (1984). Ethylene-induced microtubule reorientations:

mediation by helical arrays. *Planta* **164**, 439–447.

Sassen, M.M.A. (1965). Breakdown of the plant cell wall during the cell-fusion process. *Acta Bot. Neerl.* **14**, 165–196.

Steen, D.A. and Chadwick, A.V. (1981). Ethylene effects in pea stem tissue. Evidence of microtubule mediation. *Plant Physiol.* **67**, 460–466.

Tanimoto, E. and Masuda, Y. (1968). Effect of auxin on cell wall degrading enzymes. *Physiol. Plant.* **21**, 820–826.

Terry, M.E. and Bonner, B.A. (1980). An examination of centrifugation as a method of extracting an extracellular solution from peas, and its use for the study of indoleacetic acid-induced growth. *Plant Physiol.* **66**, 321–325.

Terry, M.E., Jones, R.L. and Bonner, B.A. (1981). Soluble cell wall polysaccharides released from pea stems by centrifugation. I. Effect of auxin. *Plant Physiol.* **68**, 531–537.

Theologis, A. (1986). Rapid gene regulation by auxin. *Annu. Rev. Plant Physiol.* **37**, 407–438.

Uhrström, I. (1969). The time effect of auxin and calcium on growth and elastic modulus in hypocotyls. *Physiol. Plant.* **22**, 271–287.

Verma, D.P.S., Maclachlan, G.A., Byrne, H. and Ewings, D. (1975). Regulation and *in vitro* translation of messenger ribonucleic acid for cellulase from auxin-treated pea epicotyls. *J. Biol. Chem.* **250**, 1019–1026.

Yoda, S. and Asida, J. (1961). Effect of auxin on the osmotic value in pea stem sections. *Nature* **192**, 577–578.

York, W.S., Darvill, A.G. and Albersheim, P. (1984). Inhibition of 2,4-dichlorophenoxyacetic acid-stimulated elongation of pea stem segments by a xyloglucan oligosaccharide. *Plant Physiol.* **75**, 295–297.

Yu, Y.B. and Yang, S.F. (1979). Auxin-induced ethylene production and its inhibition by amino-ethoxyvinylglycine and cobalt ion. *Plant Physiol.* **64**, 1074–1077.

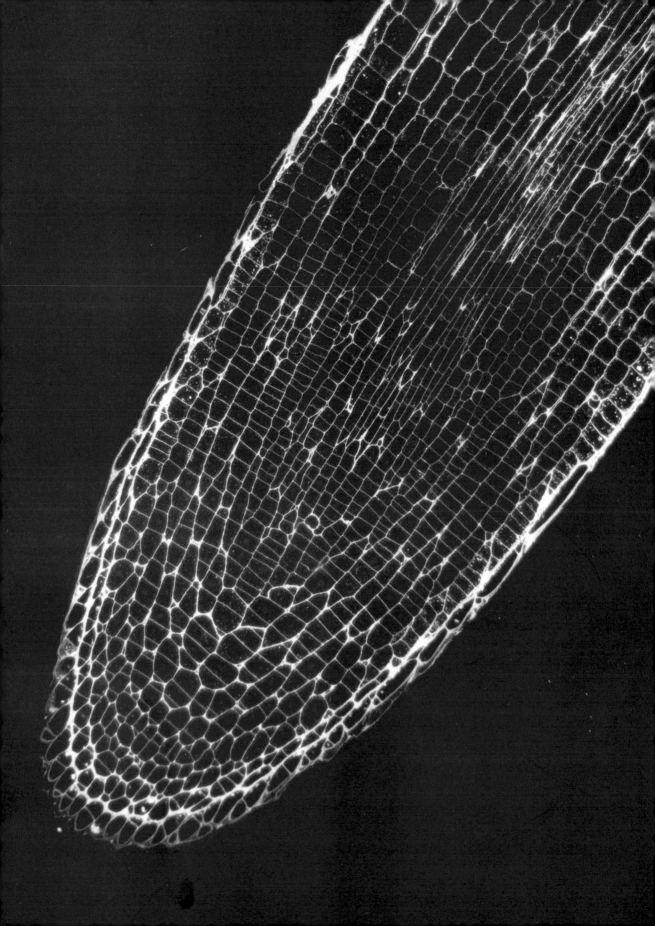

DIRECTIONAL CELL EXPANSION

Plant cells do not move relative to one another but the co-ordination of cell division and cell expansion produces a flow of cells out from the region of high mitotic activity; thus the apex advances by generating a stream of non-moving cells in its wake. In apical meristems this flow is seen within cell files which originate in the zone of cell division. As the balance between expansion and division shifts, so the size of cells changes. At the proximal end of the file where stem-cells are active, cells with a high probability of division remain small and densely cytoplasmic; further away, there is less mitotic activity and elongated, vacuolated cells occur; at the distal end of the file where growth has ceased, mature, differentiated cells form, probably with characteristic elaborations of the secondary wall. The factors that affect the balance between division and expansion therefore play an important part in morphogenesis. More will be understood about this when we have satisfactory models for cell cycle control and knowledge of the internal and external cues that affect it. Control of the division plane is discussed in a later section; the present section covers directional cell expansion.

Single cells in suspension elongate satisfactorily and so the generation of asymmetric cell shape *per se* is evidently inherent within the cell, rather than the tissue. It is the relationship between the cortical cytoskeleton and the non-random structure of the wall which ensures that cell expansion is channelled in a particular direction. Whereas the foregoing section dealt primarily with the structure of the wall, this group of chapters is more concerned with cell expansion from the cytoplasmic side. Parallelism between cortical microtubules and cellulose microfibrils was reviewed in the previous book on the plant cytoskeleton and in several subsequent reviews. It is not intended, therefore, to re-rehearse the evidence and arguments in favour of parallelism but to move on to cases either where parallelism has subsequently been shown to be exemplary, or where it can be seen to be under physiological control. Such controls are often missing or difficult to recreate for cells in culture, but understanding how they affect the shaping process is a crucial step to understanding how growth is co-ordinated *in vivo* as well as *in vitro*.

Chapters in this section are concerned with the way in which light (see also Chapter 20 by Wada and Murata), gravity and plant growth regulators influence the cytoskeleton and cell shape. In the final chapter in this section, Chapter 14, Seagull and Falconer describe what happens in the mature phase of cell growth – the formation of tracheary elements *in vitro* providing a particularly clear example of how the cytoskeleton influences the pattern of wall sculpturing.

11. REGULATION OF CELL EXPANSION

Ulrich Kutschera

Botanisches Institut, Universität Bonn,
Venusbergweg 22, D-5300 Bonn 1, Germany

Introduction

Axial plant organs such as stems or coleoptiles grow by cell enlargement, which can be defined as the irreversible deformation of the cell wall under the force of turgor pressure. The yielding of the wall leads to water uptake into the vacuole which brings about an increase in cell volume. Due to the anisotropic properties of the walls, the cells and thus the entire organ grow almost exclusively in length, whereas the diameter remains nearly constant.

Although the mechanism of cell elongation has extensively been investigated by several generations of plant physiologists, many basic questions have yet to be answered. In the present chapter the growth of single cells and multicellular plant organs is described at the biophysical level. Then, recent results on the mechanism of auxin-induced growth of excised segments and the light-induced inhibition of stem elongation of intact plant organs are summarized.

Biophysical basis of irreversible cell enlargement

The typical plant cell consists of cytoplasm bounded by two semipermeable membranes (plasmalemma, tonoplast) which surrounds a large central vacuole (Fig. 11.1A–C). The vacuole occupies most of the volume of the cell and is filled with an aqueous solution of organic and inorganic substances.

The protoplast of the plant cell is surrounded by a cell wall. The wall of the growing cell (primary wall) is a rigid structure that has the capability to undergo both elastic (reversible) and plastic (irreversible) deformation. The primary wall is composed of approximately 90% polysaccharides and 10% glycoproteins. It is a biphasic structure, consisting of relatively rigid cellulosic microfibrils embedded in a gel-like matrix. This latter is composed of hemicelluloses, pectic substances and glycoproteins.

The notion that a vacuolated plant cell can be compared with an osmometer emerged during the latter part of the nineteenth century, mainly through the work of W. Pfeffer and H. De Vries. If a cell is placed in a hypertonic solution, water is withdrawn from the vacuole and the cell becomes plasmolysed. If the external medium is replaced by a hypotonic solution, water is taken up at a rate proportional to the difference in the osmotic pressure of the vacuole π_i (MPa) and the external medium π_o (Fig. 11.1A). As a consequence of this osmotic water uptake, a hydrostatic pressure is exerted by the protoplast against the cell wall. This multidirectional hydrostatic pressure (the excess over 1 atmosphere) of the vacuolar solutes is called turgor pressure P_v (MPa) (Sachs, 1882). The water uptake leads to a volume increase of the vacuole and thus to extension of the cell wall (Fig. 11.1B). In the turgid cell the wall is elastically stretched and hence exerts an opposite counter pressure on the vacuole. This wall stress (or wall pressure P_w) is equal in magnitude to P_v and tends to force water out of the cell. In a fully turgid, non-growing cell the osmotic water influx and the amount of water that is pressed out of the cell are equal and the water potential of the cell ψ_i (MPa) is zero, i.e.

$$\psi_i = P_v - \pi_i \tag{1}$$

The Cytoskeletal Basis of Plant Growth and Form   ISBN 0–12–453770–7

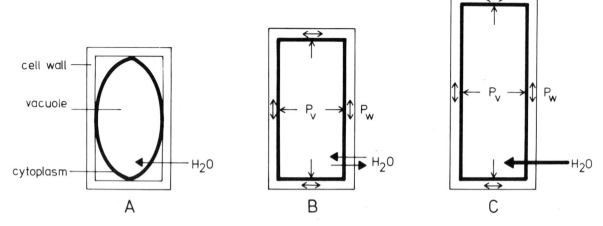

Fig. 11.1. Scheme illustrating the water uptake of a plasmolysed cell (A), a turgid, non-growing cell (B), and a turgid cell that is growing (C). P_v = turgor pressure; P_w = wall stress.

Since $P_v = \pi_i$ it follows that $\psi_i = 0$.

According to equation (1), P_v and π_i are interdependent parameters, i.e. the osmotic pressure of the cell is a measure of the maximum turgor pressure which can be exerted on the cell wall (Lockhart, 1965b).

Sachs (1882) discovered that elongation growth can only take place as long as the cells are turgid. He concluded that turgor pressure is a prerequisite for cell elongation. This observation led to the notion of turgor-driven cell elongation. Since the classical investigations of Heyn (1931, 1940) it has been generally accepted that cell elongation is caused by the irreversible (plastic) yielding of the cell wall (Lockhart, 1965a, b; Cleland, 1971; Taiz, 1984; Cosgrove, 1986; Ray, 1987).

The growth process of a single cell can thus be summarized as follows (Fig. 11.1C). Extension of the turgid cell (Fig. 11.1B) is initiated when load-bearing regions in the wall yield. This brings about a reduction in P_w (and P_v) and thus a negative ψ_i (see equation (1)). As a result, water is taken up into the vacuole and the volume of the cell increases. This influx of water is governed by the following equation:

$$dV/dt = L\Delta\psi = L(\psi_o - \psi_i) \qquad (2)$$

where dV/dt $(m^3\,s^{-1})$ is the rate of water uptake, L $(m^3\,s^{-1}\,MPa^{-1})$ is the cell membrane's

hydraulic conductance (conductivity times membrane area), and $\Delta\psi$ (MPa) is the difference in the water potential of the medium (ψ_o) and the cell (ψ_i). Since at atmospheric pressure $\psi_o = -\pi_o$, equations (1) and (2) can be combined to eliminate water potential:

$$dV/dt = L(\Delta\pi - P_v) \qquad (3)$$

where $\Delta\pi$ (MPa) is the difference between π_i and π_o.

Lockhart (1965a) contributed the idea that P_v has to exceed a certain minimum value before irreversible cell elongation occurs. He proposed the following equation:

$$dV/dt = m(P_v - Y) \qquad (4)$$

This simple function states that the growth rate (dV/dt) depends on the turgor pressure (P_v) in excess of a critical turgor Y (MPa), the 'yield threshold' (i.e. the turgor pressure below which no extension can take place) and on the yielding coefficient m $(m^3\,s^{-1}\,MPa^{-1})$.

Lockhart (1965a) defined m as a property of the cell wall expressing its rate of irreversible flow under stress and called this coefficient 'wall extensibility'. This terminology has led to some confusion in the literature since the term extensibility had already been introduced by Heyn in 1931 and defined in a different way.

Heyn (1931) determined the elastic (rever-

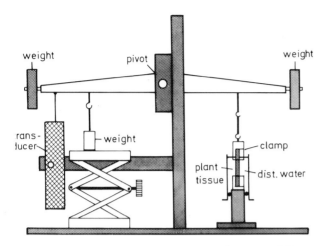

Fig. 11.2. Apparatus for measuring the extensibility of living organ segments using a linear-displacement transducer. A segment (plant tissue) was fixed between the upper and lower clamps (distance 10 mm) and submerged in distilled water. The tissue was stretched by application of a weight on the left arm of the lever.

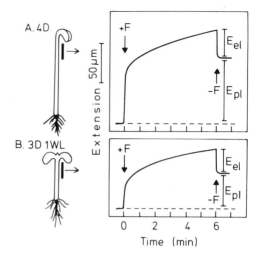

Fig. 11.3. Elastic (E_{el}) and plastic (E_{pl}) *in vivo* extensibility of the subapical region of hypocotyls from sunflower seedlings that were either grown for 4 days in darkness (A, 4D) or for 3 days in darkness and 1 day in white light (B, 3D 1WL; white light: 2.5 klux, see Kutschera, 1990). Sections, 18 mm in length, were excised from the hypocotyls and cut longitudinally into two equal halves. One segment was fixed between the clamps of the apparatus shown in Fig. 11.2, submerged in distilled water, and subjected to a constant force of 0.098 N (corresponding to a weight of 10 g) for 6 min (+ F). After removal of the force (− F) E_{el} and E_{pl} can be determined as indicated.

sible) and plastic (irreversible) longitudinal extensibility (E_{el} and E_{pl}, respectively) of the cell walls by subjecting living (or plasmolysed) oat coleoptiles to a constant external force (bending or stretching of the organ). A constant-stress extensiometer to determine E_{el} and E_{pl} is shown in Fig. 11.2 and typical measurements are given in Fig. 11.3A, B. Kutschera and Schopfer (1986a, b) have recently shown that stretching of living and frozen-thawed (killed) tissue gives essentially the same results. Thus, E_{el} and E_{pl} can be determined using fully turgid, metabolically active plant organs (*in-vivo* extensibility, Kutschera and Schopfer, 1986b). It is evident that E_{el} and E_{pl} (units: μm 10 g^{-1} 6 min^{-1}) provide no direct measurement of the yielding coefficient m as defined in equation (4). They are operational quantities which are a relative measure of the mechanical properties of the cell walls. Since, however, E_{pl} is defined as the ability of the walls to become irreversibly extended under constant stress, this parameter may be closely related to m.

Although alternative techniques to determine the mechanical properties of the cell walls are available (Taiz, 1984), as yet no method has been developed to measure m directly

(Cosgrove, 1986).

During steady growth of the cell, the rate of water influx and the rate of cell wall yielding must be equal, i.e. equations (3) and (4) can be combined (Lockhart, 1965a):

$$dV/dt = \frac{Lm}{L + m}(\Delta\pi - Y) \qquad (5)$$

Two limiting cases can be distinguished. If L is large compared to m, equation (5) reduces to equation (4), with $\Delta\pi$ substituted for P_v (yield coefficient-limited or extensibility-controlled growth). On the other hand, if L is small compared to m, equation (5) reduces to equation (3) with Y substituted for P_v (conductance, or water uptake, limited growth).

Because of water uptake during growth (Fig. 11.1C) the concentration of solutes in the vacuole (π_i) decreases. This dilution of solutes results in a corresponding reduction in $\Delta\pi$ and thus growth rate (equation (5)). In order to maintain π_i the growing cell must take up (or produce) osmotically active substances. The term 'osmoregulation' has been used to denote the processes and mechanisms by which π_i of the cell is regulated (Hellebust, 1976). It is obvious that without osmoregulation no long-term growth is possible, i.e. cell extension is dependent on the maintenance of a sufficient internal osmotic pressure.

Growth of multicellular plant organs

As pointed out above, equation (5) applies only to single isolated cells (e.g. the internode of *Nitella*).

Multicellular axial plant organs are composed of a variety of distinct types of cells, organized into tissues and tissue systems. One has to distinguish between three tissue systems: dermal, vascular and fundamental (or ground) system. The dermal tissue system of the growing stem is the epidermis. The outer wall of the epidermal cells (and in many organs also the inner epidermal wall) is much thicker than the walls of the internal tissues (see Fig. 11.4).

Moreover, the walls of the peripheral cells often contain layers of both transverse and longitudinally oriented cellulose microfibrils. In contrast, the cell walls of the fundamental tissue system (parenchyma, i.e. cortex and pith) are very thin and the orientation of cellulose microfibrils is predominantly transverse (Kutschera, 1989). The walls of the vascular tissue system (phloem, xylem) are extensible and do not have a large effect on the mechanical properties of the surrounding parenchyma tissue.

Thus, a multicellular organ consists of tissues that differ with respect to the thickness, architecture and extensibility of their cell walls. Because of the turgor pressure of the thin-walled extensible parenchyma cells, the less extensible peripheral walls are kept under longitudinal tension. The contracting force exerted by the stretched peripheral cell walls leads to a compression of the internal tissues. Hence, a mutual stress condition (tension and compression) exists between the tissues of the turgid shoot. Sachs (1882) described this mechanical property of the organ as 'tissue tension'. The term 'tissue stress', however, is more accurate since it comprises both tension (of the peripheral walls) and compression (of

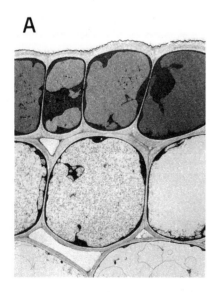 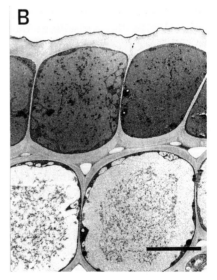

Fig. 11.4. Electron micrographs of transverse sections through the peripheral two cell layers of the subapical region of sunflower hypocotyls. The seedlings were either grown for 4 days in darkness (A) or for 3 days in darkness and for 1 day in continuous white light (B; 2.5 klux, see Kutschera, 1990). × 590; scale bar = 10 µm.

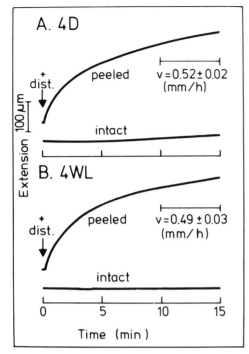

Fig. 11.5. Time course of extension of peeled and intact sections cut from sunflower hypocotyls that were either grown for 4 days in darkness (A, 4D) or for 4 days in continuous white light (B, 4WL, 2.5 klux, see Kutschera, 1990). Sections, 18 mm in length, were excised, peeled (or left intact) and fixed between the clamps of the apparatus shown in Fig. 11.2. After addition of distilled water (+ dist.) the extension of the sections (without application of a weight) was measured. The rates of length change (v) in the approximately linear range of the curves were calculated and given as means ($\pm$ SE) of six measurements each.

the internal tissues) (Kutschera, 1989).

The mutual stresses can readily be demonstrated by separation of the outer and inner cell layers of the stem (peeling of the epidermis). Upon isolation, the epidermis contracts by up to 20%, whereas the inner tissue cylinder rapidly expands owing to water uptake of the parenchyma cells (Fig. 11.5A, B). Thus, the growth process of the organ is limited by the extension of the peripheral cell walls and growth must be regarded as a co-ordinated interaction of functionally different tissues (Kutschera et al., 1987). The compressed parenchyma tissues provide the driving force for growth, whereas the yielding (plastic deformation) of the peripheral walls determines the rate of elongation. The whole organ can thus be regarded as a single, giant cell (Fig. 11.1B, C). The turgid inner tissues exert a pressure (corresponding to P_v) against the peripheral walls, which are under longitudinal and transverse tension (corresponding to P_w) (Sachs, 1882; Lockhart, 1965b; Kutschera et al., 1987; Kutschera, 1989).

Auxin-induced growth of excised sections

In order to study the mechanism of cell elongation, and hence organ growth, one has to find a way to induce (or to suppress) the growth process experimentally. One possibility is to excise sections of the growing region of axial organs and to incubate them in a solution containing the growth hormone auxin (indole-3-acetic acid, IAA). Heyn (1931, 1940) showed that the plastic extensibility (E_{pl}) of auxin-treated oat coleoptiles was about twice as high as in the control, whereas the elastic extensibility (E_{el}) was only slightly affected by the hormone. This observation led to the hypothesis that IAA induces growth by loosening of the growth-limiting peripheral cell walls of the organ (Lockhart, 1965b; Masuda and Yamamoto, 1972; Brummell and Hall, 1987; Kutschera et al., 1987). According to equation (4) a more extensible peripheral wall permits a higher rate of wall yielding and hence elongation growth. The mechanism by which IAA loosens the epidermal walls is still unknown. Three major theories have been proposed.

According to the 'gene-activation hypothesis', put forward in the 1960s by Key and others, IAA induces growth by rapid activation of certain genes, followed by the synthesis of mRNAs and proteins ('growth-limiting proteins') (Theologis, 1986). This hypothesis has been supported by the findings that IAA-induced growth does not take place if inhibitors of protein- and RNA-synthesis are added to the medium (Edelmann and Schopfer, 1989) and that IAA can rapidly increase the levels of certain mRNAs in segments of etiolated pea internodes and other organs (Theologis, 1986).

Since the auxin action that induces cell elongation is localized in the epidermis one might expect a rapid effect of IAA on the expression of specific mRNAs in the peripheral cell layer. This problem has recently been investigated using pea internodes as an experimental system.

Dietz et al. (1990) were unable to detect a rapid effect (lag-phase <1 h) of IAA on the induction of mRNAs in the epidermis and inner tissues of red-light grown pea internodes, although growth was fully established within 15 min of hormone application. In dark-grown pea internodes, however, within 30 min of IAA application effects on the expression of specific mRNAs were detectable, indicating that the methods being used were capable of detecting early effects in light-grown sections if they had occurred. Dietz et al. (1990) concluded that in red-light grown pea internodes IAA induces growth independently of the enhancement of specific mRNAs.

The 'acid-growth theory' of auxin action postulated that IAA induces growth by acidification of the cell walls, leading to wall-loosening and growth (Cleland, 1971; Taiz, 1984). This hypothesis is based on two observations. First, incubation of auxin-sensitive sections in acidic media (e.g. pH 4) results in rapid elongation growth and second, auxin causes an excretion of protons into the cell walls, which acidify the apoplastic solution to a pH of about 5. A comprehensive quantitative reinvestigation of the relationship between IAA-induced growth and proton secretion revealed, however, that these processes are mechanistically unrelated phenomena. Acid-induced wall-loosening (E_{pl}-increase) and growth does not take place at a pH of about 5. In order to stimulate growth to an extent comparable with that obtained with IAA an apoplastic pH of 3.5–4.0 is necessary. It has therefore been concluded that IAA-induced growth occurs independently of the secretion of protons (Kutschera and Schopfer, 1985a, b; Kutschera et al., 1987; Schopfer, 1989).

A third theory of IAA-induced cell wall loosening has been proposed by Ray (1962). He investigated the relationship between auxin-induced increase in cell wall material (wall synthesis) and elongation growth in oat coleoptiles and suggested a causal relationship between these processes. The recent discovery of a rapid auxin effect on the appearance of osmiophilic (electron-dense) particles at the outer epidermal wall of maize coleoptiles indicates that IAA induces the secretion of a macromolecular product into the growth-limiting peripheral cell wall (Kutschera et al., 1987; Bergfeld et al., 1988). The osmiophilic material secreted into the outer epidermal wall consists of arabinogalactan protein, which possibly acts as an epidermal wall-loosening factor in the coleoptile (P. Schopfer, personal communication). In pea internodes, auxin can rapidly increase the rate of synthesis of non-cellulosic polysaccharides in the epidermal cell walls (Kutschera and Briggs, 1987). In the epidermis of maize coleoptiles, on the other hand, a rapid auxin effect on the synthesis of wall protein (which is possibly linked to the secretion of arabinogalactan protein) was observed (Edelmann et al., 1989). These findings support the hypothesis that IAA induces growth by rapid stimulation of the secretion and incorporation of matrix material into the peripheral cell walls of the organ.

In addition to loosening of the epidermal cell walls, IAA has an effect on the orientation of newly deposited cellulose microfibrils through a corresponding change in the orientation of cortical microtubules. Bergfeld et al. (1988) have shown that in sections of maize coleoptiles, cessation of growth after depletion of endogenous IAA is correlated with an increase of longitudinally oriented microtubules and microfibrils at the inner wall surface. After addition of exogenous IAA a rapid increase of transversely oriented microtubules and microfibrils was observed. The apposition of cellulose microfibrils at the outer epidermal wall occurs independently of the initiation of IAA-mediated organ growth (Edelmann et al., 1989). It has therefore been concluded that the IAA-induced reorientation of microfibrils (Bergfeld et al., 1988) is not causally linked to the wall-loosening process. Possibly, it has a morphogenetic function and determines the mechanical anisotropy of the epidermal cell walls. Thus, the organ grows preferentially in length, in spite of the multidirectional turgor pressure of the cells (Fig. 11.1B, C).

According to equation (5) elongation growth can only take place as long as π_i, and thus $\Delta\pi$, is kept at a sufficiently high level. In the absence of external solutes, IAA-induced growth of the maize coleoptile is accompanied by a corresponding decrease of π_i, i.e. the cells behave like osmometers (Kutschera and Schopfer, 1986a). If absorbable solutes are present in the external medium (e.g. sucrose, glucose, KCl) the growth rate of the coleoptile remains nearly constant for up to 20 h due to continuous uptake of the osmotica (Stevenson and Cleland, 1981). In coleoptiles, however, exogenously applied sugars induce a strong acidification of the external medium (Kutschera and Schopfer, 1985a). Because protons can induce wall loosening and growth, excised coleoptile segments are not a suitable experimental system for the study of sugar uptake and osmoregulation during growth.

Regulation of cell elongation in the intact plant

Since the discovery of the growth hormone auxin by Went in the late 1920s it has become general practice to study the mechanism of cell elongation with excised segments, immersed in a suitable aqueous medium provided with IAA or a related growth substance. As pointed out by Brummell and Hall (1987), it is not justified to make extrapolations from results obtained on excised floating sections to the growth process in the intact plant. With excised segments, the IAA-induced loosening of an artificially stiffened cell wall (auxin-depleted sections) is studied, whereas in the intact plant no such 'minus auxin-control' exists.

One experimental approach to study the mechanism of cell elongation *in situ* is to inhibit the rapid growth of etiolated (dark-grown) seedlings by light (Mohr, 1972). It has been shown that light-induced inhibition of growth is due principally to an inhibition of cell elongation with little effect on cell number (Lockhart, 1960).

Gessner (1934) was the first to investigate the biophysical basis of the blue-light induced inhibition of growth in the hypocotyl of etiolated sunflower seedlings. After irradiation of dark-grown seedlings, a 50% inhibition of hypocotyl growth rate was observed, which was accompanied by a drastic decrease of both E_{el} and E_{pl}. Gessner (1934) concluded that blue light inhibits growth by causing a decrease in cell wall extensibility.

Later, a similar study was carried out by Lockhart (1960). After irradiation of etiolated pea seedlings for 3 h with red light a 70% inhibition of internodal growth rate was measured. Lockhart (1960) found no significant difference in the osmotic concentration (corresponding to π_i) and the water permeability of the stem cells (corresponding to L) between the irradiated and etiolated internodes. The cell wall plasticity (E_{pl}) of the stem, however, was greatly reduced as a result of prior irradiation. Lockhart (1960) concluded that visible radiation inhibits stem elongation by reducing cell wall plasticity, whereas other growth parameters (see equation (5)) are unaffected by light. Cosgrove and Green (1981) have confirmed and extended Gessner's and Lockhart's investigations using a different method to determine the yielding properties of the cell walls.

The mechanism by which light inhibits growth is unknown. It is conceivable that the synthesis of cell wall material is affected by light, which could thus bring about a change in cell wall mechanical properties and growth rate. According to the classical model of cell extension, put forward by Pfeffer (1904) and championed more recently by Burström (1979), the cell wall grows by intussusception (internal incorporation) of matrix material into the existing wall structure. This hypothesis implies a causal relationship between cell wall synthesis and elongation growth in the intact plant (Burström, 1979).

Recently, the relationship between increase in cell wall material (wall synthesis) and growth was investigated in hypocotyls of sunflower seedlings that were either grown in the dark or irradiated with continuous white light (Kutschera, 1990). The results showed that the increase in wall material was larger in the inner tissues than in the peripheral cell layers. The wall mass per length, which is a measure of the average thickness of the walls, decreased during growth in both continuous darkness and white light. This shows that wall thinning occurs during hypocotyl growth. Upon illumination of etiolated hypocotyls a 70% inhibition of growth rate was measured. The increase in wall material, however, was un-affected in both the epidermis and the inner tissues. The composition of the cell walls (cellulose, hemicellulose, pectic substances) was likewise not affected by white light. Thus, synthesis of cell wall polysaccharides and elongation growth are independent processes in the intact plant (Heyn, 1940; Mohr, 1972; Kutschera, 1990).

Because of continued synthesis, the wall mass per length increased nearly two-fold during a 24-h period of white light illumination of the etiolated hypocotyls. This resulted in a corresponding thickening of the longitudinal cell walls (Fig. 11.4A, B). The plasticity (E_{pl}) of the irradiated hypocotyls was about half that of the dark-grown control, whereas E_{el} was only slightly affected by light (Fig. 11.3A, B). These results show that the long-term effect of white light on the reduction of E_{pl} is due to the fact that cell wall synthesis continues during the inhibited growth in the light. This leads to an enhanced mechanical stability of the stem during photomorphogenesis of the seedling. Figure 11.5A, B shows that peeled sections cut from etiolated and irradiated hypocotyls rapidly extend upon addition of distilled water,

whereas in the intact (non-peeled) segment no length change occurs. This shows that the extensibility of the inner tissues is not reduced as a result of prior irradiation of the hypocotyl. Thus, the light-induced mechanical stiffening of the stem (Fig. 11.3A, B) occurs exclusively in the epidermis of the organ.

Long-term growth is only possible as long as π_i is high enough to provide a positive turgor pressure of the cells (equation (5)). Table 11.1 shows that the length of the etiolated hypo-cotyl of *Helianthus annuus* increased by approxi-mately 300% between the third and fourth day after sowing, while the diameter remained constant. The osmotic concentration of the cell sap (corresponding to π_i) decreased by only 22% during this time period. Thus, osmo-regulation occurred during growth of the etiolated stem.

Beck (1941) showed that the production of solutes in growing epidermal cells of intact etiolated sunflower hypocotyls was propor-tional to the length of the cells. This interesting observation has never been followed up. The principal osmotic substances of the sunflower hypocotyl are hexoses (glucose and fructose) and to a lesser extent organic potassium salts (McNeil, 1976). Table 11.1 shows that during growth in darkness the accumulation in soluble sugars is proportional to the length increase of the organ. Thus, growth is accompanied by a corresponding uptake (or production) of organic solutes, as originally described by Beck

Table 11.1. Length, diameter, osmotic concentration of the cell sap (OC), and amount of soluble sugars of hypocotyls cut from sunflower seedlings that were either grown for 3 days in darkness (3D), 4 days in darkness (4D) or 3 days in the dark and 1 day in white light (3D 1WL, white light: 2.5 klux, see Kutschera, 1990). Data represent means ($\pm$ SE) of 12 measurements each

Treatment	Length (mm)	Diameter (mm)	OC (mosmol kg^{-1})	Soluble sugars (mg hypocotyl^{-1})
3D	15.2 ± 0.3	1.96 ± 0.01	294 ± 3	0.74 ± 0.04
4D	45.9 ± 0.4 (+302%)	1.95 ± 0.01	229 ± 3 (−22%)	2.28 ± 0.1 (+308%)
3D 1WL	24.6 ± 0.4 (+162%)	1.95 ± 0.01	254 ± 3 (−14%)	1.21 ± 0.04 (+163%)

(1941). After irradiation of etiolated plants with white light, hypocotyl growth was greatly reduced. The osmotic concentration of the cell sap (π_i) was significantly higher than in the dark-grown plants, indicating that dilution of solutes due to water uptake was less than in the rapidly growing control. Increase in length and in soluble sugars were proportional, as in the dark-grown control plants. The data of Table 11.1 indicate that the relative maintenance of π_i (osmoregulation) during rapid hypocotyl growth in darkness is achieved by a corresponding uptake of soluble sugars into the expanding cells of the organ.

McNeil (1976) suggested that sucrose, supplied by the cotyledons via the phloem, flows into the vacuoles of the hypocotyl cells and is hydrolysed into glucose and fructose to keep the vacuolar hexose concentration high. In the stem of *Phaseolus vulgaris* the specific activity of acid invertase (which hydrolyses sucrose into hexoses) was highest in the most rapidly elongating internode and very low in internodes which had completed their elongation (Morris and Arthur, 1985). Hence, circumstantial evidence exists to suggest that this enzyme is causally involved in controlling the amount of vacuolar hexose sugars and thus π_i of the cell in the intact organ.

Acknowledgements

I wish to thank Professors Z. Hejnowicz and P.B. Green for reading the manuscript critically. Supported by the Deutsche Forschungsgemeinschaft.

References

Beck, W.A. (1941). Production of solutes in growing epidermal cells. *Plant Physiol.* **16**, 637–642.

Bergfeld, R., Speth, V. and Schopfer, P. (1988). Reorientation of microfibrils and microtubules at the outer epidermal wall of maize coleoptiles during auxin-mediated growth. *Botanica Acta* **101**, 57–67.

Brummell, D.A. and Hall, J.L. (1987). Rapid cellular responses to auxin and the regulation of growth. *Plant, Cell Environm.* **10**, 523–543.

Burström, H.G. (1979). In search of a plant growth paradigm. *Amer. J. Bot.* **66**, 98–104.

Cleland, R.E. (1971). Cell wall extension. *Annu. Rev. Plant Physiol.* **22**, 197–222.

Cosgrove, D. (1986). Biophysical control of plant cell growth. *Annu. Rev. Plant Physiol.* **37**, 377–405.

Cosgrove, D., Green, P.B. (1981). Rapid suppression of growth by blue light. Biophysical mechanism of action. *Plant Physiol.* **68**, 1447–1453.

Dietz, A., Kutschera, U. and Ray, P.M. (1990). Auxin enhancement of mRNAs in epidermis and internal tissues of the pea stem and its significance for control of elongation. *Plant Physiol.* **93**, 432–438.

Edelmann, H. and Schopfer, P. (1989). Role of protein and RNA synthesis in the initiation of auxin-mediated growth in coleoptiles of *Zea mays* L. *Planta* **179**, 475–485.

Edelmann, H., Bergfeld, R. and Schopfer, P. (1989). Role of cell-wall biogenesis in the initiation of auxin-mediated growth in coleoptiles of *Zea mays* L. *Planta* **179**, 486–494.

Gessner, F. (1934). Wachstum und Wanddehnbarkeit im *Helianthus*-Hypocotyl. *Jahrb. Wiss. Bot.* **80**, 143–168.

Hellebust, J.A. (1976). Osmoregulation. *Annu. Rev. Plant Physiol.* **27**, 485–505.

Heyn, A.N.J. (1931). Der Mechanismus der Zellstreckung. *Rec. Trav. Bot. Neerl.* **28**, 113–244.

Heyn, A.N.J. (1940). The physiology of cell elongation. *Bot. Rev.* **6**, 515–574.

Kutschera, U. (1989). Tissue stresses in growing plant organs. *Physiol. Plant* **77**, 157–163.

Kutschera, U. (1990). Cell-wall synthesis and elongation growth in hypotocyls of *Helianthus annuus* L. *Planta* **181**, 316–323.

Kutschera, U. and Briggs, W.R. (1987). Radid auxin-induced stimulation of cell wall synthesis in pea internodes. *Proc. Natl Acad. Sci. U.S.A.* **84**, 2747–2751.

Kutschera, U. and Schopfer, P. (1985a). Evidence against the acid-growth theory of auxin action. *Planta* **163**, 483–493.

Kutschera, U. and Schopfer, P. (1985b). Evidence for the acid-growth theory of fusicoccin action. *Planta* **163**, 494–499.

Kutschera, U. and Schopfer, P. (1986a). Effect of auxin and abscisic acid on cell wall extensibility in maize coleoptiles. *Planta* **167**, 527–535.

Kutschera, U. and Schopfer, P. (1986b). *In-vivo* measurement of cell-wall extensibility in maize coleoptiles. Effects of auxin and abscisic acid. *Planta* **169**, 437–442.

Kutschera, U., Bergfeld, R. and Schopfer, P. (1987). Cooperation of epidermis and inner tissues in auxin-mediated growth of maize coleoptiles. *Planta* **170**, 168–180.

Lockhart, J. (1960). Intracellular mechanism of growth inhibition by radiant energy. *Plant Physiol.* **35**, 129–135.

Lockhart, J. (1965a). An analysis of irreversible plant cell elongation. *J. Theoret. Biol.* **8**, 264–275.

Lockhart, J. (1965b). Cell extension. In *Plant Biochemistry* (eds J. Bonner and J.E. Varner), pp. 826–849. Academic Press, New York.

Masuda, Y. and Yamamoto, R. (1972). Control of auxin-induced stem elongation by the epidermis. *Physiol. Plant.* **27**, 109–115.

McNeil, D.L. (1976). The basis of osmotic pressure maintenance during expansion growth in *Helianthus annuus* hypocotyls. *Aust. J. Plant Physiol.* **3**, 311–324.

Mohr, H. (1972). *Lectures on Photomorphogenesis.* Springer, Berlin.

Morris, D.A. and Arthur, E.D. (1985). Invertase activity, carbohydrate metabolism and cell expansion in the stem of *Phaseolus vulgaris* L. *J. Exp. Bot.* **36**, 623–633.

Pfeffer, W. (1904). *Pflanzenphysiologie.* Engelmann, Leipzig.

Ray, P.M. (1962). Cell wall synthesis and cell elongation in oat coleoptile tissue. *Amer. J. Bot.* **49**, 928–939.

Ray, P.M. (1987). Principles of plant cell growth. In *Physiology of Cell Expansion During Plant Growth* (eds D.J. Cosgrove and D.P. Knievel), pp. 1–17. American Society of Plant Physiologists, Rockville, MD, USA.

Sachs, J. (1882). *Vorlesungen über Pflanzen-physiologie.* Engelmann, Leipzig.

Schopfer, P. (1989). pH-Dependence of extension growth in *Avena* coleoptiles and its implications for the mechanism of auxin action. *Plant Physiol.* **90**, 202–207.

Stevenson, T.T. and Cleland, R.E. (1981). Osmoregulation in the *Avena* coleoptile in relation to auxin and growth. *Plant Physiol.* **67**, 749–753.

Taiz, L. (1984). Plant cell expansion: regulation of cell wall mechanical properties. *Annu. Rev. Plant Physiol.* **35**, 585–657.

Theologis, A. (1986). Rapid gene regulation by auxin. *Annu. Rev. Plant Physiol.* **37**, 407–438.

12. MICROTUBULES AND THE REGULATION OF CELL MORPHOGENESIS BY PLANT HORMONES

Hiroh Shibaoka

Department of Biology, Osaka University,
Toyonaka, Osaka 560, Japan

Introduction

Unlike animals which are mobile, higher plants cannot change their location to avoid the effects of adverse environmental conditions. Instead, they alter their mode of development and assume a form that is appropriate for the environmental conditions of their habitat. The elongation of the stems of seedlings in soil is extremely rapid with almost no concomitant thickening of the stem. This type of growth results in the rapid exposure of the shoot apex to light, at which time expansion of leaves and photosynthesis can take place. Plants that bend when exposed to wind show reduced rates of stem elongation under windy conditions and thickening of stems is accelerated so that the plants become resistant to the mechanical stress imposed by the wind. In cases in which environmental factors affect the development of plants, plant hormones usually act as mediators between the reception of an environmental stimulus and the development of the plant. It has been suggested, for example, that light suppresses stem elongation by lowering the rate of transport of auxin and that mechanical stimulation causes stem thickening by inducing the synthesis of ethylene.

Plant hormones regulate plant growth in various ways. Amongst the various effects of plant hormones in response to environmental stimuli, the regulation of the direction of expansion of cells is especially important in determining the form of the plant. The direction of cell expansion is controlled by the properties of the cell wall and, in particular, by the orientation of cellulose microfibrils in the wall. The orientation of cellulose microfibrils, in turn, appears to be regulated by cortical microtubules (MTs). Thus, cortical MTs would seem to be involved in the regulation by plant hormones of the direction of cell expansion and, therefore, in the regulation by environmental factors of the form of the plant. This chapter presents the evidence that indicates the involvement of cortical MTs in mediating the actions of plant hormones in the regulation of the direction of cell expansion.

Gibberellins

The photoperiod affects not only the reproductive behaviour but also the vegetative growth of higher plants. Usually, the rate of stem elongation is higher under long-day conditions than under short-day conditions. Since long days increase levels of gibberellins and accelerate the rates of interconversion of various gibberellins, these hormones are considered to play a major role in the photoperiodic regulation of stem elongation.

Gibberellins have been shown to cause the hyperelongation of stems in various plants. Epicotyl segments of azuki bean (*Vigna angularis*) floated on a solution that contains only auxin increase their length and fresh weight at different rates; the rate of increase in fresh weight is greater than that in length (Fig. 12.1), indicating that the segments show both elongation and thickening. Segments floated on a solution that contains both auxin and gibberellin A_3 (GA_3) increase their length and fresh weight at almost the same rate (Fig. 12.1), indicating that the segments show only elongation and no thickening. Increases in length of segments are greater in the presence of GA_3 than in its absence. These observations indicate that GA_3 alters the direction of expansion of cells in azuki bean epicotyls (Shibaoka, 1972). Since the GA_3-induced promotion of elongation of segments can be reversed by MT-disrupting agents, such as colchicine (Shibaoka, 1972) and ethyl N-phenylcarbamate (Shibaoka and Hogetsu, 1977), the presence of MTs seems indispensable for the alteration of the direction of cell expansion by GA_3. The GA_3-induced promotion of elongation can also be reversed by inhibitors of the synthesis of cellulose, such as coumarin (Hogetsu et al., 1974a) and 2,6-dichlorobenzonitrile (Hogetsu et al., 1974b),

indicating that the synthesis of cellulose *de novo* is also required for the promotion of elongation by GA_3. Electron microscopic observations revealed that GA_3 causes a predominance of transverse cortical MTs in epidermal cells of azuki bean epicotyls (Shibaoka, 1974) and increases the percentage of epidermal cells with transverse cellulose microfibrils on the inner surface of the cell wall (Takeda and Shibaoka, 1981). Thus, it seems likely that gibberellins suppress the lateral expansion of cells by organizing cortical MTs such that they lie transverse to the cell axis and, as a result, newly synthesized cellulose microfibrils are laid down in the same direction. The suppression of cell swelling may facilitate the longitudinal expansion of cells.

Gibberellins also change the orientation of cortical MTs from longitudinal to transverse in seedlings of dwarf (d_5) maize (Mita and Katsumi, 1986) and of dwarf pea (Akashi and Shibaoka, 1987; Sakiyama and Shibaoka, 1990). Mesocotyls of dwarf maize seedlings have narrow elongation zones of only about 1 mm in width, while those in normal seedlings are

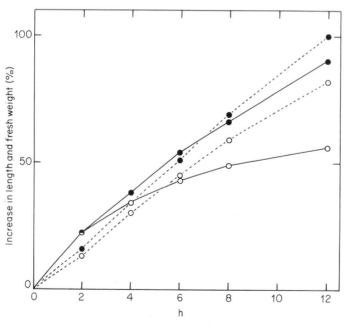

Fig. 12.1. Effects of gibberellin A_3 (GA_3) on increases in length and fresh weight of segments of azuki bean epicotyls. —○—, auxin, length; —●—, auxin + GA_3, length; – –○– –, auxin, fresh weight; – –●– –, auxin + GA_3, fresh weight. Reproduced by permission from *Plant Cell Physiol.* (Shibaoka, 1972).

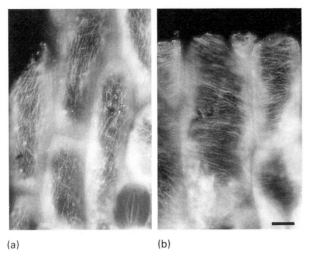

(a)	(b)

Fig. 12.2.   Effect of gibberellin A_3 (GA_3) on the orientation of cortical MTs in epidermal cells of epicotyls of the dwarf pea. a, Control; b, GA_3. Scale = 10 μm. Reproduced by permission from *Protoplasma* (Sakiyama and Shibaoka, 1990).

about 4 mm wide. In mesocotyls of dwarf maize, the distribution of epidermal cells with transverse cortical MTs is restricted within the narrow elongation zones; longitudinal MTs predominate in epidermal cells in zones where elongation is not occurring. In seedlings of normal maize, cells with transverse MTs are distributed throughout the wide elongation zones. Treatment with GA_3, which increases the width of the elongation zone in dwarf maize seedlings, increases the width of the region of cells with transverse MTs (Mita and Katsumi, 1986).

Cortical MTs in epidermal cells of dwarf pea seedlings are oriented longitudinally. GA_3 promotes elongation of epicotyls and brings about a predominance of transverse MTs (Fig. 12.2). Changes in the arrangement of cortical MTs in dwarf pea epicotyls by GA_3 have been demonstrated both by electron microscopy (Akashi and Shibaoka, 1987) and by immuno-fluorescence microscopy (Sakiyama and Shibaoka, 1990). Electron microscopic observations revealed that GA_3 changes the orientation of cellulose microfibrils in the cell wall from longitudinal to transverse (Akashi and Shibaoka, 1987). The gibberellin-induced predominance of transverse MTs seem also to be involved in the promotion by gibberellins of

elongation in dwarf maize mesocotyls and dwarf pea epicotyls.

Sawhney and Srivastava (1974, 1975) reported that colchicine inhibited the GA_3-induced elongation of lettuce hypocotyls and that prolonged treatment (72 h) with GA_3 results in a predominance of transverse MTs in cortical cells. However, since no differences in the basic structure of the cell walls and in orientation of cortical MTs and cellulose microfibrils were found between untreated and GA_3-treated seedlings within 12 h of the treatment, it was suggested that promotion by gibberellins of the elongation of lettuce hypocotyls was due to a gibberellin-induced promotion of loosening of the cell wall or of maintenance of turgor pressure within the cells (Sawhney and Srivastava, 1975).

Ethylene

Mechanical stimuli, for example, wind or touch, are known to inhibit stem elongation and to stimulate stem thickening. Ethylene is considered to be responsible for this phenomenon since plants produce ethylene in response to a mechanical stimulus, and exogenously applied ethylene often causes inhibition of stem elongation and promotion of stem thickening. Since it has been demonstrated that treatment with ethylene results in the deposition of longitudinal cellulose microfibrils on cell walls, ∮treatment with ethylene is considered to result in the cessation of stem elongation and the promotion of stem thickening by causing newly deposited cellulose microfibrils to be arranged longitudinally to the long axis of the stem∤(Eisinger, 1983).

The fact that treatment with ethylene results in the deposition of longitudinal cellulose microfibrils indicated the possibility that ethylene causes a predominance of longitudinal MTs. Steen and Chadwick (1981) demonstrated that ethylene actually brings about a predominance of longitudinal MTs in cortical cells of pea seedlings. They also found that ethylene-induced inhibition of the elongation of segments of pea epicotyls was partially alleviated by the simultaneous application of colchicine, suggesting the involvement of MTs in the ethylene-induced inhibition of the elongation of epicotyls. The effects of ethylene in altering the orientations of cortical MTs and newly deposited cellulose microfibrils from transverse to longitudinal were also demonstrated by Lang et al. (1982) in pea epicotyls.

With a view to examining whether or not depolymerization of MTs is involved in the processes of ethylene-induced reorientation of MTs, Roberts et al. (1985) examined the changes in the entire array of cortical MTs in cortical and epidermal cells of ethylene-treated pea epicotyls and mung bean hypocotyls by immunofluorescence microscopy. Ethylene causes a predominance of longitudinal MTs in both cortical and epidermal cells of pea epicotyls and mung bean hypocotyls without producing any cells in which arrays of MTs appear to have been disassembled. Each entire array of MTs seems to reorient without any intervening gross depolymerization of individual MTs. Roberts et al. (1985) also demonstrated that the change in the orientation of cortical MTs was evident as early as 30 min after treatment with ethylene when signs of stem thickening were not yet obvious to the naked eye.

Auxin

Mediation by auxin of phototropism and geotropism in shoots of higher plants is a classical example of the involvement of plant hormones in the response of plant growth to environmental conditions. Recently, it was reported that there is a difference in the orientation of cortical MTs between cells on the concave side and those on the convex side of phototropically or geotropically curved plant shoots (Nick et al., 1990). Cortical MTs adjacent to the outer tangential walls of epidermal cells of corn coleoptiles run in a wide variety of orientations. Unilateral irradiation with blue light, which causes phototropic bending, alters the orientation of cortical MTs in the epidermal cells. MTs become oriented longitudinally in cells on the irradiated concave side and transversely in cells on the shaded convex side. Changes in the orientation of cortical MTs also occur in sunflower hypocotyls concomitantly with phototropic bending (Nick et al., 1990). MTs in epidermal cells on the irradiated side of the hypocotyl are oriented longitudinally and those in cells on the shaded side are oriented transversely. MTs in cells on the concave side of geotropically curved corn coleoptiles and sunflower hypocotyls are also oriented longitudinally and those in cells on the convex side are oriented transversely (Nick et al., 1990). In an examination of the role of plant hormones in these processes, cortical MTs in epidermal cells of segments of corn coleoptiles treated with water were oriented longitudinally to the cell axis. However, longitudinal MTs in cells of segments treated with water were reoriented in a direction transverse to the cell axis when the segments were transferred to a solution of auxin (Bergfeld et al., 1988; Nick et al., 1990). Thus, as has been suggested by Nick et al. (1990), auxin seems to be involved in the changes in the orientation of cortical MTs that occur during tropic bending. The alteration in the orientation of cortical MTs occurs soon after the start of treatment with auxin; changes are appreciable within 15 min of the start of treatment and are complete within 60 min (Nick et al., 1990).

The effect of auxin in orienting cortical MTs transversely to the cell axis has also been reported in wheat coleoptiles (Volfová et al., 1977) and in oat mesocotyls (Iwata and Hogetsu, 1989). Cortical MTs in elongating mesocotyls of oat seedlings grown in the dark are arranged transversely to the cell axis. Irradiation with white light, which causes the cessation of mesocotyl elongation, alters the orientation of MTs from transverse to longitudinal. The cessation of mesocotyl elongation and the alteration of the orientation of MTs caused by light irradiation are very significantly retarded in seedlings treated with auxin. Since light irradiation has been reported to inhibit the transport of auxin from the coleoptile to the mesocotyl in oat seedlings, it seems probable that light causes the cessation of mesocotyl elongation and the alteration in the orientation of MTs by lowering the levels of auxin in mesocotyl cells (Iwata and Hogetsu, 1989).

Changes in the orientation of cortical MTs occur simultaneously with the cessation of mesocotyl elongation in epidermal cells of oat seedlings irradiated with white light, but the changes in orientation occur after the cessation of elongation in cortical cells (Iwata and Hogetsu, 1989). These results raise the question of whether the cessation of mesocotyl elongation is the result of changes in the orientation of cortical MTs or, alternatively, whether the cessation brings about the reorientation of MTs. Auxin-induced elongation of segments of azuki bean epicotyls is inhibited neither by MT-disrupting agents (Shibaoka, 1972; Shibaoka and Hogetsu, 1977) nor by inhibitors of the synthesis of cellulose (Hogetsu et al., 1974a, b). These observations indicate that auxin-induced elongation in azuki bean epicotyls involves neither the reorientation of cortical MTs nor the reinforcement of cell walls with oriented cellulose microfibrils. However, auxin-induced elongation of wheat coleoptiles was reported to be reduced by MT-disrupting agents, such as colchicine and vinblastine (Lawson and Weintraub, 1975).

Cytokinins

Cytokinins suppress stem elongation and cause stem thickening, but such effects have not been proven to be involved in the response of plant growth to the environment.

Auxin-induced elongation of segments of azuki bean epicotyls is inhibited by kinetin, one of the cytokinins. This inhibition is reversed, albeit only partially, by MT-disrupting agents, such as colchicine (Shibaoka, 1974) and ethyl N-phenylcarbamate (Shibaoka and Hogetsu, 1977) – an indication that MTs are involved in the kinetin-induced inhibition of epicotyl elongation. Electron microscopic observations revealed that treatment with kinetin results in a predominance of longitudinal MTs in epidermal cells of azuki bean epicotyls (Shibaoka, 1974). The kinetin-induced inhibition of epicotyl elongation is also reversed by inhibitors of the synthesis of cellulose, such as coumarin and 2,6-dichlorobenzonitrile (Hogetsu et al., 1974a, b). Thus, it appears that the synthesis of cellulose is also required for inhibition by kinetin of stem elongation. Kinetin appears to suppress the longitudinal expansion of cells by organizing cortical MTs in a direction longitudinal to the cell axis and, as a consequence, newly synthesized cellulose microfibrils are laid down with the same orientation. Treatment with kinetin also results in cortical MTs becoming oriented longitudinally to the cell axis in wheat coleoptile cells (Volfová et al., 1977).

Abscisic acid

Water stress causes retardation of stem elongation. Since abscisic acid (ABA) is a potent inhibitor of stem elongation and is known to accumulate in leaves subjected to water stress, it seems probable that ABA is involved in the retardation of stem elongation that is induced by water stress.

In dwarf pea epicotyls, ABA acts in the opposite way to GA$_3$ with respect to the arrangement of cortical MTs. As described above, treatment with GA$_3$, which promotes elongation of dwarf pea epicotyls, brings

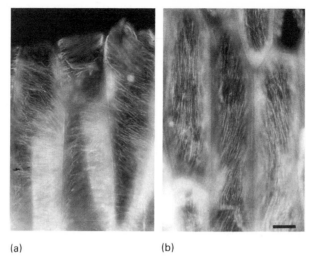

(a)   (b)

Fig. 12.3. Effect of abscisic acid (ABA) on the orientation of cortical MTs in epidermal cells of epicotyls of dwarf pea seedlings pretreated with gibberellin A_3. a, Control; b, ABA. Scale = 10 μm. Reproduced by permission from *Protoplasma* (Sakiyama and Shibaoka, 1990).

about a predominance of transverse MTs. ABA applied to GA_3-pretreated seedlings reduces the elongation rate of GA_3-treated epicotyls and eliminates the GA_3-induced predominance of transverse MTs. The net result is a predominance of longitudinal MTs (Fig. 12.3; Sakiyama and Shibaoka, 1990). Since the effects of ABA on the arrangement of cortical MTs do not become visible before the inhibition of the elongation of epicotyls by ABA becomes apparent, it is clear that the early effects of ABA on epicotyl elongation do not involve MTs. However, the predominance of longitudinal MTs brought about by treatment with ABA should have some relationship to the ABA-induced cessation of elongation of epicotyls. Longitudinally arranged MTs should orient newly deposited cellulose microfibrils longitudinally to the cell axis, and the longitudinally arranged cellulose microfibrils should suppress the longitudinal expansion of the cell. Water stress, which is known to elevate endogenous levels of ABA, inhibits increases in the elastic extensibility of the cell wall (Sakurai and Kuraishi, 1989). It is probable that the inhibition of increases in the elastic extensibility of the cell wall is caused by ABA-mediated changes in the orientation of cellulose microfibrils in the cell wall.

Jasmonic acid and related compounds; the formation of onion bulbs

Onion plants form bulbs in response to the stimulus of long-day conditions. The bulb is formed by the lateral expansion of leaf-sheath cells which expand longitudinally under short-day conditions (Heath and Holdsworth, 1948). Since the bulb is formed as a result of changes in the direction of cell expansion, it might be expected that changes in the orientation of cortical MTs are involved in this phenomenon. Mita and Shibaoka (1983) demonstrated that changes in the orientation of cortical MTs do actually occur in leaf-sheath cells of onion plants during the development of bulbs. Transverse MTs are abundant in leaf-sheath cells of onion plants grown under short-day conditions, but as day length is increased and the bulb develops, the MTs become disoriented and scattered, and finally they disappear altogether.

It has been suggested that the stimulus of long days is perceived by leaf blades, and that leaf blades exposed to the stimulus produce 'bulbing hormone' and supply it to the leaf sheaths (Heath and Holdworth, 1948). The observation that formation of onion bulbs is accompanied by a decrease in the number of cortical MTs indicates the possibility that 'bulbing hormone' is a kind of MT-disrupting agent. Recently, Koda *et al.* (1988) isolated substances that possess the ability to induce the formation of potato tubers from leaves of potato plants grown under conditions that induce the formation of tubers, i.e. short-day conditions. One of these tuber-inducing substances was identified as a glucoside of a compound related to jasmonic acid and the corresponding aglycone was named tuberonic acid (Yoshihara *et al.*, 1989). A tuber-inducing activity was also found to be associated with jasmonic acid itself and with its methyl ester (Yoshihara *et al.*, 1989), which are known to be widely distributed throughout the plant kingdom.

This result (that jasmonic acid and related compounds possess tuber-inducing activity) together with the fact that formation of potato tubers involves the lateral expansion of stolon cells, suggests the possibility that the 'bulbing hormone' of onion plants is a compound related to jasmonic acid and that such compounds have the ability to disrupt MTs. To examine such a possibility, jasmonic acid and related compounds were tested for their effects on MTs and on induction of the formation of onion bulbs. Jasmonic acid and its methyl ester were found to disrupt cortical MTs in tobacco cultured cells (Abe *et al.*, 1990), and jasmonic acid was found to induce formation of onion bulbs (Y. Koda, personal communication). These results seem to support the above-mentioned possibility. However, since methyl jasmonate disrupts MTs only in cells in the S phase in cultures of tobacco cells, it is still unclear whether induction by jasmonic acid of bulb formation is due to the disruption of cortical MTs induced by jasmonic acid.

Gibberellins seem also to be involved in the regulation of formation of onion bulbs. Unlike mature onion plants, young seedlings, having only two to three leaves, do not form bulbs even when they are grown under long-day conditions. However, such young seedlings form small bulbs when seedlings are treated with an inhibitor of the biosynthesis of gibberellins, S-3307 (Mita and Shibaoka, 1984b). GA$_3$ inhibits the induction of the formation of small bulbs by S-3307. Endogenous gibberellins seem to be suppressing bulb formation in young seedlings. S-3307 alters the orientation of cortical MTs in leaf-sheath cells from transverse to oblique or longitudinal and, conse-

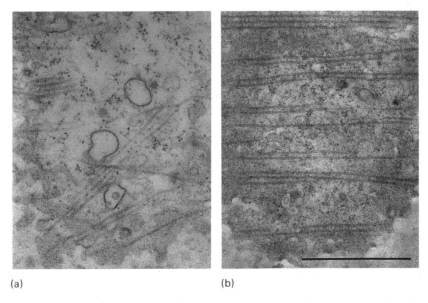

(a)   (b)

Fig. 12.4. Effect of gibberellin A$_3$ (GA$_3$) on the orientation of cortical MTs in leaf-sheath cells of the onion plant. a, Control; b, GA$_3$. Cell axis is parallel to the length of the page. Scale = 1 μm. Reproduced by permission from *Protoplasma* (Mita and Shibaoka, 1984).

quently, causes lateral expansion of leaf-sheath cells (Mita and Shibaoka, 1984b). Excision of roots alters the orientation of cortical MTs in leaf-sheath cells and causes the lateral expansion of cells, just as does treatment with S-3307 (Mita and Shibaoka, 1984a). Thus, it is tempting to speculate that roots suppress the lateral expansion of leaf-sheath cells by supplying leaf sheaths with gibberellins.

Exogenously applied GA_3 orients the cortical MTs in onion leaf-sheath cells such that they lie transverse to the cell axis (Fig. 12.4). Cortical MTs in leaf-sheath cells of onion plants grown under short-day conditions are oriented in a relatively wide range of directions, although the majority (65%) are oriented transversely or nearly transversely (60–90°) to the cell axis. However, in leaf-sheath cells of GA_3-treated plants, almost all MTs (80%) are oriented strictly transversely (80–90°) to the cell axis (Mita and Shibaoka, 1984c).

Gibberellin stabilizes cortical MTs in onion leaf-sheath cells. MT-disrupting agents, such as colchicine and cremart, disrupt cortical MTs in leaf-sheath cells and induce the formation of bulb-like swellings in the basal parts of seedlings. GA_3 protects MTs from disruption by these MT-disrupting agents, and it inhibits the swelling induced by the MT-disrupting agents (Mita and Shibaoka, 1984c). GA_3 also protects MTs in onion leaf-sheath cells from disruption by cold. Gibberellins seem to suppress bulb formation by arranging cortical MTs transversely to the cell axis and by protecting the transversely oriented MTs from disruption by 'bulbing hormone', which can be considered to be a kind of naturally occurring MT-disrupting agent.

Concluding remarks: MT-orienting mechanisms

As described above, the ability to regulate the orientation of cortical MTs has been reported to be associated with all principal plant hormones. However, the mechanism whereby plant hormones orient MTs has not been clarified. Before we try to examine the way in which plant hormones orient MTs, we should clarify the mechanism whereby the orientation of

cortical MTs is determined. To this end, the processes of repolymerization of MTs in cells with previously depolymerized MTs have often been examined in detail. In *Spirogyra* cells, the cortical MTs, which are usually arranged transversely to the cell axis, can be depolymerized by treatment with amiprophos methyl (APM). If such cells are transferred to an APM-free medium, randomly oriented MTs are repolymerized first in cortical regions of the cells and they gradually adopt a transverse orientation (Hogetsu, 1987). This result seems to indicate the presence in the *Spirogyra* cell of a factor(s) that possesses the ability to orient MTs transversely to the cell axis. Transverse organization of repolymerized MTs occurs in the cortical regions of the cells from which the bulk of the cytoplasm has been removed by centrifugation, provided that the cytoplasm has been removed before the depolymerization of pre-existing MTs and not if the cytoplasm has been removed after the depolymerization of MTs (Hogetsu, 1987). These results indicate that an MT-orienting factor(s) is associated with MTs.

Wasteneys and Williamson (1989) performed similar experiments using internode cells of *Nitella*, which usually have transverse MTs, and they found that randomly oriented MTs are repolymerized first and then increasing numbers of MTs adopt a transverse orientation. In discussing the MT-orienting mechanism, they suggest the possibility that a transverse array of MTs can be brought about by 'biased turnover' of MTs, in which non-transverse MTs turn over more rapidly than transverse MTs. Cross-bridges between MTs and the plasma membrane seem to be the most plausible candidate for the factor(s) that has the ability to reduce the rate of turnover of MTs. If there exist cross-bridges that only cross-link transverse MTs to the plasma membrane, and if transverse MTs are selectively stabilized by being cross-linked with the plasma membrane, then the degree of transverse alignment can increase.

If cross-bridges that specifically cross-link transverse MTs with the plasma membrane are responsible for the transverse arrangement of MTs, it is possible that gibberellins enhance the production of such cross-bridges. It is also possible that ABA, ethylene or cytokinin induce the production of cross-bridges that specifically cross-link longitudinal MTs. Studies on the molecular structure of cross-bridges between MTs and the plasma membrane and studies of the hormonal modifications of the structure of such cross-bridges are crucially important if we are to clarify the way in which plant hormones orient cortical MTs.

Removal of the cell wall disturbs the parallel arrays of cortical MTs in carrot cultured cells (Lloyd et al., 1980) and renders cortical MTs sensitive to cold in cultured tobacco cells (Akashi et al., 1990). The cell wall seems to be involved in the arrangement and stabilization of cortical MTs. If the cell wall is to affect the arrangement and stability of cortical MTs, some factor(s) that mediates interactions between the cell wall and MTs must be present in the plasma membrane. The observations that brief treatment of tobacco suspension culture cells with trypsin disturbs the parallel array of cortical MTs and renders cortical MTs sensitive to cold (Akashi and Shibaoka, 1991) seem to indicate that such a factor(s) is, in fact, associated with the plasma membrane and is a species of transmembrane protein. Studies on transmembrane proteins that have the ability to interact with the cell wall are also necessary for a complete description of the mechanism responsible for the orientation of MTs.

References

Abe, M., Shibaoka, H., Yamane, H. and Takahashi, N. (1990). Cell cycle-dependent disruption of microtubules by methyl jasmonate in tobacco BY-2 cells. *Protoplasma* **156**, 1–8.

Akashi, T. and Shibaoka, H. (1987). Effects of gibberellin on the arrangement of the cold stability of cortical microtubules in epidermal cells of pea internodes. *Plant Cell Physiol.* **28**, 339–348.

Akashi, T. and Shibaoka, H. (1991). Involvement of transmembrane proteins in association of cortical microtubules with the plasma membrane in tobacco BY-2 cells. *J. Cell Sci.* **98**, 169–174.

Akashi, T., Kawasaki, S. and Shibaoka, H. (1990). Stabilization of cortical microtubules by the cell wall in cultured tobacco cells. *Planta* **182**, 363–369.

Bergfeld, R., Speth, V. and Schopfer, P. (1988).

Reorientation of microfibrils and microtubules at the outer epidermal wall of maize coleoptiles during auxin-mediated growth. *Bot. Acta* **101**, 57–67.

Eisinger, W. (1983). Regulation of pea internode expansion by ethylene. *Ann. Rev. Plant Physiol.* **34**, 225–240.

Heath, O.V.S. and Holdsworth, M. (1948). Morphogenic factors as exemplified by the onion plant. *Soc. Exp. Biol. Symp.* **2**, 326–350.

Hogetsu, T. (1987). Re-formation and ordering of wall microtubules in *Spirogyra* cells. *Plant Cell Physiol.* **28**, 875–883.

Hogetsu, T., Shibaoka, H. and Shimokoriyama, M. (1974a). Involvement of cellulose synthesis in actions of gibberellin and kinetin on cell expansion. Gibberellin–coumarin and kinetin–coumarin interactions on stem elongation. *Plant Cell Physiol.* **15**, 265–272.

Hogetsu, T., Shibaoka, H. and Shimokoriyama, M. (1974b). Involvement of cellulose synthesis in actions of gibberellin and kinetin on cell expansion. 2,6-Dichlorobenzonitrile as a new cellulose-synthesis inhibitor. *Plant Cell Physiol.* **15**, 389–393.

Iwata, K. and Hogetsu, T. (1989). The effects of light irradiation on the orientation of microtubules in seedlings of *Avena sativa* L. and *Pisum sativum* L. *Plant Cell Physiol.* **30**, 1011–1016.

Koda, Y., Omer, E.A., Yoshihara, T., Shibata, H., Sakamura, S. and Okazawa, Y. (1988). Isolation of a specific potato tuber-inducing substance from potato leaves. *Plant Cell Physiol.* **29**, 1047–1051.

Lang, J.M., Eisinger, W.R. and Green, P.B. (1982). Effects of ethylene on the orientation of microtubules and cellulose microfibrils of pea epicotyl cells with polylamellate cell walls. *Protoplasma* **110**, 5–14.

Lawson, V.R. and Weintraub, R.L. (1975). Interactions of microtubule disorganizers, plant hormones, and red light in wheat coleoptile segment growth. *Plant Physiol.* **55**, 1062–1066.

Lloyd, C.W., Slabas, A.R., Powell, A.J. and Lowe, S.B. (1980). Microtubules, protoplasts and plant cell shape. An immunofluorescent study. *Planta* **147**, 500–506.

Mita, T. and Katsumi, M. (1986). Gibberellin control of microtubule arrangement in the mesocotyl epidermal cells of the d_5 mutant of *Zea mays* L. *Plant Cell Physiol.* **27**, 651–659.

Mita, T. and Shibaoka, H. (1983). Changes in microtubules in onion leaf sheath cells during bulb development. *Plant Cell Physiol.* **24**, 109–117.

Mita, T. and Shibaoka, H. (1984a). Effects of root excision on swelling of leaf sheath cells and on the arrangement of cortical microtubules in onion seedlings. *Plant Cell Physiol.* **25**, 1521–1529.

Mita, T. and Shibaoka, H. (1984b). Effects of S-3307, an inhibitor of gibberellin biosynthesis, on swelling of leaf sheath cells and on the arrangement of cortical microtubules in onion seedlings. *Plant Cell Physiol.* **25**, 1531–1539.

Mita, T. and Shibaoka, H. (1984c). Gibberellin stabilizes microtubules in onion leaf sheath cells. *Protoplasma* **119**, 100–109.

Nick, P., Bergfeld, R., Schäfer, E. and Schopfer, P. (1990). Unilateral reorientation of microtubules at the outer epidermal wall during photo- and gravitropic curvature of maize coleoptiles and sunflower hypocotyls. *Planta* **181**, 162–168.

Roberts, I.N., Lloyd, C.W. and Roberts, K. (1985). Ethylene-induced microtubule reorientations: mediation by helical arrays. *Planta* **164**, 439–447.

Sakiyama, M. and Shibaoka, H. (1990). Effects of abscisic acid on the orientation and cold stability of cortical microtubules in epicotyl cells of the dwarf pea. *Protoplasma* **157**, 165–171.

Sakurai, N. and Kuraishi, S. (1989). Growth and cell wall of dark-grown squash hypocotyls under water-stress. In *Plant Water Relations and Growth Under Stress* (eds M. Tazawa, M. Katsumi, Y. Masuda and H. Okamoto), pp. 331–338. Myu, Tokyo.

Sawhney, V.K. and Srivastava, L.M. (1974). Gibberellic acid induced elongation of lettuce hypocotyls and its inhibition by colchicine. *Can. J. Bot.* **52**, 259–264.

Sawhney, V.K. and Srivastava, L.M. (1975). Wall fibrils and microtubules in normal and gibberellic-acid-induced growth of lettuce hypocotyl cells. *Can. J. Bot.* **53**, 824–835.

Shibaoka, H. (1972). Gibberellin–colchicine interaction in elongation of azuki bean epicotyl sections. *Plant Cell Physiol.* **13**, 461–469.

Shibaoka, H. (1974). Involvement of wall microtubules in gibberellin promotion and kinetin inhibition of stem elongation. *Plant Cell Physiol.* **15**, 255–263.

Shibaoka, H. and Hogetsu, T. (1977). Effects of ethyl N-phenylcarbamate on wall microtubules and on gibberellin- and kinetin-controlled cell expansion. *Bot. Mag. (Tokyo)* **90**, 317–321.

Steen, D.A. and Chadwick, A.V. (1981). Ethylene effects in pea stem tissue. Evidence of microtubule mediation. *Plant Physiol.* **67**, 460–466.

Takeda, K. and Shibaoka, H. (1981). Effects of gibberellin and colchicine on microfibril arrangement in epidermal cell walls of *Vigna angularis* Ohwi et Ohashi epicotyls. *Planta* **151**, 393–398.

Volfová, A., Chvojka, L. and Haňkovská, J. (1977). The orientation of cell wall microtubules in wheat coleoptile segments subjected to phytohormone treatment. *Biol. Plant.* **19**, 421–425.

Wasteneys, G.O. and Williamson, R.E. (1989). Reassembly of microtubules in *Nitella tasmanica*: quantitative analysis of assembly and orientation. *Eur. J. Cell Biol.* **50**, 76–83.

Yoshihara, T., Omer, E.A., Koshino, H., Sakamura, S., Kikuta, Y. and Koda, Y. (1989). Structure of a tuber-inducing stimulus from potato leaves (*Solanum tuberosum* L.). *Agric. Biol. Chem.* **53**, 2835–2837.

13. ROLE OF THE CYTOSKELETON IN GRAVITY PERCEPTION

Andreas Sievers, Brigitte Buchen, Dieter Volkmann and Zygmunt Hejnowicz

*Botanisches Institut, Universität Bonn,
Venusbergweg 22, D-5300 Bonn 1, Germany*

Introduction

Plants live in the gravitational field of the Earth specified by the acceleration vector called gravity (1 g = 9.81 m s^{-2}). Thus, on Earth, gravity-sensitive cells are continuously stimulated. Via this stimulation the growth direction of plant organs is permanently under the control of gravity following the reaction chain: perception, transmission and response. Thus, gravity is perceived by a gravity-sensitive organ not only when its position is changed (dynamic stimulation) but also when it grows at a constant angle with respect to gravity (static stimulation). A striking example of static stimulation is the formation of compression wood on lower sides of stems in conifer trees (Westing, 1965). The physical principles for perception of gravity have often been reviewed, recently by Björkman (1988).

Gravity acts on every molecule. Two requirements, however, must be fulfilled for the susception (the physical step of perception) of dynamic stimulation: (i) gravity works on a mass M involved in gravity sensing mechanisms, (ii) the work done by gravity has to be higher than the elemental kinetic energy 1/2 kT = 2 × 10^{-21} J (20°C), to discriminate between gravity-dependent movement and thermal background. The work done by gravity on a mass M is gravity force times distance (D), hence during susception the mass M must be *displaced* by gravity. For intracellular masses involved in susception the following Minimum Condition must be fulfilled:

$$\Delta mgD > 2 \times 10^{-21} \text{ J for } D \ll \text{cell length}$$

where Δmg is the reduced weight of M (equal to the volume times difference in densities of M and the cytosol); the two vectors g and D are multiplied scalarly. (Note: The terms 'displaced', 'displacement' are used in the meaning of an actual movement from the initial equilibrium position. The term 'sedimentation' will be taken to mean the final effect of the displacement due to gravity, i.e. settling to the actual floor.)

The Minimum Condition determines the threshold for the reduced weight of M. It indicates, for example, that a cell of 20 μm in diameter is too short to fulfil the condition for mitochondria (diameter 0.5 μm, density 1.2 g cm^{-3}). If this condition is fulfilled by a mass, the gravity work may accumulate with time until the value required by the activation energy of a particular sensing mechanism is attained, because gravity-caused displacement is directional while directions of thermal motions are variable. Hence, if enough time is allowed for the action of gravity in a new direction, the Minimum Condition is not only a necessary but also a sufficient condition for dynamic gravistimulation. For example, for an amyloplast in a typical columella cell 1 s at 1 g results in the displacement which significantly exceeds that which on average might occur due to Brownian motion (Björkman, 1988). If in a cell multiple masses exist for which the Minimum Condition is fulfilled, they are all displaced in the same direction by gravity. Thus, time and multiplicity decrease the error in distinguishing a gravity-dependent displacement from the Brownian background.

Static gravistimulation is not connected with

The Cytoskeletal Basis of Plant Growth and Form   ISBN 0–12–453770–7

displacement, but with pressure or stresses (normal and/or shearing). In this case the threshold must be considered from the point of view of discrimination with respect to the pressure or stresses due to other factors, for instance turgor if the plasma membrane is involved as a sensor membrane.

Intracellular movable masses, characterized by high enough m to fulfil the Minimum Condition, and considered to be involved in sensing gravity, are called statoliths. In higher plants, amyloplasts act as statoliths. Nevertheless, in plant cells amyloplasts often occur which are gravitationally inactive (e.g. in guard, storage and secretory cells). Nuclei also fulfil the Minimum Condition, even though they are rarely considered as statoliths. Reasons for the inactivity may be immovability or lack of gravisensitivity. In the case of nuclei two additional reasons can be mentioned that exclude them as statoliths: (i) some morphogenetic functions which require a controlled angular positioning of the nucleus with respect to other organelles, i.e. it is anchored, and (ii) the fact that it is a single organelle. The cells which contain statoliths and which are involved in perception of gravity are called statocytes.

There are, however, also gravity responses which are claimed to occur without statoliths in some organs with or without starch (Westing, 1971; Caspar and Pickard, 1989) and in sporangiophores of *Phycomyces* (Dennison, 1971), as well as gravity-dependent polarity of cytoplasmic streaming (Hejnowicz *et al.*, 1985; Wayne *et al.*, 1990; Buchen *et al.*, 1991). Pickard and Thimann (1966) and Wayne *et al.* (1990) propose that the whole protoplast acts as the sensing mass leading to differential pressure and/or stress.

On the basis of the presence of statoliths the most important questions with respect to perception are: How is the work done by gravity on statoliths transferred to competent cellular structures, and what are the structures which transduce the physical stimulus into a biochemical/biophysical signal? (See reviews by Volkmann and Sievers, 1979; Sievers and Hensel, 1991.) The sedimentation of statoliths to a new position at the lateral cell wall is often considered to involve interaction of statoliths

with cortical structures such as endoplasmic reticulum (ER) cisternae, microtubules (MTs) and plasma membrane. Movement of statoliths (sliding along cytoplasmic structures) is also considered as a factor of susception, however, little attention is paid to small displacements from the initial equilibrium (Sievers and Volkmann, 1972). From our point of view this kind of displacement is of special interest because, as explained below, a small unidirectional displacement of statoliths seems to affect the interaction between statoliths and the cytoskeleton.

The cytoskeleton in statocytes

Unicellular systems

Rhizoids of the alga *Chara* are downward growing cells with a clear polar zonation with respect to the localization of organelles (Fig. 13.1; review Sievers and Schnepf, 1981). The apical and subapical zones contain thin bundles of microfilaments (MFs), the basal zone thicker ones (Fig. 13.2a, b; demonstrated by treatment with rhodamine–phalloidin; Sievers *et al.*, 1989, 1991). The apical part contains statoliths and is gravitropically responsive. In a vertically oriented rhizoid, the statoliths – compartments filled with crystallites of barium sulphate – are maintained not at the 'floor' of the cell but at a distance of 10–30 μm above. They are dynamically suspended by MFs (demonstrated by treatment with cytochalasin B (CB); Hejnowicz and Sievers, 1981), which form a network contacting and encircling the statoliths (Sievers *et al.*, 1989). When the statoliths are removed from the apex by basipetal centrifugation they are retransported with velocities increasing from 1 to 14 μm min^{-1}. This transport is mediated by actin filaments (Sievers *et al.*, 1991). Beneath the group of statoliths a spherical aggregate of anastomosing ER membranes is positioned. The integrity of this ER aggregate and of the polar organization of the whole cell is actin dependent (Bartnik and Sievers, 1988; Sievers *et al.*, 1991).

The MFs in the rhizoid apex interact dynamically with the statoliths both from the basal and

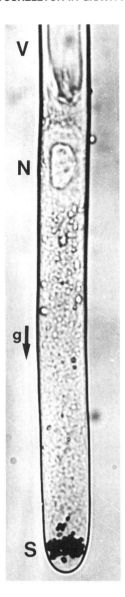

Fig. 13.1. Part of a growing *Chara* rhizoid in normal vertical orientation. The polar organization is expressed by the constant distance of the nucleus (N) and of the statoliths (S) from the vertex. g, direction of gravity; V, vacuole. Diameter of the rhizoid: 30 μm.

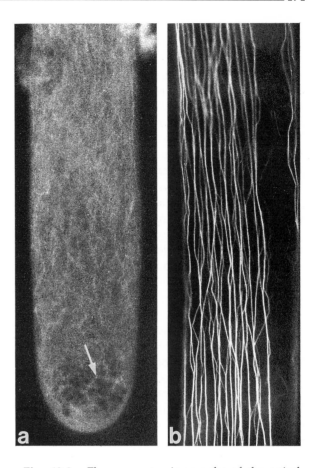

Fig. 13.2. Fluorescence micrographs of the apical with the subapical zone (a) and the basal zone (b) of a *Chara* rhizoid after staining with rhodamine–phalloidin. a, The apical and the subapical zone contain a fine network of actin filaments. Actin filaments encircle the statoliths (arrow). b, Thick, parallel oriented filament bundles are characteristic for the basal zone with cytoplasmic streaming. The bundles are interconnected by shorter and thinner filaments. Diameter of the rhizoid: 30 μm. Originals by Markus Braun.

apical side, or likely from all sides, as indicated by (i) saltations of a statolith in all directions but mostly up and down, (ii) movement of statoliths against the force of gravity in inverted rhizoids. If inverted rhizoids, which have already recovered the normal position of their statoliths, are treated with CB the statoliths move in an apical direction, i.e. in the opposite direction to the gravity vector (Hejnowicz and Sievers, 1981). This indicates that the destructive effect of CB on MFs is stronger on the basal than on the apical side of the statoliths and – when the initial balance of cytoskeletal and gravitational forces is disturbed – an acropetally directed net cytoskeletal force acts on the statoliths. Direct proof that MFs exert forces on statoliths is provided by experiments in which the position of statoliths in rhizoids, which started in a normal vertical position, was monitored during the parabolic phase of rocket flights. In this case gravity was reduced to 10^{-4} g and the weightless statoliths moved in a basipetal direction (Sievers, 1990; Volkmann et al., 1991). For MF bundles to exert a force on statoliths would necessarily mean (according to Newton's third principle) that there is an oppositely directed reaction force in the bundles, which would bring them under tension.

Microtubules are observed in the region basal to the statoliths. MT-affecting drugs did not cause statoliths to sediment, nor inhibited their saltations and their transport. Hence, MTs seem to be involved neither in positioning of the statoliths nor in graviperception (Hejnowicz and Sievers, 1981; Sievers et al., 1991).

Protonemata of the mosses Physcomitrella and Ceratodon grow upward in darkness. The apical region of the tip cell shows a polar zonation. In Physcomitrella the amyloplasts do not sediment (Jenkins et al., 1986). Both MTs and MFs occur in all zones (Doonan et al., 1985, 1988). The MTs form a three-dimensional meshwork with predominant longitudinal direction enveloping the nucleus and amyloplasts. Near the apex they converge to form a focus. MFs are abundant in the apical dome. Bundles of MFs weave amongst the amyloplasts approximately along the cell axis.

Tip cells of dark-grown protonemata of Ceratodon possess a unique longitudinal zonation: the 'apical dome' with a group of amylochloroplasts, a plastid-free zone, a zone where amylochloroplasts sediment (the only region where, during horizontal exposure, amylochloroplasts sediment), the zone containing the nucleus, amylochloroplasts, and vacuoles (Schwuchow et al., 1990). MTs are distributed throughout the cytoplasm in a mostly axial orientation. They seem to converge in the apical dome and are associated with plastids and the nucleus.

Statocytes of roots and shoots

The typical polar organization of statocytes of a Lepidium root cap (Figs 13.3 and 13.5) is characterized by the proximal positioning of

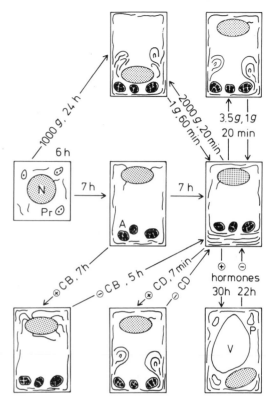

Fig. 13.3. Polar organization of statocytes. Schematic drawings of central root statocytes of Lepidium demonstrating the arrangement of the nucleus (N), ER cisternae (lines) and plastids (A, sedimented amyloplasts; P, destarched plastids; Pr, proplastids) during normal development of the root (mid line) and after different treatments. Note: (i) Physical (upper line) and chemical (lower line) effects are reversible; (ii) the cortical ER cisternae are never displaced. CB, CD, cytochalasin B, D; g, gravitational acceleration; hormones, gibberellic acid plus kinetin; V, vacuole.

the nucleus, the distal accumulation of ER cisternae with sedimented amyloplasts thereupon, while other organelles like mitochondria and dictyosomes are distributed at random (see review by Sievers and Hensel, 1991). This polar arrangement of organelles, genetically determined (Sievers et al., 1976; Volkmann et al., 1986), is achieved and maintained by means of the cytoskeleton. Such polarity is of prime importance for graviperception.

In statocytes of Lepidium roots two populations of MTs occur: one lines the anticlinal cell wall (to a lesser extent the periclinal cell wall),

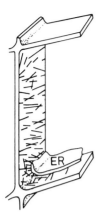

Fig. 13.4. Schematic view of a slice-like opened statocyte of *Lepidium* showing the distribution of microtubules along the cell walls. At the cell edge, between a distal ER cisterna and the plasma membrane, microtubules are arranged in a criss-cross manner. Adapted from Sievers and Hensel, 1991.

is mostly transverse in direction, and is relatively stable to colchicine; the second population is colchicine-sensitive and forms networks of crossed arrays at the distal cell edges. These MTs, in co-ordination with MFs and cross-bridging structures, stabilize the ER complex at the distal position (Fig. 13.4; Hensel, 1984a, b, 1989a). Desmotubules of the plasmodesmata additionally stabilize the cortical ER cisternae in neighbouring cells. The proximal position of the nucleus is not affected by colchicine treatment (Hensel, 1984a). MTs between the plasma membrane and an extensive network of cortical ER also exist in statocytes of *Arabidopsis* (Sack and Kiss, 1989).

The importance of MTs for the polar organization of statocytes is indicated by treatments with gibberellic acid plus kinetin, hormones which are proposed to play a role in the orientation of microtubules (Shibaoka, 1974; see also Chapter 12, this volume). The treatment results in the destruction of the polar organization of the statocyte and in the loss of amyloplastic starch (Fig. 13.3; Busch and Sievers, 1990). Immediately after the hormone treatment the roots do not respond to gravity though they still grow (Iversen, 1969; Busch and Sievers, 1990). Polarity is restored 22 h later, starch is resynthesized, and gravisensitivity is also restored.

The presence of bundles of MFs in root statocytes has been demonstrated by heavy-meromyosin decoration (Hensel, 1988), by labelling with a monoclonal anti-actin antibody (Hensel, 1986, 1989a) and by rhodamine–phalloidin (Hensel, 1989a; White and Sack, 1990). The MFs keep the nucleus in its proximal position (Hensel, 1985; Lorenzi and Perbal, 1990) and participate in the MT-mediated stabilization of the ER at the distal cell pole (Hensel, 1987). Moore and Evans (1986), referring to Moore and McClelen (1985), indicated the occurrence of MFs associated with amyloplasts and nuclei in statocytes of *Zea*.

The development of a statocyte starts with a continuous formation of ER cisternae at its proximal pole with participation of the nuclear envelope. The ER cisternae are translocated to the distal pole by a co-ordinated action of MFs (Hensel, 1985). Cytochalasin B (CB) or cytochalasin D (CD) inhibits the acropetal translocation of ER bringing about an accumulation of ER at the proximal pole (Fig. 13.3).

The polar organization of the statocytes is changed by varying the amount and direction of the gravity vector. At higher rates of centrifugation (20 min, 100–2000 g, root tip-directed) a stratification of statocytes is established according to the densities of organelles, except for the cortical ER cisternae which remain near the plasma membrane (Fig. 13.3; Sievers and Heyder-Caspers, 1983). If, after centrifugation, the roots are exposed to 1 g the ER cisternae are relocated from the lateral to the normal distal position. This process is MF dependent (Wendt *et al.*, 1987). The recovery of the distal position of the ER complex is achieved within approximately 10 min. For the same interval the latent period of the gravi-response increases (Sievers and Heyder-Caspers, 1983; Wendt and Sievers, 1986).

MFs are involved in the process of forming and maintaining the distal ER complex, by not only providing the motive force and directionality for the ER transport, but also by lifting the amyloplasts under which the ER is located. Proof that MFs exert a force on amyloplasts in a basipetal direction has been provided by a rocket experiment: reducing the gravity to 10^{-4} g (the parabolic phase of the rocket flight) caused basipetal displacement of the weight-less statoliths, as was also the case for *Chara* rhizoids (Volkmann *et al.*, 1991).

In roots growing in normal gravity, this force exerted on amyloplasts in the basipetal direction compensates (completely or partially) for the gravitational force acting on statoliths (the reduced weight of statoliths). This means that in normal vertically oriented roots the pressure of sedimented statoliths on the distal ER complex is reduced. This conclusion is supported by experiments with treatments increasing the pressure of amyloplasts on the ER complex: (i) short incubation of the roots in CD in the normal vertical position causes a displacement of the ER complex towards the anticlinal wall (except for the cortical cisterna) while amyloplasts settle to the very distal position (Fig. 13.3; Hensel, 1987); (ii) a relatively low centrifugal force (3.5 g) is sufficient for the amyloplasts to displace the ER complex (Fig. 13.3; Wendt and Sievers, 1986). Thus, when amyloplasts do press with the full weight (Δmg) on the distal ER complex its cisternae are displaced from beneath the amyloplasts (except for the cortical cisterna).

The polarity of the statocytes in roots germinated and grown on a horizontal clinostat, which never experienced normal unidirectional gravistimulation, is normal, apart from the rotation-induced random distribution of the amyloplasts (Sievers *et al.*, 1976). If, however, a short unidirectional gravistimulus (10 min of normal vertical exposure) is applied during this procedure, destructive changes develop during the following rotation: the distal ER complex disintegrates, starch is hydrolysed, and autolytic damage of the statocytes develops. The destructive changes also occur when normal vertically grown *Lepidium* roots are rotated continuously on a horizontal clinostat. Destruction of statocyte polarity can also be evoked by a longer (120 min) treatment with the Ca^{2+} ionophore A23187 (Wendt and Sievers, 1989; Sievers, 1990). Shorter treatment (15–30 min) causes the displacement of the ER complex from the distal position beneath amyloplasts towards the anticlinal cell walls (Hensel, 1989b). The reason for both effects may be the disruption of MFs caused by the elevation of the

calcium concentration (Sievers, 1990) so that the amyloplasts start to press more strongly on the ER complex. Interestingly, statocyte polarity can be disrupted by inserting an electrode to measure the transmembrane potential. This treatment can leave the statocytes with normal polarity and full gravi-electrical sensitivity, or lead alternatively to the loss of polarity (Behrens *et al.*, 1985). The latter case results in complete loss of gravi-electrical sensitivity.

The mechanical interaction between the MF network and amyloplasts is indicated by the fact that application of CB results in a considerable increase in the velocity of amyloplast sedimentation immediately after inversion (Sievers *et al.*, 1989). Direct observations of living statocytes during gravistimulation (Sack *et al.*, 1985) indicated that the amyloplasts slow down as they move to the actual floor after tilting the root. This initially high velocity can be expected if the amyloplasts interact mechanically with MFs and the previous balance of forces is disturbed. An interaction of amyloplasts with MFs, especially in the highly vacuolated statocytes of shoots, is indicated by the variety of their movements, from saltatory motion to cyclosis-type cytoplasmic streaming (Clifford and Barclay, 1980; Heathcote, 1981; Sack and Leopold, 1985; Sack *et al.*, 1986). Apart from statocytes, there is evidence from other cells, such as the alga *Mougeotia*, that interconnections exist between plastids and MFs (Wagner and Klein, 1981; Cox *et al.*, 1987).

Role of the cytoskeleton in gravi-perception

General background

The cytoskeleton is a dynamic network of filamentous proteins penetrating the cytosol (Staiger and Schliwa, 1987) and anchored at or associated with the cell membrane and/or cortical ER (Geiger, 1985; Hepler *et al.*, 1990). The cytoskeletal elements interact chemically (weak bonds) or mechanically (e.g. by encircling) with organelles. These interactions must result in a complicated state of stress (tension, compression, shear) and strain (lengthening and shortening) in cytoskeletal elements, the membranous system and the cytosol (Hameroff, 1988).

MTs, MFs and intermediate filaments are likely to be under tension in the cell. The main force which balances the tensional forces in the cytoskeleton is provided by the plasma membrane at which the cytoskeletal elements are anchored (directly and/or through the cortical ER closely associated with the plasma membrane). The plasma membrane, under tension (Wolfe and Steponkus, 1981), compresses the cytosol. The *cytoskeletal tension* is mostly due to forces generated metabolically, like shear forces generated between the actin bundle and myosin molecules which underlie protoplasmic motility (Kuroda, 1990). If there are masses in the cell which fulfil the Minimum Condition and interact with the cytoskeleton in a movable way, cytoskeletal tension must necessarily be affected by gravity; hence, there must be a gravity-dependent component of this tension. If the position of the cell changes with regard to the vertical this component changes the direction and the balance of cytoskeletal tension is disturbed.

Because of the omnipresence of the cytoskeleton in cells and of the ubiquity of mechanical interactions between the cytoskeleton and organelles, the involvement of cytoskeletal tension must be kept in mind when considering the perception of gravity. Since the cytoskeletal elements are anchored at the cortical endoplasmic reticulum and/or plasma membrane, disturbance can be transmitted from the interior of the cell to the cortical structures.

Unicellular systems

Tilting *Chara* rhizoids from the vertical to the horizontal causes sedimentation of statoliths near the plasma membrane of the lower apical wall, resulting in an asymmetric exocytosis of cell wall material via Golgi vesicles and, finally, in differential growth of the opposite flanks (see review by Sievers and Schnepf, 1981). If the statoliths are removed from the apex by basipetal centrifugation the rhizoid does not respond to gravity. However, after acropetal retransport which is mediated by actin filaments (Sievers *et al.*, 1991) the rhizoid regains its gravisensitivity. From other cell types like pollen tubes (Franke *et al.*, 1972) and secretory root cap cells (Volkmann and Czaja, 1981) it is

indicated that MFs are involved in the transport of Golgi vesicles to the cell periphery. Thus the disturbance of the MF network by gravity via statoliths should cause a disturbance of the exocytotic processes. When statoliths are displaced basipetally after inverting the rhizoid, the shape of the newly formed cell portion is changed; however, since statolith displacement was symmetrical, shape change was also symmetrical (Hejnowicz and Sievers, 1981, figs 9 and 10). This indicates that cytoskeletal tension may influence the shape of the rhizoid.

The amyloplasts in the tip cell of protonemata of *Physcomitrella* do not sediment in the gravity field. In spite of this fact they may be involved in signal transduction (Jenkins *et al.*, 1986). In the tip cell of *Ceratodon* the zone of amyloplast sedimentation is more specialized for lateral than for axial sedimentation (Walker and Sack, 1990). MTs remain closely associated with sedimented amyloplasts. The sedimentation starts within 15 min after horizontal placement, while the presentation time (threshold duration of unintermittent horizontal exposure to elicit curvature) is 12–17 min. Gravistimulated protonemata had a higher density of microtubules near the lower flank compared to the upper flank in the plastid-free zone (Schwuchow *et al.*, 1990). Treatment with oryzalin, which eliminated most of the MTs throughout the apical reigon (except for phragmoplast and spindle MTs), also eliminated the response to gravity. Schwuchow *et al.* (1990) hypothesize that a microtubule redistribution plays a role in gravitropism of the *Ceratodon* protonemata.

Statocytes of roots and shoots

The membrane potential of statocytes in *Lepidium* roots changes as early as 2 s (mean 8 s) after tilting the root (Sievers *et al.*, 1984; Behrens *et al.*, 1985). This physiological change, which itself must be somewhat delayed with respect to the susception initiating it, occurs so fast that the event can be considered to occur immediately upon tilting. Another astonishingly short threshold time is the *perception time* (T_{pc}) – the minimum duration of a stimulus provided repeatedly in an intermittent way on a clinostat (review Volkmann and Sievers, 1979). For the *Avena* coleoptile (Pickard, 1973) and also for the

Lepidium root (Kniel and Sievers, unpublished) $T_{pc} < 1$ s. Surely, T_{pc} cannot be shorter than the threshold time necessary to distinguish the gravity effect from thermal noise; however, it can be longer if there is a physiological component in the signal chain not separable from the mechanical events of susception. In *Lepidium* roots this physiological component is likely to be very short, so their T_{pc} probably represents the physical component. What mechanical event can be so fast? The initial effect of changing the orientation of a root or a shoot with respect to the vertical is the downward movement of statoliths in statocytes. It is a physical necessity that some movement of the statoliths starts immediately. The best candidate for such a minimum event seems to be a disturbance of the cytoskeletal tension caused by statoliths which start to move immediately after tilting the root (Fig. 13.5a, b); 'statoliths may trigger the transduction mechanisms via actin filaments' (Sievers *et al.*, 1989).

At a time when the existence of MF bundles in statocytes had not been documented, the effect of statolith displacement was designated as 'the sliding of the statoliths along cytoplasmic structures' (see review by Volkmann and Sievers, 1979). Relevant to this are the results

of Larsen (1965, 1969) and Iversen and Larsen (1971, 1973) concerning graviperception of *Lepidium* roots stimulated at different angles with respect to gravity, including the effect of a transient inversion (the preinversion effect) on the movement of amyloplasts and on the response. These studies have indicated that the response is due to gravity-generated motion of the amyloplasts which has a transverse component with respect to the axis of the root (asymmetric motion). The duration of the motion depends on the initial location of the amyloplasts, hence the preinversion effect. If gravistimulation is perceived via disturbance of cytoskeletal tension and is to lead to a curvature of the root, the disturbance must be asymmetric with respect to the root axis. Asymmetric disturbance of cytoskeletal tension is achieved by the transverse component of the statolith movement.

The group of hypotheses based on the sedimentation of statoliths (or even of a single 'leading statolith') to the outer flank of the cell where they could perform a signalling action seems to be inadequate because the distance travelled during T_{pc} is too small (Volkmann and Sievers, 1979). However, the conclusion that the necessary minimum for perception is fulfilled by the disturbance of the cytoskeletal tension does not preclude the sedimentation of statoliths to a new specific site as an additional step. Recently it has been possible to separate the early effect from sedimentation by using the gravitropic starchless mutant (TC7) of *Arabidopsis* (Caspar and Pickard, 1989; Kiss *et al.*, 1989), and *Nicotiana* (Kiss and Sack, 1989). Centrifugation studies by Kiss *et al.* (1989) on TC7 roots showed that the starch-free plastids of the columella cells are the most movable components. It can be calculated that the starch-free plastids fulfil the Minimum Condition and can act as statoliths. However, they do not sediment. The simplest explanation is that these presumptive statoliths are light enough to be kept far from the floor by MFs. But even such statoliths will move on tilting the root until a new steady state of cytoskeletal tension is achieved. The roots of TC7 are less sensitive to gravistimulation than those of the wild-type (WT) roots; however, the reduction

(a) 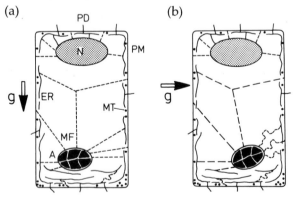 (b)

Fig. 13.5. Scheme of a central root statocyte of *Lepidium* illustrating the difference in microfilament (MF) tensions. a, Normal vertical orientation with all MFs in tension; b, horizontal orientation with asymmetrically stretched and relaxed MFs due to the gravity dependent displacement of the amyloplast (A). ER, endoplasmic reticulum; g, direction of gravity; MT, microtubule; N, nucleus; PD, plasmodesma; MF, microfilament; PM, plasma membrane.

of sensitivity – indicated by the curvature after dynamic stimulation – is much less than the reduction of the relative masses of statoliths. According to Caspar and Pickard (1989) the curvature of a root induced in the horizontal position during intervals from 2.5 to 30 min is 70% as great in TC7 as in WT, while the mass of the starch-free plastids in TC7 is only 6% of that of starch-laden ones in WT. Kiss *et al.* (1989) showed that WT roots curved in response to intermittent stimulation when the stimulation time was 10 s or less, but TC7 roots required 2 min.

In view of the fact that the statoliths at the instant of tilting the TC7 root were not sedimented, and thus could not be at a specific resting site, the group of hypotheses proposing that graviperception is due to the displacement of statoliths from their original site of sedimentation, e.g. ER membranes (Sievers and Volkmann, 1972; Volkmann and Sievers, 1979) should be ruled out when the minimum necessary for the perception is considered. This, however, does not exclude ER membranes as gravisensors where the step of signal transduction might occur.

An additional step of perception may be the sedimentation of statoliths when they are so heavy that they settle to the actual floor. The sedimentation allows qualitatively new effects of statolith movement, e.g. an asymmetrical obstructing of the supply of Golgi vesicles for exocytosis in *Chara* rhizoids, or statolith action at specific recognition sites. Lateral centrifugation, either with low or high centrifugal acceleration (Wendt *et al.*, 1987), which causes the sedimentation of amyloplasts to the centrifugal anticlinal wall where the plasma membrane is lined with the cortical ER with MTs in between, brings about a clear gravitropic curvature which surely cannot be accounted for the disturbance of cytoskeletal tension only.

The concept of two steps in perception (Lawton *et al.*, 1986) necessarily means that perception without sedimentation is qualitatively different from that operating on the basis of sedimentation. Indeed, the degree of curvature of the starchless *Arabidopsis* mutant is reduced only 30% while the gravity force exerted by plastids is reduced by as much as 94% compared with the WT (Caspar and Pickard, 1989), which is suggestive of such a qualitative difference. If there are two perception mechanisms and both operate in statocytes, there should be two perception times, a short one for the disturbance of cytoskeletal tension causing a slow response, and a longer one for the sedimentation mechanism followed by a faster response. Such an explanation may be relevant to the biphasic curve in the plot of curvature (linear) versus stimulation time (logarithmic) (Shen-Miller, 1970; Johnsson, 1971).

It should be taken into account that asymmetrical disturbance of cytoskeletal tension can be a way of perceiving dynamic gravistimulation. Sedimentation, however, is a way of perceiving long-term action of gravity in a particular direction and is thus adequate for perceiving static gravistimulation.

Gravity-dependent cytoplasmic streaming

In algal cells, different streaming velocities occur in up- and downward directions (see review by Kuroda, 1990), and they are influenced by gravity (Hejnowicz *et al.*, 1985; Wayne *et al.*, 1990; Buchen *et al.*, 1991). In *Chara* rhizoids, the velocity of streaming is faster in acropetal than in basipetal direction. This difference is due to endogenous motive shearing forces generated by MF bundles in the cortical cytoplasm (Fig. 13.2b; Hejnowicz *et al.*, 1985) and enhanced by gravity in normal vertical orientation of the cell. Experiments under 10^{-4} g in rockets show that both streaming velocities increased, although the basipetal streaming rate increased more than the acropetal one (Buchen *et al.*, 1991).

In normal vertically oriented *Nitellopsis* internodal cells, the downward streaming velocity is faster than the upward rate (Wayne *et al.*, 1990). Upon horizontal exposure, both streaming velocities become equal. In contrast to the rhizoid, no steady endogenous component exists. The effect of gravity is much higher than that which can result from vectorial addition of the gravity force to the shearing motive forces generated at the interphase ecto/endoplasm where the MF bundles are located. This means that there must be a mechanism which amplifies the effect of gravity on streaming velocities in *Nitellopsis*.

Different densities of the medium in which the cells are kept and the external Ca^{2+} and K^+ concentrations affect the graviresponsiveness of the streaming. The voltage-dependent Ca^{2+} and K^+ blockers affect the ratio of the velocities in the vertical position. Removal of one or both cell ends by ligation, or treatment of the ends with UV light, completely eliminates the gravity-induced polarity of cytoplasmic streaming (Wayne *et al.*, 1990). Based on these data we propose that gravity causes a change of shear stress between the plasma membrane and the cell wall in the internodal cell of *Nitellopsis*. That is, there must be a reaction force in the ectoplasm to the force which moves the endoplasm. We infer that the ectoplasm is restrained in a normal non-ligated cell along the lateral plasma membrane. The branches of the MF bundle system which run into the region beneath the plasma membrane (Williamson *et al.*, 1986) may be relevant to this. The plasma membrane, in turn, must be restrained along the cell wall. The reaction force can induce a shear stress in these restraining connections which would affect physiological and electrochemical activities of the plasma membrane. In the horizontal position, full symmetry of the peripheral shear stress exists with respect to the up- and downward streaming. However, in the vertical position, the gravity acting on the protoplast results in a force which must affect the shear stresses in the plasma membrane asymmetrically on both sides, so that it enhances the stress on one side and decreases it on the opposite side. A decrease of the stress enhances the streaming velocity; in consequence the plasma membrane activity and the streaming velocities can be changed in an asymmetric way. The model sketched above is consistent with all observations of Wayne *et al.* (1990) and is quantitatively realistic. It is based on the disturbance of the cytoskeletal tension and the shear stress in the plasma membrane. We also assume that in *Chara* rhizoids a release of cytoskeletal tension enhances the streaming velocity on the change to microgravity. It has been indicated (Hejnowicz *et al.*, 1985) that, in

very long cells, the actin bundles themselves have mass large enough to be affected by gravity, and that this effect may be transmitted to the cytoskeletal elements which stabilize the actin bundles in the ectoplasm. The Minimum Condition indicates that the bundles may act as gravity-sensing masses if they are >25 µm. Quantitatively, this hypothesis is realistic considering the activation energy for tension-dependent ionic channels.

Sporangiophores of *Phycomyces* with a large vacuole show longitudinal channels of protoplasmic streaming in the cortical cytoplasm (Dennison, 1971). Dennison (1971) suggested that lateral displacement of the vacuole in tilted sporangiophores may be involved in the long-term graviresponse, making the layer of cytoplasm thicker on the lower side. A disturbance of the cytoskeletal tension may be involved both in this long-term response, as well as in the faster response triggered by the physical deformation of the cell.

The cytoskeleton in transduction

A review of the common transduction pathways relevant to graviresponse is provided by Björkman (1988). How does a disturbance of cytoskeletal tension generate the transduction step (first biochemical/biophysical signal)? If the cytoskeletal elements are attached to stretch sensitive ion channels either in the cortical ER or in the plasma membrane, or even in the amyloplast membrane, the disturbance can cause asymmetric stimulation of the channels and directional ion transport. Stretch activated channels in plant cells have been described for the cell membrane of tobacco protoplasts (Falke *et al.*, 1988) and in guard cells of *Commelina* (Schroeder and Hedrich, 1989). The activation energy for channel opening is quite high (Björkman, 1988), thus the energy transmitted by the cytoskeleton must be focused on the channels, i.e. the cytoskeleton should act as a part of an amplifying system (Volkmann and Sievers, 1991). Such a possibility exists for sensing the disturbance of the cytoskeleton due to gravity-induced movement of statoliths, or due to gravitational stresses in the cell wall system as postulated by Edwards and Pickard (1987). A question remains whether stretch-activated channels allow dramatic changes in ion fluxes, including calcium which appears to be involved in transduction (Björkman and Leopold, 1987). It may be relevant that a Ca^{2+} pump in the ER of the *Lepidium* root has been proposed as the trigger for rapid changes in membrane potential following gravistimulation (Sievers *et al.*, 1984; Sievers, 1990).

Realization of graviresponse by the cytoskeleton

The typical graviresponse of plant organs depends on differential elongation rates on opposite sides of the organ. There are many aspects of this differential growth; we are interested in those which involve the cytoskeleton. One aspect has been described by Nick *et al.* (1990) who have shown that differential growth in the horizontally positioned sunflower hypocotyl is correlated with the reorientation of microtubules underlying the outer wall of the upper epidermis – from transverse to longitudinal – whereas the orientation in the lower epidermis maintained their original transverse direction. Studies on the role of the change in MT orientation in the regulation of cell elongation rates (pea stem), showed that the change in MT orientation is not a consequence of elongation decline but rather its cause (Laskowski, 1990). Hush *et al.* (1990) point out the importance of mechanical stresses in cell walls. We wish to point out the importance of the distribution of stresses in the cytoskeleton as a fundamental factor controlling the orientation of MTs.

Concluding remarks

The cytoskeleton, especially MFs, appears to be the factor in graviperception which participates in sensing the gravity-dependent displacement of masses when the initial balance is disturbed by changing the gravity direction or magnitude. The experimental data indicate that the disturb-

ance of cytoskeletal tension transmits this displacement to the membranous system of the cell, initiating the chain of events which leads to graviresponse. The change of cytoskeletal tension might thus be an important event for gravity-dependent phenomena. However, we can take another look at this change; gravity-dependent movement of statoliths allows cytoskeletal tension to change and thus enables *in vivo* experimental studies of cytoskeleton mechanics in cell biology. Statoliths and gravity are natural means by which the cytoskeleton can be mechanically affected and from this point of view gravity perception constitutes a way of looking into the cellular machinery.

Acknowledgements
This work was supported by Bundesminister für Forschung und Technologie, Bonn, and Ministerium für Wissenschaft und Forschung, Düsseldorf, ('AGRAVIS') and by Deutsche Forschungsgemeinschaft, Bonn.

References

Bartnik, E. and Sievers, A. (1988). In-vivo observations of a spherical aggregate of endoplasmic reticulum and of Golgi vesicles in the tip of fast-growing *Chara* rhizoids. *Planta* **176**, 1–9.

Behrens, H.M., Gradmann, D. and Sievers, A. (1985). Membrane-potential responses following gravistimulation in roots of *Lepidium sativum* L. *Planta* **163**, 463–472.

Björkman, T. (1988). Perception of gravity by plants. *Adv. Bot. Res.* **15**, 1–41.

Björkman, T. and Leopold, A.C. (1987). An electric current associated with gravity sensing in maize roots. *Plant Physiol.* **84**, 841–846.

Buchen, B., Hejnowicz, Z., Braun, M. and Sievers, A. (1991). Cytoplasmic streaming in *Chara* rhizoids: Studies in a reduced gravitational field during parabolic flights of rockets. *Protoplasma*, in press.

Busch, M.B. and Sievers, A. (1990). Hormone treatment of roots causes not only a reversible loss of starch but also of structural polarity in statocytes. *Planta* **181**, 358–364.

Caspar, T. and Pickard, B.G. (1989). Gravitropism in a starchless mutant of *Arabidopsis*. *Planta* **177**, 185–197.

Clifford, P.E. and Barclay, G.F. (1980). The sedimentation of amyloplasts in living statocytes of the dandelion flower stalk. *Plant Cell Environ.* **3**, 381–386.

Cox, G., Hawes, C.R., van der Lubbe, L. and Juniper, B.E. (1987). High-voltage electron microscopy of whole, critical-point dried plant cells. 2. Cytoskeletal structures and plastid motility in *Selaginella. Protoplasma* **140**, 173–186.

Dennison, D.S. (1971). Gravity receptors in *Phycomyces.* In *Gravity and the Organism* (eds S.A. Gordon and M.J. Cohen), pp. 65–71. University of Chicago Press, Chicago.

Doonan, J.H., Cove, D.J. and Lloyd, C.W. (1985). Immunofluorescence microscopy of microtubules in intact cell lineages of the moss, *Physcomitrella patens. J. Cell Sci.* **75**, 131–147.

Doonan, J.H., Cove, D.J. and Lloyd, C.W. (1988). Microtubules and microfilaments in tip growth: evidence that microtubules impose polarity on protonemal growth in *Physcomitrella patens. J. Cell Sci.* **89**, 533–540.

Edwards, K.L. and Pickard, B.G. (1987). Detection and transduction of physical stimuli in plants. In *The Cell Surface in Signal Transduction* (eds E. Wagner, H. Grepping and B. Millet), NATO ASI Series vol. H 12, pp. 40–66. Springer-Verlag, Berlin.

Falke, L.C., Edwards, K.L., Pickard, B.G. and Misler, S. (1988). A stretch-activated anion channel in tobacco protoplasts. *FEBS Lett.* **237**, 141–144.

Franke, W.W., Herth, W., Van der Woude, W.J. and Morré, D.J. (1972). Tubular and filamentous structures in pollen tubes: Possible involvement as guide elements in protoplasmic streaming and vectorial migration of secretory vesicles. *Planta* **105**, 317–341.

Geiger, B. (1985). Microfilament-membrane interaction. *TIBS* **10**, 456–461.

Hameroff, S.R. (1988). Coherence in the cytoskeleton: implications for biological information processing. In *Biological Coherence and Response to External Stimuli* (ed. H. Fröhlich), pp. 242–265. Springer-Verlag, Berlin.

Heathcote, D.G. (1981). The geotropic reaction and statolith movements following geostimulation of mung bean hypocotyls. *Plant Cell Environ.* **4**, 131–140.

Hejnowicz, Z. and Sievers, A. (1981). Regulation of the position of statoliths in *Chara* rhizoids. *Protoplasma* **108**, 117–137.

Hejnowicz, Z., Buchen, B. and Sievers, A. (1985). The endogenous difference in the rates of acropetal and basipetal cytoplasmic streaming in *Chara* rhizoids is enhanced by gravity. *Protoplasma* **125**, 219–229.

Hensel, W. (1984a). A role of microtubules in the polarity of statocytes from roots of *Lepidium sativum* L. *Planta* **162**, 404–414.

Hensel, W. (1984b). Microtubules in statocytes from roots of cress (*Lepidium sativum* L.). *Protoplasma* **119**, 121–134.

Hensel, W. (1985). Cytochalasin B affects the structural polarity of statocytes from cress roots (*Lepidium sativum* L.). *Protoplasma* **129**, 178–187.

Hensel, W. (1986). Demonstration of microfilaments in statocytes of cress roots. *Naturwissenschaften* **73**, 510.

Hensel, W. (1987). Cytodifferentiation of polar plant cells: formation and turnover of endoplasmic reticulum in root statocytes. *Exp. Cell Res.* **172**, 377–384.

Hensel, W. (1988). Demonstration by heavy-meromyosin of actin microfilaments in extracted cress (*Lepidium sativum* L.) root statocytes. *Planta* **173**, 142–143.

Hensel, W. (1989a). Tissue slices from living root caps as a model system in which to study cytodifferentiation of polar cells. *Planta* **177**, 296–303.

Hensel, W. (1989b). Effects of elevated calcium and of calmodulin-antagonists upon the cytodifferentiation of statocytes in the root cap of cress. *J. Plant Physiol.* **134**, 460–465.

Hepler, P.K., Palevitz, B.A., Lancelle, S.A., McCauley, M.M. and Lichtscheidl, I. (1990). Cortical endoplasmic reticulum in plants. *J. Cell Sci.* **96**, 355–373.

Hush, J.M., Hawes, C.R. and Overall, R.L. (1990). Interphase microtubule re-orientation predicts a new cell polarity in wounded pea roots. *J. Cell Sci.* **96**, 47–61.

Iversen, T.-H. (1969). Elimination of geotropic responsiveness in roots of cress (*Lepidium sativum*) by removal of statolith starch. *Physiol. Plant.* **22**, 1251–1262.

Iversen, T.-H. and Larsen, P. (1971). The starch statolith hypothesis and the optimum angle of geotropic stimulation. *Physiol. Plant.* **25**, 23–27.

Iversen, T.-H. and Larsen, P. (1973). Movement of amyloplasts in the statocytes of geotropically stimulated roots. The pre-inversion effect. *Physiol. Plant.* **28**, 172–181.

Jenkins, G.I., Courtice, G.R.M. and Cove, D.J. (1986). Gravitropic responses of wild-type and mutant strains of the moss *Physcomitrella. Plant Cell Environ.* **9**, 637–644.

Johnsson, A. (1971). Investigations of the geotropic curvature of the *Avena* coleoptile. I. The geotropic response curve. *Physiol. Plant.* **25**, 35–42.

Kiss, J.Z. and Sack, F.D. (1989). Reduced gravitropic sensitivity in roots of a starch-deficient mutant of *Nicotiana sylvestris*. *Planta* **180**, 123–130.

Kiss, J.Z., Hertel, R. and Sack, F.D. (1989). Amyloplasts are necessary for full gravitropic sensitivity in roots of *Arabidopsis thaliana*. *Planta* **177**, 198–206.

Kuroda, K. (1990). Cytoplasmic streaming in plant cells. *Int. Rev. Cytol.* **121**, 267–307.

Larsen, P. (1965). Geotropic responses in roots as influenced by their orientation before and after stimulation. *Physiol. Plant.* **18**, 747–765.

Larsen, P. (1969). The optimum angle of geotropic stimulation and its relation to the starch statolith hypothesis. *Physiol. Plant.* **22**, 469–488.

Laskowski, M.J. (1990). Microtubule orientation in pea stem cells: a change in orientation follows the initiation of growth rate decline. *Planta* **181**, 44–52.

Lawton, J.R., Juniper, B.E. and Hawes, C.R. (1986). Ultrastructural changes in statocytes in grass nodes after short periods of gravistimulation: Do statoliths fall through the vacuole? Evidence from thick section high voltage EM. *J. Exp. Bot.* **37**, 693–704.

Lorenzi, G. and Perbal, G. (1990). Actin filaments responsible for the location of the nucleus in the lentil statocyte are sensitive to gravity. *Biol. Cell* **68**, 259–263.

Moore, R. and Evans, M.L. (1986). How roots perceive and respond to gravity. *Am. J. Bot.* **73**, 574–587.

Moore, R. and McClelen, C.E. (1985). Root gravi-responsiveness and columella cell structure in carotenoid-deficient seedlings of *Zea mays*. *Ann. Bot.* **56**, 83–90.

Nick, P., Bergfeld, R., Schäfer, E. and Schopfer, P. (1990). Unilateral reorientation of microtubules at the outer epidermal wall during photo- and gravitropic curvature of maize coleoptiles and sunflower hypocotyls. *Planta* **181**, 162–168.

Pickard, B.G. (1973). Geotropic response patterns of the *Avena* coleoptile. I. Dependence on angle and duration of stimulation. *Can. J. Bot.* **51**, 1003–1021.

Pickard, B.G. and Thimann, K.V. (1966). Geotropic response of wheat coleoptiles in absence of amyloplast starch. *J. Gen. Physiol.* **49**, 1065–1086.

Sack, F.D. and Kiss, J.Z. (1989). Rootcap structure in wild type and in a starchless mutant of *Arabidopsis*. *Am. J. Bot.* **76**, 454–464.

Sack, F.D. and Leopold, A.C. (1985). Cytoplasmic streaming affects gravity-induced amyloplast sedimentation in maize coleoptiles. *Planta* **164**, 56–62.

Sack, F.D., Suyemoto, M.M. and Leopold, A.C. (1985). Amyloplast sedimentation kinetics in gravistimulated maize roots. *Planta* **165**, 295–300.

Sack, F.D., Suyemoto, M.M. and Leopold, A.C. (1986). Amyloplast sedimentation and organelle saltation in living corn columella cells. *Am. J. Bot.* **73**, 1692–1698.

Schroeder, J.I. and Hedrich, R. (1989). Involvement of ion channels and active transport in osmoregulation and signaling of higher plant cells. *TIBS* **14**, 187–192.

Schwuchow, J., Sack, F.D. and Hartmann, E. (1990). Microtubule distribution in gravitropic protonemata of the moss *Ceratodon*. *Protoplasma* **159**, 60–69.

Shen-Miller, J. (1970). Reciprocity in the activation of geotropism in oat coleoptiles grown on clinostats. *Planta* **92**, 152–163.

Shibaoka, H. (1974). Involvement of wall microtubules in gibberellin promotion and kinetin inhibition of stem elongation. *Plant Cell Physiol.* **15**, 255–263.

Sievers, A. (1990). Transduction of the gravity signal in plants. In *Signal Perception and Transduction in Higher Plants* (eds R. Ranjeva and A.M. Boudet), NATO ASI Ser. H 47, pp. 297–306. Springer-Verlag, Berlin.

Sievers, A. and Hensel, W. (1991). Root cap: structure and function. In *Plant Root: The Hidden Half* (eds Y. Waisel, U. Kafkafi and A. Eshel). pp. 53–74. Dekker, New York.

Sievers, A. and Heyder-Caspers, L. (1983). The effect of centrifugal accelerations on the polarity of statocytes and on the graviperception of cress roots. *Planta* **157**, 64–70.

Sievers, A. and Schnepf, E. (1981). Morphogenesis and polarity of tubular cells with tip growth. In *Cytomorphogenesis in Plants* (ed. O. Kiermayer), pp. 265–299. Springer-Verlag, Wien.

Sievers, A. and Volkmann, D. (1972). Verursacht differentieller Druck der Amyloplasten auf ein komplexes Endomembransystem die Geoperzeption in Wurzeln? *Planta* **102**, 160–172.

Sievers, A., Volkmann, D., Hensel, W., Sobick, V. and Briegleb, W. (1976). Cell polarity in root statocytes in spite of simulated weightlessness. *Naturwissenschaften*, **63**, 343.

Sievers, A., Behrens, H.M., Buckhout, T.J. and Gradmann, D. (1984). Can a Ca^{2+} pump in the endoplasmic reticulum of the *Lepidium* root be the trigger for rapid changes in membrane potential after gravistimulation? *Z. Pflanzenphysiol.* **114**, 195–200.

Sievers, A., Kruse, S., Kuo-Huang, L.-L. and Wendt, M. (1989). Statoliths and microfilaments in plant cells. *Planta* **179**, 275–278.

Sievers, A., Kramer-Fischer, M., Braun, M. and Buchen, B. (1991). The polar organization of the growing *Chara* rhizoid and the transport of statoliths are actin dependent. *Botanica Acta* **104**, 103–109.

Staiger, C.J. and Schliwa, M. (1987). Actin localization and function in higher plants. *Protoplasma* **141**, 1–12.

Volkmann, D. and Czaja, A.W.P. (1981). Reversible inhibition of secretion in root cap cells of cress after treatment with cytochalasin B. *Exp. Cell Res.* **135**, 229–236.

Volkmann, D. and Sievers, A. (1979). Graviperception in multicellular organs. In *Encyclopedia of Plant Physiology*, N.S., vol. 7: *Physiology of Movements* (eds W. Haupt and M.E. Feinleib), pp. 573–600. Springer-Verlag, Berlin.

Volkmann, D. and Sievers, A. (1991). Forschung unter reduzierter Schwerkraft: Bedeutung für die Gravitationsbiologie. *Naturwissenschaften*, in press.

Volkmann, D., Behrens, H.M. and Sievers, A. (1986). Development and gravity sensing of cress roots under microgravity. *Naturwissenschaften* **73**, 438–441.

Volkmann, D., Buchen, B., Tewinkel, M., Hejnowicz, Z. and Sievers, A. (1991). Oriented movement of statoliths studied in a reduced gravitational field during parabolic flights of rockets. *Planta*, in press.

Wagner, G. and Klein, K. (1981). Mechanism of chloroplast movement in *Mougeotia*. *Protoplasma* **109**, 169–185.

Walker, L.M. and Sack, F.D. (1990). Amyloplasts as possible statoliths in gravitropic protonemata of the moss *Ceratodon purpureus*. *Planta* **181**, 71–77.

Wayne, R., Staves, M.P. and Leopold, A.C. (1990). Gravity-dependent polarity of cytoplasmic streaming in *Nitellopsis*. *Protoplasma* **155**, 43–57.

Wendt, M. and Sievers, A. (1986). Restitution of polarity in statocytes from centrifuged roots. *Plant Cell Environ.* **9**, 17–23.

Wendt, M. and Sievers, A. (1989). The polarity of statocytes and the gravisensitivity of roots are dependent on the concentration of calcium in statocytes. *Plant Cell Physiol.* **30**, 929–932.

Wendt, M., Kuo-Huang, L.-L. and Sievers, A. (1987). Gravitropic bending of cress roots without contact between amyloplasts and complexes of endoplasmic reticulum. *Planta* **172**, 321–329.

Westing, A.H. (1965). Formation and function of compression wood in gymnosperms. *Bot. Rev.* **31**, 381–480.

Westing, A.H. (1971). A case against statoliths. In *Gravity and the Organism* (eds S.A. Gordon and M.J. Cohen), pp. 97–101. University of Chicago Press, Chicago.

White, R.G. and Sack, F.D. (1990). Actin microfilaments in presumptive statocytes of root caps and coleoptiles. *Am. J. Bot.* **77**, 17–26.

Williamson, R.E., Perking, J.L. and McCurdy, D.W. (1986). Production and use of monoclonal antibodies to study the cytoskeleton and other components of the cortical cytoplasm of *Chara*. *Eur. J. Cell Biol.* **41**, 1–8.

Wolfe, J. and Steponkus, P.L. (1981). The stress–strain relation of the plasma membrane of isolated plant protoplasts. *Biochim. Biophys. Acta* **643**, 663–668.

14. *IN VITRO* XYLOGENESIS

Robert W. Seagull[1] and Marcia M. Falconer[2]

[1] *USDA/ARS, Southern Regional Research Center,*
1100 Robert E. Lee Boulevard, PO Box 19687, New Orleans, LA 70179, USA
[2] *Department of Biology, University of Ottawa,*
Ottawa, Ontario, Canada K1N 6N5

Introduction

One of the most studied morphogenetic processes in plants is xylem cell differentiation. This process offers a number of useful features which make it an ideal model system for the study of differentiation. Some of these features are as follows: (i) differentiation is terminal, thus eliminating the possibility of tracheary elements differentiating into some other types of cell; (ii) xylem cell differentiation is easily detected microscopically, using phase contrast or polarized light microscopy; (iii) differentiation follows a well-defined series of events which can be studied by a variety of biochemical and microscopical techniques; (iv) various manipulations of intact tissues or the use of tissue-culture techniques can induce the spontaneous formation of tracheary elements.

Patterns of cell development

The xylem tissue of the plant is composed of cells of various shapes and sizes. Primary xylem is normally produced through meristematic activity in shoot and root tips. Secondary xylem is generated by the activity of the vascular cambium (Esau, 1960). Xylem tissue (either primary or secondary) includes tracheary elements – both tracheids and vessel members – fibre cells, and parenchyma (Esau, 1960).

Functional only after cell death, the cell walls of tracheary and vessel elements form the hollow conductive tubing for water and dissolved nutrients. The cell walls of xylem tissue are thickened and reinforced with bands of cellulose microfibrils enabling them to withstand the pressures of the surrounding tissues and the hydrostatic pressure of conducting water. The extent of thickening and the patterns of wall bands varies within the plant and thus must be under developmental control.

The formation of tracheary elements follows a precise developmental sequence (Torrey *et al.*, 1971; Barnett, 1981). Enlargement of the pro-vascular cells is followed by the deposition of the secondary cell wall. The extent and pattern of cell wall deposition varies depending on the type (primary or secondary) of tracheary element being formed. The process of wall lignification follows that of cellulose synthesis in the secondary cell wall. Cessation of wall deposition is followed by cell senescence. Autolysis of the cytoplasm and dissolution of unlignified cell walls results in the final formation of the hollow tube network of the vascular tissue.

Protoxylem and metaxylem

Primary xylem can be broadly categorized into protoxylem and metaxylem, based on the location within the plant organ and cell wall organization. Protoxylem is found in the elongating portion of a plant organ (root or shoot) and contains only tracheary elements embedded in parenchyma. Cell wall thickening patterns are in the form of annular rings or helical bands. The more complex tissue, metaxylem, is found in the non-elongating regions of organs and is composed of fibre cells, in addition to tracheary elements and parenchyma. It contains cells that are generally shorter, with thicker walls in reticulate, scalariform or pitted wall patterns.

The separation of protoxylem and metaxylem is not distinct with respect to the timing of development or the location in the organ. Manipulation of cell extension rates during development can induce increases or decreases

The Cytoskeletal Basis of Plant Growth and Form   ISBN 0–12–453770–7

ounts of proto- and metaxylem ging the locations of these cell an organ (O'Brien, 1981). These ᵔ indicate that final wall pattern can ᵔced by the characteristics of cell ᵔn before xylem differentiation occurs. An extensive examination of 1350 species of angiosperms (Bierhorst and Zamora, 1965) indicated that there is a transition between proto- and metaxylem, with annular and spiral band patterns being transformed into scalariform and pitted wall thickenings.

The study of xylogenesis has been primarily concerned with the development of xylem cells, both tracheary elements and vessels, and focused on several major questions: (i) the origin of the xylem tissue from its cellular precursors; (ii) the nature of the stimuli controlling its origin (i.e. the role of hormones, mitosis and DNA synthesis); and (iii) the events which distinguish the tracheary element as a distinct cell type (i.e. the mechanism of wall synthesis and pattern formation. In this chapter the authors will focus on the question of the events leading to tracheary element formation and more specifically the cytoplasmic control of cell wall development. For detailed reviews of the other major areas of study in xylogenesis see the reviews written by Torrey *et al.* (1971), Roberts (1976), Dodds (1981), Barnett (1981), Savage (1983), Bengochea *et al.* (1983), Fukuda and Komamine (1985), Aloni (1987a, b) and Sugiyama and Komamine (1987).

Early observations of xylogenesis

Cytoplasmic patterning
Research on the mechanism(s) of tracheary element formation has been going on for over a hundred years. As reported by Sinnott and Bloch (1945), early researchers observed changes in the cytoplasm that predicted differentiation into tracheary elements. Crüger (1855) showed that in living cells, later destined to be tracheary elements, actively streaming strands of denser cytoplasm occupy the positions where wall thickenings will be laid down. Dipple (1867) extended these observations by reporting that the granular bands of streaming

cytoplasm remain intact during plasmolysis and still reflect the band pattern in the cell wall.

Our understanding of the cytoplasmic organization and the changes that occurred within the cell during the formation of tracheary elements did not change appreciably from the days of Crüger and Dipple until the introduction of aldehyde fixation and electron microscopy. It soon became clear that one of the first recognized changes in cell morphology was the clustering of cortical microtubules into bands that predict the precise location and orientation of the secondary wall microfibrils of the secondary wall bands. These initial studies and subsequent ones form a cornerstone in our understanding of the role of cytoskeletal elements in plant cell morphogenesis (for reviews see Hepler and Palevitz, 1974; Gunning and Hardham, 1982; Seagull, 1990a).

The role of the cytoskeleton
All of the early work on xylogenesis and the role of cytoskeletal elements was done in intact plant systems (Cronshaw and Bouck, 1965; Pickett-Heaps and Northcote, 1966; Pickett-Heaps, 1967; Robards, 1968; Barlow, 1969; Robards and Kidwai, 1972) or in tissues induced to form tracheary elements due to wounding (Hepler and Newcomb, 1964; Roberts and Baba, 1968). This can be summarized into a few key observations:

(1) Clusters of cortical microtubules, closely associated with the plasma membrane, predict the location and orientation of the cellulose microfibrils of the secondary cell wall (Hepler and Newcomb, 1964; Wooding and Northcote, 1964; Cronshaw and Bouck, 1965; Esau *et al.*, 1966).

(2) Microtubules change their patterns of distribution during wall band formation, being evenly positioned across the forming band early in formation, but restricted to the edges of thickenings later in wall deposition (Pickett-Heaps and Northcote, 1966; Goosen-De Roo, 1973a, b, c).

(3) Disruption of the microtubule arrays with depolymerizing agents results in formation of an abnormal secondary cell wall

(Cronshaw, 1967; Hepler and Fosket, 1971). (4) Microtubules are involved in determining the location and orientation of secondary wall microfibrils, but they are *not* involved in the *synthesis* of cell wall components (Chrispeels, 1972; Hogetsu *et al.*, 1974).

Advances in technology

Two major advancements in technology have had a tremendous impact on the study of xylogenesis. The first was the introduction of immunocytochemical and immunofluorescence techniques to the study of the plant cytoskeleton (reviewed by Lloyd, 1987). These techniques allowed three-dimensional examination of cytoskeletal elements and provided information which previously was obtained only through the very labour-intensive technique of serial section reconstruction of electron micrographs. In addition, the ease of observation resulted in the rapid examination of orders of magnitude more cells, thus facilitating the generation of quantitative, as well as qualitative, data. This increase in observations of cells at various stages of development has led to a better appreciation of the dynamics of cytoskeletal arrays.

The second major advance was the introduction of an *in vitro* xylogenesis system, i.e. the induction of tracheary element development in tissue culture (Kohlenbach and Schmidt, 1975; Fukuda and Komamine, 1985; Fukuda, 1989a). Unlike intact systems, tissue culture systems tend to have relatively more tracheary elements at various stages of development than do intact tissues and these tracheary elements can be easily monitored microscopically.

In vitro xylogenesis

Culture systems

There are several *in vitro* culture techniques that have been used to study xylogenesis (for review see Bengochea *et al.*, 1983; Fukuda and Komamine, 1985) but the one which has received the widest attention is the *Zinnia* mesophyll system established by Kohlenbach and Schmidt (1975) and subsequently developed and improved by Fukuda and

Komamine (1980a, b, 1982, 1983) and Sugiyama *et al.* (1986). The specific advantages of this system include: (i) formation of tracheary elements from single cells, thus eliminating possible complications due to cell–cell interactions and facilitating uniform cell exposure to medium; (ii) cell differentiation occurs rapidly (within 48 h of culturing); (iii) cells are grown under defined conditions, thus facilitating manipulations and assessment of the effects of culture modifications; (iv) the high degree of synchrony of tracheary element development and the large numbers of cells that differentiate (routinely 30% and up to 60% under ideal conditions) facilitates biochemical studies; and (v) this system develops all the types of xylem cells found in intact systems as well as undergoing developmental transitions from proto- to metaxylem.

This system has been used to study various aspects of tracheary element differentiation, including hormonal control (Burgess and Linstead, 1984; Church and Galston, 1988a, b), changes in wall composition (Fukuda and Komamine, 1982; Ingold *et al.*, 1988), the role of DNA metabolism and cell division (Fukuda and Komamine, 1980b, 1981; Kohlenbach and Schopke, 1981; Thelen and Northcote, 1989), presumed cellulose synthase in the plasma membrane (Haigler and Brown, 1986), the role of calcium in xylogenesis (Roberts and Haigler, 1989, 1990) and the role of the cytoskeleton in wall development (Falconer and Seagull, 1985a, b, 1986, 1988; Fukuda, 1987, 1989b; Iwasaki *et al.*, 1988; Kobayashi *et al.*, 1987, 1988).

In this chapter we would like to highlight major new observations, made using the *Zinnia* system, which extend our understanding of the role of cytoskeletal elements in xylogenesis. We will not cover the role that the *Zinnia* system has played in elucidating other aspects of xylogenesis, such as the pre-programming of cells for differentiation, roles of hormones, wall lignification and cell senescence. These aspects have been reviewed previously (Fukuda and Komamine, 1985; Fukuda, 1989a).

The role of cytoskeletal elements

Co-localization of microtubules and microfibrils
Study of the *Zinnia* xylogenesis system, using immunocytochemical and fluorescence techniques, not only confirms the earlier EM studies on intact systems, but also extends these observations to illustrate the dynamics of cytoskeletal elements during differentiation and the possible mechanism involved in the reorganization of the microtubules.

Staining of wall material with calcofluor white (Tinopal LPW) has allowed us to identify and examine large numbers of cells in the early stages of xylogenesis, when wall bands are not detectable by other light microscopy techniques (Falconer and Seagull, 1985a). Early development of tracheary elements can be sub-divided into three categories: (i) microtubules grouped into bands without evident secondary walls; (ii) groups of microtubules subtending wall material visible using calcofluor white; and (iii) a complex microtubule pattern reflected by well-developed wall thickenings detected by phase contrast and polarization optics. Cells in the first two categories would appear undifferentiated when viewed using non-fluorescence techniques. In all of these categories the cells are alive, thus we are not including the later developmental stages of wall lignification (although in late category three cells there is some wall lignification evident), protoplast autolysis and end plate dissolution. Our categories would fit into the phase III sequence of differentiation outlined for *Zinnia* by Fukuda and Komamine (1985).

Monitoring both the cell wall deposition (calcofluor staining) and microtubule organization (indirect immunofluorescence) allows us to follow progressions in development, using wall staining patterns to indicate previous microtubule arrays (Falconer and Seagull, 1988). It is clear from our observations that within a single cell there can be several stages of microtubule organization and wall deposition (Falconer and Seagull, 1985a) and a progression of development, with clusters of microtubules forming and then splaying out (Falconer and Seagull, 1988). At all times, the deposition of wall material follows the change in microtubule arrays. These transitions in the organization of the microtubule arrays result in the conversion of wall patterns from discrete bands to scalariform to pitted patterns, just as is seen in the transition from protoxylem to metaxylem (Bierhorst and Zamora, 1965).

Control of wall patterns
The observations of co-localization of microtubules with wall microfibrils and the fact that the grouping of microtubules precedes the deposition of localized wall patterns strongly supports the concept of microtubules controlling the location and orientation of wall bands in developing tracheary elements. This does not answer the question of what determines the location of the secondary wall bands but merely shifts the emphasis to what defines the location and orientation of the groups of microtubules.

A central question in the development of cell wall thickenings is: 'How do microtubules change from a dispersed to a clustered array?' It was proposed by Gunning *et al.* (1978) that microtubule turnover was responsible for the establishment of new microtubules in clustered arrays, through the function of microtubule organizing centres (MTOCs) or nucleating sites (NS). The *Zinnia* system has provided information that is consistent with the hypothesis that lateral association of existing microtubules can result in the formation of the observed clustered arrays (Falconer and Seagull, 1985b; Roberts *et al.*, 1985). Stabilization of microtubules with taxol does not prevent the formation of clustered arrays. In fact, taxol-induced clustering of microtubules resulted in earlier formation and increased numbers of tracheary elements (Falconer and Seagull, 1985b). These observations indicate that: (i) microtubule turnover is not essential for the development of wall thickenings; (ii) early induction of microtubule clustering results in wall band formation, thus the machinery which deposits wall components in discrete bands may be essentially the same as that which deposits material for dispersed wall synthesis; and (iii) one of the earliest steps in xylogenesis must be the bundling of microtubules since the induction of bundling results in early wall band formation.

The formation of localized wall deposits appears to be directly influenced by the presence of a non-uniform array of microtubules. Wall thickenings can be introduced in non-xylogenic cell types if the distribution of microtubule arrays is modified by drug treatment. Treatment with taxol results in the synthesis of wall thickenings in suspension culture cells of *Vicia hajastana* (Weerdenburg and Seagull, 1988) and cotton fibres (Seagull, 1989b, 1990b). Partial removal of microtubule arrays in cotton by exposure to oryzalin or colchicine, giving a non-uniform distribution of microtubules, results in the production of wall thickenings (Seagull, 1989b, 1990b). While the production of a non-uniform distribution of microtubules in these systems is not normal, once established, non-uniform wall deposition follows. These observations may indicate that transitions in the microtubule arrays are pivotal points in the differentiation of tracheary elements.

Work with the *Zinnia* system has also provided information about the regulation of wall pattern formation during xylogenesis. We have established a relationship between cell shape and the development of specific wall band patterns (Falconer and Seagull, 1986). Whether a cell will produce a banded or a webbed pattern of wall thickenings (for definitions see Falconer and Seagull, 1986) appears to be dependent on the microtubule organization before differentiation starts. Parallel arrays of microtubules generate banded wall patterns while randomly organized microtubule arrays generate webbed wall patterns (Fig. 14.1). The wall thickening patterns can also be related to cell shape. Elongated cells or portions of cells generate banded wall patterns, while spherical cells or portions of cells generate webbed wall patterns. Using pulse treatments with microtubule disrupting agents changes the microtubule organization, cell morphology and subsequent wall band patterns. Only cells which had changed shape (from elongate to rounded) in the presence of microtubule disrupting agents produced random arrays of microtubules upon recovery from the drug. These cells subsequently produced webbed wall patterns. If cells did not change morpho-

logy during drug exposure, due to minimal cell expansion, then the returning microtubules were organized into parallel arrays and subsequent wall thickenings were organized into band patterns.

The exact relationship between cell shape, microtubule pattern and wall band pattern remains obscure since one cannot change microtubule pattern without changing cell shape, or vice versa. Changes in both microtubule organization and cell shape occur when cells are treated with microtubule disrupting agents. It is possible that cell shape is the primary controlling mechanism for wall band patterns, with shape influencing microtubule organization and ultimately wall pattern. Cells which do not change shape during drug treatment produce parallel arrays of microtubules upon removal of the drug. These observations indicate that MTOCs or NS may exist at the plasmalemma and may be involved in determining the organization of the returning array of microtubules (Falconer *et al.*, 1988).

Regulation of xylogenesis
There are dramatic changes in the microtubule populations that occur before any detectable changes are seen in the cell wall. These changes must be brought about through modifications in the chemistry or kinetics of the microtubule arrays. The formation of clustered arrays of microtubules is essential to form proper wall development and the process of MT bundling is an early step in xylogenesis. Extensive turnover of the microtubule population is not essential for xylogenesis and it appears that the dispersed arrays of microtubules in undifferentiated cells can be clustered to form the patterns essential to proper wall deposition (Falconer and Seagull, 1985b).

What are the factors in the cell that control the organization and dynamics of the microtubule population? From the literature we know that microtubules can be induced to bundle through the activity of microtubule associated proteins (MAPs) (for review see Gunning and Hardham, 1982; Olmsted, 1986). Recently, microtubule bundling proteins have been isolated from carrot cells (Cyr and Palevitz, 1989; see also Cyr, Chapter 5, this

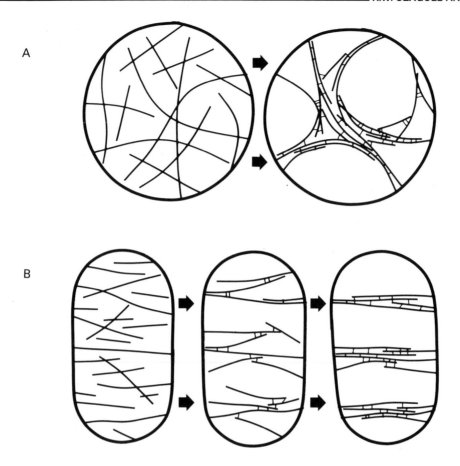

Fig. 14.1. Model of how lateral association of microtubules would result in the formation of webbed and banded patterns of wall thickenings (based on the data from Falconer and Seagull, 1985b, 1986). A, When randomly organized arrays of microtubules are laterally associated they tend to form circular patterns of bundles. B, Lateral association of parallel arrays of microtubules tends to result in the formation of parallel bundles. In either case, the final microtubule pattern could be modified by factors such as the extent of lateral association and the length of individual microtubules.

volume). Such proteins have not been reported in the *Zinnia* culture system and it would be of interest to screen the *Zinnia* system with antibodies against the bundling proteins from carrot. Two differentiation specific proteins have been identified from the *Zinnia* system (Fukuda and Komamine, 1983); however, their relationship to the proteins isolated from carrot or their possible interaction with microtubules remains unknown.

Changes in the level of tubulin (relative to total protein) and the numbers of microtubules have been documented for the early culture stages in the *Zinnia* system (Fukuda and Komamine, 1985; Fukuda, 1987, 1989b). Some of this increase in tubulin may be related to the transformation of the cells from leaf mesophyll to suspension culture, however immature tracheary elements do exhibit approximately twice the number of microtubules as non-differentiated cells at the same age (Fukuda, 1987). This specific increase in tubulin content requires both protein and mRNA synthesis. Multiple tubulin isotypes have been found in other systems and it is believed that they may be involved in defining microtubule stability and function (for review see Seagull, 1989a). Electrophoresis patterns did not illustrate various tubulin isotypes in the *Zinnia* system (Fukuda, 1987). Further experiments are necessary to determine if indeed there is only a single isotype of tubulin in the *Zinnia* system or that multiple isotypes exist but have not been detected. Given the expanding literature on tubulin isotypes and post-translational modification of tubulin, it seems reasonable that the *Zinnia* system would show some tubulin diversity.

Starting after 2 days in culture, there is an increase in the rates of tubulin synthesis and degradation, thus the level of tubulin does not alter at this time (Fukuda, 1987, 1989b). Tubulin levels do not change during the rearrangement of microtubules from axially oriented to transversely oriented, nor do they change when microtubules are clustered into groups; indicating that net tubulin synthesis does not play an essential role in these processes. The observed increase in microtubule numbers occurs after the increase in tubulin content, rather than during the increase, indicating a possible change in the equilibrium between free and polymerized tubulin (Fukuda, 1987). This change in the equilibrium may reflect the synthesis of stabilizing components (such as MAPs) for the microtubules; or a change in some other cytoplasmic components, such as ion balance, that could modify microtubule stability.

Calcium ions have significant effects on cell morphogenesis and the function of the cytoskeleton (Wick, 1988). Changes in $[Ca^{2+}]$ appear to influence the development of tracheary elements. Inhibition of calcium uptake or decreases in the amount of available Ca^{2+} inhibit xylogenesis (Roberts and Haigler, 1989, 1990). These authors show that differentiation of tracheary elements appears to involve at least two Ca^{2+} regulated steps: one at the induction of differentiation (i.e. the initiation of the cultures), and the other just prior to wall band formation. It is not yet known if Ca^{2+} regulation is directly related to cytoskeletal function.

The role of actin in xylogenesis

The importance of actin-like microfilaments in a variety of plant cell processes has become evident in recent years (for review see Seagull,

1989a). Improvements in preservation techniques have shown microfilaments to be associated with microtubules in all stages of the cell cycle and in various developmental processes. Some of the earliest work on xylogenesis (for review see Sinnott and Bloch, 1945) indicated that cytoplasmic streaming and thus presumably microfilaments were somehow involved in the pre-patterning of wall band location. More recently, microfilaments have been shown to be associated with the clustered arrays of microtubules during wall thickening in tracheary elements of *Azolla* (Heath and Seagull, 1982). The possible involvement of microfilaments in xylogenesis, using the *Zinnia* system, has been examined by Kobayashi *et al.* (1987, 1988). Their studies indicate that microfilaments are not involved in the lateral association of microtubules to form bundles, since exposure to the microfilament disrupting agent cytochalasin B does not prevent the formation of localized wall thickenings. This agent does, however, inhibit the normal reorientation of microtubule arrays from axial to transverse, and so wall bands are abnormally oriented axially. Staining of microfilaments indicates that they are located between the bundles of microtubules, rather than within them. Microfilament patterns in plant cells are very labile and subject to disruption during staining procedures, resulting in poor preservation and possibly inaccurate interpretation of results. The actin arrays that exist between the bundled microtubules often appear very amorphous, rather than fibrillar (Kobayashi *et al.*, 1989). In addition, actin cables appear somewhat fragmented. We have analysed microfilament arrays in the *Zinnia* system and have found more prominent actin cables and actin staining associated with the microtubule bundles (Fig. 14.2). These apparent discrepancies in actin localization are most likely the result of differences in preservation. Only further experimentation and observations will clarify whether actin is or is not associated with the bundles of microtubules. The inhibitor studies of Kobayashi *et al.* indicate that if actin is associated with the microtubules it is not involved in the lateral association of these elements into bundles. Actin has been proposed to function

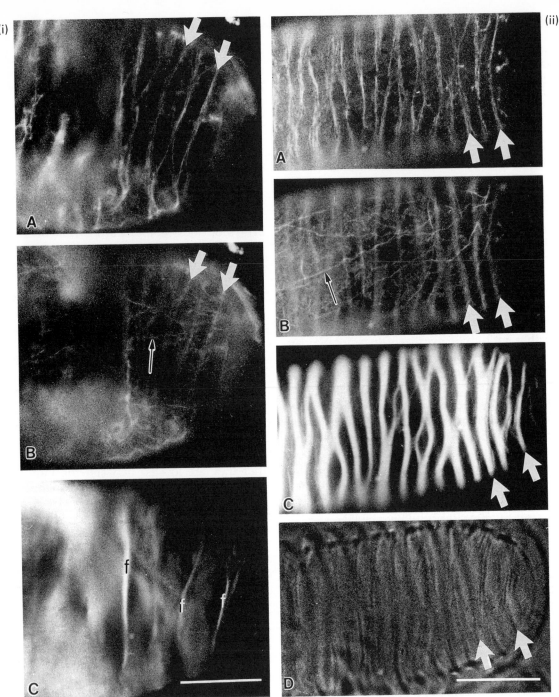

Fig. 14.2. Indirect immunofluorescent observations of developing tracheary elements. Microtubules (A) were visualized using antibodies, actin patterns (B) were visualized with rhodamine-labelled phalloidin, wall patterns were visualized using the fluorescent dye Tinopal LPW (C) or with phase contrast (D). Scale bar = 20 μm. (i) A tracheary element, early in development, with no secondary wall thickening visible with Tinopal staining (C), yet microtubules are clustered into bundles (A, arrows). Actin (B) is organized into bundles running perpendicular to the microtubules (small arrow) and parallel to the microtubules (large arrow). Wall staining (C) shows an even distribution of wall material. Some folding of the wall is evident (f). (ii) A tracheary element, late in development, with secondary wall deposition clearly evident with wall staining (C) or with phase contrast microscopy (D). Microtubules (A) and microfilaments (B) are clearly associated with the wall thickenings (large arrows). Cables of actin also run transversely (small arrows) to the bundles of microtubules.

with microtubules to produce ordered wall microfibrils deposition (Seagull, 1989a). Only further experimentation on the *Zinnia* system will clarify this point.

Future directions

There are many unanswered questions regarding the involvement of cytoskeletal elements in the formation of localized wall thickenings during xylogenesis. As discussed above, the *Zinnia* system may represent one of the best experimental systems to answer many of these questions.

The use of direct immunofluorescence techniques has expanded our understanding of how microtubules reorganize during xylogenesis and the timing of such events (Fukuda, 1989a). These studies have been restricted by the limited number of probes with which this system can be studied. The structural elements of the cytoskeleton have been examined using antibodies to tubulin and fluorescently labelled phalloidin. The regulatory proteins (i.e. those involved in controlling cytoskeletal organization and function) however have not been examined, even at a superficial level. For the most part this is the direct result of a lack of probes which bind to cytoskeletal components other than tubulin and actin. There are a number of laboratories currently active in the isolation of cytoskeletal associated proteins from a variety of plant sources (for review see Chapters 1–5, this volume). The identification and characterization of such proteins in the *Zinnia* system may clarify the mechanisms of microtubule aggregation and subsequent splaying out during the transition of wall patterns from banded to pitted (Falconer and Seagull, 1988).

There are very few proteins which appear to be specifically synthesized during xylogenesis (Fukuda and Komamine, 1983). The proteins that were identified appeared at a time when microtubules were undergoing dramatic reorganization, thus it is possible that these proteins play a role in differentiation-specific changes in the cytoskeleton (Fukuda, 1989b). Differential distribution of such proteins could explain the observed non-synchronous development of wall thickenings within individual cells (Falconer and Seagull, 1985a, 1988).

The detailed study of cytoskeletal regulatory proteins during xylogenesis may provide clues as to the induction of xylogenesis. A specific set of hormones and environmental conditions is essential for the induction of tracheary elements (Fukuda and Komamine, 1985; Fukuda, 1989a). Free $[Ca^{2+}]$ appears to be an important factor at several stages of xylogenesis (Roberts and Haigler, 1989, 1990). We know that these factors influence the plant cytoskeleton (Seagull, 1989a), yet the precise sequence of events leading to the observed changes in cytoskeletal organization is unknown. The synchrony and relatively high level of differentiation of the *Zinnia* system may facilitate the study of the target molecules and secondary messengers that lead to cytoskeletal changes.

While immunocytochemical techniques have facilitated the study of the changes in microtubule arrays, we now have to complement these observations with electron microscopic studies to determine with greater detail how the cytoskeleton achieves its final organization and function. Fukuda and co-workes (Fukuda, 1987; Fukuda and Kobayashi, 1989) illustrated increases in tubulin content and microtubule number that immediately precede wall thickening. Somehow the clustering of these microtubules results in the clustering of the cellulose synthesizing complexes on the plasma membrane (Haigler and Brown, 1986). We know that microtubules and newly synthesized wall material are both closely associated with the plasma membrane. We do not know the mechanism for this interaction, whether via protein cross-bridges, ionic interactions or via some other process. The improved ultrastructural detail of cell wall–plasma membrane–cytoskeletal interactions offered by rapid freezing and freeze-substitution techniques (Lancelle *et al.*, 1987) should provide ultrastructural evidence concerning the mechanism of cytoskeletal involvement. In addition these techniques should provide alternative preservation methods for the examination of the microfilament arrays, thus facilitating the assessment of their function, if any, in the development of wall thickenings.

References

Aloni, R. (1987a). Differentiation of vascular tissue. *Ann. Rev. Plant Physiol.* **38**, 179–204.

Aloni, R. (1987b). The induction of vascular tissues by auxin. In *Plant Hormones and their Role in Plant Growth and Development* (ed. P.J. Davies), pp. 363–374. Martinus Nijhoff, Boston.

Barlow, P.W. (1969). Differences in response to colchicine by differentiating xylem cells in root of *Pisum. Protoplasma* **68**, 79–83.

Barnett, J.R. (1981). Current research into tracheary element formation. *Commentaries in Plant Science* **11**, 161–174.

Bengochea, T., Harry, G.I. and Dodds, J.H. (1983). Cytodifferentiation of xylem cells *in vitro*: an overview. *Histochem. J.* **15**, 411–418.

Bierhorst, D.W. and Zamora, P.M. (1965). Primary xylem element associations of angiosperms. *Amer. J. Bot.* **52**, 657–710.

Burgess, J. and Linstead, P. (1984). *In vitro* tracheary element formation: Structural studies and the effect of triiodobenzoic acid. *Planta* **160**, 481–489.

Chrispeels, M.J. (1972). Failure of colchicine or cytochalasin to inhibit protein secretion by plant cells. *Planta* **108**, 283–287.

Church, D.L. and Galston, A.W. (1988a). Kinetics of determination of the differentiation of isolated mesophyll cells of *Zinnia elegans* to tracheary elements. *Plant Physiol.* **88**, 92–96.

Church, D.L. and Galston, A.W. (1988b). Hormonal induction and anti-hormonal inhibition of tracheary element differentiation in *Zinnia* cell cultures. *Phytochemistry* **27**, 2435–2439.

Cronshaw, J. (1967). Tracheid differentiation in tobacco pith cultures. *Planta* **72**, 72, 78–90.

Cronshaw, J. and Bouck, G.B. (1965). The fine structure of differentiating xylem elements. *J. Cell Biol.* **24**, 415–431.

Cruger, H. (1855). Zur Entwicklungsgeschichte der Zellenwand. *Botan. Zeit.* **13**, 617–629.

Cyr, R.J. and Palevitz, B.A. (1989). Microtubule-binding proteins from carrot. 1. Initial characterization and microtubule bundling. *Planta* **177**, 245–260.

Dipple, L. (1867). Die Entstehung der wandstandl-gen Protoplasmastronchen. *Abhandl Naturforsch. Gellesch. Halle* **10**, 53–68.

Dodds, J.H. (1981). The role of the cell cycle and cell division in xylogenesis. In *Xylem Cell Development* (ed. J.R. Barnett), pp. 153–162. Chambers Green, Tunbridge Wells, Kent.

Esau, K. (1960). *Anatomy of Seed Plants*, pp. 75–94. J. Wiley, New York.

Esau, K., Cheadle, V.I. and Gill, R.H. (1966). Cytology of differentiating tracheary elements. III. Structures associated with cell surfaces. *Amer. J. Bot.* **53**, 765–771.

Falconer, M. and Seagull, R.W. (1985a). Immuno-fluorescent and calcofluor white staining of developing tracheary elements in *Zinnia elegans* L. suspension cultures. *Protoplasma* **125**, 190–198.

Falconer, M. and Seagull, R.W. (1985b). Xylogenesis in tissue culture: Taxol effects on microtubule reorganization and lateral association in differentiating cells. *Protoplasma* **128**, 157–166.

Falconer, M. and Seagull, R.W. (1986). Xylogenesis in tissue culture. II. Microtubules, cell shape and secondary wall patterns. *Protoplasma* **133**, 140–148.

Falconer, M. and Seagull, R.W. (1988). Xylogenesis in tissue culture. III. Continuing wall deposition during tracheary element development. *Protoplasma* **144**, 10–16.

Falconer, M.M., Donaldson, G. and Seagull, R.W. (1988). MTOCs in higher plant cells: an immuno-fluorescent study of microtubule assembly sites following depolymerization by APM. *Protoplasma* **144**, 46–55.

Fukuda, H. (1987). A change in tubulin synthesis is the process of tracheary element differentiation and cell division of isolated *Zinnia* mesophyll cells. *Plant Cell Physiol.* **28**, 517–528.

Fukuda, H. (1989a). Cytodifferentiation in isolated single cells. *Bot. Mag. Tokyo* **102**, 491–501.

Fukuda, H. (1989b). Regulation of tubulin degradation in isolated *Zinnia* mesophyll cells in culture. *Plant Cell Physiol.* **30**, 243–252.

Fukuda, H. and Kobayashi, H. (1989). Dynamic organization of the cytoskeleton during tracheary-element differentiation. *Develop. Growth and Differ.* **31**, 9–16.

Fukuda, H. and Komamine, A. (1980a). Establishment of an experimental system for the tracheary element differentiation from single cells isolated from the mesophyll of *Zinnia elegans. Physiol. Plant.* **52**, 57–60.

Fukuda, H. and Komamine, A. (1980b). Direct evidence of cytodifferentiation to tracheary elements without intervening mitosis in a culture of single cells isolated from the mesophyll of *Zinnia elegans. Plant Physiol.* **65**, 61–64.

Fukuda, H. and Komamine, A. (1981). Relationship between tracheary element differentiation and DNA synthesis in single cells isolated from the mesophyll of *Zinnia elegans* – Analysis of inhibitors of DNA synthesis. *Plant Cell Physiol.* **22**, 41–49.

Fukuda, H. and Komamine, A. (1982). Lignin synthesis and its related enzymes as markers of tracheary-element differentiation in single cells isolated from the mesophyll of *Zinnia elegans*. *Planta* **155**, 423–430.

Fukuda, H. and Komamine, A. (1983). Changes in the synthesis of RNA and protein during tracheary element differentiation in single cells isolated from the mesophyll of *Zinnia elegans*. *Plant and Cell Physiol.* **24**, 603–614.

Fukuda, H. and Komamine, A. (1985). Cytodifferentiation. In *Cell Culture and Somatic Cell Genetics* (ed. I.K. Vasil), vol. 2, pp. 149–212. Academic Press, London.

Goosen-De Roo, L. (1973a). The fine structure of the protoplast in primary tracheary elements of the cucumber after plasmolysis. *Acta Bot. Neerl.* **22**, 467–485.

Goosen-De Roo, L. (1973b). The relationship between cell organelles and cell wall thickenings in primary tracheary elements of cucumber. I. Morphological aspects. *Acta Bot. Neerl.* **22**, 279–300.

Goosen-De Roo, L. (1973c). The relationship between cell organelles and cell wall thickenings in primary tracheary elements of the cucumber. II. Quantitative aspects. *Acta Bot. Neerl.* **22**, 301–320.

Gunning, B.E.S. and Hardham, A.R. (1982). Microtubules. *Ann. Rev. Plant Physiol.* **33**, 651–698.

Gunning, B.E.S., Hardham, A.R. and Hughes, J. (1978). Evidence for initiation of microtubules in discrete regions of the cell cortex of *Azolla* root-tip cells, and an hypothesis on the development of cortical arrays of microtubules. *Planta* **143**, 161–179.

Haigler, C.H. and Brown, R.M. Jr (1986). Transport of rosettes from the Golgi apparatus to the plasma membrane in isolated mesophyll cells of *Zinnia elegans* during differentiation to tracheary elements in suspension culture. *Protoplasma* **134**, 111–120.

Heath, I.B. and Seagull, R.W. (1982). Oriented microfibrils and the cytoskeleton: A critical comparison of models. In *The Cytoskeleton in Plant Growth and Development* (ed. C.W. Lloyd), pp. 163–182. Academic Press, London.

Hepler, P.K. and Fosket, D.E. (1971). The role of microtubules in vessel member differentiation in *Coleus*. *Protoplasma* **72**, 213–236.

Hepler, P.K. and Newcomb, E.H. (1964). Microtubules and fibrils in the cytoplasm of *Coleus* cells undergoing secondary wall deposition. *J. Cell Biol.* **20**, 529–533.

Hepler, P.K. and Palevitz, B.A. (1974). Microtubules and microfilaments. *Ann. Rev. Plant Physiol.* **25**, 309–362.

Hogetsu, T., Shibaoka, H. and Shimokoriyama, H. (1974). Involvement of cellulose synthesis in the actions of gibberellin and kinetin on cell expansion.

2,6-Dichlorobenzonitrile as a new cellulose-synthesis inhibitor. *Plant Cell Physiol.* **15**, 389–393.

Ingold, E., Sugiyama, M. and Komamine, A. (1988). Secondary cell wall formation. Changes in cell wall constituents during the differentiation of isolated mesophyll cells of *Zinnia elegans* to tracheary elements. *Plant Cell Physiol.* **29**, 295–303.

Iwasaki, T., Fukuda, H. and Shibaoka, H. (1988). Inhibition of cell division and DNA synthesis by gibberellin in isolated *Zinnia* mesophyll cells. *Plant Cell Physiol.* **27**, 717–724.

Kobayashi, H., Fukuda, H. and Shibaoka, H. (1987). Reorganization of actin filaments associated with the differentiation of tracheary elements in *Zinnia* mesophyll cells. *Protoplasma* **138**, 69–71.

Kobayashi, H., Fukuda, H. and Shibaoka, H. (1988). Interrelationship between the spatial disposition of actin filaments and microtubules during the differentiation of tracheary elements in cultured *Zinnia* cells. *Protoplasma* **143**, 29–37.

Kohlenbach, H.W. and Schmidt, B. (1975). Cytodifferezierung in Form einer direkten Umwandllung isolierter Mesophyllzellen zu Tracheiden. *Z. Pflanzenphysiol.* **75**, 369–374.

Kohlenbach, H.W. and Schopke, C. (1981). Cytodifferentiation to tracheary elements from isolated mesophyll protoplasts of *Zinnia elegans*. *Naturwissenschaften* **68**, 576–577.

Lancelle, S.A., Cresti, M. and Hepler, P.K. (1987). Ultrastructure of the cytoskeleton in freeze-substituted pollen tubes on *Nicotiana alata*. *Protoplasma* **140**, 141–150.

Lloyd, C.W. (1987). The plant cytoskeleton: The impact of fluorescence microscopy. *Ann. Rev. Plant Physiol.* **38**, 119–139.

O'Brien, T.P. (1981). The primary xylem. In *Xylem Development* (ed. J.R. Barnett), pp. 14–46. Chambers Green, Tunbridge Wells, England.

Olmsted, J.B. (1986). Microtubule-associated proteins. *Ann. Rev. Cell Biol.* **2**, 421–457.

Pickett-Heaps, J.D. (1967). The effects of colchicine on the ultrastructure of dividing plant cells, xylem wall differentiation and distribution of cytoplasmic microtubules. *Dev. Biol.* **15**, 206–236.

Pickett-Heaps, J.D. and Northcote, D.H. (1966). Relationship of cellular organelles to the formation and development of the plant cell wall. *J. Exp. Bot.* **17**, 20–26.

Robards, A.W. (1968). On the ultrastructure of differentiating secondary xylem in willow. *Protoplasma* **65**, 449–464.

Robards, A.W. and Kidwai, P. (1972). Microtubules and microfibrils in xylem fibers during secondary cell wall formation. *Cytobiologie* **6**, 1–21.

Roberts, A.W. and Haigler, C.H. (1989). Rise in chlorotetracycline fluorescence accompanies tracheary element differentiation in suspension cultures of *Zinnia*. *Protoplasma* **152**, 37–45.

Roberts, A.W. and Haigler, C.H. (1990). Tracheary-element differentiation in suspension-cultured cells of *Zinnia* requires uptake of extracellular Ca^{2+}. *Planta* **180**, 502–509.

Roberts, K., Burgess, J., Roberts, I. and Linstead, P. (1985). Microtubule rearrangements during plant cell growth and development: an immunofluorescent sudy. In *Botanical Microscopy* (ed. A.W. Robards), pp. 263–284. Oxford University Press, Oxford.

Roberts, L.W. (1976). *Cytodifferentiation in Plants: Xylogenesis as a Model System*. Cambridge University Press, London.

Roberts, L.W. and Baba, S. (1968). IAA-induced xylem differentiation in the presence of colchicine. *Plant and Cell Physiol.* **9**, 315–321.

Savage, R.A. (1983). The role of plant hormones in higher plant cellular differentiation. I. A critique. *Histochem. J.* **15**, 437–445.

Seagull, R.W. (1989a). The plant cytoskeleton. *Crit. Rev. Plant Sci.* **8**, 131–167.

Seagull, R.W. (1989b). The role of the cytoskeleton during oriented microfibril deposition. II. Microfibril disposition in cells with disrupted cytoskeletons. In *Cellulose and Wood, Chemistry and Technology* (ed. C. Schuerch), pp. 811–825. Wiley Interscience, New York.

Seagull, R.W. (1990a). The role of the cytoskeletal elements in organized wall microfibril deposition. In *Biosynthesis and Biodegradation of Cellulose and Cellulosic Materials* (eds C.H. Haigler and P. Weimer), pp. 143–163. Marcel Dekker, New York.

Seagull, R.W. (1990b). The effects of microtubule and microfilament disrupting agents on cytoskeletal arrays and wall deposition in developing cotton fibers. *Protoplasma* **159**, 44–59.

Seagull, R.W., Falconer, M.M. and Weerdenburg, C.A. (1987). Microfilaments: Dynamic arrays in higher plant cells. *J. Cell Biol.* **104**, 995–1004.

Sinnott, E.W. and Bloch, R. (1945). The cytoplasmic basis of intercellular patterns in vascular differentiation. *Amer. J. Bot.* **32**, 151–156.

Sugiyama, M. and Komamine, A. (1987). Relationship between DNA synthesis and cytodifferentiation to tracheary elements. *Oxford Surveys of Plant Molecular and Cell Biology* **4**, 343–346.

Sugiyama, M., Fukuda, H. and Komamine, A. (1986). Effects of nutrient limitation and γ-radiation on tracheary element differentiation and cell division in single mesophyll cells of *Zinnia elegans*. *Plant Cell Physiol.* **27**, 601–606.

Thelen, M.P. and Northcote, D.H. (1989). Identification and purification of a nuclease from *Zinnia elegans* L.: a potential molecular marker for xylogenesis. *Planta* **179**, 181–195.

Torrey, J.G., Fosket, D.E. and Hepler, P.K. (1971). Xylem formation: a paradigm of cytodifferentiation in higher plants. *Amer. Scient.* **59**, 338–352.

Weerdenburg, C.A. and Seagull, R.W. (1988). The effects of taxol and colchicine on microtubule and microfibril arrays in elongating plant cells in culture. *Can. J. Bot.* **66**, 1797–1816.

Wick, S.M. (1988). Immunolocalization of tubulin and calmodulin in meristematic plant cells. In *Calcium Binding Proteins* (ed. M.P. Thompson), pp. 21–45. CRC Press, Boca Raton.

Wooding, F.B.P. and Northcote, D.H. (1964). The development of the secondary wall of the xylem of *Acer pseudoplatanus*. *J. Cell Biol.* **2**, 327–337.

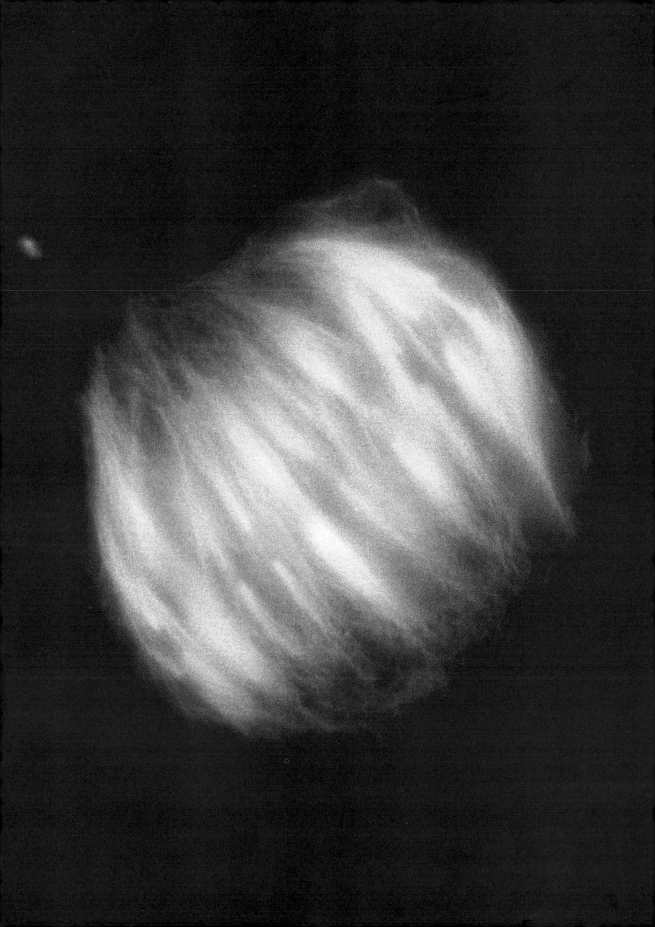

MITOSIS

Traditionally, mitosis is the highpoint of the cell cycle: the visible and dramatic culmination of the invisible processes of interphase. From the morphogenetic point of view, however, mitosis gives the appearance of being sandwiched between two important and linked processes to which it seems unrelated. That is, the division plane is predicted in the alignment of premitotic cytoskeletal structures, is enacted by the postmitotic phragmoplast, but is not necessarily anticipated by the alignment of the spindle. Because of the inextricable relationship between premitosis and cytokinesis in terms of cytoplasmic division, they are discussed in the later section on division plane alignment and so it is only for organizational convenience that nuclear division is covered out of sequence. Of course, nuclear and cytoplasmic division are normally co-ordinated and some of the experiments on whether they are obligatorily linked or separable processes are described later. One of the crucial aspects of cell partitioning about which little is known is how cells are prompted into division. Empirically, combinations of plant growth regulators are known to induce cells to divide *in vitro* but, rationally, hardly anything at all is known about the mechanism. It is for this reason that studies on mitotic regulation are so important in plant biology. Although fibrous structural proteins are the main theme of this book there are signalling processes, involving ions and enzymes, which must orchestrate and mark the transitions between the various premitotic and mitotic phases. This is the subject of Chapter 16 by Wolniak who also reviews evidence for conserved cell cycle regulatory genes, the description of which will have an enormous impact on plant biology.

The process of mitosis in higher plants is essentially similar to that in more intensively studied animal cells. There are, however, distinctive differences. For example, in the former, there is no centriole during a life cycle that does not involve the use of a basal body, necessary for forming cilia or flagella. Perhaps it is only a minor difference that higher plant spindles have no centrioles at their poles for this is no impediment to cell division. The spindle poles may not be so tightly focused but the microtubule nucleating material, whatever it is, is evidently still able to count and segregate itself into two opposing poles which, as in the case of *Haemanthus* endosperm, can be many micrometres broad. It is at other stages of the cytoskeletal cycle that the absence of a definable microtubule organizing centre(s) forms problems, at least for plant cell biologists, certainly not the cell. Other than the poles (where there is strong but still indirect evidence for believing that MT nucleating material is located) there is probably no other location where it is agreed that such stuff must reside. Hence there is a problem in discussing the continuity between the arrays; how one MT array succeeds another and where the MTs are nucleated. In this section, Lambert, Vantard, Schmit and Stoeckel (Chapter 15) review such problems for mitosis as well as discussing the distinctive features of plant cell division. The question of where MTs are initiated is also dealt with in chapters by Palevitz, Wick, and Lloyd (Chapters 4, 17 and 18 respectively) for it is a problem that affects interphase and preprophase as well as division.

15. MITOSIS IN PLANTS

Anne-Marie Lambert, Marylin Vantard, Anne-Catherine Schmit and Herrade Stoeckel

Institut de Biologie Moléculaire des Plantes, Centre National de la Recherche Scientifique, 12 Rue du Général Zimmer, 67084 Strasbourg Cedex, France

Introduction

The aim of this review is to focus on concepts that emerged from recent studies of mitosis in higher plant cells, mostly developed by means of new experimental approaches and functional assays.

In this chapter the authors will restrict their discussion to selected events that they consider to be crucial in understanding singularities of the plant mitotic cell:

(1) The onset of mitosis and spindle formation.
(2) Kinetochore structure and function.
(3) Actin interaction with the mitotic spindle.
(4) Origin of the phragmoplast.
(5) Mitotic control and perspectives.

The mechanisms that regulate the entry into mitosis, the chromosome movements and, later, the exit of mitosis, remain poorly understood in higher plant cells.

Difficulties that are encountered in studying the molecular mechanisms of plant microtubule dynamics *in vivo*, at selected stages of the cell cycle, are due to inherent properties of the plant cell: (i) the cell wall renders microinjection and protein incorporation difficult, and (ii) synchronous cell cultures are hard to obtain. After protoplast isolation, mitosis is, in general, coupled to regeneration of the cell wall leading to the development of callus and microcolonies and seldom leads to a population of cycling, isolated cells.

Citation of numerous original data will unfortunately not be possible here. Recent reviews have been written on the plant cytoskeleton (Staiger and Schliwa, 1987; Lloyd, 1989; Seagull, 1989) and the plant mitotic spindle (Baskin and Cande, 1990). Basic concepts of mitosis were also brought up to date (Nicklas, 1988; Wolniak, 1988; Cande, 1989, 1990; McIntosh and Koonce, 1989; Mitchison, 1989; Sato *et al.*, 1989; Brinkley, 1990). These reviews provide references and detailed sources of information.

The onset of mitosis in higher plants

In yeasts, fungi, as in animal cells, the centrosome is known to be a key regulator of the spatial and temporal organization of the mitotic spindle. Duplication and further separation of centrosomes, as well as drastic changes of their microtubule nucleation activity and microtubule turnover, are well documented (Mazia, 1987; Cande, 1990). Higher plants have no characterized centrosomes although nucleation and dynamics of their microtubules suggest that they may possess cell-cycle-dependent MTOC (microtubule organizing centre) activities that are under spatial and temporal regulation.

In most higher plant cells, the onset of mitosis is characterized by two major and successive events, involving different microtubule populations: (i) formation of the preprophase-band (PPB), and (ii) development of the bipolar spindle. Recent data indicate that microtubules radiating from the nucleus take part in the PPB formation (Flanders *et al.*, 1990).

The function and organization of the PPB are reviewed in Chapters 17 and 18 and will not be discussed here. However, as far as mitosis itself is concerned, the question remains whether these two transitory cytoskeletal architectures are under dependent or independent controls of G2-M phase transition. In naturally

The Cytoskeletal Basis of Plant Growth and Form   ISBN 0–12–453770–7

wall-free higher plant cells, such as in *Haemanthus* endosperm, spindle formation occurs without a PPB. This indicates that assembly and function of the mitotic spindle may be controlled independently of the dynamics of cortical microtubule arrays.

MTOC activity of the plant nuclear surface and spindle formation

Cytoplasmic microtubules of higher plant cells radiate from the surface of the nucleus toward the cell cortex. This is one of the striking singularities of the plant cytoskeleton, as compared with other cell types. Therefore if the activity of the plant nuclear surface as a potential MTOC is correct and if this activity is cell-cycle dependent, it represents one of the key factors in the control of the mitotic onset.

Recent studies (Adoutte *et al.*, 1985) using the incorporation of *Paramecium* axonemal tubulin into lysed *Haemanthus* endosperm cells revealed sites of microtubule assembly both at the nuclear envelope and at the distal end of

endogenous plant microtubules (Vantard *et al.*, 1990). The new data obtained by means of a functional assay strongly support the hypothesis that the whole plant nuclear surface acts as a microtubule nucleation site. Sites of microtubule anchoring at the nuclear envelope were detected by electron microscopy (Lambert, 1980) or by anti-tubulin labelling (De Mey *et al.*, 1982; Bajer and Mole-Bajer, 1982; Wick, 1985a, b) and were confirmed by confocal microscopy (Fig. 15.1). They correspond to multiple nucleation centres uniformly dispersed on the nucleus, with electron-dense material resembling pericentriolar components, usually associated with them. Human autoantibodies that stain pericentriolar nucleating material in animal cells (Calarco-Gillam *et al.*, 1983) decorate the periphery of the nucleus and the spindle poles of higher plant cells (Clayton *et al.*, 1985; Wick, 1985a, b), although the significance of this cross-reactivity has been recently questioned (Harper *et al.*, 1989).

At the onset of mitosis, when intranuclear chromosomes condense, tubulin incorporation increases on the nuclear surface of *Haemanthus*

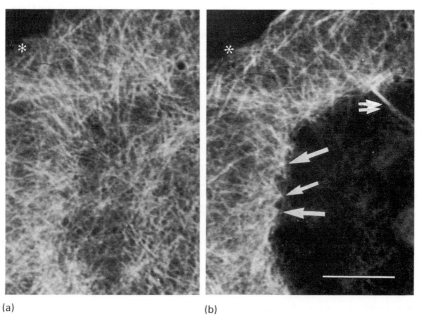

(a)   (b)

Fig. 15.1. Confocal microscopy of *Haemanthus* endosperm cell labelled with anti-tubulin antibodies: optical sections above (a) and through (b) the nucleus. Multiple sites of microtubule anchorage are seen on the nuclear surface (arrows). One thin microtubule bundle penetrates into the nucleus (double arrow). Scale bar = 10 μm.

endosperm cells in prophase, suggesting a cell-cycle-dependent control of its nucleation capacity. It should be investigated if these events are accompanied by phosphorylation of centrosomal-type proteins around the plant nucleus, as this has been demonstrated in animal interphase cell extracts where mitotic centrosome activity is induced by addition of cdc2 kinase (Verde et al., 1990).

In Haemanthus endosperm cells, the assembly of the bipolar acentriolar spindle around the nucleus is achieved by a transitory convergence of microtubules, forming aster-like centres, which progressively lead to the formation of spindle poles (Schmit et al., 1983). This involves increased microtubule interactions that might be mediated by selective microtubule associated proteins (MAPs). Calmodulin that is concentrated at the centriolar polar regions of animal cells is found in these prophase microtubule centres (Vantard et al., 1985), suggesting that calmodulin-mediated mechanisms may control these events in higher plants.

Altogether, these observations suggest that the diffuse MTOC-like components on the plant nuclear surface may change in their spatial distribution in a cell-cycle-dependent manner. Such changes may be reflected later in the broad and diffuse poles of the higher plant spindle at metaphase. Comparable events occur in animal cells and have been described as a 'flexible' centrosome (Mazia, 1987).

If the higher plant cell has a constant tubulin content, an increase in polymer assembly may lead to a slower rate of microtubule polymerization or to shorter, and/or more dynamic microtubules as prophase progresses. This is the case during spindle formation. Long interphase microtubules that elongated toward the cell periphery, as indicated by exogenous tubulin incorporation at their distal ends (Vantard et al., 1990), are replaced by straight, shorter microtubules forming the two half spindles around the nucleus. Although no systematic studies of the dynamic instability of higher plant microtubules have yet been performed in living cells, our present views strongly suggest that comparable regulatory mechanisms occur in spite of significant differences of the microtubule nucleation processes.

Actin in interphase–mitosis transition

The molecular mechanisms that control the F-actin redistribution and dynamics during the formation of the mitotic spindle in higher plants remain obscure.

In walled, vacuolated cells, radial strands of F-actin seem to have a strategic role in prepro-phase band organization where they intermingle with microtubules, as well as during the premitotic development of the phragmo-some (Lloyd and Traas, 1988; Flanders et al., 1990, and Lloyd, Chapter 18).

The onset of mitosis in cultured animal cells is characterized by the disassembly of actin stress fibres and rounding up of the cells. However, in other cell types, such as leucocytes, cytoplasmic actin seems to be involved in the positioning and motility of the centrosomes (Euteneuer and Schliwa, 1985). Particular observations on higher plant cells during pre-mitosis indicate a progressive reorganization of the cytoplasmic F-actin network at the cell cortex and into the transvacuolar strands in the centre of the cell (Lloyd, 1988, 1989; Mole-Bajer et al., 1988; Schmit and Lambert, 1987). It is unknown if, and how, this spatial redistribution of the plant cytoplasmic microfilaments is related to the microtubule dynamics, during spindle formation, and if these events can be correlated with the inhibition of cytoplasmic streaming that has been reported at prophase (Otâ, 1961; Mineyuki et al., 1983).

The plant kinetochore: structure and function

The sudden breakdown of the nuclear envelope allows the capture of spindle microtubules by the kinetochores. However, investigations using confocal microscopy indicate that thin microtubule bundles penetrate earlier into the prophase nucleus, suggesting that the disorganization of the plant nuclear envelope is gradual before its final and sudden rupture. It is unknown if lamina-like components are involved. As a consequence the attachment of microtubules to the kinetochores during prophase–prometaphase transition is asynchronous and may depend on the spatial distribution of pre-kinetochores within the nucleus. The spectacular jumps of individual chromosomes, as observed in microcinematography, reveals the fast dynamics of microtubules during this anchorage. It will be important to know if microtubule-capture occurs when kinetochore components are being activated through phosphorylation at the time of the nuclear envelope breakage which is known to be under *cdc2*-kinase control (Murray and Kirschner, 1989).

Experimental breakthroughs have yielded entirely new insights into the function of kinetochores as mitotic motors (Mitchison, 1989). Centromeric/kinetochore proteins appear as active links between centromeric-specific DNA sequences and spindle microtubules. Most data obtained so far come from yeast and mammalian chromosomes. Plant kinetochore components remained unidentified until the recent discovery that human autoantibodies from CREST patients, reacting with centromeric antigens of mammalian chromosomes (Brinkley, 1990), recognize kinetochores of the higher plant *Haemanthus* (Mole-Bajer *et al.*, 1990). Western blotting revealed common epitopes in centromeric polypeptides of HeLa and *Haemanthus* cell extracts. This indicates a striking conservation of centrosome/kinetochore components in both animal and plant kingdoms, although the molecular mass of the polypeptides is different. Two proteins of 65 kD and 135 kD were stained in *Haemanthus* cell extracts using antibodies that recognize HeLa centromeric peptides of 52 kD and 80 kD and several minor bands of 140–200 kD. These studies also highlight the composite structure of the higher plant kinetochore that was characterized up to now by its quasi homogeneous 'ball-cup' organization. Four granules per centromere are detected in the G2 nucleus, while multiple or at least two distinct subunits are resolved within each sister kinetochore in prometaphase (Mole-Bajer *et al.*, 1990). This configuration leads to a double kinetochore fibre anchored at each sister-kinetochore. Such organization is also detected by use of confocal microscopy, as illustrated in Fig. 15.2. It is unknown if this corresponds to the half chromatid concept predicted by Bajer and

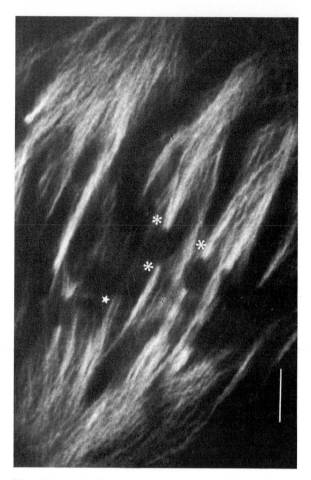

Fig. 15.2. Confocal microscopy of a prometaphase spindle (*Haemanthus* endosperm cell – anti-tubulin labelling). The double structure of each sister kinetochore and double kinetochore fibre is detected (marks). Intermingling of neighbouring fibres can be followed on serial optical sections. Polar regions are diffuse. Scale bar = 10 μm.

Östergren (1961). Microtubule interactions between neighbouring kinetochores during prometaphase contribute to the interdependent and asynchronous movements of chromosomes (Lambert and Bajer, 1975; Bajer, 1987).

Such composite kinetochore structure may be extended in the case of diffuse centromeric chromosomes of several plant species such as *Luzula* (review, Godward, 1985) and may reflect the compound organization that has been proposed as a functional model for mammalian chromosomes (Brinkley, 1990).

Sister kinetochore separation and anaphase

The process of chromosome segregation involves the bipolar orientation of sister-kinetochores, and further kinetochore-fibre disassembly. Sudden separation of sister-kinetochores, resulting in a quasi synchronous chromosome splitting is microtubule-independent and not related to spindle forces as demonstrated in microtubule depleted cells after long exposure to colchicine (Mole-Bajer, 1969; Bajer and Mole-Bajer, 1972; Lambert, 1980). An increase in cytosolic calcium seems to be involved as a signal (Keith *et al.*, 1985). Recent studies with iontophoretically injected Ca^{2+} into *Tradescantia* stamen hair cells during anaphase provided new and direct evidence that a Ca^{2+} increase might regulate chromosome motion during the whole process of anaphase (Zhang *et al.*, 1990). The mechanism by which an increase of Ca^{2+} to the micromolar level induces an acceleration of chromosome movement is not known (Wolniak, 1988; see also Chapter 16, this volume). It may facilitate microtubule disassembly, but may also activate dynein and/or kinesin which were found at the kinetochore and within the spindle, although their valid identification in the higher plant mitotic cells remains debatable.

Selective distribution of calmodulin along plant kinetochore fibres (Vantard *et al.*, 1985; Wick *et al.*, 1985) suggests that kinetochore–microtubule dynamics in anaphase might be controlled by a Ca^{2+}-calmodulin-dependent process. This could involve particular MAPs, bound to the kinetochore microtubules, regulating their selective disassembly (Lambert and Vantard, 1986).

Such mechanisms that might involve a calcium gradient within the two half spindles during anaphase, would allow microtubule disassembly at the kinetochore, as it is now assumed, and at the same time permit polar assembly of new microtubules that elongate toward the equator and invade the interzone (De Mey *et al.*, 1982; Bajer and Mole-Bajer, 1982).

Structure and dynamics of the plant kinetochore fibre

Metaphase and anaphase dynamics of microtubules (review Baskin and Cande, 1990) have revealed peculiarities of the higher plant mitotic cells. The kinetochore fibre has a complex structure. Two different microtubule populations intermingle: stable microtubules which are cold and drug resistant are anchored at the kinetochore and form the core of the fibre that is surrounded by numerous skew and labile non-kinetochore microtubules. This leads to the 'Fir-tree' pattern described in *Haemanthus* (Bajer and Mole-Bajer, 1986) and *Allium* mitotic cells (Palevitz, 1988) for example. As a result, the plant kinetochore fibre is connected to the poles through intermingling non-kinetochore microtubules. Whether such organization is structurally and/or functionally comparable to aster and kinetochore microtubule interactions of animal cells is unknown. In spite of these structural differences the kinetics of the polewards chromosome movements have comparable features to those of animal cells. In anaphase, the chromosomal speed is about $1-2$ μm min^{-1}, with variations. Several models suggest that labile, lateral interactions between microtubules contribute to the chromosome motion of higher plants (Bajer, 1987). The discovery, in animal cells, that microtubule disassembly occurs primarily at the plus ends while still attached to the kinetochores, localizes the force production at the kinetochore itself, which represents a 'mitotic motor' (Mitchison, 1989; Sluder, 1990). It is highly probable that these mechanisms occur in most eukaryotic cells, as suggested by experiments using microinjection of exogenous tubulin (Gorbsky *et al.*, 1987; Geuens *et al.*, 1989). Further investigations on the functional properties of the plant kinetochore are, however, one of present priorities in this field. Studies using microtubule inhibitors (Morejohn *et al.*, 1987) or UV irradiation suggest however that the integrity of the whole kinetochore fibre is needed for chromosome motion within the higher plant spindle. During anaphase, the growth of microtubules from the poles towards the equator, as described above in *Haemanthus* endosperm cells, contributes to develop antagonist forces that maintain the poles apart.

Actin interactions with the mitotic spindle

As we have stressed previously, the cytoplasmic plant actin network remains as a permanent elastic cage around the microtubular spindle during the whole process of mitosis and cytokinesis (Schmit and Lambert, 1987, 1988; Traas *et al.*, 1987; Mole-Bajer *et al.*, 1988; Lloyd, 1989) as well as in meiosis (Traas *et al.*, 1989). Microfilaments are also detected inside the spindle (Forer, 1984) although their activity is highly debated (Forer, 1988) as chromosome movements are not arrested by actin inhibitors such as the cytochalasins. Recent investigations by microinjection of anti-actin antibodies (Mole-Bajer *et al.*, 1988) and of fluorescent phalloidin (Schmit and Lambert, 1990) into living mitotic cells of *Haemanthus*, did not inhibit chromosome motion. While chromosomes are moving to the poles, the actin microfilaments labelled by injected phalloidin are stretched on the surface of the microtubular

spindle (Fig. 15.3). At the poles, where no forces are developed, the F-actin remains as a network, as in interphase. One can assume that active interactions are developed between actin filaments and microtubules, particularly at the interface between the mitotic spindle and the cytoplasmic cell cortex. This cellular domain is of utmost importance in the metabolic exchanges between cytoplasm and the mitotic spindle, and deserves particular attention. In spite of a large body of evidence confirming that F-actin is an important cytoskeletal component of the plant mitotic cell, we can only speculate on its role.

Spindle–phragmoplast relations, the phragmoplast equator as a MTOC and a site of actin assembly?

Cytokinesis will not be included here. However, the origin of the phragmoplast will be discussed in relation to the mitotic spindle, taking into account recent investigations using new experimental approaches.

Phragmoplast architecture is highly complex

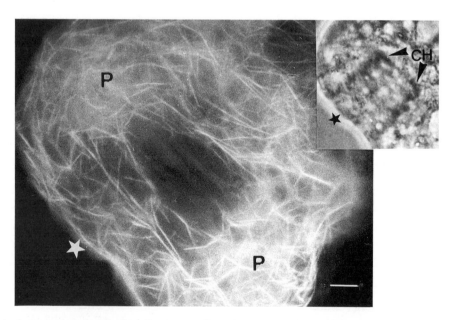

Fig. 15.3. Actin interaction with the mitotic spindle in living anaphase (*Haemanthus* endosperm cell – inset). Fluorescent phalloidin was microinjected in metaphase. During chromosome (CH) migration, F-actin labelled microfilaments are stretched on the spindle surface and remain as a network at the poles (P). Scale bar = 10 μm.

and its origin is controversial. Two or even three main hypotheses are currently debated. Phragmoplast microtubules could be: (i) the remnants of half-spindle polar microtubules that invade the interzone in late anaphase and would be stabilized at their plus interdigitating ends; (ii) the result of new tubulin assembly at the equator during telophase, including also, (iii) new polymerization of microtubules from the nuclear surface of reforming nuclei (see reviews by Hepler and Wolniak, 1984; Seagull, 1989; Baskin and Cande, 1990).

Successful incorporation of exogenous tubulin into lysed *Haemanthus* endosperm cells in late anaphase–telophase indicates intense microtubule assembly at the equator (Vantard *et al.*, 1990). Homogeneous tubulin incorporation is also detected at the distal ends of endogenous microtubules which form the typical barrel-shaped phragmoplast. Previous indirect observations on the restoration of birefringence after UV irradiation (Inoué, 1964), together with immunocytochemical investigations and detection in electron-microscopy of dense equatorial material that resembles MTOC (Lambert and Bajer, 1972) strongly favour the model in which phragmoplast microtubules are the result of new tubulin assembly. The equatorial region of the phragmoplast could be considered as a MTOC. In that case, phragmoplast microtubules may be nucleated either at their plus ends, anchored at the equator, according to polarity data (Euteneuer *et al.*, 1982) or may correspond to a specific microtubule population of opposite polarity to that of spindle microtubules. The mechanisms that control phragmoplast formation might be independent or uncoupled from those that control microtubule function in chromosome movement. Phragmoplast and cell plate development between non-sister nuclei during cellularization of syncytia, for example, and in cell fragments, favour these views (Bajer and Mole-Bajer, 1982, 1986; De Mey *et al.*, 1982).

Another exciting point is the intense accumulation of actin within the phragmoplast (see reviews by Lloyd, 1988, 1989). This has been investigated recently *in vivo* by using double microinjection of differently labelled phalloidin into the same mitotic cell, at subsequent stages of anaphase (Schmit and Lambert, 1990). This new technique permits the dynamics of the actin network to be visualized, and suggests that actin may be newly assembled within the phragmoplast domain in telophase. According to these observations, the future cell plate domain is an active site of cytoskeletal assembly, which may be associated with the formation of the new cell membrane and cell wall. Heavy meromyosin labelling indicates also that microfilament polarity is opposite on two sides of the cell plate, in isolated phragmoplasts, suggesting the possible involvement of actin in the vesicle transport toward the cell plate (Kakimoto and Shibaoka, 1988). Whether actin that is present in the interphase network, the preprophase band, around the mitotic spindle, and in the phragmoplast corresponds to specific isotypes remains entirely unknown and should undoubtedly be the subject of further investigations.

Control of mitosis and perspectives

Difficulties in obtaining synchronous cultures of isolated higher plant cells represent an enormous handicap in the present development of genetic analysis. By contrast, yeast and fungi appear as key models.

As p34 *cdc*2 kinase homologues have been found in all eukaryotic cells studied so far, from yeast to HeLa cells, it is highly probable that comparable mechanisms are present in higher plants.

One priority is now to identify and experiment on *cdc*2 regulation of plant microtubule dynamics and centrosome-like activity, as recently reported on *Xenopus* cell extracts (Verde *et al.*, 1990). It would be exciting to identify changes in cyclin or cyclin-like proteins during anaphase progression and to investigate potential dependence or independence in microtubule nucleation/dynamics in the phragmoplast. Another major perspective is the understanding of the plant kinetochore activity. Cloning and sequencing of the yeast centromere genes now offers the possibility of identifying centromeric DNA sequences that might have comparable functions in chromosomes of higher organisms and particularly of higher plants (Bloom *et al.* 1989).

The presence of MAPs such as kinesin and dynein, which function as mechanochemical ATPases for microtubule-based movements, remains to be proved in the higher plant spindle and particularly at the kinetochores.

On the other hand, binding sequences for MAPs on tubulins also indicate striking similarities with repeats detected on particular centromeric protein sequences, suggesting that MAPs may also link micrótubules to the kinetochore (see review by Brinkley, 1990).

Cell models, including lysed cells or isolated spindles appear to be powerful in the investigation of chromosome movement (see review by Cande, 1989) and in assays on exogenous tubulin incorporation (Vantard *et al.*, 1990). The use of a microinjected 'reporter' tubulin in the living cells should soon bring new insight in the dynamics of microtubules in higher plants.

Developmentally regulated expression of tubulin and actin genes in higher plants is a fast-growing field (Silflow *et al.*, 1987; Lloyd, 1989; McLean *et al.*, 1990). More information is needed to understand the differential activity of microtubule and F-actin populations, particularly in the mitotic cells.

These perspectives represent a large field of investigation where molecular and structural approaches should be, as far as possible, tightly bound to studies of the living cell.

Acknowledgements

We are grateful to Dr Jan De Mey (EMBL, Heidelberg) for stimulating discussions. Drs André Adoutte, Nicolette Levilliers, Anne-Marie Hill shared with us the results on *Paramecium* tubulin incorporation. We thank Monique Wehr and Christine Peter for their help in preparing the manuscript. This work was supported by the Centre National de la Recherche Scientifique and the Ministère de la Recherche et de l'Enseignement Supérieur (87 TO 233).

References

Adoutte, A., Claisse, M., Maunoury, R. and Beisson, J. (1985). Tubulin evolution: ciliate-specific epitopes are conserved in the ciliary tubulin of Metazoa. *J. Mol. Evol.* **22**, 220–229.

Bajer, A.S. (1987). Substructure of the kinetochore and reorganization of kinetochore microtubules during early prometaphase in *Haemanthus* endosperm. *Eur. J. Cell Biol.* **43**, 23–34.

Bajer, A.S. and Mole-Bajer, J. (1972). Spindle dynamics and chromosome movements. *Int. Rev. Cyt.* **Suppl. 3**, 1–273.

Bajer, A.S. and Mole-Bajer, J. (1982). Asters, poles, and transport properties within spindle-like microtubule arrays. In *Cold Spring Harbor Symp. Quant. Biol.* **46**, 263–283.

Bajer, A.S. and Mole-Bajer, J. (1986). Reorganization of microtubules in endosperm cells and cell fragments of the higher plant *Haemanthus*. *J. Cell Biol.* **102**, 263–281.

Bajer, A.S. and Östergren, G. (1961). Centromere-like behavior of non-centromeric bodies. *Hereditas* **47**, 563–598.

Baskin, T.I. and Cande, W.Z. (1990). The structure and function of the mitotic spindle in flowering plants. *Ann. Rev. Plant Physiol. Plant Mol. Biol.* **41**, 277–315.

Bloom, K., Hill, A., Kenna, M. and Saunders, M. (1989). The structure of a primitive kinetochore. *Trends Biochem. Sci.* **14**, 223–227.

Brinkley, B.R. (1990). Centromeres and kinetochores: integrated domains on eukaryotic chromosomes. *Curr. Opin. Cell Biol.* **2**, 446–452.

Calarco-Gillam, P.D., Siebert, M.C, Hubble, R., Mitchison, T. and Kirschner, M. (1983). Centrosome development in early mouse embryos as defined by an autoantibody against pericentriolar material. *Cell* **35**, 621–629.

Cande, W.Z. (1989). Mitosis *in vitro*. In *Mitosis, Molecules and Mechanisms* (eds J. Hyams and B.R. Brinkley), pp. 303–326. Academic Press, London.

Cande, W.Z. (1990). Centrosomes: composition and reproduction. *Curr. Opin. Cell Biol.* **2**, 301–305.

Clayton, L., Black, C.M. and Lloyd, C.W. (1985). Microtubule nucleating sites in higher plant cells identified by an autoantibody against pericentriolar material. *J. Cell Biol.* **101**, 319–324.

De Mey, J., Lambert, A.M., Bajer, A.S., Moeremans, M. and De Brabander, M. (1982). Visualization of microtubules in interphase and mitotic plant cells of *Haemanthus* endosperm with the immunogold staining method. *Proc. Natl Acad. Sci. U.S.A.* **79**, 1898–1902.

Euteneuer, U. and Schliwa, M. (1985). Evidence for an involvement of actin in the positioning and motility of centrosomes. *J. Cell Biol.* **101**, 96–103.

Euteneuer, U., Jackson, W.T. and McIntosh, J.R. (1982). Polarity of spindle microtubules in *Haemanthus* endosperm. *J. Cell Biol.* **94**, 644–653.

Falconer, M.M., Donaldson, G. and Seagull, R.W. (1988). MTOCs in higher plant cells: an immuno-fluorescent study of microtubule assembly sites following depolymerization by APM. *Protoplasma* **144**, 46–55.

Flanders, D.J., Rawlins, D.J., Shaw, P.J. and Lloyd, C.W. (1990). Nucleus associated microtubules help determine the division plane of plant epidermal cells: avoidance of four-way junctions and the role of cell geometry. *J. Cell Biol.* **110**, 1111–1122.

Forer, A. (1984). Does actin produce the force that moves a chromosome to the pole during anaphase? *Can. J. Biochem. Cell Biol.* **63**, 595–598.

Forer, A. (1988). Do anaphase chromosomes chew their way to the pole or are they pulled by actin? *J. Cell Sci.* **91**, 449–453.

Geuens, G., Hill, A.M., Levilliers, N., Adoutte, A. and De Brabander, M. (1989). Microtubule dynamics investigated by microinjection of *Paramecium* axonemal tubulin: Lack of nucleation of proximal assembly of microtubules at the kinetochore during prometaphase. *J. Cell Biol.* **108**, 939–953.

Godward, M.B.E. (1985). The kinetochore. *Int. Rev. Cyt.* **94**, 77–105.

Gorbsky, G.J., Sammak, P.S. and Borisy, G.G. (1987). Microtubule dynamics in chromosome motion visualized in living anaphase cells. *J. Cell Biol.* **104**, 9–18.

Harper, J.D.I., Mitchison, J.M., Williamson, R.E. and John, P.C.L. (1989). Does the autoimmune serum 5051 specifically recognize microtubule organizing centres in plant cells? *Cell Biol. Int. Rep.* **13**, 471–483.

Hepler, P.K. and Palevitz, B.A. (1986). Metabolic inhibitors block anaphase A *in vivo. J. Cell Biol.* **102**, 1995–2005.

Hepler, P.K. and Wolniak, S.M. (1984). Membranes in the mitotic apparatus: their structure and function. *Int. Rev. Cyt.* **90**, 169–238.

Inoué, S. (1964). Organization and function of the mitotic spindle. In *Primitive Motile Systems in Cell Biology* (eds R.D. Allen and H. Kamiya), pp. 549–598. Academic Press, New York.

Kakimoto, T. and Shibaoka, H. (1988). Cytoskeletal ultrastructure of phragmoplast-nuclei complexes isolated from cultured tobacco cells. *Protoplasma* **Suppl. 2**, 95–103.

Keith, H., Ratan, R., Maxfield, F., Bajer, A.S., Shelanski, M.C. (1985). Local cytoplasmic calcium gradients in living mitotic cells. *Nature,* (Lond.) **316**, 848–850.

Lambert, A.M. (1980). The role of chromosomes in anaphase trigger and nuclear envelope activity in spindle formation. *Chromosoma* **76**, 295–308.

Lambert, A.M. and Bajer, A.S. (1972). Dynamics of spindle fibers and microtubules during anaphase and phragmoplast formation. *Chromosoma* **39**, 101–144.

Lambert, A.M. and Bajer, A.S. (1975). Fine structure dynamics of the prometaphase spindle. *J. Micro. Biol. Cell.* **23**, 191–194.

Lambert, A.M. and Vantard, M. (1986). Calcium and calmodulin as regulators of chromosome movement during mitosis in higher plants. In *Molecular and Cellular Aspects of Calcium in Plant Development* (ed. A.J. Trewavas), NATO AJJ series, Life Sciences, pp. 175–183. Plenum, New York.

Lloyd, C.W. (1988). Actin in plants. *J. Cell Sci.* **90**, 185–192.

Lloyd, C.W. (1989). The plant cytoskeleton. *Curr. Opin. Cell Biol.* **1**, 30–35.

Lloyd, C.W. and Traas, J.A. (1988). The role of F-actin in determining the division plane of carrot suspension cells. Drug studies. *Development* **102**, 211–222.

Mazia, D. (1987). The chromosome cycle and the centrosome cycle in the mitotic cycle. *Int. Rev. Cyt.* **100**, 49–92.

McIntosh, J.R. and Koonce, M.P. (1989). Mitosis. *Science* **246**, 622–628.

McLean, B.G., Eubanks, S. and Meagher, R.B. (1990). Tissue-specific expression of divergent actins in soybean root. *The Plant Cell* **2**, 335–344.

Mineyuki, Y., Yamada, M., Takagi, M., Wada, M. and Furuya, M. (1983). A digital image processing technique for the analysis of particle movements. Its application to organelle movements during mitosis in *Adiantum capillus veneris* protonemata. *Plant and Cell Physiol.* **24**, 225–234.

Mitchison, T.J. (1989). Mitosis: basic concepts. *Curr. Opin. Cell Biol.* **1**, 67–74.

Mole-Bajer, J. (1969). Fine structural studies of apolar mitosis. *Chromosoma* **26**, 427–448.

Mole-Bajer, J., Bajer, A.S. and Inoué, S. (1988). Three dimensional localization and redistribution of F-actin in higher plant mitosis and cell plate formation. *Cell Motil. Cytosk.* **10**, 217–228.

Mole-Bajer, J., Bajer, A.S., Zinkowski, R.P., Balczon, R.D. and Brinkley, B.R. (1990). Autoantibodies from a patient with scleroderma CREST recognized kinetochores of the higher plant *Haemanthus. Proc.*

Natl Acad. Sci. U.S.A. **87**, 3599–3603.

Morejohn, L.C., Bureau, T.E., Mole-Bajer, J., Bajer, A.S. and Fosket, D.E. (1987). Oryzalin, a dinitro-aniline herbicide, binds to plant tubulin and inhibits microtubule polymerization *in vitro*. *Planta* **172**, 252–264.

Murray, A.W. and Kirschner, M.W. (1989). Dominoes and clocks: The union of two views of the cell cycle. *Science* **246**, 614–621.

Nicklas, R.B. (1988). The forces that move chromosomes in mitosis. *Ann. Rev. Biophys. Biophys. Chem.* **17**, 431–439.

Otâ, T. (1961). The role of cytoplasm in cytokinesis of plant cells. *Cytologia* **26**, 428–447.

Palevitz, B.A. (1988). Microtubular firtrees in mitotic spindles of onion roots. *Protoplasma* **142**, 74–78.

Pickett-Heaps, J.D. (1969). The evolution of the mitotic apparatus: an attempt at comparative ultrastructural cytology in dividing plant cells. *Cytobios* **3**, 257–280.

Sato, H., Kobayashji, A. and Itoh, J. (1989). Molecular basis of physical and chemical probes for spindle assembly. *Cell Struct. Func.* **14**, 1–34.

Schmit, A.C. and Lambert, A.M. (1987). Characterization and dynamics of cytoplasmic F-actin in higher plant endosperm cells during interphase, mitosis and cytokinesis. *J. Cell Biol.* **105**, 2157–2166.

Schmit, A.C. and Lambert, A.M. (1988). Plant actin filament and microtubule interactions during anaphase–telophase transition: effects of antagonist drugs. *Biol. Cell* **64**, 309–319.

Schmit, A.C. and Lambert, A.M. (1990). Microinjected fluorescent phalloidin *in vivo* reveals the F-actin dynamics and assembly in higher plant mitotic cells. *The Plant Cell* **2**, 129–138.

Schmit, A.C., Vantard, M. and Lambert, A.M. (1983). Aster-like microtubule centers establish spindle polarity during interphase–mitosis transition in higher plant cells. *Plant Cell Rep.* **2**, 285–288.

Seagull, R.W. (1989). The plant cytoskeleton. In *Critical Reviews in Plant Sciences* **8**, 131–167.

Silflow, C.D., Oppenheimer, D.G., Kopczak, S.D., Ploense, S.E. and Ludwig, S.R. (1987). Plant tubulin genes: structure and differential expression during development. *Dev. Genet.* **8**, 435–460.

Sluder, G. (1990). Functional properties of kinetochores in animal cells. *Curr. Opin. Cell Biol.* **2**, 23–27.

Staiger, C.J. and Schliwa, M. (1987). Actin localization and function in higher plants. *Protoplasma* **141**, 1–12.

Traas, J.A., Doonan, J.H., Rawlins, D.J., Shaw, P.J., Watts, J. and Lloyd, C.W. (1987). An actin network is present in cytoplasm throughout the cell cycle of carrot cells and associates with the dividing nucleus. *J. Cell Biol.* **105**, 387–395.

Traas, J.A., Burgain, S., De Vaulx, R. (1989). The organization of the cytoskeleton during meiosis in eggplant (*Solanum melongena* L.): microtubules and F-actin are both necessary for coordinated meiotic division. *J. Cell Sci.* **92**, 541–550.

Vantard, M., Lambert, A.M., De Mey, J., Picquot, P. and Van Eldik, L. (1985). Characterization and immunocytochemical distribution of calmodulin in higher plant endosperm cells: localization in the mitotic apparatus. *J. Cell Biol.* **101**, 488–499.

Vantard, M., Levilliers, N., Hill, A.M., Adoutte, A. and Lambert, A.M. (1990). Incorporation of *Paramecium* axonemal tubulin into higher plant cells reveals functional sites of microtubule assembly. *Proc. Natl Acad. Sci. U.S.A.* **87**, 8825–8829.

Verde, F., Labbé, J.C., Dorée, M. and Karsenti, E. (1990). Regulation of microtubules dynamics by cdc2 protein kinase in cell-free extracts of *Xenopus* eggs. *Nature* **343**, 233–238.

Wick, S.M. (1985a). The higher plant mitotic apparatus: redistribution of microtubules, calmodulin and microtubule initiation material during its establishment. *Cytobios* **43**, 285–294.

Wick, S.M. (1985b). Immunofluorescence microscopy of tubulin and microtubule arrays in plant cells. III. Transition between mitotic, cytokinetic and interphase microtubule arrays. *Cell Biol. Int. Rep.* **9**, 357–371.

Wick, S.M., Muto, S. and Duniec, J. (1985). Double immunofluorescence labeling of calmodulin and tubulin in dividing plant cells. *Protoplasma* **126**, 198–206.

Wolniak, S.M. (1988). The regulation of mitotic spindle function. *Biochem. Cell Biol.* **66**, 490–514.

Zhang, D.H., Callaham, D.A. and Hepler, P.K. (1990). Regulation of anaphase chromosome motion in *Tradescantia* stamen hair cells by calcium and related signaling agents. *J. Cell Biol.* **111**, 171–182.

16. PATTERNS OF REGULATION DURING MITOSIS

Stephen M. Wolniak

Department of Botany and Center for Agricultural Biotechnology, University of Maryland, College Park, MD 20742, USA

Introduction

In this chapter the regulation of progression through the major morphological transitions during mitosis, namely nuclear envelope breakdown and anaphase onset, is described. The cell division cycle kinase whose catalytic subunit is known as p34^{cdc2} probably plays important roles in initiating entry into mitosis from interphase and for progression through nuclear envelope breakdown. The data describing p34^{cdc2} kinase come almost exclusively from other organisms, but there are early indications that it is a likely regulatory component in plant cells. After nuclear envelope breakdown, mitotic progression appears to be modulated by changes in cytosolic calcium activity. Calcium activity increases may be random temporally, but appear to be followed by increases in protein phosphorylation, some the result of calcium-stimulated kinase activity. Periods of protein phosphorylation appear to be followed by episodes of protein dephosphorylation. A major purpose of this chapter is to point out directions for future experimentation on mitotic regulation in plant cells.

Progression through the cell cycle and mitotic transitions.

The somatic cell cycle in plants and animals is commonly broken into interphase and mitosis. Interphase consists of a period of nuclear DNA replication (S, for synthesis) that is both preceded and followed by intervals when no replication occurs (G$_1$ before S and G$_2$ after S). Mitosis (M) usually occupies but a small portion of the cycle and for the purpose of this discussion, includes cytokinesis. The focus of this chapter is on the regulation of entry into, and progression through, the M-phase of the cell cycle.

Entry into the M-phase of the cell cycle is defined morphologically by the condensation of the replicated chromosomes within the nuclear envelope, during mitotic prophase. Shortly thereafter, during mid to late prophase, the nucleolus disperses. The mitotic spindle (McDonald, 1989; Baskin and Cande, 1990), comprising a more-or-less well-focused biconical array of microtubules becomes organized around the nucleus during late prophase. Nuclear envelope breakdown occurs and is followed by chromosomal congression to the metaphase plate. Metaphase culminates with the simultaneous splitting of sister chromatids at anaphase onset. Nuclear envelope breakdown and anaphase onset, two of the major morphological transitions during mitosis, are probably the manifestation of many distinct biochemical processes (Wolniak, 1988). An important facet of both, however, is that they appear to be spontaneous, pancellular events, and as such, are likely candidates to be under the regulatory control of signalling cascades (Harris, 1977).

An important event that precedes and apparently causes nuclear envelope breakdown both *in vivo* and *in vitro* is the phosphorylation of a class of nuclear proteins, known collectively as the lamins (Gerace and Blobel, 1980; Burke and Gerace, 1986; Gerace and Burke, 1988; Heald and McKeon, 1990; Peter *et al.*, 1990b; Ward and Kirschner, 1990). *In vitro* experiments indicate that lamin phosphorylation at specific serine and threonine residues is necessary and sufficient for nuclear envelope

The Cytoskeletal Basis of Plant Growth and Form   ISBN 0–12–453770–7

breakdown (Burke and Gerace, 1986; Suprynowitz and Gerace, 1986). The signalling cascade that culminates in lamin phosphorylation is complex, involving at least three different types of protein kinases (Shenoy et al., 1989; Maller, 1990; Ward and Kirschner, 1990). A cytosolic calcium activity rise has been demonstrated as a regulatory step preceding nuclear envelope breakdown in sea urchins (Steinhardt and Alderton, 1988).

At the end of metaphase, the sister chromatids unwind from each other, and separate synchronously. Sister chromatid separation can occur spontaneously at all chromosomes even in the absence of an organized mitotic apparatus (Mole-Bajer, 1958; Lambert, 1980) and once this event occurs, the chromosomes move apart by a short distance and then fail to separate any further. Thus, the signal for separation is independent of kinetochore attachments to spindle microtubules, but further chromosomal movement is not. Chromosomal separation in the absence of spindle microtubules suggests that anaphase onset involves the release of a 'brake' that holds the chromosomes together during metaphase while the anaphase motor is running. Among possible candidates for this metaphase brake are one or more centromeric proteins, identified immunologically, and known as CENP-A through CENP-D (for CENtromeric Protein) (Pluta et al., 1990) and at least as two species of INCENP (for INner CENtromeric Protein) that are present in or on metaphase kinetochores in a variety of animal (Cooke et al., 1987, 1990; Pluta et al., 1990) and plant (Mole-Bajer et al., 1990) cells. Some of these antigens move with the kinetochores as they migrate with their chromosomes to the spindle poles during anaphase, while others (Cooke et al., 1987, 1990; Pankov et al., 1990; Pluta et al., 1990) remain at the former metaphase plate during anaphase, and ultimately reside in the spindle midzone as the cell enters cytokinesis. Antigens in this latter group reside in the core of the centromere and between the chromatids and remain tenaciously attached to them during metaphase but not during anaphase (Cooke et al., 1987), a condition that would be necessary for a metaphase brake component. The mechanism for their release

from the kinetochore or the chromosome may underlie the mechanism for sister chromatid separation at anaphase onset.

Checkpoints in mitotic regulation

The notion of a regulatory control point (Van'tHof and Kovacs, 1972) has long been recognized as a timepoint in a multi-stepped process when the cell undergoes a non-reversible shift (i.e. a 'point of no return' (Mazia, 1961)) that commits it to enter the next phase of that process. The cell moves into a new physiological state, responding differently to the same stimuli from cells that have not passed through the switch point. Numerous control points (also described at various times as transition points, restriction points or switch points) have been determined through the pharmacological manipulation of cells at various points in their cycles. The classical manifestations of a block at a transition point is the observation of an extended arrest following a short inhibitory treatment (Mitchison, 1971). In a well-defined system, it is also possible to induce precocious entry into different phases of mitosis through the early induction of regulatory cascades (Chen and Wolniak, 1987a; Wolniak, 1987; Larsen and Wolniak, 1990). Key to the recognition that precocious entry is occurring, or that progression is anomalously fast, is a thorough knowledge of normal rates of progression, such as the extensive temporal database that has been developed for mitosis in stamen hair cells of the spiderwort plant, Tradescantia virginiana (Wolniak, 1988).

Hartwell (1978) was the first to propose that genetic manipulation or mutational analysis could delineate rate-limiting gene products essential for a process as complex as the cell cycle through specific deletions or alterations manifested in conditional mutations. The cessation of the cell cycle at a particular point in a conditional mutant grown under restrictive conditions indicates the essential participation of that anomalous or nonexistent gene product in the process and identifies that control point as a 'checkpoint' (Hartwell, 1978; Hartwell and Weinert, 1989). An important checkpoint occurs at the G_2/M transition in fission yeast, where some of the gene products involved in the process have been identified.

Regulatory cascades and mitotic progression

p34[cdc2] kinase: a major player in cell cycle regulation

During the past several years, a number of studies have demonstrated an increase in the activity of a specific kinase at one or more points during the cell cycle and during mitosis (Nurse and Bissett, 1981; Beach et al., 1982; Simanis and Nurse, 1986; Labbe et al., 1988). Increases in this kinase activity have been linked to entry into mitosis from interphase, chromatin condensation, nucleolar dispersion, nuclear envelope breakdown and the exit from mitosis. Homologues for the kinase have been found throughout the eukaryotic realm, from yeast to humans, and the activity of this kinase promotes meiotic and mitotic activity in animal eggs and embryos. The name of this kinase, p34[cdc2], is derived from an apparent molecular weight for the catalytic subunit of 34 kD in a wide variety of organisms and from its initial description as a conditional mutation in fission yeast, *Schizosaccharomyces pombe*. This cell division cycle or *cdc*, mutation, known as *cdc*2, fails to progress through two checkpoints in its cell cycle at the restrictive temperature, namely, G_1/S and G_2/M (Nurse and Bissett, 1981; Beach et al., 1982).

Our thinking about cell cycle regulation has been altered profoundly by the discovery, activity and ubiquity of p34[cdc2] and its homologues, whose activity appears to be necessary at least for entry into M from G_2 in all organisms that have been analysed thus far (Draetta and Beach, 1989; Lohka, 1989; Marx, 1989; Moreno et al., 1989; Norbury and Nurse, 1989; Lewin, 1990; Reed and Wittenburg, 1990). All p34[cdc2] homologues appear to contain a 16 amino-acid motif, *EGVPSTAIREISLLKE* (known as *PSTAIR*) (Lee and Nurse, 1987; Norbury and Nurse, 1989). Other, related proteins, containing as much as 14/16 of the *PSTAIR* sequence may also be involved in cell cycle regulation, but in functions far removed from that of p34[cdc2] (Norbury and Nurse, 1989). In

several lower and higher plants, a p34[cdc2] homologue has been found (John et al., 1989; Feiler and Jacobs, 1990), though the details of its kinase activity are less well described. Homologues of the p34 kinase may be a universal component involved in cell cycle regulation throughout the eukaryotic realm.

The activity of p34[cdc2] kinase is dependent upon its association with one or more regulatory subunits, in a multimeric complex, where it phosphorylates serine or threonine residues on its substrates, primarily in motifs flanked by basic amino acid residues (Shenoy et al., 1989; Ward and Kirschner, 1990). Among the regulatory components associated with p34[cdc2] is the group of polypeptides that have become known as the cyclins (Evans et al., 1983; Swenson et al., 1986; Luca and Ruderman, 1989). The name cyclin was originally coined for a set of polypeptides that appeared during interphase and disappeared during mitosis in embryonic animals. Now, cyclin genes are known to encode several types of proteins whose molecular weights range from approximately 45 to 65 kD that are distinguished by type on the basis of a conserved sequence, known as a 'cyclin box' (Hadwiger et al., 1989; Minshull et al., 1989a). While the specific functions for all of the cyclin-isotypes are not known, the so-called A- and B-type cyclins bind to p34[cdc2] and regulate its activity during mitosis (Whitfield et al., 1990). In yeast, cyclin homologues apparently serve in a similar capacity with p34[cdc2] during M. The 63-kD product of *cdc*13 is a cyclin that is necessary for the initiation of M (Booher and Beach, 1987; Booher et al., 1989). A 55-kD peptide in extracts from peas appears to be reversibly associated with p34[cdc2] in a direct relationship with its kinase activity; this peptide may also be a cyclin (Feiler and Jacobs, 1990). It appears that cyclin binding confers substrate specificity on p34[cdc2] kinase (Norbury and Nurse, 1989; Moreno and Nurse, 1990; Whitfield et al., 1990), possibly to direct the activity of p34[cdc2] for different processes (Lewin, 1990).

The proteolytic degradation of cyclin is necessary for p34[cdc2] kinase inactivation (Minshull et al., 1989b). In *Drosophila*, A-type cyclin is degraded during prometaphase while

B-type cyclin is degraded later, at or near the onset of anaphase (Whitfield *et al.*, 1990). In yeast (Moreno *et al.*, 1989), some clams (Murray, 1989), and human cells (Pines and Hunter, 1989), the timing of B-type cyclin degradation also appears to coincide with the onset of anaphase, while in some echinoderms, the timing of cyclin degradation is less clear (Neant *et al.*, 1989). The signal for cyclin degradation seems to come from $p34^{cdc2}$ itself, and apparently involves cyclin phosphorylation. The post-translational modifications of cyclin are followed by its dissociation from $p34^{cdc2}$, its proteolytic degradation, and a concomitant loss of $p34^{cdc2}$ kinase activity (Draetta *et al.*, 1989). The loss of activity in some way triggers the exit from mitosis.

In its association with cyclin, a homologue of the yeast $p34^{cdc2}$ kinase is *the* important enzymatic component in M-phase promoting factor (MPF) (Arion *et al.*, 1988), a cytoplasmic extract that triggers the release from meiotic arrest in amphibian and echinoderm oocytes and acts as the active factor in promoting cell cycle progression in early animal embryogenesis (Dunphy *et al.*, 1988; Gautier *et al.*, 1988; Maller *et al.*, 1989). Active MPF comprises $p34^{cdc2}$ kinase and at least one species of cyclin (Dunphy *et al.*, 1988; Draetta *et al.*, 1989; Murray and Kirschner, 1989b; however, see Norbury and Nurse, 1989; Minshull *et al.*, 1989b). The accumulation of cyclin during the embryonic cell cycle (Murray and Kirschner, 1989b) contrasts sharply with the approximately constant level of $p34^{cdc2}$ kinase present in the cytosol (Draetta and Beach, 1988; Norbury and Nurse, 1989).

Although cyclin association with $p34^{cdc2}$ is followed by an increase in the activity of the enzyme (Draetta and Beach, 1988, 1989; Murray and Kirschner, 1989a, b; Norbury and Nurse, 1989), the binding of an additional factor, encoded by $cdc25^+$, is necessary for full kinase activation (Moreno *et al.*, 1989; Ducommun *et al.*, 1990). This gene product is a phosphoprotein of rather wide-ranging molecular weight (depending on the organism from which it is obtained) that was first discovered in yeast (Ducommun *et al.*, 1990), that is expressed primarily during G_2 (Sadhu *et al.*,

1990), and that appears to be involved in the dephosphorylation of $p34^{cdc2}$ at specific tyrosine residues, for maximal kinase activation (Draetta *et al.*, 1988b; Booher *et al.*, 1989; Gould and Nurse, 1989; Moreno *et al.*, 1989; Morla *et al.*, 1989; Ducommun *et al.*, 1990; Lewin, 1990). The dephosphorylation of $p34^{cdc2}$ may be under the control of a specific phosphatase (Booher and Beach, 1989; Cyert and Thorner, 1989; Morla *et al.*, 1989).

The role of $p34^{cdc2}$ in the G_1/S transition is not well described (Lewin, 1990), but mutational analysis suggests that certain classes of other cyclin-like proteins, known as *CLN1-CLN3* in fission yeast, are rate-limiting and specific for G_2/S progression (Hadwiger *et al.*, 1989; Richardson *et al.*, 1989). These polypeptides appear to bind to $p34^{cdc2}$ and activate its kinase activity (Wittenberg *et al.*, 1990). It is unclear if $p34^{cdc2}$ functions at multiple control points in the cell cycle of other organisms.

$p34^{cdc2}$ kinase acts on specific substrates during prophase

There are several known substrates for $p34^{cdc2}$ at the G_2/M transition (see review by Moreno and Nurse, 1990) that have been discerned from studies on animal and yeast cells, namely, histone H1 (Arion *et al.*, 1988; Langan *et al.*, 1989; Meijer *et al.*, 1989), the nucleolar proteins nucleolin and NO-38 (Peter *et al.*, 1990a), some of the phosphorylation sites on specific lamins (Peter *et al.*, 1990b; Heald and McKeon, 1990; Ward and Kirschner, 1990), and microtubule-associated-protein-2 (MAP-2) (see Shenoy *et al.*, 1989). Histone H1 phosphorylation has been linked to chromatin condensation (Bradbury *et al.*, 1974). The phosphorylation of nucleolin and NO-38 have been linked with nucleolar dispersion (Peter *et al.*, 1990a), and the phosphorylation of lamins is necessary for nuclear envelope breakdown (Gerace and Burke, 1988).

In animals, $pp60^{c-src}$ is a promiscuous tyrosine kinase that exhibits high enzymatic activity during the M-phase of the cell cycle (Chackalaparampil and Shalloway, 1988; Morgan *et al.*, 1989), and appears to be a substrate for $p34^{cdc2}$ (Morgan *et al.*, 1989; Shenoy *et al.*, 1989) as well as protein kinases A and C (Yaciuk *et al.*, 1989). Once activated, $pp60^{c-src}$ participates in a phosphorylation cascade involving a variety of substrates, some of which serve as kinases on their own after being phosphorylated. The cascade has been linked to the phosphorylation of MAP-2 protein kinase (Ray and Sturgill, 1988; Anderson *et al.*, 1990), an enzyme that may be indirectly involved in microtubule stability and, through S6 kinase II, in the phosphorylation of lamin C (Shenoy *et al.*, 1989; Ward and Kirschner, 1990).

The addition of the p34^{cdc2}/cyclin complex to interphase extracts containing centrosomes and microtubule protein results in changes in microtubule stability (Verde *et al.*, 1990). The microtubules in the extract become associated with the centrosomes and then exhibit a transient phase of rapid growth followed by rapid shrinkage to a uniform size distribution. At steady-state, the microtubules assemble and shrink rapidly. These changes *in vitro* resemble spindle assembly *in vivo* during late prophase and prometaphase, but it is not known if p34^{cdc2} kinase activity is specific for the effects observed, or if its role in the activation of the spindle microtubule organizing centres is indirect. Nevertheless, the phosphorylation of centrosomal components as it is detected with the anti-phosphoprotein antibody, MPM-2, appears to be a necessary prerequisite for normal spindle organization in mammalian cells (Centonze and Borisy, 1990).

While p34^{cdc2} kinase activity has been linked to chromatin condensation, nucleolar disassembly, nuclear envelope breakdown and spindle assembly through the phosphorylation of specific substrates, the conditional disruption of a type 1 phosphatase in *Aspergillus* results in an increase of phosphoantigen staining in the spindle plaque and in the nucleolus (Doonan and Morris, 1989). Because spindle structure is anomalous and the nucleolus fails to disperse during prophase, the phosphatase may play necessary dephosphorylation roles for the normal completion of both nucleolar dispersion and spindle formation. Thus, co-ordinated waves of phosphorylation and dephosphorylation may regulate mitotic progression.

There is *every* reason to believe that the p34^{cdc2} kinase plays a regulatory role for mitosis in the plant cell. In binding assays with immunological probes, a 34-kD antigen has been found to be present in *Arabidopsis*, oats, and in dividing cells of *Chlamydomonas* (John *et al.*, 1989). In a broader survey employing both genetic and immunological assays, a 34-kD peptide that closely resembles p34^{cdc2} is present in a variety of higher plants, and as part of a high molecular weight complex, it possesses histone H1 kinase activity (Feiler and Jacobs, 1990). There is a high level of homology between this gene product from peas and those translated from yeast and human p34^{cdc2} transcripts (Feiler and Jacobs, 1990) but the details of p34^{cdc2} regulation at specific points in the cell cycle may differ from those in animal and fungal cells (John *et al.*, 1989).

Higher order regulation of p34^{cdc2} kinase

Several additional factors act as regulators for p34^{cdc2} kinase activity. Like p34^{cdc2}, these factors were initially discovered through mutation analysis in yeast (Draetta *et al.*, 1988a). One of these genes, *wee1*, is not required for mitosis, but its overexpression in yeast results in cells that progress through the cell cycle at a slower rate than normal (Russell and Nurse, 1987a), resulting in daughter cells that are larger than those of wild type strains (Hagan *et al.*, 1990). Under-expression of the *wee1* gene product results in small, rapidly dividing cells. The *wee1* gene product resembles a protein kinase and apparently acts as an inhibitor of mitosis. *nim1* is a gene that also encodes a kinase-like protein, but apparently stimulates cell cycle progression (Russell and Nurse, 1987b). The products of *nim1* and *wee1* work in opposite directions with p34^{cdc2} and cell cycle progression.

cdc25, and its homologue in budding yeast, *M1H1* (Russell *et al.*, 1989), produces a gene product that can also stimulate the initiation of mitosis and thereby offset the effects of the *wee1* gene product to a limited extent. Its inactivation results in slower division rates in larger cells (Hagan *et al.*, 1990), while its overexpression can result either in progression through the cell cycle that is more rapid than wild type strains, or in cell death. It has been suggested that the *cdc25* gene product may act as a 'docking factor' that targets a tyrosine phosphatase to the p34[cdc2]/cyclin complex for kinase activation (Ducommun *et al.*, 1990). In later stages in the development of *Drosophila*, another gene product, known as *string*, and homologous with *cdc25* in yeast (Edgar and O'Farrell, 1989) may supplant cyclin to activate p34[cdc2] kinase (Gould and Nurse, 1989).

The p34[cdc2] regulator known as *suc1* (Hayles *et al.*, 1986) encodes a small protein of *c.* 13 kD in fission yeast (Brizuela *et al.*, 1987; Booher *et al.*, 1989) and humans (Draetta *et al.*, 1987; Draetta and Beach, 1989). It is uniformly abundant during the cell cycle and can form a complex with p34[cdc2] that exhibits very high kinase activity *in vivo* (Draetta and Beach, 1989). The disruption of the *suc1* gene results in cell cycle arrest. The *suc1* gene product may somehow be involved in the process by which the cell exits from mitosis. Undoubtedly, other factors will emerge as activators or suppressors for the enzyme and consequently, for cell cycle regulation. None of these factors has yet been described in plant cells.

The distribution of p34[cdc2] during mitosis

During interphase, immunological localizations of p34[cdc2] in mammalian cells place the kinase in the cytosol, distributed uniformly, and within the nucleus in a punctate pattern (Bailly *et al.*, 1989), while other antibodies in yeast (Booher *et al.*, 1989) and in other mammalian cells (Riabowol *et al.*, 1989) describe a largely intranuclear distribution for the antigen. In higher eukaryotic cells, from prophase through the onset of anaphase, p34[cdc2] apparently resides in or around the spindle pole regions before it becomes situated in a cluster of vesicles around the spindle midzone (Bailly *et al.*, 1989; Riabowol *et al.*, 1989). In yeast, the distribution of p34[cdc2] remains intranuclear during early portions of mitosis (Booher *et al.*, 1989). Later, and coincident with cyclin (*cdc13*) degradation, the intranuclear distribution of anti-p34[cdc2] antibodies is lost (Alfa *et al.*, 1989).

The centrosomal distribution of p34[cdc2] correlates with a possible function, namely the phosphorylation of centrosomal proteins for the construction and stabilization of the spindle (Centonze and Borisy, 1990). However, p34[cdc2] is not the only kinase to be specifically associated with the centrosomal region of the spindle during mitosis. Ohta and co-workers (1990) have found that the calcium/calmodulin-dependent protein kinase II is a spindle component in mammalian cells that co-localizes with the centrosomes during metaphase and anaphase and is present in the midzone during telophase. Its function in the centrosome has been linked to microtubule stability, through the phosphorylation of a 62-kD spindle protein (Dinsmore and Sloboda, 1988, 1989). An alternative demonstration of the requirement of p34[cdc2] for progression into mitosis comes from the microinjection of an antibody directed against the kinase into mammalian cells, an experiment that results in an arrest at the G_2/M transition (Riabowol *et al.*, 1989).

Because p34[cdc2] kinase is active when it is part of a multimeric complex, antibodies directed against these factors might be expected to co-localize with the kinase at certain points during the cell cycle. In double-labelling experiments, antibodies directed against the kinase and cyclin-like proteins are identically distributed, up to the point where kinase activity is lost (Alfa *et al.*, 1989; Booher *et al.*, 1989). The overlap between the distributions of p13[suc1] and p34[cdc2] antibodies is most pronounced late in mitosis, after the distribution of the latter becomes centred in the spindle midzone into a punctate array (Bailly *et al.*, 1989).

Other regulators and mitotic progression

p34[cdc2] kinase may function at multiple points during the cell cycle and mitosis. Different roles for p34[cdc2] may be possible through the associations of the kinase with different cyclins and other regulators (Richardson *et al.*, 1989; Lewin, 1990). Pharmacological studies on cell cycle and mitotic progression point to other kinases such as protein kinase C (Nishizuka, 1986, 1989; Huang, 1989; O'Brian and Ward, 1989; Levin *et al.*, 1990) and perhaps calcium/calmodulin-dependent protein kinase (Colbran *et al.*, 1989) or calcium-dependent but calmodulin-independent kinases (Harmon *et al.*, 1987; Polya *et al.*, 1989) as likely regulators during mitosis (Wolniak, 1988). In budding yeast, the recent discovery of a protein kinase C gene, *PKC1*, whose conditional deletion results in a G_2-stage-specific arrest of the cell cycle, with a simultaneous arrest of cell growth (Levin *et al.*, 1990), suggests that protein kinase C plays an integral role in the regulation of both the cell growth and cell division cycles. Because of the phenotype observed, this gene product could function *upstream* of the G_2/M activation of $p34^{cdc2}$, and possibly be part of a cascade responsible for $p34^{cdc2}$ activation for its role during mitosis.

In *Aspergillus nidulans*, the product of the *nimA* gene, a probable kinase, is necessary for progression through G_2/M transition. Significantly, its conditional disruption results in the arrest of both cell division and cell growth, while its overexpression results in an *acceleration* of cell division and cell growth (Osmani *et al.*, 1987, 1988a, b). In animal cells, various tyrosine-kinases (Vila and Weber, 1988; Morgan *et al.*, 1989; Shenoy *et al.*, 1989; Peaucellier *et al.*, 1990) and cyclic-nucleotide-dependent kinases (Browne *et al.*, 1990) are also probably involved in mitotic progression. Because neither of these classes of kinases is known in higher plants, any model for mitotic regulation in plants incorporating protein phosphorylation must be restricted accordingly.

Calcium plays a role in the regulation of mitotic progression

The likely involvement of calcium in mitosis (Harris, 1977; Hepler, 1977) comes from observations that show the sensitivity of the spindle microtubules to its presence (Salmon and Segall, 1980; Kiehart, 1981), and the existence of a membrane-based calcium buffering system in and around the spindle, presumably for its stabilization (Hepler, 1980; Wolniak *et al.*, 1980, 1983; Kiehart, 1981). Pharmacological perturbations of calcium activity result in dramatically altered rates of mitotic progression (see review by Wolniak, 1988). The intentional buffering of intracellular calcium activity in echinoderm embryos results in mitotic arrest (Silver, 1989), while the induction of mitosis in mammalian cells, through treatment with various mitogens, is typically followed by elevations in cytosolic calcium activity (McNeil *et al.*, 1985; Byron and Villereal, 1989).

A large body of pharmacological evidence points to calcium as a likely regulator of mitotic progression in stamen hair cells from *Tradescantia virginiana*. Stamen hair cells exhibit remarkable temporal precision and clarity of chromosome behaviour during mitosis, providing the highest known level of temporal precision for studies on mitotic progression (Hepler, 1985; Wolniak and Bart, 1985a; also see review by Wolniak, 1988). We have utilized this precision as a bioassay for pharmacological studies involving the perturbation of extracellular or cytosolic calcium (Wolniak and Bart, 1985a, b; Larsen *et al.*, 1989), polyphosphoinositide metabolism (Wolniak, 1987; Chen and Wolniak, 1987b), or protein kinase C activity (Larsen and Wolniak, 1990). A rise in cytosolic calcium activity is necessary for anaphase in stamen hair cells, and at least some of this calcium enters the cytosolic compartment from the extracellular space (Hepler, 1985; Wolniak and Bart, 1985a, b; Chen and Wolniak, 1987a; Wolniak, 1988). Hepler and Callaham (1987) have shown that a rise in cytosolic calcium activity follows anaphase onset in stamen hair cells and correlates directly with the velocity of chromosome separation during anaphase (Zhang *et al.*, 1990). We have found that multiple mitotic transitions are blocked or induced by identical inhibitor or elicitor treatments, administered at different times during mitosis. These results suggest that oscillatory shifts in the activities of the same elicitors at different times are necessary to trigger the different events we call nuclear envelope breakdown, anaphase onset and cell plate vesicle aggregation (see review by Wolniak, 1988).

Polyphosphoinositide metabolism is a logical pathway for the control of a rapidly oscillating regulatory cascade involving calcium (Abdel-Latif, 1986; Berridge, 1984, 1988, 1989) and culminating in anaphase (Whitman and Cantley, 1988; Wolniak, 1988). In animals and probably many kinds of plants (Boss, 1989; Memon *et al.*, 1989a,b; Rincon *et al.*, 1989; also see review by Einspahr and Thompson, 1990), the hydrolysis of phosphatidylinositol 4,5-bisphosphate results in the production of inositol 1,4,5-trisphosphate and a 1,2-diacylglycerol, which elicit the release of sequestered calcium and the activation of protein kinase C, respectively. In plants and certain fungi, the major pool of exchangeable calcium may be within the vacuole (Schumaker and Sze, 1987; Cornelius *et al.*, 1989; Alexandre and Lassalles, 1990; Einspahr and Thompson, 1990). Kinase activation is followed by the phosphorylation of specific substrates. Autophosphorylation of protein kinase C at specific residues both sustains its activity after calcium homeostasis and controls its specificity (Flint *et al.*, 1990). Protein kinase C homologues have been identified through consensus hybridization from dicot plant cDNA libraries, though their regulatory domains appear to differ from those in animals (Lawton *et al.*, 1989). Other calcium-dependent (diacylglycerol-independent (Harmon *et al.*, 1987; Blowers and Trewavas, 1988)) and phospholipid dependent (Martiny-Baron and Scherer, 1989) protein kinases have been isolated and characterized from plants. These types of enzymes could play regulatory roles in a regulatory cascade for mitotic progression.

Lithium chloride is, amongst other things, an inhibitor of polyphosphoinositide cycling through its blockade of inositol 1-monophosphatase (Berridge *et al.*, 1989), which results in an inhibition of phosphoinositide breakdown. The specific reversal of a lithium-blocked process by myo-inositol post-treatment is indicative of polyphosphoinositide involvement in the regulation of that process. Mitotic arrest induced by lithium is reversible in sea urchin embryos (Forer and Sillers, 1987) and stamen hair cells (Wolniak, 1987) by treatment with myo-inositol, but not scyllitol, its stereo-isomer. Calcium or 1,2-dioctanoylglycerol reverse metaphase arrest in stamen hair cells while the 1,3-isomer of the lipid is insufficient for reversal, results suggesting that a lipid-dependent factor, possibly a protein kinase, is activated in the anaphase cascade (Wolniak, 1987). Neomycin, an aminoglycoside antibiotic, through its inhibition of phospholipase C, blocks polyphosphoinositide cycling (Prentki *et al.*, 1986; Tysnes *et al.*, 1987). Its addition to

stamen hair cells results in metaphase arrest that is partially reversible by the addition of 1,2-dioctanoylglycerol, or by calcium, but not by 1,3-dioctanoylglycerol or magnesium, providing an independent link between lipid-dependent kinase activation and anaphase (Larsen *et al.*, 1991). Treatment with clindamycin, a related antibiotic with no known effect on phospholipase C, results in mitotic progression that is identical to that in untreated control cells (Larsen *et al.*, 1991).

The author and co-workers tested whether a rise in cytosolic calcium is sufficient for sister chromatid separation or if the calcium rise could be supplanted by the artificial activation of protein kinase C. They have shown (Larsen *et al.*, 1989) that metaphase arrest induced by the calcium chelator quin2 is reversible by post-treatment with the potent protein kinase C activator, 1,2-dioctanoylglycerol, but not by its inactive 1,3-isomer. They then added very low levels of 1,2-dioctanoylglycerol (Ebanks *et al.*, 1989) to otherwise untreated cells at known times during mitosis (Larsen and Wolniak, 1990) and found that progression was accelerated *or retarded* significantly as a function of the time of its addition to the cells, with sharply-defined (<2 min) starting limits for promotion or inhibition. 1,2-Dioctanoylglycerol accelerated mitotic progression when added at any point between 10 min prior to nuclear envelope breakdown, up to *c.* 15 min after nuclear envelope breakdown. Remarkably, addition of 1,2-dioctanoylglycerol, starting at any point 18–23 min after nuclear envelope breakdown *delayed* anaphase onset over 55 min. None of these treatments affected the rate of subsequent progression or the rate of chromosomal separation. The addition of 1,2-dioctanoylglycerol at any point starting 25–29 min after nuclear envelope breakdown resulted in a small reduction in the metaphase transit time, and a significant acceleration in the start of cell plate vesicle aggregation without a change in the rate of chromosome separation (*c.* 1.3 μm min^{-1}). In controls, 1,3-dioctanoylglycerol was totally without effect on mitotic progression. The author and co-workers propose that low levels of 1,2-dioctanoylglycerol hyperactivate a protein kinase (C) that, in turn, phosphorylates mitotically-important substrates in stamen hair cells. During intervals when progression is promoted by 1,2-dioctanoylglycerol, triggered phosphorylation appears to be necessary for further mitotic progression. The time-dependency of 1,2-dioctanoylglycerol addition for mitotic progression or arrest indicates intervals when the phosphorylation of certain mitotic substrates is the prerequisite for progression while at other times, the inappropriate or untimely phosphorylation of these same substrates alters serial pathways in a fashion manifested as arrest. Alternatively, during the treatment period where delayed anaphase onset is observed, the cell may have activated a phosphatase (Booher and Beach, 1989; Cyert and Thorner, 1989; Doonan and Morris, 1989; Ohkura et al., 1989) as part of its regulatory cascade that offsets the kinase hyperactivation induced.

The differences in response by the cells at a function of the time of treatment with these drugs is indicative of changes in regulatory states within the cell at different stages of mitosis. The time-dependency of the treatments on subsequent mitotic progression appears to affect regulatory cascades at points prior to or after they have exerted their effects on progression. The differences in progression that result from identical treatments initiated just a few minutes apart delineate control points for mitotic regulation where the same elicitors are involved in the entry into different phases of the process.

Cytosolic calcium activity determinations in mitotic cells

In contrast to pharmacological evidence suggesting the involvement of cytosolic calcium activity changes in the anaphase cascade, direct calcium measurements in living cells provide a thoroughly equivocal view of changes during metaphase (see reviews by Wolniak, 1988; Hepler, 1989), leading to the conclusion that there is no clear-cut correlation between calcium rises and progression into anaphase (Ratan et al., 1988; Tombes and Borisy, 1989; Kao et al., 1990). The lack of uniformity during metaphase and anaphase among cell types may

be indicative of multiple physiological strategies for the regulation of a process common to virtually all eukaryotic cells. Nuclear envelope breakdown, however, is apparently preceded by a cytosolic calcium rise in a variety of organisms (Poenie et al., 1985, 1986; Steinhardt and Alderton, 1988; Kao et al., 1990).

Oscillations in cytosolic calcium activity occur frequently, albeit at irregular intervals in animals (Ratan et al., 1988; Byron and Villereal, 1989; Tombes and Borisy, 1989). Seemingly random oscillations of cytosolic calcium activity could serve as a portion of the trigger necessary for anaphase, with the cell not progressing until a number of these pulses were processed in a frequency-encoded signal (Rapp, 1987; Berridge, 1989). Short, localized fluxes of calcium in the cytosol (Ratan et al., 1988; Tombes and Borisy, 1989; Kao et al., 1990) could go undetected or be reported as small, but gradual rises in activity when measured in experiments designed for large-scale amplitude determinations. Gradual increases in cytosolic calcium occur prior to and during anaphase (Tombes and Borisy, 1989), while brief transients have been reported in other cells using similar experimental methods (Poenie et al., 1985, 1986; Ratan et al., 1988; Tombes and Borisy, 1989). The situation in plants may be equally complex during late prophase or late metaphase, where its analysis has proven to be difficult (Keith et al., 1985), but simpler during anaphase where a clear calcium-amplitude rise follows chromosome separation (Hepler and Callaham, 1987).

The duration of a cytosolic calcium level shift is a function of both its magnitude and the activity of the homeostatic pumps involved in its reduction to resting levels. The slope of the calcium gradient over some distance, rather than its magnitude may have regulatory significance (Hepler, 1989), where gradual pancellular rises could produce steep calcium gradients in regions near sites of active calcium transport (Hepler and Wolniak, 1984). Localized gradients created by gradual increases in cytosolic calcium could be dampened by the chelating agents used to make calcium determinations (Hepler, 1989), thus altering the signal during its measurement. Moreover, UV-illumination used in calcium detection could, in and of itself, supplant calcium rises in regulatory cascades by activating protein kinase C (Matsui and DeLeo, 1990; also see Larsen and Wolniak, 1990).

Regulatory cascades involving p34^{cdc2}- and calcium-dependent pathways

Regulatory cascades during mitosis may involve a variety of elicitors and operate largely through the phosphorylation and dephosphorylation of proteins

It is reasonable to hypothesize that mitotic progression after nuclear envelope breakdown involves waves of protein phosphorylation and dephosphorylation that are triggered by oscillations in the cytosolic calcium activity. These changes may occur at rather irregular intervals (Wolniak, 1988; Silver, 1989), a result of shifts in calcium channel activity, with cation movement into the cytosol from both internal and external pools. Tombes and Borisy (1989) have shown that shifts in calcium activity may occur irregularly in mammalian cells and that the shifts may modulate mitotic progression, but do not serve as the sole signals for the transitions. Recently, Kao and co-workers (1990) showed with 'caged' calcium, that a rise in cytosolic calcium activity apparently precedes nuclear envelope breakdown in Swiss 3T3 fibroblasts, possibly for calcium-dependent phosphorylation of lamins at specific residues (Ward and Kirschner, 1990). Brief, localized changes in cytosolic calcium activity during metaphase could go undetected, especially if they occurred some minutes prior to sister chromatid separation. Experiments employing caged calcium should be repeated in a strict temporal context, where the metaphase transit time is a known quantity, and where the elevation of calcium levels in the cell can be manipulated at a series of time points during prometaphase and metaphase.

It seems likely that a cyclic pattern of phosphorylation is followed by an interval of dephosphorylation (Booher and Beach, 1989; Vandre and Borisy, 1989; Cyert and Thorner,

1989). Significant dephosphorylation of specific substrates (i.e. MPM-2 antigens (Davis et al., 1983)) precedes or accompanies anaphase onset (Vandre et al., 1984; Millar et al., 1987; Vandre and Borisy, 1989). In fission yeast, two genes, known as dis2 and sds21, encode similar 37-kD type 1 protein phosphatase homologues (Ohkura et al., 1989). The disruption of both genes results in the failure of sister chromatid separation. Specific phosphatase activity is also necessary for the completion of anaphase in Aspergillus in the form of the gene product of bimG11$^+$ (Doonan and Morris, 1989).

Linkages between p34^{cdc2} and calcium in mitotic regulation

Moreno and co-workers (1989) intentionally eliminated calcium-activated kinase activities from their assays to demonstrate the fluctuation of cdc-kinase activity in isolates from different phases of the cell cycle. This elimination demonstrates the times and extents of cdc-kinase activities, but fails to exclude calcium-dependent regulatory pathways that may play significant roles in mitotic progression. Patel and co-workers (1989), using ammonia-activated sea urchin eggs, established a link between calcium, polyphosphoinositide cycling, and p34^{cdc2} kinase for entry into M that may be fundamental to all eukaryotes. An elevation in cytosolic calcium activity is modulated by polyphosphoinositide cycling, and is necessary for the phosphorylation of cyclin, which, in turn, binds to and activates p34^{cdc2} kinase, in a step necessary for the G$_2$/M transition (Patel et al., 1989).

If p34^{cdc2} kinase, maximally active in its complex with cyclin, is part of the trigger for sister chromatid separation, then an increase in its abundance in the metaphase cell might promote precocious entry into anaphase. The microinjection of p34^{cdc2} kinase at various times during the cell cycle, either alone or complexed with cyclin B, stimulated changes in cell shape (towards roundness), in the organization of the cytoskeleton (towards a shift in actin filament distribution and a reduction in the numbers of interphase microtubules) and in chromatin structure (towards condensation) (Lamb et al., 1990). However, neither spindle formation nor the promotion of entry into anaphase was observed (Lamb *et al.*, 1990). The failure of these injections to induce anaphase indicates that p34^{cdc2} cyclin is not limiting for anaphase, or that the signal comprises more than one elicitor. Alternatively, p34^{cdc2} is not directly involved in sister chromatid separation at all, but functions upstream of anaphase in setting the stage for chromatid separation to occur.

p34^{cdc2} kinase, either alone, or as part of a multimeric complex, may alter calcium levels in the cytosol. Recently, Picard and co-workers (1990) microinjected the p34 motif *PSTAIR* into unfertilized starfish oocytes and observed cortical vesicle exocytosis and an elevation of the fertilization envelope, events usually associated with fertilization, and intimately linked with an increase in cytosolic calcium activity (Steinhardt *et al.*, 1977; Gilkey *et al.*, 1978). Measurements of cytosolic calcium activity showed that *PSTAIR*, but not similar sequences, elevated cytosolic calcium levels transiently above 1 μM within 1 min of injection. The co-injection of the calcium chelator EGTA and *PSTAIR* suppressed cortical vesicle break-down, a result suggesting that *PSTAIR* activates the cortical events through calcium. *PSTAIR* injection was sufficient to induce rises in cytosolic calcium levels, even after calcium depletion from the medium or calcium influx blockades, apparently through the mobilization of intracellular calcium stores. The co-injection of p13^{suc1} with *PSTAIR* showed that the rise in cytosolic calcium was independent of p34^{cdc2} kinase activity. The linkage between p34^{cdc2} and cytosolic calcium activity as affected by *PSTAIR* may be the crucial connection between the p34^{cdc2} and calcium-based pathways of cell cycle regulation.

Future directions

It is necessary to determine the role of p34^{cdc2} in the control of mitotic progression in all kinds of eukaryotes. In the current wave of cell cycle research, plants are the only major group of eukaryotic organisms not widely represented. We should also focus on interacting regulatory pathways involving p34^{cdc2} homologues and calcium. It is unlikely that a single regulatory switch controls the multifaceted processes of cell cycle progression, mitosis and cytokinesis. The known checkpoints for progression require specific gene products to be available and active, while some of the transition points require certain physiological conditions to be met. It seems reasonable to propose that p34^{cdc2} functions in concert with a variety of other kinases and with calcium to initiate and then sustain progression through mitosis. Waves of protein phosphorylation are co-ordinated with waves of dephosphorylation. Cyclins and other regulating factors are probably present and functional in plant cells. It is expected that more similarities than differences will emerge between plants and other eukaryotes.

Ongoing advances in our understanding of mitotic regulation

Since this article was written, there have been several important developments that provide evidence for the involvement of p34^{cdc2} in mitotic regulation in plant cells. The apparent p34^{cdc2} homologues from alfalfa (Hirt *et al.*, 1991), rice (Hata, 1991) *Arabidopsis* (Ferreira *et al.*, 1991) and corn (Colasanti *et al.*, 1991) have been isolated, and characterized. All display high homology with human and yeast p34^{cdc2}. Insertion of the alfalfa sequence into a temperature-sensitive fission yeast *cdc*2 mutant resulted in the full restoration of cell cycle activity at the restrictive temperature, suggesting functional homology for the gene in the higher plant (Hirt *et al.*, 1991). The abundance of the p34^{cdc2} homologue in different organs of alfalfa was also measured by northern blot analysis (Hirt *et al.*, 1991). Similar experiments were performed with corn homologues (Colasanti *et al.*, 1991). Higher transcript levels were found in shoots and suspension culture cells than in roots. Cultured alfalfa cells showed fluctuations in the levels of this transcript during somatic embryogenesis. Immunological studies show that the distribution of p34^{cdc2} in wheat seed-lings correlates directly with cell division activity in basal leaf meristems (John *et al.*, 1990).

During differentiation of these leaf cells, levels of p34^{cdc2} labelling decline significantly (John *et al.*, 1990). Thus, it appears that a p34^{cdc2} kinase gene is present in higher plants, and its product is most abundant in meristematic tissues. Some efforts are now being directed towards the regulation of p34^{cdc2} in higher plants. A recent report (John *et al.*, 1991) shows that a homologue of the p34^{cdc2} regulator known as p13^{suc1} is present in wheat. Its distribution, determined immunologically, appears to be ubiquitous in the plant. Its involvement in mitosis in higher plants is not yet known.

Acknowledgements

I am grateful for support for this work from the National Science Foundation (grant number DCB 87-00422) and from the Maryland Agricultural Experiment Station (Research Grant Number J-003, and Hatch Project MD-J-136; Contribution Number 8237). I am also grateful to Drs Paul Larsen, John C. Watson, Eliot Herman, Kathy Kamo and Albert Ades and to Mr Douglas Mills for listening patiently to my slowly evolving ideas.

References

Abdel-Latif, A.A. (1986). Calcium-mobilizing receptors, polyphosphoinositides, and the generation of second messengers. *Pharmacol. Rev.* **38**, 227–272.

Alexandre, J. and Lassalles, J.P. (1990). Effect of D-*myo*-inositol 1,4,5-*tris*phosphate on the electrical properties of the red beet vacuole membrane. *Plant Physiol.* **93**, 837–840.

Alfa, C.E., Booher, R., Beach, D.H. and Hyams, J.S. (1989). Fission yeast cyclin: subcellular localisation and cell cycle regulation. *J. Cell Sci.* **Suppl. 12**, 9–19.

Anderson, N.G., Maller, J.L., Tonks, N.K. and Sturgill, T.W. (1990). Requirement for integration of signals from two distinct phosphorylation pathways for activation of MAP kinase. *Nature (Lond.)* **343**, 651–653.

Arion, D., Meijer, L., Brizuela, L. and Beach, D. (1988). *cdc*2 is a component of the M phase-specific histone H1 kinase: evidence for identity with MPF. *Cell* **55**, 371–378.

Bailly, E., Doree, M., Nurse, P. and Bornens, M. (1989). p34^{cdc2} is located in both nucleus and cytoplasm; part is centrosomally associated at G$_2$/M and enters vesicles at anaphase. *EMBO (Eur. Mol. Biol. Organ.) J.* **8**, 3985–3995.

Baskin, T.I. and Cande, W.Z. (1990). The structure and function of the mitotic spindle in flowering plants. *Ann. Rev. Plant Physiol. Mol. Biol.* **41**, 277–315.

Beach, D.H., Durkacz, B. and Nurse, P.M. (1982). Functionally homologous cell cycle control genes in budding and fission yeast. *Nature* **300**, 706–709.

Berridge, M.J. (1984). Inositol phosphate and diacylglycerol as second messengers. *Biochem. J.* **220**, 345–360.

Berridge, M.J. (1988). Inositol lipids and calcium signalling. *Proc. R. Soc. Lond. Ser. B* **234**, 359–378.

Berridge, M.J. (1989). Cell signalling through cytoplasmic calcium oscillations. In *Cell to Cell Signalling: from Experiments to Theoretical Models* (ed. A. Goldbeter), pp. 449–459. Academic Press, New York.

Berridge, M.J., Downes, C.P. and Hanley, M.R. (1989). Neural and developmental actions of lithium: a unifying hypothesis. *Cell* **59**, 411–419.

Blowers, D.P. and Trewavas, A.J. (1988). Phosphatidylinositol kinase activity of a plasma membrane-associated calcium-activated protein kinase from pea. *FEBS Lett.* **238**, 87–89.

Booher, R. and Beach, D. (1987). Interaction between *cdc13*$^+$ and *cdc2*$^+$ in the control of mitosis in fission yeast; dissociation of the G$_1$ and G$_2$ roles of the *cdc2*$^+$ protein kinase. *EMBO (Eur. Mol. Biol. Organ.) J.* **6**, 3441–3447.

Booher, R. and Beach, D. (1989). Involvement of a type 1 protein phosphatase encoded by *bws1*$^+$ in fission yeast mitotic control. *Cell* **57**, 1009–1016.

Booher, R., Alfa, C.E., Hyams, J.S. and Beach, D.H. (1989). The fission yeast *cdc2/cdc13/suc1* protein kinase: regulation of catalytic activity and nuclear localization. *Cell* **58**, 486–497.

Boss, W.F. (1989). Polyphosphoinositide metabolism: its relation to signal transduction in plants. In *Second Messengers in Plant Growth and Development* (eds W.F. Boss and D.J. Morre), pp. 29–56. Alan R. Liss, New York.

Bradbury, E.M., Inglis, R.J. and Matthews, H.R. (1974). Control of cell division by very lysine-rich histone (H1) phosphorylation. *Nature (Lond.)* **247**, 257–261.

Brizuela, L., Draetta, G. and Beach, D. (1987). p13^{suc1} acts in the fission yeast cell division cycle as a component of the p34^{cdc2} protein kinase. *EMBO (Eur. Mol. Biol. Organ.) J.* **6**, 3507–3514.

Browne, C.L., Bower, W.A., Palazzo, R.E. and Rebhun, L.I. (1990). Inhibition of mitosis in fertilized sea urchin eggs by inhibition of the cyclic AMP-dependent protein kinase. *Exp. Cell Res.* **188**, 122–128.

Burke, B. and Gerace, L. (1986). A cell free system to study reassembly of the nuclear envelope at the end of mitosis. *Cell* **44**, 639–652.

Byron, K.L. and Villereal, M.L. (1989). Mitogen-induced $[Ca^{2+}]_i$ changes in individual human fibroblasts. *J. Biol. Chem.* **264**, 18234–18239.

Centonze, V.E. and Borisy, G.G. (1990). Nucleation of microtubules from mitotic centrosomes is modulated by a phosphorylated epitope. *J. Cell Sci.* **95**, 405–411.

Chackalaparampil, I. and Shalloway, D. (1988). Altered phosphorylation and activation of pp60^{c-src} during fibroblast mitosis. *Cell* **52**, 801–810.

Chen, T.-L.L. and Wolniak, S.M. (1987a). Mitotic progression in stamen hair cells of *Tradescantia* is accelerated by treatment with ruthenium red and Bay K-8644. *Eur. J. Cell Biol.* **45**, 16–22.

Chen, T.-L.L. and Wolniak, S.M. (1987b). Lithium induces cell plate dispersion during cytokinesis in *Tradescantia*. *Protoplasma* **141**, 56–63.

Colasanti, J., Tyers, M. and Sundaresan, V. (1991). Isolation and characterization of cDNA clones encoding a functional p34^{cdc2} homologue from *Zea mays*. *Proc. Natl Acad. Sci. USA* **88**, 3377–3381.

Colbran, R.J., Schworer, C.M., Hashimoto, Y., Fong, Y.-L., Rich, D.P., Smith, M.K. (1989). Calcium/calmodulin-dependent protein kinase II. *Biochem. J.* **258**, 313–325.

Cooke, C.A., Heck, M.M.S. and Earnshaw, W.C. (1987). The inner centromere protein (*INCENP*) antigens: movement from inner centromere to midbody during mitosis. *J. Cell Biol.* **105**, 2053–2067.

Cooke, C.A., Bernat, R.L. and Earnshaw, W.C. (1990). *CENP-B*: a major human centromere protein located beneath the kinetochore. *J. Cell Biol.* **110**, 1475–1488.

Cornelius, G., Gebauer, G. and Techel, D. (1989). Inositol *tris*phosphate induces calcium release from *Neurospora crassa* vacuoles. *Biochem. Biophys. Res. Comm.* **162**, 852–856.

Cyert, M.S. and Thorner, J. (1989). Putting it on and taking it off: phosphoprotein phosphatase involvement in cell cycle regulation. *Cell* **57**, 891–893.

Davis, F.M., Tao, T.Y., Fowler, S.K. and Rao, P.N. (1983). Monoclonal antibodies to mitotic cells. *Proc. Natl Acad. Sci. U.S.A.* **80**, 2926–2930.

Dinsmore, J.H. and Sloboda, R.D. (1988). Calcium and calmodulin-dependent phosphorylation of a 62 kD protein induces microtubule depolymerization in sea urchin mitotic apparatuses. *Cell* **53**,

769–780.

Dinsmore, J.H. and Sloboda, R.D. (1989). Micro-injection of antibodies to a 62 kD mitotic apparatus protein arrests mitosis in dividing sea urchin embryos. *Cell* **57**, 127–134.

Doonan, J.H. and Morris, N.R. (1989). The *bimG* gene of *Aspergillus nidulans*, required for completion of anaphase, encodes a homolog of mammalian phosphoprotein phosphatase 1. *Cell* **57**, 987–966.

Draetta, G. and Beach, D. (1988). Activation of *cdc2* protein kinase during mitosis in human cells: cell cycle-dependent phosphorylation and subunit rearrangement. *Cell* **54**, 17–26.

Draetta, G. and Beach, D. (1989). The mammalian *cdc2* protein kinase: mechanisms of regulation during the cell cycle. *J. Cell Sci. Suppl.* **12**, 21–27.

Draetta, G., Brizuela, L., Potashkin, J. and Beach, D. (1987). Identification of p34 and p13, human analogs of the cell cycle regulators of fission yeast encoded by *cdc2*$^+$ and *suc1*$^+$. *Cell* **50**, 319–325.

Draetta, G., Brizuela, L., Moran, B. and Beach, D. (1988a). Regulation of the vertebrate cell cycle by the *cdc2* protein kinase. *Cold Spring Harb. Symp. Quant. Biol.* **53**, 195–201.

Draetta, G., Piwnica-Worms, H., Morrison, D., Druker, B., Roberts, T. and Beach, D. (1988b). *cdc2* is a major cell-cycle regulation tyrosine kinase substrate. *Nature (Lond.)* **336**, 738–744.

Draetta, G., Luca, F., Westendorf, J., Brizuela, L., Ruderman, J. and Beach, D. (1989). *cdc2* protein kinase is complexed with both cyclin A and B: evidence for proteolytic inactivation of MPF. *Cell* **56**, 829–838.

Ducommun, B., Draetta, G., Young, P. and Beach, D. (1990). Fission yeast *cdc25* is a cell-cycle regulated protein. *Biochem. Biophys. Res. Comm.* **167**, 301–309.

Dunphy, W.G., Brizuela, L., Beach, D. and Newport, J. (1988). The *Xenopus cdc2* protein is a component of MPF, a cytoplasmic regulator of mitosis. *Cell* **54**, 423–431.

Ebanks, R., Roifman, C., Mellors, A. and Mills, G.B. (1989). The diacylglycerol analogue, 1,2,-*sn*-dioctanoylglycerol, induces an increase in cytosolic free Ca^{2+} and cytosolic acidification of T lymphocytes through a protein kinase C-independent process. *Biochem. J.* **258**, 689–698.

Edgar, B.A. and O'Farrell, P.H. (1989). Genetic control of cell division patterns in the *Drosophila* embryo. *Cell* **57**, 177–187.

Einspahr, K.J. and Thompson, G.A., Jr (1990). Transmembrane signaling via phosphatidylinositol 4,5-bisphosphate hydrolysis in plants. *Plant Physiol.* **93**, 361–366.

Evans, T., Rosenthal, E.T., Youngblom, J., Distel, D. and Hunt, T. (1983). Cyclin: a protein specified by maternal mRNA in sea urchin eggs that is destroyed at each cleavage division. *Cell* **33**, 389–396.

Feiler, H.S. and Jacobs, T.W. (1990). Cell division in higher plants: a *cdc2* gene, its 34-kDa product, and histone H1 kinase activity in pea. *Proc. Natl Acad. Sci. U.S.A.* **87**, 5397–5401.

Ferreira, P.C.G., Hemerly, A.S., Villarroel, R., van Montagu, M. and Inze, D. (1991). The *Arabidopsis* functional homolog of the p34^{cdc2} protein kinase. *Plant Cell* **3**, 531–540.

Flint, A.J., Paladini, R.D. and Koshland, D.D., Jr (1990). Autophosphorylation of protein kinase C at three separated regions of its primary sequence. *Science* (Wash., DC) **249**, 408–411.

Forer, A. and Sillers, P.J. (1987). The role of the phosphatidylinositol cycle in mitosis in sea urchin zygotes. Lithium inhibition is overcome by *myo*-inositol but not by other cyclitols or sugars. *Exp. Cell Res.* **170**, 42–55.

Gautier, J., Norbury, C., Lohka, M., Nurse, P. and Maller, J. (1988). Purified maturation-promoting factor contains the product of a *Xenopus* homolog of the fission yeast cell cycle control gene *cdc2*$^{+}$. *Cell* **54**, 433–439.

Gerace, L. and Blobel, G. (1980). The nuclear envelope lamina is reversibly depolymerized during mitosis. *Cell* **19**, 277–287.

Gerace, L. and Burke, B. (1988). Functional organization of the nuclear envelope. *Ann. Rev. Cell Biol.* **4**, 335–374.

Gilkey, J.C., Jaffe, L.F., Ridgway, E.B. and Reynolds, G.T. (1978). A free calcium wave traverses the activating egg of the medaka, *Oryzias latipes*. *J. Cell Biol.* **76**, 448–466.

Gould, K.L. and Nurse, P. (1989). Tyrosine phosphorylation of the fission yeast *cdc2*$^{+}$ protein kinase regulates entry into mitosis. *Nature* (Lond.) **342**, 39–45.

Hadwiger, J.A., Wittenberg, C., Richardson, H.E., deBarros Lopes, M. and Reed, S.I. (1989). A family of cyclin homologs that control the G$_1$ phase in yeast. *Proc. Natl Acad. Sci. U.S.A.* **86**, 6255–6259.

Hagan, I.M., Riddle, P.N. and Hyams, J.S. (1990). Intramitotic controls in the fission yeast *Schizosaccharomyces pombe*: the effect of cell size on spindle length and the timing of mitotic events. *J. Cell Biol.* **110**, 1617–1621.

Harmon, A.C., Putnam-Evans, C. and Cormier, M.J. (1987). A calcium-dependent but calmodulin-independent protein kinase from soybean. *Plant.*

Physiol. **83**, 830–837.

Harris, P. (1977). Triggers, trigger waves, and mitosis: a new model. In *Monographs in Cell Biology* (eds J.R. Jeder, E.D. Buetow, I.L. Cameron, G.M. Padilla and A.M. Zimmerman), pp. 75–104. Academic Press, New York.

Hartwell, L.H. (1978). Cell division from a genetic perspective. *J. Cell Biol.* **77**, 627–637.

Hartwell, L.H. and Weinert, T.A. (1989). Checkpoints: controls that ensure the order of cell cycle events. *Science* (Wash., DC) **246**, 629–634.

Hata, S. (1991). cDNA cloning of a novel *cdc2*$^{+}$/ *CDC28*-related protein kinase from rice. *FEBS Lett.* **279**, 149–152.

Hayles, J., Aves, S. and Nurse, P. (1986). *suc1*$^{+}$ is an essential gene involved in both the cell cycle and growth in fission yeast. *EMBO* (*Eur. Mol. Biol. Organ.*) *J.* **5**, 3373–3379.

Heald, R. and McKeon, F. (1990). Mutations of phosphorylation sites in lamin A that prevent nuclear lamina disassembly in mitosis. *Cell* **61**, 579–589.

Hepler, P.K. (1977). Membranes in the spindle apparatus: their possible role in the control of microtubule assembly. In *Mechanism and Control of Cell Division* (eds T. Rost and E.M. Gifford, Jr), pp. 212–232. Dowden, Hutchinson and Ross, Stroudsberg, PA.

Hepler, P.K. (1980). Membranes in the mitotic apparatus of barley cells. *J. Cell Biol.* **86**, 490–499.

Hepler, P.K. (1985). Calcium restriction prolongs metaphase in dividing *Tradescantia* stamen hair cells. *J. Cell Biol.* **100**, 1363–1368.

Hepler, P.K. (1989). Calcium transients during mitosis: observations in flux. *J. Cell Biol.* **109**, 2567–2574.

Hepler, P.K. and Callaham, D.A. (1987). Free calcium increases during anaphase in dividing stamen hair cells of *Tradescantia*. *J. Cell Biol.* **105**, 2137–2143.

Hepler, P.K. and Wolniak, S.M. (1984). Membranes in the mitotic apparatus: their structure and funtion. *Inter. Rev. Cytol.* **90**, 169–238.

Hirt, H., Pay, A., Gyorgyy, J., Bako, L., Nemeth, K., Bogre, L. *et al.* (1991). Complementation of a yeast cell cycle mutant by an alfalfa cDNA encoding a protein kinase homologous to p34^{cdc2}. *Proc. Natl Acad. Sci. U.S.A.* **88**, 1636–1640.

Huang, K.-P. (1989). The mechanism of protein kinase C activation. *Tren. Neurosci.* **12**, 425–432.

John, P.C.L., Sek, F.J. and Lee, M.G. (1989). A homolog of the cell cycle control protein p34^{cdc2} participates in the division cycle of *Chlamydomonas* and a similar protein is detectable in higher plants and remote taxa. *Plant Cell* **1**, 1185–1193.

John, P.C.L., Sek, F.J., Carmichael, J.P. and McCurdy, D.W. (1990). p34^{cdc2} homologue level, cell division, phytohormone responsiveness and cell differentiation in wheat leaves. *J. Cell Sci.* **97**, 627–630.

John, P.C.L., Sek, F.J. and Hayles, J. (1991). Association of the plant p34^{cdc2}-like protein with p13^{suc1}: implications for control of cell division cycles in plants. *Protoplasma* **161**, 70–74.

Kao, J.P.Y., Alderton, J.M., Tsien, R.Y. and Steinhardt, R.A. (1990). Active involvement of Ca^{2+} in mitotic progression of Swiss 3T3 fibroblasts. *J. Cell Biol.* **111**, 183–196.

Keith, C.H., Ratan, R., Maxfield, F.R., Bajer, A.S. and Shelanski, M.L. (1985). Local cytoplasmic gradients in living mitotic cells. *Nature* (Lond.) **316**, 848–850.

Kiehart, D.P. (1981). Studies on the *in vivo* sensitivity of spindle microtubules to calcium ions and evidence for a vesicular calcium-sequestering system. *J. Cell Biol.* **88**, 604–617.

Labbé, J.-C., Picard, A., Karsenti, E. and Dorée, M. (1988). An M-phase-specific protein kinase of *Xenopus* oocytes: partial purification and possible mechanism of its periodic activation. *Develop. Biol.* **127**, 157–169.

Lamb, N.J.C., Fernandez, A., Watrin, A., Labbe, J.-C. and Cavadore, J.-C. (1990). Microinjection of p34^{cdc2} kinase induces marked changes in cell shape, cytoskeletal organization, and chromatin structure in mammalian fibroblasts. *Cell* **60**, 151–165.

Lambert, A.-M. (1980). The role of chromosomes in anaphase trigger and nuclear envelope activity in spindle formation. *Chromosoma* (Berl.) **76**, 295–308.

Langan, T.A., Gautier, J., Lohka, M., Hollingsworth, R., Moreno, S., Nurse, P. *et al.* (1989). Mammalian growth-associated H1 histone kinase: a homolog of *cdc2$^+$/CDC28* protein kinases controlling mitotic entry in yeast and frog cells. *Mol. Cell. Biol.* **9**, 3860–3868.

Larsen, P.M. and Wolniak, S.M. (1990). 1,2-Dioctanoylglycerol accelerates or retards mitotic progression in *Tradescantia* stamen hair cells, depending on the time of its addition. *Cell Motil. Cytoskelet.* **16**, 190–203.

Larsen, P.M., Chen, T.-L.L. and Wolniak, S.M. (1989). Quin2-induced metaphase arrest in stamen hair cells can be reversed by 1,2-dioctanoylglycerol, but not by 1,3-dioctanoylglycerol. *Eur. J. Cell Biol.* **48**, 212–219.

Larsen, P.M., Chen, T.-L.L. and Wolniak, S.M. (1991). Neomycin reversibly disrupts mitotic progression in stamen hair cells of *Tradescantia*. *J. Cell Sci.*, **98**, 159–168.

Lawton, M.A., Yamamoto, R.T., Hanks, S.K. and Lamb, C.J. (1989). Molecular cloning of plant transcripts encoding protein kinase homologs. *Proc. Natl Acad. Sci. U.S.A.* **86**, 3140–3144.

Lee, M. and Nurse, P. (1987). Complementation used to clone a human homologue of the fission yeast cell cycle control gene *cdc2$^+$*. *Nature* (Lond.) **327**, 31–35.

Lehner, C.F. and O'Farrell, P.H. (1990). The roles of *Drosophila* cyclins A and B in mitotic control. *Cell* **61**, 535–547.

Levin, D.E., Fields, F.O., Kunisawa, R., Bishop, J.M. and Thorner, J. (1990). A candidate protein kinase C gene, *PKC1*, is required for the *S. cerevisiae* cell cycle. *Cell* **62**, 213–224.

Lewin, B. (1990). Driving the cell cycle: M phase kinase, its partners, and substrates. *Cell* **61**, 743–752.

Lohka, M.J. (1989). Mitotic control by metaphase-promoting factor and *cdc* proteins. *J. Cell Sci.* **92**, 131–135.

Luca, F.C. and Ruderman, J.V. (1989). Control of programmed cyclin destruction in a cell-free system. *J. Cell Biol.* **109**, 1895–1909.

Maller, J.L. (1990). *Xenopus* oocytes and the biochemistry of cell division. *Biochemistry* **29**, 3157–3166.

Maller, J.L., Gautier, J., Langan, T. and Lohka, M.J. (1989). Maturation-promoting factor and the regulation of the cell cycle. *J. Cell Sci. Suppl.* **12**, 53–63.

Martiny-Baron, G. and Scherer, G.F.E. (1989). Phospholipid-stimulated protein kinase in plants. *J. Biol. Chem.* **264**, 18052–18059.

Marx, J.L. (1989). The cell cycle coming under control. *Science* (Wash., DC) **245**, 252–255.

Matsui, M.S. and DeLeo, V.A. (1990). Induction of protein kinase C activity by ultraviolet radiation. *Carcinogenesis* **11**, 229–234.

Mazia, D. (1961). Mitosis and the physiology of cell division. In *The Cell*, vol. III. (eds J. Brachet and A.E. Mirsky), pp. 77–412. Academic Press, New York.

McDonald, K. (1989). Mitotic spindle ultrastructure and design. In *Mitosis: Molecules and Mechanisms* (eds J.S. Hyams and B.R. Brinkley), pp. 1–38. Academic Press, New York.

McNeil, P.L., McKenna, M.P. and Taylor, D.L. (1985). A transient rise in cytosolic calcium follows stimulation of quiescent cells with growth factors and is inhibitable with phorbol myristate acetate. *J. Cell Biol.* **101**, 372–379.

Meijer, L., Arion, D., Golsteyn, R., Pines, J., Brizuela, L., Hunt, T. *et al.* (1989). Cyclin is a component of the sea urchin egg M-phase specific histone H1 kinase. *EMBO (Eur. Mol. Biol. Organ.) J.* **8**, 2275–2282.

Memon, A.R., Chen, Q. and Boss, W.F. (1989a). Inositol phospholipids activate plasma membrane ATPase in plants. *Biochem. Biophys. Res. Comm.* **162**, 1295–1301.

Memon, A.R., Rincon, M. and Boss, W.F. (1989b). Inositiol trisphosphate metabolism in carrot (*Daucus carota* L.) cells. *Plant Physiol.* **91**, 477–480.

Millar, S.E., Freeman, M. and Glover, D.M. (1987). The distribution of 'mitosis specific' antigen during *Drosophila* development. *J. Cell Sci.* **87**, 95–104.

Minshull, J., Blow, J.J. and Hunt, T. (1989a). Translation of cyclin mRNA is necessary for extracts of activated *Xenopus* eggs to enter mitosis. *Cell* **56**, 947–956.

Minshull, J., Pines, J., Golsteyn, R., Standart, N., Mackie, S., Colman, A. *et al.* (1989b). The role of cyclin synthesis, modification and destruction in the control of cell division. *J. Cell Sci. Suppl.* **12**, 77–97.

Mitchison, J.M. (1971). *The Biology of the Cell Cycle.* Cambridge University Press, Cambridge.

Mole-Bajer, J. (1958). Cine-micrographic analysis of *c*-mitosis in endosperm. *Chromosoma* (Berl.) **9**, 322–358.

Mole-Bajer, J., Bajer, A.S., Zinkowski, R.P., Balczon, R.D. and Brinkley, B.R. (1990). Autoantibodies from a patient with scleroderma CREST recognized kinetochores of the higher plant *Haemanthus*. *Proc. Natl Acad. Sci. U.S.A.* **87**, 3599–3603.

Moreno, S. and Nurse, P. (1990). Substrates for p34[cdc2]: *in vivo veritas? Cell* **61**, 549–551.

Moreno, S., Hayles, J. and Nurse, P. (1989). Regulation of p34[cdc2] protein kinase during mitosis. *Cell* **58**, 361–372.

Morgan, D.O., Kaplan, J.M., Bishop, J.M. and Varmus, H.E. (1989). Mitosis-specific phosphorylation of pp60[c-src] by p34[cdc2] associated protein kinase. *Cell* **57**, 775–786.

Morla, A.O., Draetta, G., Beach, D. and Wang, J.Y.J. (1989). Reversible tyrosine phosphorylation of *cdc2*: dephosphorylation accompanies activation during entry into mitosis. *Cell* **58**, 193–203.

Murray, A.M. (1989). Cyclin synthesis and degradation and the embryonic cell cycle. *J. Cell Sci. Suppl.* **12**, 65–76.

Murray, A.M. and Kirschner, M.W. (1989a). Dominoes and clocks: the union of two views of the cell cycle. *Science* (Wash., DC) **246**, 614–621.

Murray, A.M. and Kirschner, M.W. (1989b). Cyclin synthesis drives the early embryonic cell cycle. *Nature* (Lond.) **339**, 275–280.

Neant, I., Charbonneau, M. and Guerrier, P. (1989). A requirement for protein phosphorylation in regulating the meiotic and mitotic cell cycles in echinoderms. *Develop. Biol.* **132**, 304–314.

Nishizuka, Y. (1986). Studies and perspectives on protein kinase C. *Science* (Wash., DC) **233**, 305–312.

Nishizuka, Y. (1989). Studies and prospectives of the protein kinase C family for cellular regulation. *Cancer* **63**, 1892–1903.

Norbury, C.J. and Nurse, P. (1989). Control of the higher eukaryote cell cycle by p34[cdc2] homologues. *Biochim. Biophys. Acta* **989**, 85–95.

Nurse, P. and Bissett, Y. (1981). Gene required in G_1 for commitment to cell cycle and in G_2 for control of mitosis in fission yeast. *Nature* (Lond.) **292**, 558–560.

O'Brian, C.A. and Ward, N.E. (1989). Biology of the protein kinase C family. *Canc. Metastas. Rev.* **8**, 199–214.

Ohkura, H., Kinoshita, N., Miyatani, S., Toda, T. and Yanagida, M. (1989). The fission yeast *dis2*[+] gene required for chromosome disjoining encodes one of two putative type 1 protein phosphatases. *Cell* **57**, 997–1007.

Ohta, Y., Ohba, T. and Miyamoto, E. (1990). Ca^{2+}/calmodulin-dependent protein kinase II: localization in the interphase nucleus and the mitotic apparatus of mammalian cells. *Proc. Natl Acad. Sci. U.S.A.* **87**, 5341–5345.

Osmani, S.A., May, G.S. and Morris, N.R. (1987). Regulation of the mRNA levels of *nimA*, a gene required for the G_2/M transition in *Aspergillus nidulans*. *J. Cell Biol.* **104**, 1495–1504.

Osmani, S.A., Engle, D.B., Doonan, J.H. and Morris, N.R. (1988a). Spindle formation and chromatin condensation in cells blocked at interphase by mutation of negative cell cycle control gene. *Cell* **52**, 241–251.

Osmani, S.A., Pu, R.T. and Morris, N.R. (1988b). Mitotic induction and maintenance by overexpression of a G_2-specific gene that encodes a potential protein kinase. *Cell* **53**, 237–244.

Pankov, R., Lemieux, M. and Hancock, R. (1990). An antigen located in the kinetochore region in metaphase and on polar microtubule ends in the midbody region in anaphase, characterized using a monoclonal antibody. *Chromosoma (Berl.)* **99**, 95–101.

Patel, R., Twigg, J., Crossley, I., Golsteyn, R. and Whitaker, M. (1988). Calcium-induced chromatin condensation and cyclin phosphorylation during chromatin condensation cycles in ammonia-activated sea urchin eggs. *J. Cell Sci. Suppl.* **12**, 129–144.

Peaucellier, G., Anderson, A.C. and Kinsey, W.H. (1990). Protein tyrosine phosphorylation during meiotic divisions of starfish oocytes. *Develop. Biol.* **138**, 391–399.

Peter, M., Nakagawa, J., Dorée, M., Labbé, J.-C. and Nigg, E.A. (1990a). Identification of major nucleolar proteins as candidate mitotic substrates of *cdc2* kinase. *Cell* **60**, 791–801.

Peter, M., Nakagawa, J., Doree, M., Labbe, J.-C. and Nigg, E.A. (1990b). *In vitro* disassembly of the nuclear lamina and M phase-specific phosphorylation of lamins by *cdc2* kinase. *Cell* **61**, 591–602.

Picard, A., Cavadore, J.-C., Lory, P., Bernengo, J.-C., Ojeda, C. and Dorée, M. (1990). Microinjection of a conserved peptide sequence of p34^{cdc2} induces a Ca^{2+} transient in oocytes. *Science* (Wash., DC) **247**, 327–329.

Pines, J. and Hunter, T. (1989). Isolation of a human cyclin cDNA: evidence for cyclin mRNA and protein regulation in the cell cycle and for interaction with p34^{cdc2}. *Cell* **58**, 833–845.

Pluta, A.F., Cooke, C.A. and Earnshaw, W.C. (1990). Structure of the human centromere at metaphase. *TIBS* **15**, 181–185.

Poenie, M., Alderton, J., Tsien, R.Y. and Steinhardt, R.A. (1985). Changes of free calcium levels with stages of the cell division cycle. *Nature* (Lond.) **315**, 147–149.

Poenie, M., Alderton, J., Tsien, R.Y. and Steinhardt, R.A. (1986). Calcium rises abruptly and briefly throughout the cell at the onset of anaphase. *Science* (Wash., DC) **233**, 886–889.

Polya, G.M., Morrice, N. and Wettenhall, R.E.H. (1989). Substrate specificity of wheat embryo calcium-dependent protein kinase. *FEBS Lett.* **253**, 137–140.

Prentki, M., Deeney, J.T., Matshinsky, F.M. and Joseph, S.K. (1986). Neomycin, a specific drug to study the inositol-phospholipid signalling system? *FEBS Lett.* **197**, 285–288.

Rapp, P.E. (1987). Why are so many biological systems periodic? *Prog. Neurobiol.* **29**, 261–273.

Ratan, R.R., Shelanski, M.L. and Maxfield, F.R. (1988). Long-lasting and rapid calcium changes during mitosis. *J. Cell Biol.* **107**, 993–999.

Ray, L.B. and Sturgill, T.W. (1988). Insulin-stimulated microtubule-associated protein kinase is phosphorylated on tyrosine and threonine *in vivo*. *Proc. Natl Acad. Sci. U.S.A.* **85**, 3753–3757.

Reed, S.I. and Wittenberg, C. (1990). Mitotic role for the *CDC28* protein kinase of *Saccharomyces cerevisiae*. *Proc. Natl Acad. Sci. U.S.A.* **87**, 5697–5701.

Riabowol, K., Draetta, G., Brizuela, L., Vandre, D. and Beach, D. (1989). The *cdc2* kinase is a nuclear

protein that is essential for mitosis in mammalian cells. *Cell* **57**, 393–401.

Richardson, H.E., Wittenburg, C., Cross, F. and Reed, S.I. (1989). An essential G$_1$ function for cyclin-like proteins in yeast. *Cell* **59**, 1127–1133.

Rincon, M., Chen, Q. and Boss, W. (1989). Characterization of inositol phosphates in carrot (*Daucus carota* L.) cells. *Plant Physiol.* **89**, 126–132.

Russell, P. and Nurse, P. (1987a). Negative regulation of mitosis by *wee1*, a gene encoding a protein kinase homolog. *Cell* **49**, 559–567.

Russell, P. and Nurse, P. (1987b). The mitotic inducer *nim1$^+$* functions in a regulatory network of protein kinase homologs controlling the initiation of mitosis. *Cell* **49**, 569–576.

Russell, P., Moreno, S. and Reed, S.I. (1989). Conservation of mitotic controls in fission and budding yeasts. *Cell* **57**, 295–303.

Sadhu, K., Reed, S.I., Richardson, H. and Russell, P. (1990). Human analog of fission yeast *cdc25* mitotic inducer is predominantly expressed in G$_2$. *Proc. Natl Acad. Sci. U.S.A.* **87**, 5139–5143.

Salmon, E.D. and Segall, R.R. (1980). Calcium-labile mitotic spindles isolated from the sea urchin egg (*Lytechinus variegatus*). *J. Cell Biol.* **86**, 355–365.

Schumaker, K.S. and Sze, H. (1987). Inositol 1,4,5-trisphosphate releases Ca^{2+} from vacuolar membrane vesicles of oat root. *J. biol. Chem.* **262**, 3944–3946.

Shenoy, S., Choi, J.-K., Bagrodia, S., Copeland, T.D., Maller, J.L. and Shalloway, D. (1989). Purified maturation promoting factor phosphorylates pp60^{c-src} at the sites phosphorylated during fibroblast mitosis. *Cell* **57**, 763–774.

Silver, R.B. (1989). Nuclear envelope breakdown and mitosis in sand dollar embryos is inhibited by microinjection of calcium buffers in a calcium-reversible fashion, and by antagonists of intracellular Ca^{2+} channels. *Develop. Biol.* **131**, 11–26.

Simanis, V. and Nurse, P. (1986). The yeast cell cycle control gene *cdc2* of fission yeast encodes a protein kinase potentially regulated by phosphorylation. *Cell* **45**, 261–268.

Steinhardt, R.A. and Alderton, J. (1988). Intracellular free calcium rise triggers nuclear envelope breakdown in the sea urchin embryo. *Nature* (Lond.) **332**, 364–366.

Steinhardt, R.A., Zucker, R. and Schatten, G. (1977). Intracellular calcium release at fertilization in the sea urchin egg. *Develop. Biol.* **58**, 185–196.

Suprynowitz, F.A. and Gerace, L. (1986). A fractionated cell-free system for analysis of prophase nuclear disassembly. *J. Cell Biol.* **103**, 2073–2082.

Swenson, K.L., Farrell, K.M. and Ruderman, J.V.

(1986). The clam embryo protein cyclin A induces entry into M phase and the resumption of meiosis in *Xenopus* oocytes. *Cell* **47**, 861–870.

Tombes, R.M. and Borisy, G.G. (1989). Intracellular free calcium in mammalian cells: anaphase onset is calcium modulated, but is not triggered by a brief transient. *J. Cell Biol.* **109**, 627–636.

Tysnes, O.-B., Verhoeven, A.J.M. and Holmsen, H. (1987). Neomycin inhibits agonist-stimulated polyphosphoinositide metabolism and response in human platelets. *Biochem. Biophys. Res. Comm.* **144**, 454–462.

Vandre, D.D. and Borisy, G.G. (1989). Anaphase onset and dephosphorylation of mitotic phosphoproteins occur concomitantly. *J. Cell Sci.* **94**, 245–258.

Vandre, D.D., Davis, F.M., Rao, P.N. and Borisy, G.G. (1984). Phosphoproteins are components of mitotic microtubule organizing centers. *Proc. Natl Acad. Sci. U.S.A.* **81**, 4439–4443.

Van't Hof, J. and Kovacs, C.J. (1972). Mitotic cycle regulation in the meristem of cultured roots: the principal control point hypothesis. In *The Dynamics of Meristem Cell Populations* (eds Miller, M.W. and Keuhnert, C.C.), pp. 15–32. Plenum, New York.

Verde, F., Labbé, J.-C., Dorée, M. and Karsenti, E. (1990). Regulation of microtubule dynamics by *cdc2* protein kinase in cell-free extracts of *Xenopus* eggs. *Nature* (Lond.) **343**, 233–238.

Vila, J. and Weber, M.J. (1988). Mitogen-stimulated tyrosine phosphorylation of a 42-kD cellular protein: evidence for a protein kinase-C requirement. *J. Cell Physiol.* **135**, 285–292.

Ward, G.E. and Kirschner, M.W. (1990). Identification of cell cycle-regulated phosphorylation sites on nuclear lamin C. *Cell* **61**, 561–577.

Whitfield, W.G.F., Gonzalez, C., Maldonado-Codina, G. and Glover, D.M. (1990). The A- and B-type cyclins of *Drosophila* are accumulated and destroyed in temporally distinct events that define separable phases of the G_2-M transition. *EMBO (Eur. Mol. Biol. Organ.) J.* **9**, 2563–2572.

Whitman, M. and Cantley, L. (1988). Phosphoinositide metabolism and the control of cell proliferation. *Biochim. Biophys. Acta* **948**, 327–344.

Wittenberg, C., Sugimoto, K. and Reed, S.I. (1990). G_1-specific cyclins of *S. cerevisiae*: cell cycle periodicity, regulation by mating pheromone, and association with the $p34^{cdc2}$ protein kinase. *Cell* **62**, 225–237.

Wolniak, S.M. (1987). Lithium alters mitotic progression in stamen hair cells of *Tradescantia* in a time-dependent and reversible fashion. *Eur. J. Cell Biol.* **44**, 286–293.

Wolniak, S.M. (1988). The regulation of mitotic spindle function. *Biochem. Cell Biol.* **66**, 490–514.

Wolniak, S.M. and Bart, K.M. (1985a). The buffering of calcium with quin2 reversibly forestalls anaphase onset in stamen hair cells of *Tradescantia. Eur. J. Cell Biol.* **39**, 33–40.

Wolniak, S.M. and Bart, K.M. (1985b). Nifedipine reversibly arrests mitosis in stamen hair cells of *Tradescantia. Eur. J. Cell Biol.* **39**, 273–277.

Wolniak, S.M., Hepler, P.K. and Jackson, W.T. (1980). Detection of the membrane-calcium distribution during mitosis in *Haemanthus* endosperm with chlorotetracycline. *J. Cell Biol.* **87**, 23–32.

Wolniak, S.M., Hepler, P.K. and Jackson, W.T. (1983). Ionic changes in the mitotic apparatus at the metaphase/anaphase transition. *J. Cell Biol.* **96**, 598–605.

Yaciuk, P., Choi, J.-K. and Shalloway, D. (1989). Mutation of amino acids in $pp60^{c-src}$ that are phosphorylated by protein kinase C and A. *Mol. Cell. Biol.* **9**, 2453–2463.

Zhang, D.H., Callaham, D.A. and Hepler, P.K. (1990). Regulation of anaphase chromosome motion in *Tradescantia* stamen hair cells by calcium and related signaling agents. *J. Cell Biol.* **111**, 171–182.

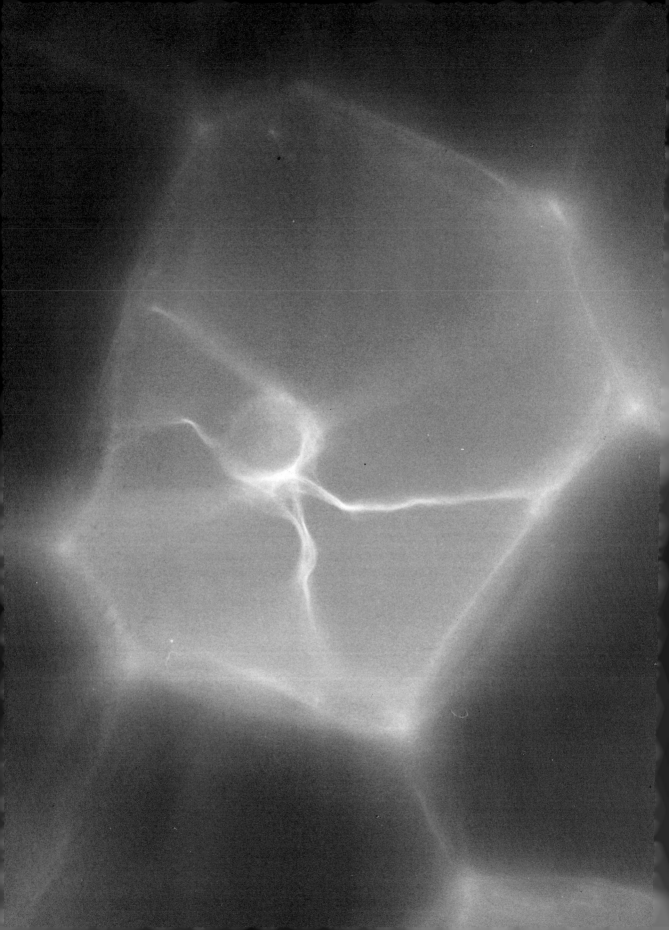

DETERMINATION OF THE DIVISION PLANE

Low power micrographs of sectioned plants show a network of cell walls, the pattern of which depends almost entirely upon the alignment of cell plates across dividing cells. All walls in a tissue, except for the outer ones without a neighbour, were at one time a cell plate and although some walls may be selectively stretched by the effects of directional cell expansion during the non-dividing phase, it is during cytoplasmic division that basic tissue patterns are established. Transverse sections of roots and shoots reveal a histology in which adjacent cells can have distinctly different morphologies and functions. The source of this differentiation is best seen in longitudinal sections where adjacent cells can be traced along files towards the apical region: it is in the apical region that progenitors divide longitudinally to initiate new files. These watershed divisions initiate the differentiation process and increase the width of the growing axis, whereas products of transverse division (placing new cells into pre-existing files) potentially increase length.

Regulating the division plane is the central shaping process during programmed development. It is seen during embryogenesis and the repetitive production of modular organs, such as leaves, during post-embryonic growth but is also expressed during unpredicted (although not necessarily unprogrammed) events as when quiescent cells are stimulated to divide to seal a wound.

In addition to the *plane* of division, the actual site of division is important. When cells divide, the new cell plate is positioned so that it is off-set from the wall separating two cells in the adjacent file. This staggering of the joints produces the pattern which many commentators have likened to the bonding of bricks and the placement of partitions in liquid foams. It is during cell formation that tissue texture is expressed. However, the partitioning of cellular space is predicted before mitosis. In large, vacuolated cells, premitotic cytoplasmic strands that tether the central nucleus have been known for some considerable time to come together in a phragmosome which anticipates the division plane. A later discovery was that the cortical preprophase band of microtubules – certainly more widely known than the phragmosome – is also a forerunner of the division plane. Common to both is the need to consider how the spatial information expressed by premitotic structures is retained through space and time until the cell plate finally unites with the mother wall. The phragmosome and the preprophase band (PPB) in higher plant cells are the subject of two chapters in this section and one of these discusses how the arrangement of the premitotic cytoskeleton may give rise to the characteristic bonding pattern of cells in tissues. Division plane alignment in moss and fern protonemata is discussed in the chapters by Doonan and by Wada and Murata in the context of the life cycle of filamentous organisms in the next section. Moss protonemata contain no PPBs although their cross walls are carefully positioned relative to the light source, and so it is important to appreciate that there may be several different mechanisms for aligning the division plane. Some are discussed in this section by Brown and Lemmon (Chapter 19). During sporogenesis, as opposed to histogenesis, there is no PPB and these authors describe alternative spatial influences on quadripartitioning of the spore.

It is encouraging that the division-aligning cytoskeletal mechanisms have been identified in filamentous plants as well as in the more complex tissues of higher plants. Moreover, these are not mechanisms in isolation for we can begin to see how light and cellular dimensions influence the division plane. That is, not only during the programmed stages of histogenesis but also during the subsequent response to environmental factors.

17. THE PREPROPHASE BAND

Susan M. Wick

Department of Plant Biology, University of Minnesota,
220 BSC, 1445 Gortner Avenue, St Paul, MN 55108, USA

Introduction

The preprophase band (PPB) was first described a quarter of a century ago by Pickett-Heaps and Northcote (1966a, b) as a band of cortical microtubules (MTs) that forms before mitosis in plant cells. Often, though not always, the PPB is positioned so that the plane or curved surface it defines at least roughly bisects the nucleus. Although the PPB disappears as the mitotic spindle is established, the memory of its placement is retained, and at cytokinesis the new cell plate is guided to fuse with the parent cell walls along the line delineated previously by the PPB. In this respect, the PPB can be conceptualized to be a reflection of a cell's commitment to divide in a particular orientation or position. Thus, while one might be tempted to think of PPB formation as a mitotic phenomenon due to its proximity in time and space to the mitotic apparatus, it is appropriate to consider it as a part of cytokinesis, as it is one of the manifestations that a cell has established a division site (Gunning, 1982; Gunning and Wick, 1985). Under certain conditions, it is possible to uncouple PPB formation from nuclear events that are required prior to mitosis (Mineyuki et al., 1988b), as discussed on p. 240.

Spatial aspects of the PPB were reviewed extensively by Gunning (1982) in the first edition of this book as part of a chapter on the cytokinetic apparatus, and more briefly by Gunning and Wick (1985). In the intervening years, researchers have begun to address some of the questions raised in those reviews and fill in some of the missing pieces of information. The aim of this chapter will be to concentrate primarily on new information that has accrued since the previous works were written.

Occurrence of preprophase bands

With respect to the occurrence of PPBs, the current state of knowledge allows some generalizations to be made. First, it appears that PPBs are strictly a phenomenon of vegetative cells: they do not occur in any meiotic plant cells (Brown and Lemmon, 1989a, b; Staiger and Cande, 1990, and earlier references therein). Generative cell division likewise does not involve a PPB (Palevitz and Cresti, 1989). Second, evolutionarily, PPBs are found first in land plants (i.e. in the five-kingdom system, the Kingdom Plantae, which includes bryophytes and vascular plants). They are, however, not found in every cell type. Cells of the filamentous protonemata of the mosses *Funaria* and *Physcomitrella* – the divisions of which show precise spatial regulation – do not form a PPB (Schmeidel et al., 1981; Doonan et al., 1987). On the other hand, the fern *Adiantum* does have a PPB in its protonematal cells (Wada et al., 1980; Murata and Wada, 1989). For reasons not understood, Jenni et al. (1990) could not see PPBs in fern prothalli using immunofluorescence microscopy, although they could clearly visualize them with electron microscopy. Doonan et al. (1987) postulate that the lack of a PPB in *Physcomitrella* protonemata could be attributed to the observation that these cells have primarily endoplasmic MTs, with few cortical MTs that could be reorganized into a PPB. Another explanation is needed to account for the lack of a PPB in *Funaria* protonemata, which have a cortical MT system as well as a set of endoplasmic MTs (Schmiedel et al., 1981).

PPBs are found in some moss cells once the gametophyte switches over to formation of buds and leafy shoots, and thus three-dimensional growth (Schmiedel et al., 1981; Doonan *et al.*, 1987). However, some oddities are found among the bryophytes. Epidermal cells of the *Funaria* sporophyte have asymmetrically shaped PPBs (Sack and Paolillo, 1985). Some cells of the thallus of the liverwort *Marchantia* have incomplete PPBs that do not form a closed ring (Apostolakos and Galatis, 1985a), while the hornwort *Phaeoceros* has in its sporocyte cells a PPB that makes a complete ring, but which is highly asymmetrical in form (Brown and Lemmon, 1988), with tightly banded MTs on one side of the cell and MTs that are splayed apart on the other side.

Attempts to find a PPB in the dividing cell that gives rise to the *Funaria* stomate were unsuccessful (Sack and Paolillo, 1985). These stomates are binucleate and single-celled as a result of cytokinesis that begins normally but is never completed. If a PPB is truly missing from these cells, they fall into the general category of cells in which lack of a PPB is correlated with absence of cell plate fusion with the parent walls. Another example in this category is the division of archesporial cells in the sporocyte capsule of bryophytes. Products formed here, the spore mother cells, are separate entities, not contiguous with parent tissue, and there is no PPB predicting the placement of this division (Busby and Gunning, 1988; Gambardella *et al.*, 1990). One very well characterized cell type that lacks a PPB is endosperm at the stage when it is without a cellulosic wall. (See Gunning (1982) for references and for other examples of cell divisions in this category, and for a table that compiles reports of PPBs in bryophytes and vascular plants.)

A third generalization about the occurrence of PPBs is that all types of cell divisions within tissues involve PPBs (Pickett-Heaps and Northcote, 1966b; Gunning *et al.*, 1978). That is, as long as a division meets the criterion of producing a cell plate that grows to meet and fuse with parent walls, the cell preparing for that division will display a PPB regardless of the orientation of the division: whether the division is symmetrical or asymmetrical; whether one or both daughter cells will proceed to differentiate without further division, as in stomatal cells (Pickett-Heaps and Northcote, 1966b); whether daughter cells are destined to

become parts of different tissues within which they will undergo further divisions, as in the production of new files of cells in a meristem (formative divisions, Gunning *et al.*, 1978); or whether daughter cells remain as part of the same tissue and cell type as the parent (proliferative divisions, Gunning *et al.*, 1978).

The studies mentioned above on which this conclusion is based dealt with wheat leaf epidermis and *Azolla* root meristems, both of which exhibit a high degree of organization, with ordered files of cells and predictable patterns of cell division. Because the question of whether the presence of PPBs correlates with the degree of morphogenetic order was raised (Gunning, 1982), it is of interest to determine whether cells from less organized tissues also form PPBs. A tissue that shows less order than either of the above systems is dicot mesophyll spongy parenchyma. There clearly are mechanisms operating to limit divisions within these cells that would result in an increase in the number of layers of spongy parenchyma. However, divisions that result in insertion of new walls anticlinal to leaf surfaces do not follow a specific pattern. Thus these cells do not appear to be subject to as many constraints to divide in a particular orientation in order to function properly, as do, for instance, cells of grass epidermis dividing to form a stomatal subsidiary cell. Yet cells of spongy parenchyma, with its lower degree of morphogenetic order, likewise indicate with a PPB their commitment to a proliferative division (Wick *et al.*, 1989).

A few recent studies have commented on a related question, which is whether (excluding cell types such as wall-less endosperm, which never have PPBs) every cell that is judged to be in preprophase or prophase on the basis of cell position, nuclear position, or state of chromosome condensation has a PPB or the remnants of one that is being replaced by a mitotic spindle. In cultured cells of *Spartina* (Hogan, 1988) and tobacco (Katsuta *et al.*, 1990) and in cells of grass leaf epidermis (Cho and Wick, 1989) and onion cotyledon epidermis (Mineyuki and Palevitz, 1990), all cells preparing for mitosis exhibit a PPB.

The data on *Spartina* and tobacco cells bring

us to a final topic regarding the occurrence of PPBs, which is their presence in various suspension culture cells. Initial attempts to find PPBs in cultured cells were unsuccessful, results that were consistent with the idea that the PPB is a marker for cells about to undergo divisions giving rise to morphogenetic order. That is, if division orientation in cultured cells is not crucial, then perhaps preparation of a particular cortical division site is not necessary, and a PPB, which reflects preparation of a division site, would not be formed. Several reports of PPBs in suspension culture cells indicate that this is not the case (Falconer and Seagull, 1985; Gorst et al., 1986; Simmonds, 1986; Doonan et al., 1987; Kakimoto and Shibaoka, 1987; Hogan, 1988; Wang et al., 1989a, b; Katsuta et al., 1990).

Some of the cell lines used for these experiments grew as files, i.e. exhibited one-dimensional organization of growth (Gorst et al., 1986; Hogan, 1988) and others were capable of undergoing embryogenesis under suitable culture conditions, or were recently derived from such cultures or from organized plant tissues (Falconer and Seagull, 1985; Gorst et al., 1986). On the basis of the information, one might conclude that it is not very surprising that PPBs were found in some of these cultures, since filamentous chains of cells that are part of an intact plant, such as uniseriate hairs, are known to display PPBs (Busby and Gunning, 1980). Also, cells of the embryogenic lines might be expected to retain a memory for PPB formation, even though growth conditions do not promote embryo formation at a given time. However, others of these cell cultures were unable to form anything but unorganized colonies, and even these were found to contain PPBs (Gorst et al., 1986; Simmonds, 1986; Wang et al., 1989a), although sometimes they were few and irregular in appearance (Gorst et al., 1986). Thus, as in the mesophyll cell example mentioned above, apparent reduction in, or lack of, order among cells does not automatically indicate division without prior preparation of a division site as visualized by PPB formation.

Attempts have been made to determine whether cell cultures that are capable of regenerating ordered groupings of cells tend to display PPBs more regularly than do cultures that have lost this ability. For samples in which there are noticeable changes in nuclear configuration before spindle formation, the most direct approach is to examine all preprophase–prophase cells for PPBs, as was done by Hogan (1988) and Katsuta et al. (1990). Another approach taken has been to compare the percentage of cells with a PPB with the percentage that are in mitosis or cytokinesis (Gorst et al., 1986; Wang et al., 1989a; Traas et al., 1990). Rigorous interpretation of figures obtained this way requires knowing the duration of the PPB stage relative to that of mitosis or cytokinesis for each cell line studied, for the particular growth conditions of the experiment.

Spatial relationship between PPB site, nucleus, and ultimate division site in cells within tissues

PPB and nucleus

Under normal circumstances, the PPB and nucleus show a close association. A typical PPB is ring-shaped and planar, and is commonly aligned with the nuclear equator (Fig. 17.1a). Sometimes the PPB is acentric with respect to the nucleus (examples in Galatis and Mitrakos, 1979; Wick and Duniec, 1984; Eleftheriou, 1985b), especially if there are space restrictions in an asymmetrical cell division. In grass stomatal subsidiary cell mother cells (SMCs), the PPB is not planar, but rather forms a three-dimensional array that cradles a portion of the nucleus (see Fig. 17.5).

Endoplasmic MTs linking the nucleus to the PPB and other regions of the cortex in cells within tissues have been noted often (Wick and Duniec, 1983 and earlier references therein; Tiwari et al., 1984; Bakhuizen et al., 1985; Gunning and Wick, 1985; Mineyuki et al., 1989; Brown and Lemmon, 1990; Flanders et al., 1990; Mineyuki and Palevitz, 1990; see also Fig. 17.1b). In some cells, interaction between the PPB site and the nucleus or its associated MTs causes deformation of the nucleus in the orientation of the PPB (Galatis, 1982). Decreased movement of small organelles in the region between the PPB and nucleus (Mineyuki et al., 1984) and nuclear resistance to centrifugal

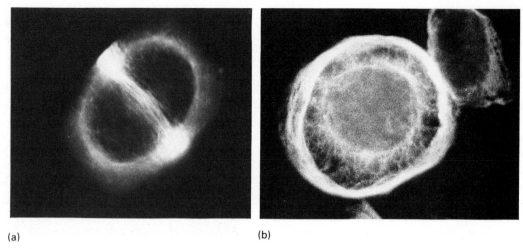

(a)   (b)

Fig. 17.1. Tubulin immunofluorescence microscopy of isolated cells of onion root tips. a, Side view of a cell reveals cortical PPB microtubules aligned with the nuclear equator and microtubules around the nucleus (× 1450). b, An end-on view of a preprophase cell shows microtubules radiating from the nuclear surface towards the cortex; the PPB is seen as a bright cortical ring (× 1300). Reproduced from the *Journal of Cell Biology* (1984) **99**, 63s–69s by copyright permission of the Rockefeller University Press.

displacement away from the PPB region (Pickett-Heaps, 1969a; Galatis *et al.*, 1984b) suggest a gelation or increased tensile strength of the cytoplasm, such as might result from the presence of a fibrous network. In grass SMCs, this increased stabilization of the nuclear position continues through mitosis (Pickett-Heaps, 1969a). Transvacuolar strands connecting the nucleus and PPB in suspension culture cells contain actin microfilaments as well as MTs (Lloyd and Traas, 1988; Katsuta *et al.*, 1990). No such actin links have yet been found in most cells within tissues (Cho and Wick, 1990; McCurdy and Gunning, 1990; Mineyuki and Palevitz, 1990), although they are seen in vacuolate leaf epidermal cells (Goodbody and Lloyd, 1990). Also there is a dense aggregation of actin filamens in SMCs in the region bounded by the curved PPB (Cho and Wick, 1990).

Various relationships between endoplasmic MTs and those of the PPB and mitotic array have been proposed. Suggestions have been made that the nuclear envelope nucleates MTs linking the nucleus and PPB (Bakhuizen *et al.*, 1985; Flanders *et al.*, 1990) or even ultimately contributes MTs to the PPB (Flanders *et al.*, 1990). Interaction of MTs emanating from

opposite polar organizers at the nuclear surface may be responsible for defining the PPB site in liverwort cells (Brown and Lemmon, 1990). Wang *et al.* (1989a) claim that perinuclear MTs do not appear in soybean cultures until the PPB is already well formed, which suggests that such a mechanism is not operating here. The alternate view, that the PPB might contribute MTs to the nuclear surface via endoplasmic MTs, is discussed by Brown and Lemmon (1988); earlier work concerning this hypothesis is covered by Wick and Duniec (1984).

The normally close association between nucleus and PPB appears not to be a prerequisite for establishment of the cortical division site. In some divisions, the PPB is formed before the nucleus migrates into its final position relative to the PPB (for examples see Gunning *et al.*, 1978; Mineyuki and Palevitz, 1990). Especially in the case of the SMC division during formation of grass stomatal complexes, there is convincing evidence that messages regarding establishment of the division site are transcellular, and do not rely on nuclear positioning within the cell preparing for division. The SMC will make a PPB adjacent to the guard mother cell that is inducing it to divide even if the nucleus has been displaced away from it by centrifugation (Galatis *et al.*, 1984b) or by cytochalasin B treatment (Cho and Wick, 1990), or if there is a cell wall (Galatis *et al.*, 1984a) or wall stubs (Apostolakos and Galatis, 1987) that prevent the nucleus from approaching its normal position, partially cradled within the space outlined by the PPB. In SMCs that subtend two or more guard mother cells in the same file, multiple PPBs can form, each adjacent to one of the inducing cells, even though there is a single nucleus (Galatis *et al.*, 1983). On the other hand, cells made binucleate by means of caffeine treatment often form a single PPB at their next division (Pickett-Heaps, 1969b; Apostolakos and Galatis, 1987; Wang *et al.*, 1989b; Wick and Duniec, unpublished observations). Thus, the number or position of nuclei is not correlated with the number or placement of PPBs.

PPB and the ultimate division site in the cell cortex

Evidence continues to mount that, under normal circumstances, the PPB accurately marks the cortical site that will be bisected by the new cell plate when it reaches the parent cell walls at cytokinesis (Gunning *et al.*, 1978; Galatis and Mitrakos, 1979; Busby and Gunning, 1980; Galatis *et al.*, 1982; Eleftheriou and Tsekos, 1982; Brown and Lemmon, 1984, 1988; Eleftheriou, 1985b). Electron microscopy studies apparently led some to conclude that the PPB in grass SMCs is in the shape of a ring that is bent only slightly or not at all in periclinal planes, and thus does not accurately reflect the curve of the subsidiary cell wall that is laid down (Pickett-Heaps and Northcote, 1966b; Galatis *et al.*, 1983). Subsequent work using immunfluorescence microscopy, which allows visualization of a much greater portion of cell depth in a single image, indicates that the SMC PPB shows some variation in degree of curvature, but that it does appear to predict the placement of the future cell wall (Cho and Wick, 1989; Cleary and Hardham, 1989; Wick *et al.*, 1989; see also Fig. 17.2). (Because Cleary and Hardham used sectioned material, it is not clear whether their fig. 18 is a true paradermal

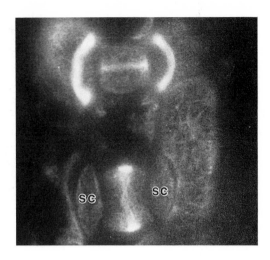

Fig. 17.2. Segment of rye leaf epidermis containing developing stomatal complexes. A lens-shaped subsidiary cell (sc) is visible to either side of the guard mother cell at the bottom, which has a longitudinal PPB; in the developmentally younger cells above these, PPBs mark the forthcoming division of each of the SMCs. Note that the curved PPBs accurately predict the placement of the wall that outlines the subsidiary cells. The upper guard mother cell is still in interphase and shows a characteristic transverse microtubule band (× 1200). Reproduced from *Cell Biology International Reports* (1989) **13**, 95–106, by copyright permission of Academic Press, New York.

view, or whether this is at a slight angle to paradermal, which could account for the PPB appearing to lie right at the SMC lateral wall.) Some *Triticum* species have subsidiary cells that appear triangular in medial section with electron microscopy, but these cells assume the normal lens shape in more cortical regions (Galatis *et al.*, 1984a); whatever distorts the curve of the wall in interior planes apparently does not interfere with the cell plate reaching and bisecting the premarked division site, which is in the SMC cortex.

Apostolakos *et al.* (1990) find that tubulin–colchicine complexes often accumulate at the division site, and attribute this to the ability of the site to nucleate, attract and/or stabilize MTs. Attempts are underway to define whatever it is that endows the division site with its characteristic abilities to support PPB formation and cell plate fusion, but so far the search has yielded few clues. Calmodulin is apparently not concentrated here (Wick *et al.*, 1985). By the time of cytokinesis, the site previously marked by the PPB apparently contains factors that assist in the final step of normal cell plate insertion, the process of making the plate taut and flat (Mineyuki and Gunning, 1990). Experimentally inducing the cell plate to grow to an area of the cortex other than the division site prevents this flattening of the cell plate. It is not yet known whether any other properties of a misplaced cell wall are abnormal.

An interesting situation is seen in some dividing thallus cells of *Marchantia*. Here, incomplete PPBs form, while other specialized sites in the cell cortex induce formation of MT bundles (Apostolakos and Galatis, 1985a). At cytokinesis, the cell plate can assume an unusual shape as it fuses with the previous PPB sites, as well as with other cortical regions, possibly those previously occupied by the other MT bundles (Apostolakos and Galatis, 1985b). Thus it seems that the PPB site may not be the only cortical region at which MTs can be organized and stabilized prior to mitosis, and towards which a phragmoplast can be attracted. (See also Apostolakos and Galatis (1987) regarding MT-organizing and phragmoplast-attracting characteristics of the rims of wall strips formed in the presence of caffeine.)

However important premitotic preparation of the division site may be, it is clear that preparation alone is not sufficient to guarantee precise division control. Several studies indicate that the site previously occupied by the PPB can exert an attractive force on the edges of the growing phragmoplast, provided that the edges come within a short distance (no more than a few micrometres) of it. The force of attraction can perhaps best be appreciated when the distance between the division site and the phragmoplast is experimentally increased, and the cell plate must move along the parent wall or grow an extra distance to reach it (Ôta, 1961; Galatis *et al.*, 1984b; Gunning and Wick, 1985; Apostolakos and Galatis, 1987). The cortical division site is also able to attract rafts of new cell plate that become free-floating after depolymerization of MTs with colchicine (Galatis *et al.*, 1984b).

Because the phragmoplast begins its growth between the telophase nuclei, the mitotic apparatus needs to be placed near the pre-determined division site and oriented in such a way that the growing phragmoplast can reach the 'zone of attraction' (Galatis *et al.*, 1983, 1984a, b; Apostolakos and Galatis, 1987; Venverloo and Libbenga, 1987; Cho and Wick, 1990).

It is worth noting that the final PPB site is the one that predicts division orientation and placement. In vacuolate cells, PPB formation is coincident with consolidation of transvacuolar strands, which radiate from the nucleus, into a phragmosome, a raft of cytoplasm that lies in the division plane and has the PPB at its rim (Gunning, 1982; Venverloo, 1990). Apparently, in response to wounding, an occasional cell that has already begun phragmosome and PPB formation can replace these with a new phragmosome and PPB of a different orientation, which then reflects the alignment of the subsequent division plane (Venverloo, 1990).

Relationship of nucleus to site of phragmoplast fusion with parent walls

This last observation about one of the prerequisites for attaining correct final disposition of the cell plate brings the discussion back to the nucleus, and to consideration of what is responsible for the morphogenetically crucial role of positioning it relative to the PPB site. As seen on p. 235, the nucleus apparently does not usually influence PPB placement. Can the PPB or associated cytoskeletal elements control placement and orientation of the nucleus? Several reports that nuclear migration can occur before PPB formation argue against a direct role of the PPB in nuclear migration (Pickett-Heaps and Northcote, 1966b; Pickett-Heaps, 1969c; Venverloo and Libbenga, 1987; Brown and Lemmon, 1988a; Flanders *et al.*, 1990; Katsuta *et al.*, 1990). However, in view of the evidence for physical linkage between PPB and nucleus, one might suspect that once the nucleus is in position, the PPB-associated cytoskeleton could be instrumental in defining spindle pole location. (See discussion in Wick and Duniec, 1984.) In support of this hypothesis, recent studies indicate that spindle poles appear always to form along a line orthogonal to that of the PPB, even in cells in which the spindle will be oblique later in mitosis (Mineyuki *et al.*, 1988a; Cho and Wick, 1989; Cleary and Hardham, 1989).

We have already seen that something stabilizes the SMC nucleus to resist displacement throughout preprophase and mitosis (Pickett-Heaps, 1969a). Ôta (1961) centrifuged mitotic stamen hair cells of *Tradescantia* (which, like SMCs, are relatively large) with enough force to displace the spindle. He found that it would not only return to the normal position relative to the predicted division site, but also would realign its axis to the correct one if it had been disrupted. That is, there appeared to be a mechanism for assuring the correct position and orientation of the nucleus throughout mitosis.

One possibility is that preprophase endoplasmic MTs perform this function until they disappear at the beginning of mitosis, and that there is a second system operating during mitosis. This might be particularly evident in large cells that have considerable space into which the nucleus could go astray if not tethered close to the correct division site. Indeed, Venverloo and Libbenga (1987) have evidence from observing individual vacuolate cells exposed to inhibitors that MTs are responsible for positioning the nucleus centrally, whereas actin is required to hold the nucleus in its normal position during mitosis. In cells of grass leaf epidermis one intriguing candidate for the second nuclear tethering system is the actin that faces spindle poles during mitosis (Cho and Wick, 1990, 1991). In SMCs, this takes the form of a dense accumulation of actin that surrounds the SMC nucleus on the side facing the inducing guard mother cell. Actin in the guard mother cell faces the poles at the time of spindle formation and remains located along the lateral edges while the spindle rotates and separates chromosomes diagonally, and while daughter nuclei, with the growing phragmoplast between them, rotate back to the original position of the prophase spindle poles. Actin is further implicated in a nuclear tethering and orienting role during mitosis in guard mother cells and SMCs by the observation that

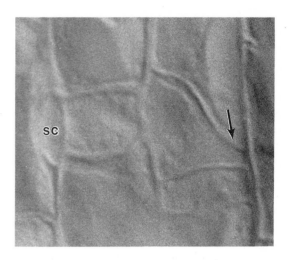

Fig. 17.3. Aberrant subsidiary cell formation in rye leaf epidermis exposed to cytochalasin B. Instead of curving to form a normal lens-shaped cell next to the guard mother cell, the subsidiary cell wall on the right (arrow) has fused to the anticlinal wall distal to the guard mother cell. A normal subsidiary cell (sc) is seen to the left (× 1260). Figure courtesy of Soon-Ok Cho.

division in both is spatially disrupted by cytochalasins (Cho and Wick, 1990 and references therein; see also Fig. 17.3). Correct placement of the cell plate in asymmetric divisions of root tip cells is also inhibited by cytochalasin D (Mineyuki and Palevitz, 1990).

In both the guard mother cell and in the centrifuged stamen hair cell, the supposed tethering system appears to have some elastic properties and/or the ability to re-establish connections across a gap. Whether actin at spindle poles is a general feature and whether its organization is dependent on PPB formation remains to be determined. Lloyd and Traas (1988) and Katsuta *et al.* (1990) postulate that, in cultured cells, one function of the PPB and its associated MTs in transvacuolar strands is to guide placement of actin in the cytoplasmic strands between the nucleus and cortex, thus ensuring proper positioning of the nucleus.

There are numerous examples of naturally occurring and experimentally induced situations in which incorrect placement and alignment of the spindle leads to dissociation of the site of cell plate fusion with parent walls and the cortical site previously occupied by the PPB. Drug-induced nuclear displacement leads to abnormal wall orientation in vacuolate epidermal cells dividing in response to wounding (Venverloo and Libbenga, 1987). If centrifugation of SMCs is severe enough, complete recovery does not occur, and the cell plate may reach neither or only one of the correct division sites facing the guard mother cell (Galatis et al., 1984b). A similar result is found when cytochalasin B treatment is used (Cho and Wick, 1990). Division abnormalities also occur if a SMC is induced to divide by several guard mother cells simultaneously, in which case the single nucleus cannot be properly oriented relative to each, and the phragmoplast cannot fuse at all the PPB sites (Galatis et al., 1983, 1984a). The reverse situation is caused by CIPC treatment, which leads to formation of multiple nuclear fragments and thereby multiple phragmoplasts that fuse with many cortical regions in addition to the PPB site (Clayton and Lloyd, 1984; Tiwari et al., 1984; Gunning and Wick, 1985; Lloyd and Traas, 1988). A telephose nucleus positioned directly in front of the division site can also prevent normal cell plate orientation (Galatis et al., 1984a).

Development of the PPB

Transition from interphase MT array to PPB
The scheme of MT deployment as a PPB replaces the interphase MT array was first elucidated in root meristem cells by correlating progressive changes in nucleoli and in chromatin condensation with changes in MT distribution (Wick and Duniec, 1983, 1984; Gunning and Wick, 1985). Since then, the process has been examined in numerous other systems, and while there are some variations unique to certain types of division, a general pattern is evident. In some cell types, development of canals around the nucleoli and condensing chromatin serves as an additional marker to correlate nuclear and PPB stages (Eleftheriou, 1985a, b).

A PPB is first recognized as a wide band of cortical MTs that usually is superseded by a narrower band. Sometimes double bands are

formed, especially in elongated cells at early PPB stages. Whereas PPB narrowing occurs in some cultured cells (Katsuta et al., 1990), in others the band tends to remain wider than those found in most cells within tissues, and it is sometimes wider on one side (Gorst et al., 1986; Simmonds, 1986; Hogan, 1988). Guard mother cells of some species likewise tend to be fairly broad at maturity (e.g. Busby and Gunning, 1980; Galatis, 1982; Galatis et al., 1982; Cho and Wick, 1989; Mullinax and Palevitz, 1989), although narrower ones are formed in others (Mineyuki et al., 1988a; Cleary and Hardham, 1989). Whether or not the PPBs of fern protonemata narrow appears to depend on the light conditions in which the cultures are grown (Murata and Wada, 1989). On the other hand, in Datura epidermal cells, the course of PPB development seems to be a function of cell dimensions: PPBs of elongated cells narrow with maturity but those of isodiametric cells, which show more irregular division patterns, do not narrow (Flanders et al., 1990).

Near the time that the PPB is first distinguishable, two other sets of MTs can be found: perinuclear MTs and endoplasmic MTs that radiate from the nuclear area towards the cortex. Before PPB MTs appear in wheat protophloem, perinuclear MTs aligned parallel to the anticipated PPB are visible (Eleftheriou, 1985b), while in soybean cultures derived from protoplasts, perinuclear MTs are found only after the PPB is well developed (Wang et al., 1989a). Perinuclear MTs gradually assume the alignment of spindle MTs and focus towards two polar regions, which always are aligned orthogonal to the PPB (or, in the case of the curved SMC PPB, aligned parallel to a line in the perinuclear cortex that bisects the PPB, forming equal angles to either side). As mentioned on p. 237, this relationship holds even for spindles that will rotate later in mitosis to become diagonal (Mineyuki et al., 1988a; Cho and Wick, 1989; Cleary and Hardham, 1989). Eventually the PPB MTs disappear, and in cells without a large vacuole, the nucleus-to-cortex MTs do as well.

In many cells, both the interphase array and the PPB are transverse to the longest axis of the

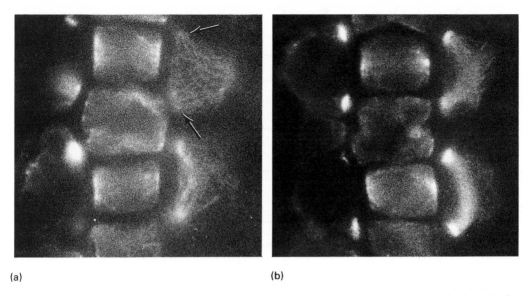

(a)   (b)

Fig. 17.4. Stages in PPB formation in SMCs of rye epidermis. a,b, Intersecting arrays of microtubules appear to fan out from two regions (arrows) on the paradermal faces of each SMC, become progressively bundled (lower SMC of a, upper one of b) and eventually form the curved line of the mature PPB (lower SMC of b) (× 1600). Reproduced from the *Journal of Cell Science* (1989) **92**, 581–594 by copyright permission of The Company of Biologists.

cell, a situation that has led to speculation about whether or not PPB formation and narrowing could occur by a condensation or 'bunching up' of existing MTs (Pickett-Heaps, 1969a; Wick and Duniec, 1983; Lloyd and Traas, 1988; Flanders *et al.*, 1990). However, in cells preparing for morphogenetically critical divisions, establishment of the proper division plane often involves redirecting cellular polarity from being parallel to tissue and organ polarity to being perpendicular or oblique to it. Here, the trend is for cells to reorient their interphase array first, and then proceed to make a PPB that reflects the new polarity.

In guard mother cells of grasses, there is a stage at which both interphase arrays are present; because at this stage each array is wider at the cell edges than in the centre of the periclinal face, their combined image is that of a cross with widened tips or a radial array along the periclinal faces of the cell (Galatis, 1982; Cho and Wick, 1989; Cleary and Hardham, 1989; Mullinax and Palevitz, 1989; see also Palevitz, Chapter 4, this volume). Even-

tually the more transverse MTs disappear and the longitudinal band consolidates its MTs somewhat to form the PPB. In contrast, the onion guard mother cell interphase array is comprised of randomly arranged MTs that become reorganized first into a wide, and then a narrower PPB (Mineyuki *et al.*, 1989). When pea root cells respond to a nearby wound by re-entering the division cycle and dividing in planes that are parallel to the cut surface, they first reorient their interphase MTs into arrays parallel to the wound, and then form PPBs in the new orientation (Hush *et al.*, 1990). PPB development in SMCs also starts with an altered interphase array. Here, the new array is first seen as MTs radiating from two foci along the cell edge facing the file of cells that contains guard mother cells (Cho and Wick, 1989). Apparent interaction between MTs fanning out from each focal point eventually results in consolidation of the curved PPB (Figs 17.4a, b and 17.5). This mode of PPB development suggests that the band is formed of two sets of antiparallel MTs.

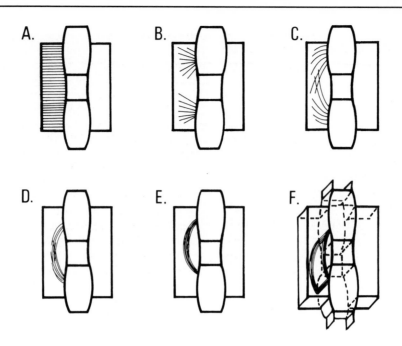

Fig. 17.5. Diagrammatic representation of PPB development in grass SMCs. Transverse interphase cortical microtubules (A) are replaced by two radiating arrays of microtubules in the paradermal cortex (B). Interaction of microtubules within and between these arrays (C, D) appears to give rise to the curved, narrow band of microtubules visible in paradermal view (E). Focusing through the cell indicates that MTs in the paradermal cortex near the leaf surface are aligned like those in the paradermal cortex facing the leaf interior. F shows the three-dimensional form of the mature PPB. Details of how the anticlinal portions of the SMC PPB are formed are not yet available.

Timing of PPB appearance

The temporal aspects of PPB formation have been investigated by Mineyuki *et al.* (1988b) in onion and by Gunning and Sammut (1990) in wheat. In root tip cells of normally grown onion seedlings, all cells with a PPB are found to have a G2 level of DNA, indicating that they have finished the S phase of the cell cycle. Pulse-chase experiments reveal that some cells produce a PPB very shortly after finishing S, although not all cells may follow this timetable. The PPB in some wheat root tip cells likewise can begin to form at the end of S-phase. With increasing time after S, one begins to see cells with a narrower, more mature PPB and finally the stage at which PPB MTs begin to disappear (Gunning and Sammut, 1990). If onion seedlings are exposed to inhibitors of DNA synthesis, the PPB and nuclear cycles can be partially disengaged, and some cells form a PPB before finishing S phase (Mineyuki *et al.*, 1988b); however, in cultured tobacco cells, one

of the same inhibitors hinders PPB formation as well as progression of the nuclear cycle (Katsuta *et al.*, 1990). According to the breakdown of the onion root tip cell cycle given by González-Fernández *et al.* (1971), G2 occupies about 17% of total cell cycle time, mitosis/cytokinesis takes about 12% of the cycle, and nearly half of mitosis is spent in prophase. If the PPB generally forms early in G2 and persists through at least part of prophase, as it does in several species, it is not surprising that cells with PPBs are so readily found in meristematic tissues.

Another aspect of PPB timing has been examined in cells that had left the division cycle but have been induced to divide again in response to wounding. Pea root cells can establish a PPB within about a day of wounding (Hush *et al.*, 1990); *Nautilocalyx* epidermal cells may take three to four days to establish a phragmosome and PPB, but if a second wound is inflicted before division, reorientation of the division plane, visualized as formation of a phragmosome parallel to the new wound, can occur in as little as seven hours (Venverloo, 1990).

Actin in the PPB

Several studies note a band of actin filaments co-aligned with the PPB of MTs, both in meristematic cells (Palevitz, 1987; McCurdy and Gunning, 1990; Mineyuki and Palevitz, 1990) and in cultured cells (Kakimoto and Shibaoka, 1987; Traas et al., 1987; Lloyd and Traas, 1988; Katsuta et al., 1990). Realignment of cortical actin, first in a wide band and then in a narrower band corresponding to the phragmosome (and thus, presumably, coincident with the PPB), is an early response to wounding in leaf epidermis of Tradescantia (Goodbody and Lloyd, 1990). However, actin has not been found associated with PPBs of grass leaf epidermal cells, although cortical actin can be visualized in these cells (Cho and Wick, 1990, 1991). When present, the actin band is usually wider than the MT band, and the degree of banding varies considerably from one experimental system to another. PPB MTs must be intact in order for the actin band to form (Katsuta et al., 1990; McCurdy and Gunning, 1990; Mineyuki and Palevitz, 1990). Integrity of actin is not required for formation of the PPB (Cho and Wick, 1990; Katsuta et al., 1990; Mineyuki and Palevitz, 1990), but recent evidence indicates that in root tip cells, actin is required for normal narrowing of the PPB (Mineyuki and Palevitz, 1990). Clearly, much more needs to be done to ascertain the universality of the presence of actin in the PPB, and to determine its role in establishment of the division site.

Acknowledgements

I thank my past and current colleagues Jadwiga Duniec, Brian Gunning, Yoshi Mineyuki, and Soon-Ok Cho for challenging and stimulating discussions. Soon-Ok Cho and Kris Kirkeby helped with preparation of Fig. 17.5. The writing of this chapter was supported in part by USDA grant 89–37261–4910 and NSF DCB/8918547.

References

Apostolakos, P. and Galatis, B. (1985a). Studies on the development of the air pores and air chambers of Marchantia paleacea. III. Microtubule organization in preprophase–prophase initial aperture cells – formation of incomplete preprophase microtubule bands. Protoplasma 128, 120–135.

Apostolakos, P. and Galatis, B. (1985b). Studies on the development of the air pores and air chambers of Marchantia paleacea. IV. Cell plate arrangement in initial aperture cells. Protoplasma 128, 136–146.

Apostolakos, P. and Galatis, B. (1987). Induction, polarity and spatial control of cytokinesis in some abnormal subsidiary cell mother cells of Zea mays. Protoplasma 140, 26–42.

Apostolakos, P., Galatis, B., Katsaros, C. and Schnepf, E. (1990). Tubulin conformation in microtubule-free cells of Vigna sinensis. An immunofluorescent and electron microscope study. Protoplasma 154, 132–143.

Bakhuizen, R., van Spronsen, P.C., Sluiman-den Hertog, F.A.J., Venverloo, C.J. and Goosen-de Roo, L. (1985). Nuclear envelope radiating microtubules in plant cells during interphase mitosis transition. Protoplasma 128, 43–51.

Brown, R.C. and Lemmon, B.E. (1984). Plastid apportionment and preprophase microtubule bands in monoplastidic root meristem cells of Isoetes and Selaginella. Protoplasma 123, 95–103.

Brown, R.C. and Lemmon, B.E. (1988). Preprophasic microtubule systems and development of the mitotic spindle in hornworts (Bryophyta). Protoplasma 143, 11–21.

Brown, R.C. and Lemmon, B.E. (1989a). Minispindles and cytoplasmic domains in microsporogenesis of orchids. Protoplasma 148, 26–32.

Brown, R.C. and Lemmon, B.E. (1989b). Morphogenetic plastid migration and microtubule organization during megasporogenesis in Isoetes. Protoplasma 152, 136–147.

Brown, R.C. and Lemmon, B.E. (1990). Polar organizers mark division axis prior to preprophase band formation in mitosis of the hepatic Reboulia hemisphaerica (Bryophyta). Protoplasma 156, 74–81.

Busby, C.H. and Gunning, B.E.S. (1980). Observations on pre-prophase bands of microtubules in uniseriate hairs, stomatal complexes of sugar cane, and Cyperus root meristems. Eur. J. Cell Biol. 21, 214–223.

Busby, C.H. and Gunning, B.E.S. (1988). Establishment of plastid-based quadripolarity in spore mother cells of the moss Funaria hygrometrica. J. Cell Sci. 91, 117–126.

Cho, S.-O. and Wick, S.M. (1989). Microtubule orientation during stomatal differentiation in grasses. *J. Cell Sci.* **92**, 581–594.

Cho, S.-O. and Wick, S.M. (1990). Distribution and function of actin in the developing stomatal complex of winter rye (*Secale cereale* cv. Puma). *Protoplasma* **157**, 154–164.

Cho, S.-O. and Wick, S.M. (1991). Actin in the developing stomatal complex of winter rye: a comparison of actin antibodies and Rh-phalloidin labeling of control and CB-treated tissues. *Cell Motil. Cytoskeleton* **19**, 25–36.

Clayton, L. and Lloyd, C.W. (1984). The relationship between the division plane and spindle geometry in *Allium* cells treated with CIPC and griseofulvin: an anti-tubulin study. *Eur. J. Cell Biol.* **34**, 248–253.

Cleary, A.L. and Hardham, A.R. (1989). Microtubule organization during development of stomatal complexes in *Lolium rigidum*. *Protoplasma* **149**, 67–81.

Doonan, J.H., Cove, D.J., Corke, F.M.K. and Lloyd, C.W. (1987). Pre-prophase band of microtubules, absent from tip-growing moss filaments, arises in leafy shoots during transition to intercalary growth. *Cell Motil. Cytoskeleton* **7**, 138–153.

Eleftheriou, E.P. (1985a). Abundance of microtubules in preprophase bands of some *Triticum* species. *Planta* **163**, 175–182.

Eleftheriou, E.P. (1985b). Microtubules and root protophloem ontogeny in wheat. *J. Cell Sci.* **75**, 165–179.

Eleftheriou, E.P. and Tsekos, I. (1982). Development of protophloem in roots of *Aegilops comosa* var. *thessalica*. I. Differential divisions and preprophase bands of microtubules. *Protoplasma* **113**, 110–119.

Falconer, M.M. and Seagull, R.W. (1985). Immunofluorescent and Calcofluor white staining of developing tracheary elements in *Zinnia elegans* L. suspension cultures. *Protoplasma* **125**, 190–198.

Flanders, D.J., Rawlins, D.J., Shaw, P.J. and Lloyd, C.W. (1990). Nucleus-associated microtubules help determine the division plane of plant epidermal cells: avoidance of four-way junctions and the role of cell geometry. *J. Cell Biol.* **110**, 1111–1122.

Galatis, B. (1982). The organization of microtubules in guard cell mother cells of *Zea mays*. *Can. J. Bot.* **60**, 1148–1166.

Galatis, B. and Mitrakos, K. (1979). On the differential divisions and preprophase microtubule bands involved in the development of stomata of *Vigna sinensis* L. *J. Cell Sci.* **37**, 11–37.

Galatis, B., Apostolakos, P., Katsaros, C. and Loukari, H. (1982). Pre-prophase microtubule band and local wall thickening in guard cell mother cells of some Leguminosae. *Ann. Bot.* **50**, 779–791.

Galatis, B., Apostolakos, P. and Katsaros, C. (1983). Synchronous organization of two preprophase microtubule bands and final cell plate arrangement in subsidiary cell mother cells of some *Triticum* species. *Protoplasma* **117**, 24–39.

Galatis, B., Apostolakos, P. and Katsaros, C. (1984a). Positional inconsistency between preprophase microtubule band and final cell plate arrangement during triangular subsidiary cell and atypical hair cell formation in two *Triticum* species. *Can. J. Bot.* **62**, 343–359.

Galatis, B., Apostolakos, P. and Katsaros, C. (1984b). Experimental studies on the function of the cortical cytoplasmic zone of the preprophase microtubule band. *Protoplasma* **122**, 11–26.

Gambardella, R. and Alfano, F. (1990). Monoplastidic mitosis in the moss *Timmiella barbuloides* (Bryophyte). *Protoplasma* **156**, 29–38.

González-Fernández, A., Giménez-Martín, G. and de la Torre, C. (1971). The duration of the interphase periods at different temperatures in root tip cells. *Cytobiologie* **3**, 367–371.

Goodbody, K.C. and Lloyd, C.W. (1990). Actin filaments line up across *Tradescantia* epidermal cells, anticipating wound-induced division planes. *Protoplasma* **157**, 92–101.

Gorst, J., Wernicke, W. and Gunning, B.E.S. (1986). Is the preprophase band of microtubules a marker of organization in suspension cultures? *Protoplasma* **134**, 130–140.

Gunning, B.E.S. (1982). The cytokinetic apparatus: its development and spatial regulation. In *The Cytoskeleton in Plant Growth and Development* (ed. C.W. Lloyd), pp. 229–292. Academic Press, London.

Gunning, B.E.S. and Wick, S.M. (1985). Preprophase bands, phragmoplasts, and spatial control of cytokinesis. *J. Cell Sci. Suppl.* **2**, 157–179.

Gunning, B.E.S. and Sammut, M. (1990). Rearrangements of microtubules involved in establishing cell division planes start immediately after DNA synthesis and are completed just before mitosis. *Plant Cell* **2**, 1273–1282.

Gunning, B.E.S., Hardham, A.R. and Hughes, J.E. (1978). Preprophase bands of microtubules in all categories of formative and proliferative cell division in *Azolla* roots. *Planta* **143**, 145–160.

Hogan, C.J. (1988). Preprophase bands in a suspension culture of the monocot *Spartina pectinata*. *Exp. Cell Res.* **175**, 216–222.

Hush, J.M., Hawes, C.R. and Overall, R.L. (1990).

Interphase microtubule re-orientation predicts a new cell polarity in wounded pea roots. *J. Cell Sci.* **96**, 47–61.

Jenni, V., Cattelan, H. and Roos, U.-P. (1990). Immunofluorescence and ultrastructure of mitosis and cell division in the fern *Athrium filix-femina*. *Bot. Helv.* **100**, 101–119.

Kakimoto, T. and Shibaoka, H. (1987). Actin filaments and microtubules in the preprophase band and phragmoplasts of tobacco cells. *Protoplasma* **140**, 151–156.

Katsuta, J., Hashiguchi, Y. and Shibaoka, H. (1990). The role of the cytoskeleton in positioning of the nucleus in premitotic tobacco BY-2 cells. *J. Cell Sci.* **95**, 413–422.

Lloyd, C.W. and Traas, J.A. (1988). The role of F-actin in determining the division plane of carrot suspension cells. Drug studies. *Development* **102**, 211–221.

McCurdy, D.W. and Gunning, B.E.S. (1990). Reorganization of cortical actin microfilaments and microtubules at preprophase and mitosis in wheat root-tip cells: a double label immunofluorescence study. *Cell Motil. Cytoskeleton* **15**, 76–87.

Mineyuki, Y. and Gunning, B.E.S. (1990). A role for preprophase bands of microtubules in maturation of new cell walls, and a general proposal on the function of preprophase band sites in cell division in higher plants. *J. Cell Sci.* **97**, 527–537.

Mineyuki, Y. and Palevitz, B.A. (1990). Relationship between preprophase band organization, F-actin and the division site in *Allium*: Fluorescence and morphometric studies on cytochalasin-treated cells. *J. Cell Sci.* **97**, 283–295.

Mineyuki, Y., Takagi, M. and Furuya, M. (1984). Changes in organelle movement in the nuclear region during the cell cycle of *Adiantum* protonema. *Plant Cell Physiol.* **25**, 297–308.

Mineyuki, Y., Marc, J. and Palevitz, B.A. (1988a). Formation of the oblique spindle in dividing guard mother cells of *Allium*. *Protoplasma* **147**, 200–203.

Mineyuki, Y., Wick, S.M. and Gunning, B.E.S. (1988b). Preprophase bands of microtubules and the cell cycle: kinetics and experimental uncoupling of their formation from the nuclear cycle in onion root-tip cells. *Planta* **174**, 518–526.

Mineyuki, Y., Marc, J. and Palevitz, B.A. (1989). Development of the preprophase band from random cytoplasmic microtubules in guard mother cells of *Allium cepa* L. *Planta* **178**, 291–296.

Mullinax, J.B. and Palevitz, B.A. (1989). Microtubule reorganization accompanying preprophase band formation in guard mother cells of *Avena sativa* L. *Protoplasma* **149**, 89–94.

Murata, T. and Wada, M. (1989). Re-organization of microtubules during preprophase band development in *Adiantum* protonemata. *Protoplasm* **151**, 73–80.

Ôta, T. (1961). The role of cytoplasm in cytokinesis of plant cells. *Cytologia* **26**, 428–447.

Palevitz, B.A. (1987). Actin in the preprophase band of *Allium cepa*. *J. Cell Biol.* **104**, 1515–1519.

Palevitz, B.A. and Cresti, M. (1989). Cytoskeletal changes during generative cell division and sperm formation in *Tradescantia virginiana*. *Protoplasma* **150**, 54–71.

Pickett-Heaps, J.D. (1969a). Preprophase microtubules and stomatal differentiation; some effects of centrifugation on symmetrical and asymmetrical cell division. *J. Ultrastruct. Res.* **27**, 24–44.

Pickett-Heaps, J.D. (1969b). Preprophase microtubule bands in some abnormal mitotic cells of wheat. *J. Cell Sci.* **4**, 397–420.

Pickett-Heaps, J.D. (1969c). Preprophase microtubules and stomatal differentiation in *Commelina cyanea*. *Aust. J. Biol. Sci.* **22**, 375–391.

Pickett-Heaps, J.D. and Northcote, D.H. (1966a). Organization of microtubules and endoplasmic reticulum during mitosis and cytokinesis in wheat meristems. *J. Cell Sci.* **1**, 109–120.

Pickett-Heaps, J.D. and Northcote, D.H. (1966b). Cell division in the formation of the stomatal complex of the young leaves of wheat. *J. Cell Sci.* **1**, 121–128.

Sack, F.D. and Paolillo, D.J. Jr (1985). Incomplete cytokinesis in *Funaria* stomata. *Amer. J. Bot.* **72**, 1325–1333.

Schmiedel, G., Reiss, H.-D. and Schnepf, E. (1981). Associations between membranes and microtubules during mitosis and cytokinesis in caulonema tip cells of the moss *Funaria hygrometrica*. *Protoplasma* **108**, 173–190.

Simmonds, D.H. (1986). Prophase bands of microtubules occur in protoplast cultures of *Vicia hajastana* Grossh. *Planta* **167**, 469–472.

Staiger, C.J. and Cande, W.Z. (1990). Microtubule distribution in *dv*, a maize meiotic mutant defective in the prophase to metaphase transition. *Dev. Biol.* **138**, 231–242.

Tiwari, S.C., Wick, S.M., Williamson, R.E. and Gunning, B.E.S. (1984). Cytoskeleton and integration of cellular function in cells of higher plants. *J. Cell Biol.* **99**, 63s–69s.

Traas, J.A., Doonan, J.H., Rawlins, D.J., Shaw, P.J., Watts, J. and Lloyd, C.W. (1987). An actin network is present in the cytoplasm throughout the cell cycle of carrot cells and associates with the dividing nucleus. *J. Cell Biol.* **105**, 387–395.

Traas, J.A., Renaudin, J.P., Teyssendier de la Serve, B. (1990). Changes in microtubular organization mark the transition to organized growth during organogenesis in *Petunia hybrida*. *Plant Sci.* **68**, 249–256.

Venverloo, C.J. (1990). Regulation of the plane of cell division in vacuolated cells. II. Wound-induced changes. *Protoplasma* **155**, 85–94.

Venverloo, C.J. and Libbenga, K.R. (1987). Regulation of the plane of cell division in vacuolated cells. I. The function of nuclear positioning and phragmosome formation. *J. Plant Physiol.* **131**, 267–284.

Wada, M., Mineyuki, Y., Kadota, A. and Furuya, M. (1980). The changes of nuclear position and distribution of circumferentially aligned cortical microtubules during the progression of cell cycle in *Adiantum* protonemata. *Bot. Mag. Tokyo* **93**, 237–245.

Wang, H., Cutler, A.J. and Fowke, L.C. (1989a). High frequencies of preprophase bands in soybean protoplast cultures. *J. Cell Sci.* **92**, 575–580.

Wang, H., Cutler, A.J. and Fowke, L.C. (1989b). Preprophase bands in cultured multinucleate soybean protoplasts. *Protoplasma* **150**, 110–116.

Wick, S.M. and Duniec, J. (1983). Immunofluorescence microscopy of tubulin and microtubule arrays in plant cells. I. Preprophase band development and concomitant appearance of nuclear envelope-associated tubulin. *J. Cell Biol.* **97**, 235–243.

Wick, S.M. and Duniec, J. (1984). Immunofluorescence microscopy of tubulin and microtubule arrays in plant cells. II. Transition between the pre-prophase band and the mitotic spindle. *Protoplasma* **122**, 45–55.

Wick, S.M., Muto, S. and Duniec, J. (1985). Double immunofluorescence labeling of calmodulin and tubulin in dividing plant cells. *Protoplasma* **126**, 198–206.

Wick, S.M., Cho. S.-O. and Mundelius, A.R. (1989). Microtubule deployment within plant tissues: fluorescence studies of sheets of intact mesophyll and epidermal cells. *Cell Biol. Int. Rep.* **13**, 95–106.

18. CYTOSKELETAL ELEMENTS OF THE PHRAGMOSOME ESTABLISH THE DIVISION PLANE IN VACUOLATED HIGHER PLANT CELLS

Clive W. Lloyd

Department of Cell Biology, John Innes Institute,
John Innes Centre for Plant Science Research,
Colney Lane, Norwich NR4 7UH, UK

Sinnott and Bloch frequently stated that students of cell division neglected vacuolated cells to concentrate instead on more 'typical', densely cytoplasmic meristematic cells. The latter are certainly more convenient to fix, section, make cell squashes and there is also a higher likelihood of encountering division figures. However, large vacuolated cells can be induced to divide by wounding the tissue, and the great advantage they then possess is that the nucleus migrates into the centre of the cell to divide, suspended by transvacuolar strands that, unlike the situation in meristematic cells, are openly and visibly involved in anchoring the nucleus. Furthermore, Sinnott and Bloch (1940) observed that these strands gradually aggregated, anastomosing into a diaphragm which occupied the position of the future wall. Since this phragmosome was established from early prophase, long before the cell plate formed within it, the phragmosome effected and predicted the division plane rather than anything associated with the 'rolling nucleus'. Later electron microscopic work on the equally important and predictive preprophase band of microtubules (MTs) diverted attention to the cortex, and transvacuolar elements were largely neglected. Consequently, there is comparatively little literature on the phragmosome to review. However, in the latter part of the 1980s, fluorescence microscopy studies on vacuolated cells established that microtubular and actin filaments help tether the premitotic nucleus and construct the phragmosome. A satisfactory first description can now be pieced together for how the cytoskeletal elements participate in this critical process. The expectation is that mechanisms visible in vacuolated cells will prove to be relevant to division plane alignment in other higher plant cells.

The cytological observations of Sinnott and Bloch

In their 1940 study of cytokinesis in large, strongly vacuolated cells Sinnott and Bloch stated that the plane of the next division, and the exact position of the future cell wall, are indicated by the distribution of cytoplasm much earlier than the plane can be deduced from nuclear orientation. In normal tissue, the cross walls between cells in a file are off-set to avoid aligning with cross walls in neighbouring files. After wounding, however, new division walls are deposited parallel to the wound and are not usually off-set. Division planes in such tissue are therefore predictable. Nuclei were observed by Sinnott and Bloch to migrate into the centre of the vacuole on transvacuolar strands, and by prophase these cytoplasmic strands had begun to fuse. Fusion of strands produced a more or less continuous phragmosome which anticipated the future division

The Cytoskeletal Basis of Plant Growth and Form   ISBN 0–12–453770–7

plane by aligning parallel to the wound. At this stage other single strands not in the phragmosomal plane could be seen to pass up and down from the nucleus, at right angles to the future division plane. They termed this configuration (transverse phragmosome and perpendicular apical/basal strands) a Maltese cross. However, the terminology is not entirely appropriate since the phragmosome defines a two-dimensional sheet whilst the apical/basal elements are one-dimensional strands. At a later stage, the phragmoplast was observed to grow out within the phragmosome, depositing the cell plate which eventually fused at the point where the cytoplasmic diaphragm had already joined the mother wall prior to division. The significance of the phragmosome, therefore, is that it is a visible expression of polarity, expressed premitotically.

In later work (1941a) this general method of division of vacuolated cells in secondary, wound meristems was found to be essentially identical throughout a range of plants and cell types. In vacuolated dividing cells of elongating shoot tips, the phragmosome was seen to bifurcate where its path would have otherwise caused it to align with a neighbouring cross wall. In traumatic tissue, phragmosomes need not, however, exhibit such avoidance (Sinnott and Bloch, 1941b). Normal plant tissue in section can resemble a foam or a honeycomb, the key feature of which is that only three walls in a plane meet at a point. The quasi-hexagonal packing of un-wounded plant cells depends upon a mechanism whereby the premitotic transvacuolar strands that contribute to the phragmosome avoid contacting an existing three-way junction. This problem is at the heart of cell and tissue geometry. Division plane alignment affects not only the shape of daughter cells but the relative positions of the walls in multicellular tissue. Placement of cross walls had been suggested to be influenced by physical forces much as liquid films respond to surface tension. The key insight of Sinnott and Bloch is that they recognized that wall alignment required a study of the forces acting on the phragmosome from the beginning and not just upon the growing wall that followed it.

Electron microscopy of the preprophase band of MTs in vacuolated cells

The first indication that the cytoskeleton is involved in the division plane was Pickett-Heaps and Northcote's (1966) discovery of the preprophase band (PPB) of MTs in young wheat leaf cells. This circumferential band accurately forecasts where the cell plate will fuse with the maternal wall, but disappears by metaphase. The predictive ability of the PPB has subsequently been confirmed for a wide variety of divisions: symmetrical, asymmetrical, curved, formative (i.e. a division giving rise to a new cell type), proliferative, unprogrammed (i.e. quiescent cells stimulated to divide by wounding) or programmed (reviewed in Gunning, 1982). There has been a considerable amount of work devoted to the PPB, almost all of it performed on densely cytoplasmic cells in which a phragmosome is not visible. As a result, attention focused upon the properties of the cortex in determining the division plane. The work of the Leiden laboratory showed, however, that the phragmosome and the PPB are not unrelated. In leaf explants of *Nautilocalyx lynchii*, the process of explantation stimulates the large, vacuolated epidermal cells to divide, chiefly periclinally, and the nucleus migrates along transvacuolar strands into the centre of the cell, 10–20 h before metaphase (Venverloo et al., 1980). Phragmosome formation (i.e. the aggregation of strands) starts some hours after nuclear migration but at least 2 h prior to the disappearance of the nucleolus. Electron microscopic studies showed that a band of MTs (i.e. the PPB) occurred at the junction between phragmosome and cortex. In wounded pea tissue with precisely aligned divisions in adjacent cells, Hardham and McCully (1982) also saw bands of MTs at the cortex of the predicted division planes (although no phragmosomes were reported). Other electron microscopy work from the Leiden laboratory indicated that phragmosomes themselves contain MTs (and bundles of microfilaments) (Goosen-de Roo et al., 1984). Bakhuizen et al. (1985) proposed that MTs which radiated from the nucleus into the transvacuolar strands had a role in mobilizing the nucleus, then stabilizing it in the division plane. This established that both phragmosome and PPB contained MTs although it was not clear how (if at all) the two devices were related, nor which of the two formed first. In a later drug study on leaf explants of *N. lynchii* Venverloo and Libbenga (1987) concluded that the two structures formed concomitantly. They also found that anti-MT drugs (colchicine, oryzalin), or the anti-actin agent cytochalasin B (CB), inhibited phragmosome formation and nuclear positioning, implying that actin and MTs played a part in the transvacuolar aspect of division plane establishment.

Fluorescent phallotoxins locate actin filaments in the division plane

Fluorescent phallotoxins have played an important part in clarifying the distribution of actin filaments during the plant cell's cycle. Actin filaments have been known for some time to exist in interphase cells but it is now clear that they do not disappear from cells at the onset of division, as was once thought. In onion root tip cells, F-actin can be stained in the phragmoplast (Clayton and Lloyd, 1985), and in the preprophase band (Palevitz, 1987). During mitosis, too, F-actin cages the nucleus and occurs within the higher plant cell's nucleus (see Lambert *et al.*, Chapter 15, this volume).

One of the reasons why the discovery of different F-actin arrays has been episodic could be the sensitivity of some classes of filaments, in some cell types, to aldehyde fixation; hence actin may not always be detectable under standard conditions used for preparing cells for MT immunofluorescence. However, because of their low molecular weight, fluorescent phallotoxins can be introduced into unfixed, walled plant cells along with low levels of detergent and dimethyl sulphoxide (DMSO). This non-fixation method demonstrated that a complex network of cytoplasmic actin filaments and cables was present throughout the cell cycle of carrot suspension cells (Traas *et al.*, 1987). This was seen when cells were extracted first, before adding rhodamine–phalloidin (RhPh); it was

seen when RhPh was added at lower concentrations (10^{-8} M) than known to affect actin polymerization; the networks were also seen when RhPh was introduced into cells by electroporation. In the latter method, pulses of direct current induce transient holes to open in the plasma membrane, allowing RhPh to enter the cell without the need for DMSO and detergent, although both methods yielded identical observations. Using aldehydes as fixatives, only relatively thick cables of F-actin could be seen at the cortex of carrot cells, as well as cables that ran deeper in the cytoplasm and in transvacuolar strands. But the non-fixation methods also showed that a very fine class of actin filaments occurred at the cell cortex, transverse to the cell's long axis, parallel to the cortical MTs. During mitosis, in carrot suspension cells, most of the large actin cables were seen to disappear but a meshwork of bundles remained in the cytoplasm. During preprophase, actin filaments co-distributed with the MTs of the PPB but in addition, these filaments connected with other actin strands outside the division plane. During metaphase, the nucleus could be seen to be caged by actin filaments, the spindle itself contained F-actin, and actin strands connected the nucleus to the cortex. Similarly during cytokinesis, the RhPh-stained phragmoplast was not seen as an isolated structure but to be continuous with other cytoplasmic F-actin-containing strands that ran through the cell.

This actin-based view of the cell cycle is therefore different in several significant respects to that of the established MT cycle. A key feature is that the central mitotic and cytokinetic apparatus are not free-floating but connected to a system of transvacuolar and cortical actin cables that remains throughout division. The MTs and F-actin of the PPB disappear by metaphase, but actin cables remain in the division plane to tether the nucleus. Employing a different approach in which lysine pretreatment was used to stabilize F-actin, Kakimoto and Shibaoka (1987) also reported that actin filaments paralleled the MTs of the PPB in BY-2 tobacco suspension cells. Actin filaments were also seen between the central phragmoplast and the cell cortex. A subsequent study on carrot suspension cells (Lloyd and Traas, 1988) concentrated upon actin in the division plane. At metaphase two classes of nucleus-associated actin filaments could be discerned: filaments between nucleus and cortex, transverse to the cell's long axis (i.e. the phragmosome) and other filaments that radiate from the spindle poles (cf. the apical–basal strands of Sinnott and Bloch). Cytokinesis in carrot suspension cells is often eccentric, producing a gap within which it was to be seen that actin strands connect the outgrowing phragmoplast to the cortical division site.

The significance, therefore, of the phragmosomal actin filaments, which radiate between nucleus and cortex throughout division, is that they, rather than any cortical structure, provide the memory of a division plane determined in the premitotic phase. The PPB and the phragmoplast never coincide in time, but the phragmosomal actin records the plane they both occupy, guiding the cytokinetic apparatus out to the former PPB site.

A model for the interaction between actin and MTs in establishing the division plane has been proposed (Lloyd and Traas, 1988; see also Lloyd, 1989). In this, the central, premitotic nucleus is tethered by actin strands that span the vacuole and when at the cortex, some of the finer filaments parallel the cortical microtubules. Bunching of the cortical MTs to form the PPB is suggested to sweep the cortical actin

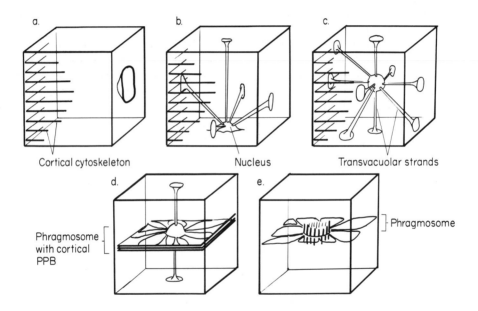

Fig. 18.1. Model for the involvement of the cytoskeleton in division plane formation. a, In the quiescent, vacuolated cell, the nucleus is contained in the cortical cytoplasm. The interphase cortical cytoskeleton (only partially represented) contains MTs and actin, with thicker actin cables (not shown) in the sub-cortex. b, In response to wounding, the nucleus migrates towards the wound. The nucleus is conspicuously associated with actin filaments which appear to be essential for this traumatotactic (phase I) migration. c, In preparation for mitosis, the nucleus migrates (phase II) into the centre of the vacuole suspended by actin-containing strands. The strands also contain nucleus-associated microtubules. d, The cortical cytoskeleton becomes reorganized to form a preprophase band containing MTs and actin filaments. As part of this process, most of the strands radiating from the nucleus concomitantly realign so that they come to lie in the division plane. These fused strands constituting the phragmosome contain actin filaments and MTs. e, By metaphase, the cytoplasmic MTs have depolymerized, leaving phragmosomal actin filaments to mark the division plane. These filaments continue out into the cortex. Later the phragmoplast expands within the phragmosome until the cell plate fuses with the side walls.

into the band and to re-align the nucleus-radiating strands into the plane circumscribed by the PPB. Actin strands not reorganized into the PPB remain between the nucleus and end walls, constituting the apical–basal (or polar plasmic) strands perpendicular to the phragmosome. Subsequently, however, microtubules have been seen in vacuolated cells to form a continuous link between the nucleus and the PPB and the model is updated to accommodate this (see Fig. 18.1).

Microtubules in the division plane

Electron microscopic studies of dividing vacuolated cells indicated that the phragmosome contains microtubules. This was observed by Goosen-de Roo et al. (1984) for long narrow cambial cells of the ash, Fraxinus excelsior L., and by Bakhuizen et al. (1985) for large highly vacuolated Nautilocalyx lynchii cells. Microtubules radiated from the nuclear envelope into the transvacuolar strands. In small, non-vacuolated cells of Pisum sativum, microtubules were also observed between the nuclear envelope and the PPB, suggesting that this potentially new class of MTs was not peculiar to vacuolated cells. Despite the difficulty of routinely observing elements between nucleus and cortex, in cells where that gap is narrow, immunofluorescence observations of favourably presented cells have provided occasional hints that MTs might actually bridge nucleus and PPB (see Palevitz, Chapter 4; and Wick, Chapter 17, this volume). Similarly, although it is difficult to establish that the fragmentary MTs observed in EM studies are actually continuous, such data have encouraged the idea that the premitotic nucleus is anchored to the cortex – and to the PPB zone in particular – by MTs.

In a recent study, Flanders et al. (1990) performed optical sectioning and computer-aided image reconstruction of anti-tubulin-labelled Datura stramonium epidermal cells to confirm that hypothesis. The gap between nucleus and cortex in these large, vacuolated cells, is 10 μm or more, allowing it to be seen unambiguously that the central nucleus is anchored by transvacuolar strands containing thick bundles of MTs. Figure 18.2 presents a stereopair of reconstituted confocal microscope sections through a Nautilocalyx lynchii epidermal cell. This illustrates the periclinal PPB, stained with anti-tubulin, and the MTs that radiate to it from the central nucleus. Katsuta et al. (1990) have also convincingly shown for BY-2 tobacco suspension cells that MTs connect the nuclear surface to the PPB. In elongated Datura stem epidermal cells possessing broad, probably still-forming PPBs, the MT bundles radiate from the nucleus: most of the bundles contact the cortex within a broad zone defined by the broad PPB, whilst others pass away, perpendicularly, towards the end walls. Where the PPB is tight (i.e. mature) the radiating strands are equally tightly confined to the same plane, with only the polar strands passing out of the phragmosomal plane. This applies for both transversely and longitudinally-dividing cells. Instead of describing it as a two-dimensional Maltese cross, this conformation may be better likened to a wheel, with the nucleus at the hub, the PPB as the rim, the radiating MTs forming the spokes, with the polar strands forming the axle. Because broad PPBs become tighter as chromatin becomes more condensed it is possible to see this as a temporal progression: as the cortical MTs 'bunch up', the nucleus-associated strands adopt shallower angles to the transverse division plane as they are drawn into the phragmosome (see Figs 18.1 and 18.3). According to this interpretation, the two structures – PPB and phragmosome – form simultaneously. The nucleus-associated MTs are almost certainly newly-polymerized for they are not seen in premitotic cells. Since they connect with the PPB in forming the division plane, it follows that the PPB itself contains new MTs and is not merely formed by a congregation of old interphase MTs. If this proves to be the case it would be formally possible for a new PPB to be superimposed, at divergent angles, to the existing cortical MT array. But since this has not been convincingly demonstrated to occur, it would appear that when they contact the cortex, nascent nucleus-associated MTs intercalate and share the direction of existing cortical MTs as the latter reorganize.

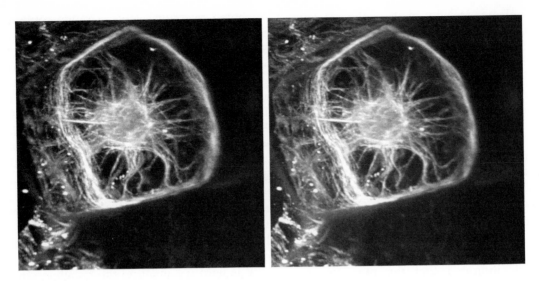

Fig. 18.2. Stereopair of the microtubules within a periclinal phragmosome of a *Nautilocalyx lynchii* epidermal cell. The large epidermal cells (approximately 50 μm diameter) in an explant divide parallel to the cut surface. Here, an anti-tubulin labelled phragmosome has been optically sectioned by confocal laser microscopy and then the stack of images reconstructed by computer methods. In this periclinal phragmosome, the nucleus is clearly connected to cortical PPB microtubules by radial MT bundles. (From K.C. Goodbody, C.J. Venverloo and C.W. Lloyd, in preparation.)

It is interesting that the nucleus-associated MTs do not immediately grow out along a predetermined plane, but that the phragmosomal plane is gradually negotiated or 'hunted' within a broad PPB zone. This prompts two major questions. What powers this cytoskeletal rearrangement, microtubular or actin-based motors? What factors influence the precise alignment of the phragmosomal strands?

In elongated cells, strands running to the end walls, or out of the greater zone within which the PPB forms, remain to constitute the polar strands, roughly perpendicular to the phragmosome. Most other strands appear to form the phragmosome. An interesting exception, noted by Sinnott and Bloch (1941b), is that where the phragmosome would otherwise have co-aligned with the cross wall between neighbouring cells, it splits (see Fig. 18.3). Flanders *et al.* (1990) report such behaviour for MTs: both the nucleus-radiating MT strands and the PPB itself, split to avoid contacting existing three-way wall junctions.

Apart from re-emphasizing the relationship between the cytoplasmic strands and the place-

ment of the PPB, we should ask why the wall on the other side of a junction with a cross wall becomes a prohibited area for the cytoskeleton as the division plane forms. During interphase, cortical MTs themselves exhibit no great tendency to avoid such regions. One explanation, based more on behaviour of the premitotic transvacuolar strands, was proposed by Flanders *et al.* (1990). The patterns eventually adopted by the nucleus-associated strands could be mimicked by the patterns of soap bubble walls held in a flexible hexagonal frame, or by springs radiating from a central disc, their other ends free to slide by means of Teflon collars along the edge of a hexagonal frame. By distorting the frame into different shapes, such analogue models demonstrate the general point that elements under tension adopt a minimal path if free to move. Such paths tend to be perpendicular to side walls, tend to be the shortest route and therefore avoid the corners of the frame which, as distant points from the centre, represent longest routes. When the frames are made into rectangular hexagons, the introduction of only a

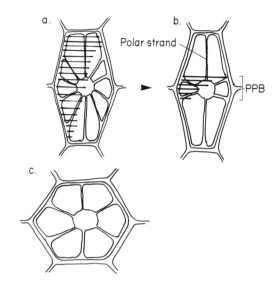

Fig. 18.3. Radial, nucleus-associated strands in *Datura stramonium* epidermal cells. This summarizes the image reconstruction observations of Flanders *et al.* (1990) on cells of differing shape. In elongated stem epidermal cells (a) bundles of MTs radiate from the central nucleus. In such long cells, the PPBs are initially broad (20–30 µm) but become tighter (transition to b). The radial strands become reorganized as the cortical band tightens until most of these transvacuolar MT bundles eventually accumulate within a transverse cytoplasmic bar (that is, the phragmosome). Strands connected to the end walls appear not to participate in this realigning process involving the side walls and remain as perpendicular, polar strands. The division plane is therefore transverse to the cell's long axis. Cytoplasmic strands split rather than contact existing three-way junctions. In the isodiametric cells (c) it was observed that the radial MT strands did not reorganize into a dominant transverse bar encircled by the condensing PPB. PPB formation in such cells was unusual in that bands were broad and some were angled, like the hands of a clock. Such cells divided at variable angles to the stem axis. (Reproduced from Lloyd, 1991.)

slight convex angle between two long sides causes bubble walls or springs to move to avoid this vertex; the elements split either side of the angle, much as the phragmosomal strands split to avoid contacting a three-way junction. This suggests that in elongated cells, strands radiating from the nucleus will (pro-

vided they are under tension and free to move relative to the cortex) tend to move to a position(s) transverse to the long axis. During early preprophase, however, when the band is broad and still forming, microtubule-containing strands in elongated *Datura* cells adopt non-minimal configurations as they pass obliquely away to distant sites of long side walls (diagrammed in Fig. 18.3). The presence of cortical interphase MTs at those sites may well serve to anchor these early, premitotic, transvacuolar strands in long paths, but as the cortical MTs reorganize to form the PPB, most of the strands increasingly conform to minimal configurations according to minimal path principles. That is, against a background of reorganizing (depolymerizing?) cortical MTs, the initially oblique radial strands will become freed to move into the phragmosomal plane. Realignment of the radial, actin-containing strands could therefore simultaneously reorganize the cortical cytoskeleton into a tighter band, particularly since PPB MTs appear to emerge from these radial strands (Flanders *et al.*, 1990).

Consistent with this is the recent finding (Mineyuki and Palevitz, 1990) that cytochalasin D (CD) causes PPBs in *Allium* root tip cells to be two to three times broader than in controls. An alternative possibility – that increased lateral interaction between cortical MTs tightens the band – seems to be discounted by the finding that previously narrowed bands rewiden in the presence of CD (Mineyuki and Palevitz, 1990). Although the latter authors were studying densely cytoplasmic meristematic cells the implications from vacuolated cells are that the radial actin component may play a central part in selecting the division plane.

What is interesting is that isodiametric, nonorthogonal cells (which tend towards regular hexagons in section) are often seen to contain a star-like arrangement of premitotic strands (Fig. 18.3c). This persists until the formation of the PPB, although PPBs seem weak in this class of cells – often being angled and quite broad. Unlike the elongated cells, in which most strands eventually define the phragmosomal plane, other strands remain in these nonorthogonal cells perhaps because they are already in minimal path configurations. However, despite the apparent irregularity, and the failure to form a classical 'Maltese cross' where the majority of the cytoplasmic strands are in the phragmosome, the cells demonstrably divide, although the tissue they form is an irregular patchwork. Another contributory factor to the failure of these cells to form a tight PPB could be the disposition of MTs in the interphase array. During interphase, in elongated *Datura* stem epidermal cells, MTs can be followed, by optical sectioning, from wall to wall indicating that they form a transverse array that encircles the cell (Flanders *et al.*, 1989). The MTs are therefore topologically predisposed to form a circumferential band (i.e. the PPB). The non-orthogonal, isodiametric cells present a different picture: MTs are generally parallel to one another on any one cell face but there are faces (generally the outer epidermal wall) where MTs of divergent orientation spill over from neighbouring faces, forming a criss-cross effect. In such cases, where MTs are not wrapped with uniform alignment around a common axis, it may be geometrically impossible for the MTs to bunch up to form a tight, hoop-like PPB. From examining these cells it is difficult not to conclude that MT organization is influenced by cellular geometry.

The relative contributions of MTs and actin filaments to constructing the division plane: drug studies

Before strands anchoring the premitotic nucleus were discovered to contain actin and MTs, the existence of these cytoskeletal elements was implied by drug studies. However, in those studies in which mitosis is induced by wounding, it is important to appreciate that the initial traumatotactic nuclear migration (phase I) is distinct from premitotic nuclear migration that follows (phase II) (see Fig. 18.1). Indeed, nuclei several ranks away from the wound may undergo phase I migration to a wall on the wound side, although only the nearest few ranks may subsequently undergo division

involving premitotic nuclear migration into the centre of the cell (phase II). Traumatotactic nuclear migration is sensitive to CB but not to colchicine (Schnepf and von Traitteur, 1973). The cytoplasmic strands that radiate from the nucleus adjacent to the wound wall in *Tradescantia* leaf epidermal cells have subsequently been shown to contain F-actin by Goodbody and Lloyd (1990) who similarly found that this phase could be inhibited by CD.

In protoplasts derived from BY-2 tobacco suspension cells, nuclei also adopt a central location, supported by transvacuolar strands (Katsuta and Shibaoka, 1988). This location is maintained in the presence of the anti-MT drug, propyzamide, but not in the presence of CB, indicating that actin filaments alone can tether the nucleus as appears to be the case for the nuclei during traumatotactic migration.

Following the traumatotactic phase, nuclei in wounded tissue undergo another migration to the centre of the cell in preparation for mitosis. This premitotic migration is a normal feature of division in unwounded, vacuolated cells. Using leaf explants of *N. lynchii*, Venverloo and Libbenga (1987) dissected the various contributions of actin filaments and MTs to phragmosome formation. Treatment with colchicine inhibited nuclear migration to the centre of the cell, although once in such a central location nuclei were not displaced by colchicine treatment. However, migration to the centre was not inhibited by CB. So, whereas traumatotactic migration involves actin but not MTs, premitotic nuclear migration appears to involve MTs. Katsuta *et al.* (1990) have reexamined nuclear migration in BY-2 suspension cells, employing double fluorescence methods to visualize actin and MTs. They report that the premitotic, transvacuolar strands contain both cytoskeletal elements but that migration is not inhibited by cytochalasin treatment. This again indicates a role for MTs in premitotic migration. However, as Lloyd and Traas (1988) found for carrot suspension cells, actin filaments are not completely destroyed by CD; this can also be seen to be the case for BY-2 cells and so it cannot be concluded that MTs have an exclusive role in migration of the premitotic nucleus. Since simultaneous treatment with CD and propyzamide reduced the percentage of cells with a central nucleus to below 1%, it is still possible that both elements, perhaps necessarily in concert, play a part in this process.

If traumatotaxis represents phase I movement, and premitotic nuclear migration phase II, then movement of strands into the phragmosome represents a third phase. Again, Venverloo and Libbenga (1987) found that anti-MT and anti-actin drugs inhibited phragmosome formation, as well as nuclear positioning. Abnormally placed cell walls resulted from such treatment. Observations on chloroisopropyl phenylcarbamate (CIPC)-treated carrot suspension cells are relevant to this (Lloyd and Traas, 1988). CIPC induces tri- and multi-polar spindles although it has been shown to have no effect on the PPB. Split, three-way phragmoplasts successfully grow out between the three groups of chromatin separated on tri-polar spindles, and actin filaments can be seen to connect the leading edges of the abnormal phragmoplasts to sites that could not have been predicted by a planar PPB. Apart from demonstrating the remarkable plasticity of the phragmoplast this shows that cytokinesis can proceed, albeit abnormally, where the cytoskeleton is disorganized. This, however, seems to depend on the continued presence of actin filaments which, unlike MTs, remain in the cytoplasm after metaphase (Traas et al., 1987; Lloyd and Traas, 1988).

The polar strands may have a role in positioning the nucleus. These strands contain actin filaments (Lloyd and Traas, 1988) and MTs (Flanders et al., 1990; Katsuta et al., 1990). Centrifugation of dividing nuclei in Tradescantia stamen hairs can snap polar strands, in which case the pole wobbles violently (Ôta, 1961). In Nautilocalyx epidermal cells, weakening of these strands is reported (Venverloo and Libbenga, 1987) to cause displacement of the nucleus within the phragmosome. CD treatment of carrot suspension cells cause the spindle axis to tilt (Lloyd and Traas, 1988) but since phragmosomal actin as well as actin in polar strands was depolymerized by CD it is not clear which class played the more important part.

In summarizing this section, it would now appear that whereas traumatotactic nuclear migration involves actin, premitotic nuclear migration and phragmosome formation involve MTs as well. Further work is needed to establish whether the polar strands play a part in influencing the attitude of the spindle relative to the phragmosome, and whether they help to maintain the position of the nucleus within the phragmosomal plane. Following metaphase, however, cytoplasmic MTs disappear from the scene and actin filaments guide the phragmoplast out to the cortex. While isolating phragmoplasts, Kakimoto and Shibaoka (1988) observed that nuclei were still bound to these structures in the reported absence of MTs and actin. It is conceivable therefore that a third element may be involved in nuclear anchorage (but see Shaw et al. on intermediate filaments, Chapter 6, this volume).

Supracellular forces affecting division plane alignment

Elongated cells tend to divide transversely, perpendicular to the long wall, and this general observation has been encapsulated in various rules or laws. Similarly, cell plates tend to avoid aligning with cross walls between neighbours. Both broad conclusions indicate that the mechanisms for division plane alignment is sensitive to the shape of the mother cell and to immediate neighbours. Yet, division planes appear to be co-ordinated over much larger areas to produce organized tissue. Some ideas on co-ordination throughout tissue fields are covered by Green and Selker, Chapter 22, this volume. When tissue is wounded, adjacent cells are stimulated to divide parallel to the injury. Almost certainly there are factors at work found only in wounded tissue, but the co-ordination of division planes involved in this might provide useful clues to the process in unwounded tissue. Sinnott and Bloch (1945) cut notches out of Coleus internodes and observed the redifferentiation of surrounding pith cells to reconstitute phloem and xylem strands. Dividing cells contained phragmosomes, and these formed parallel to the course of the future vascular strand (parallel to the edge of the cut) and not to the original axis of elongation. Similarly, elongated stem epidermal cells of *Tradescantia albovittata*, which occur in regular files, can be induced to divide against the grain, longitudinally, by making a longitudinal razor slit (Goodbody and Lloyd, 1990). In leaf epidermis of this plant, wound divisions form a flat ellipse around a razor slit and perfectly encircle a single cell punctured with a microneedle.

These examples amply demonstrate that factors in the tissue field can override existing patterns of division and considerations of individual cell shape. In such traumatic tissue, phragmosomes do not avoid other phragmosomes in adjacent cells but tend to lie directly opposite them (Sinnott and Bloch, 1941a, b; 1945). This is not to say that a phragmosome in one cell aligns with a pre-existing wall between cells in a neighbouring file, but that sets of phragmosomes, developing synchronously in response to the wound, can align across tissue regardless of cell boundaries. Staining wounded *Tradescantia* leaves with rhodamine–phalloidin, Goodbody and Lloyd (1990) found that transvacuolar actin cables lined up across groups of cells, parallel to the wound, before the planar array of phragmosomal actin accumulated in that position. These premitotic actin cables focused upon a common point upon the separating wall. That is, lines (i.e. actin cables) as well as planes (i.e. phragmosomes and PPBs) can pass across groups of cells. Two intriguing, and interlinked, features of wound divisions are therefore: the formation of four-way .instead of three-way junctions; and the common alignment of these wound walls over large fields. Discussing these in turn: the *minimal path hypothesis* (Flanders *et al.*, 1990) depends upon the prior existence of a vertex to be avoided by cytoskeletal elements free to hunt a shorter path between nucleus and cortex in setting up the phragmosome. In normal, unwounded tissue, cell plates tend to attach to mother walls at 90° (see Fig. 18.4i). Subsequent lack of expansion of that plate for one cell cycle, whilst other mature walls continue to expand, causes the adjacent cell to buckle where the plate attaches (Korn, 1980) as

illustrated in Fig. 18.4ii, iii. What was a straight wall in the neighbouring cell becomes faceted during this process: a 90° attachment of a cell plate later bends to three 120° (perhaps only locally) angles as the non-dividing neighbour gains an extra face and a vertex. Subsequent division of the neighbouring cell would not yield touching cell plates where there is a vertex (on a long path from the central nucleus (Fig. 18.4iv)) to be avoided. However, where neighbouring cells divide simultaneously – as in wounded tissue – there is no such prior faceting of common walls along the prospective division line, and so the wound division plane, as in Fig. 18.4v, can pass from cell to cell, forming four-way junctions across the tissue (Goodbody and Lloyd, 1990).

The fact that premitotic actin cables can 'meet' at a common point on a separating wall could be taken to indicate that the cytoskeleton reads changes printed onto separating walls by gradients passing out from the wound. For instance, wound factors could cause field changes in cytoskeletal anchoring points. The precision of alignment observed between actin cables in *T. albovittata* leaves indicates, however, that cables in adjacent cells not only arrive in the same plane, but can focus upon the same point. Sinnott and Bloch (1941b) questioned whether transvacuolar strands could read gradients with such precision and they hinted instead at special points between cells. They suggested that if phragmosomes carried on, at their junctions with the side walls, an exchange of nutrients with the adjacent cell, then the region where an adjacent transverse wall attaches would be a blind spot to be avoided. A current restatement of this might be that transvacuolar elements, and actin cables in particular, preferentially align at plasmodesmata or some such specialized foci shared by cells. This would account for a co-alignment of phragmosomes in wound tissue which has developed from co-alignment of premitotic strands, and avoidance in normal tissue of three-way junctions where sharing is blocked by a previously attached cell plate.

There are therefore three hypotheses for avoidance of four-way junctions: cytoplasmic strands may tend not to make stable attach-

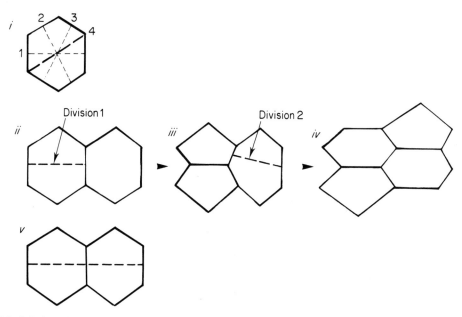

Fig. 18.4. Model for the avoidance of four-way junctions in asynchronously dividing cells and their formation in synchronously dividing cells. i, Isotropically expanding epidermal cells divide in planes 1, 2 and 3, but not 4 (which would form four-way junctions with neighbouring cells). According to the minimal path hypothesis (Flanders *et al.*, 1990), planes 1, 2 and 3 are the minimal paths selected by elements under tension and which are free to glide relative to the cortex. Such planes are stable and contact the mid-edge perpendicularly. Plane 4, which joins vertices, represents a longer path avoided by tensile elements. ii–iv, These three figures show the effect of two asynchronous divisions. First (ii) the left-hand cell divides transversely. The cell plate contacts the common wall perpendicularly. According to Korn (1980), the new wall does not expand for one generation whilst all other walls continue to do so; this causes the common wall (iii) to buckle. This introduces a vertex and an extra facet in the neighbouring cell. This vertex represents a feature to be avoided when the right-hand cell divides. In the final cell pattern produced by these two divisions (iv), the two new walls are not in contact but staggered, separated by an intervening facet. v, This illustrates the effect of two simultaneous divisions, as occurs following wounding. When one cell does not divide significantly ahead of the other, there is no pre-buckling of the common wall; there is no vertex to be avoided by the division plane aligning mechanisms and the divisions need not be off-set.

ments at vertices because they are distant points to be avoided by tensile strands, or because a cell–cell aligning system is blocked where three walls meet, or because three-way junctions are areas which are chemically hostile to cytoskeletal attachment.

The second issue is why a particular line is taken by wound divisions across tissue, paralleling the wound, disregarding normal features of cell geometry. In cells induced by wounding to differentiate into xylem elements, Sinnott and Bloch (1945) saw that the characteristic wall thickenings and, importantly, the cytoplasmic banding that preceded them, were directly opposite one another in adjacent cells around the wound. The complex pattern thus set up passed across cell boundaries, resembling 'lines of force in a magnetic field'. Microtubules are now known to bunch up in patterns that anticipate the secondary thickening of xylem (see Seagull and Falconer, Chapter 14, this volume). A recent paper by Hush *et al.* (1990) confirms that MTs begin to re-align, parallel to the wound, by 5 h after making a cut in pea roots. Ethylene – a wound hormone – rapidly induces MTs in unwounded pea and mung bean epidermal cells to reorientate (Roberts *et al.*, 1985) and is a likely agent of physiological control. However, a list of stimuli for cytoskeletal reorganization would include, in addition to ethylene: other hormones, calcium gradients, changes in cellular dimensions and wall angles, sealing of plasmodesmata and altered connectivity, altered ionic currents and electrical fields (see Hush *et al.*, 1990), and realignment of stress paterns.

Recent advances have identified the cytoskeletal elements that set the division plane. Next, it will be necessary to determine which (if any) of these field forces the cytoskeleton sets its compass by. This is particularly important in unwounded tissue in which the co-ordination of cytoskeletal behaviour is essential for producing the ordered growth of tissues and organs.

References

Bakhuizen, R., Van Spronsen, P.C., Sluiman-Den Hertog, F.A.J., Venverloo, C.J. and Goosen-de Roo, L. (1985). Nuclear envelope radiating microtubules in plant cells during interphase mitosis transition. *Protoplasma* **128**, 43–51.

Clayton, L. and Lloyd, C.W. (1985). Actin organization during the cell cycle in meristematic plant cells. *Exp. Cell Res.* **156**, 231–238.

Flanders, D.J., Rawlins, D.J., Shaw, P.J. and Lloyd, C.W. (1989). Computer-aided 3-D reconstruction of interphase microtubules in epidermal cells of *Datura stramonium* reveals principles of array assembly. *Development* (Camb.) **106**, 531–541.

Flanders, D.J., Rawlins, D.J., Shaw, P.J. and Lloyd, C.W. (1990). Nucleus-associated microtubules help determine the division plane of plant epidermal cells: avoidance of 4-way junctions and the role of cell geometry. *J. Cell Biol.* **110**, 1111–1122.

Goodbody, K. and Lloyd, C.W. (1990). Actin filaments line-up across *Tradescantia* epidermal cells, anticipating wound-induced division planes. *Protoplasma* **157**, 92–101.

Goosen-de Roo, L., Bakhuizen, R., Van Spronsen, P. and Libbenga, K.R. (1984). The presence of extended phragmosomes containing cytoskeletal elements in fusiform cambial cells of *Fraxinus excelsior* L. *Protoplasma* **122**, 145–152.

Gunning, B.E.S. (1982). The cytokinetic apparatus: its development and spatial regulation. In *The Cytoskeleton in Plant Growth and Development* (ed. C.W. Lloyd), pp. 229–292. Academic Press, London.

Hardham, A.R. and McCully, M.E. (1982). Repro-gramming of cells following wounding in Pea (*Pisum sativum* L.) roots. I. Cell division and differentiation of new vascular elements. *Protoplasma* **112**, 143–151.

Hush, J.M., Hawes, C.R. and Overall, R.L. (1990). Interphase microtubule re-orientation predicts a new cell polarity in wounded pea roots. *J. Cell Sci.* **96**, 47–61.

Kakimoto, T. and Shibaoka, H. (1987). Actin filaments and microtubules in the preprophase band and phragmoplast of tobacco plant. *Protoplasma* **140**, 151–156.

Katsuta, J. and Shibaoka, H. (1988). The roles of the cytoskeleton and the cell wall in nuclear positioning in tobacco BY-2 cells. *Plant Cell Physiol.* **29**, 403–413.

Katsuta, J., Hashiguchi, Y. and Shibaoka, H. (1990). The role of the cytoskeleton in positioning of the nucleus in premitotic BY-2 cells. *J. Cell Sci.* **95**, 413–422.

Korn, R.W. (1980). The changing shape of plant cells: transformation during cell proliferation. *Ann. Bot.* **46**, 649–666.

Lloyd, C.W. (1989). The plant cytoskeleton. *Current Opinion in Cell Biology* **1**, 30–35.

Lloyd, C.W. (1991). How does the cytoskeleton read the laws of geometry in aligning the division plane of plant cells? *Development, Suppl.* **1**, 55–66.

Lloyd, C.W. and Traas, J.A. (1988). The role of F-actin in determining the division plane of carrot suspension cells. Drug Studies. *Development* **102**, 211–222.

Mineyuki, Y. and Palevitz, B.A. (1990). Relationship between preprophase band organization, F-actin and the division site in *Allium*. *J. Cell Sci.* **97**, 283–295.

Ôta, T. (1961). The role of cytoplasm in cytokinesis of plant cells. *Cytologia* (Tokyo) **25**, 297–308.

Palevitz, B.A. (1987). Actin in the preprophase band of *Allium cepa*. *J. Cell Biol.* **104**, 1515–1519.

Pickett-Heaps, J.D. and Northcote, D.H. (1966). Organization of microtubules and endoplasmic reticulum during mitosis and cytokinesis in wheat meristems. *J. Cell Sci.* **1**, 109–120.

Roberts, I.N., Lloyd, C.W. and Roberts, K. (1985). Ethylene-induced microtubule reorientations: mediation by helical arrays. *Planta* **164**, 439–447.

Schnepf, E. and von Traitteur, R. (1973). On the traumatotactic movement of nuclei in *Tradescantia* leaves. *Z. Pflanzenphysiol.* **69**, 181–184.

Sinnott, E.W. and Bloch, R. (1940). Cytoplasmic behaviour during division of vacuolate plant cells. *Proc. Natl Acad. Sci. U.S.A.* **26**, 223–227.

Sinnott, E.W. and Bloch, R. (1941a). Division in

vacuolate plant cells. *Am. J. Bot.* **28**, 225–232.

Sinnott, E.W. and Bloch, R. (1941b). The relative position of cell walls in developing plant tissues. *Am. J. Bot.* **28**, 607–617.

Sinnott, E.R. and Bloch, R. (1945). The cytoplasmic basis of intercellular patterns in vascular differentiation. *Am. J. Bot.* **32**, 151–156.

Traas, J.A., Doonan, J.D., Rawlins, D.J., Shaw, P.J., Watts, J. and Lloyd, C.W. (1987). An actin network is present in the cytoplasm throughout the cell cycle of carrot cells and associates with the dividing nucleus. *J. Cell Biol.* **105**, 387–395.

Venverloo, C.J. and Libbenga, K.R. (1987). Regulation of the plane of cell division in vacuolated cells. I. The function of nuclear positioning and phragmosome formation. *J. Plant Physiol.* **131**,. 267–284.

Venverloo, C.J., Hovenkamp, P.H., Weeda, A.J. and Libbenga, K.R. (1980). Cell division in *Nautilocalyx* explants. I. Phragmosome, preprophase band and plane of division. *Z. Pflanzenphysiol.* **100**, 161–174.

19. THE CYTOKINETIC APPARATUS IN MEIOSIS: CONTROL OF DIVISION PLANE IN THE ABSENCE OF A PREPROPHASE BAND OF MICROTUBULES

Roy C. Brown and Betty E. Lemmon

*Department of Biology, University of Southwestern Louisiana,
Lafayette, LA 70504-2451, USA*

Introduction

The quadripartitioning of a sporocyte into a tetrad of predictably arranged spores is a splendid example of precise geometrical cell division. Comparative studies of the phenomenon of cytoplasmic quadripartitioning have provided insight into the fundamental nature of the cytokinetic apparatus in plants. As defined by Gunning (1982) the cytokinetic apparatus is a complex cytoskeletal system that functions both in determination of plane of division and completion of the process of cell plate formation.

In histogenesis, the cytokinetic apparatus includes two microtubule arrays that are temporally and structurally distinct. The preprophase band of microtubules (PPB), which is organized as part of the cytoplasmic preparation for mitosis in walled plant cells, is one of the first structural indications that a cell is preparing to divide (Wick and Duniec, 1983). The orientation of the PPB clearly marks the future plane of division which, in most cases, will be at right angles to the spindle. The second component of the cytokinetic apparatus, the phragmoplast, develops in late anaphase. The phragmoplast is a complex of cytoskeleton, endoplasmic reticulum, and vesicles associated with cell-plate deposition. The phragmoplast microtubule system is composed of two sets of microtubules of opposite polarity (for review, Gunning, 1982). At first associated with the interzonal spindle, the microtubules of the phragmoplast spread centrifugally into the cortical cytoplasm as the cell plate expands, eventually completing division by joining with parental walls at sites previously occupied by the PPB.

In sporogenesis, the cytokinetic apparatus is fundamentally different from that described above for histogenesis. Sporocytes embarking on the determinant morphogenetic pathway leading to spore tetrads are isolated from the influence of neighbouring cells by a callosic or mucopolysaccharide wall, and division planes seem to be endogenously controlled. No PPB marks the division sites, and cytokinesis usually involves the formation of wall around the spores which is contiguous with, or inside of, the sporocyte wall. Determination of division plane appears to result from reorganization of the cytoplasm into spore domains. The concept of cytoplasmic domains has been invoked to account for partitioning of the cytoplasm in other coenocytic systems, both plant and animal (Menzel, 1986; Warn, 1986; Goff and Coleman, 1987). Such cytoplasmic domains function as spheres of activity within the common cytoplasm. The microtubular cytoskeleton in

The Cytoskeletal Basis of Plant Growth and Form   ISBN 0–12–453770–7

sporocytes is predictive of the spore domains and plays a key role in cytokinesis.

Spore domains can be established either prior to, or after, nuclear division. Premeiotic establishment of division quadripolarity is typical of the simple land plants, whereas postmeiotic determination of division plane is typical of higher plants (ferns and seed plants). The only simple plants reported to lack pre-division quadripolarity are members of the subclass Marchantiidae (Bryophyta: Hepaticae).

Premeiotic determination of division plane

Quadrilobing of the sporocyte

Quadrilobing of the cytoplasm in sporogenesis of bryophytes provides clear evidence that division polarity for the entire meiotic process can be determined from the onset of meiosis. The spore domains anticipate the eventual distribution of the four nuclei resulting from the second meiotic division. Interestingly, the fundamental quadripolarity of the sporocyte as exemplified by cytoplasmic lobing is also reflected in prepatterning of the exine and the placement of apertures in young spores of the tetrad (Brown and Lemmon, 1988a).

Lobing of the cytoplasm in meiotic prophase is most extreme in hepatics of the Jungermanniidae (Fig. 19.1) and in bryopsid mosses, especially those of the Polytrichidae, in which the cleaveage furrows reach nearly to the nucleus lying in the small central portion of the sporocyte. Lobing, which begins early in prophase and results in pronounced definition of the future spores, appears to be independent of nuclear position. The sporocyte nucleus regularly undergoes a dramatic two-way migration during meiotic prophase (Brown and Lemmon, 1987a, 1988a, 1989a; Busby and Gunning, 1988a). In early prophase, the nucleus moves from the centre of the sporocyte to the periphery, and then in midprophase it returns to the centre. Although this unusual nuclear migration is seldom mentioned, it is so frequently illustrated that we assume it to be a characteristic feature of sporogenesis in plants. No correlation has been shown between the pattern of the cytoplasmic lobing and the path

of nuclear migration. A nucleus that happens to lie in one of the lobes, rather than between lobes, migrates into the central portion of the cytoplasm before development of the meiotic spindle (Brown and Lemmon, 1988a). Although no specific arrangement of the cytoskeleton marks the path of nuclear migration, perturbation of microtubules prevents the nucleus from returning to the centre in monoplastidic sporocytes of the moss *Funaria* (Busby and Gunning, 1989).

Organization of the cytoskeleton in meiotic prophase reflects the establishment of division quadripolarity. In the polyplastidic hepatics, microtubules emanate from a centrosomal-like focus in each of the four cytoplasmic lobes (Fig. 19.1). These four cones of microtubules impinge on the central nucleus and distort it toward the four lobes. In the monoplastidic mosses (Fig. 19.2) and hornworts, microtubules emanating from a single plastid located in each of the lobes result in a similar distortion of the nucleus toward the four lobes. Development of the unique quadripolar microtubule system of meiotic prophase, which is eventually transformed into the meiotic spindle, is best understood in monoplastidic meiosis.

Plastid polarity in monoplastidic meiosis

In monoplastidic cells, the single plastid is intimately involved with the process of cell division (for review, Brown and Lemmon, 1990a). Monoplastidic sporocytes exhibit the phenomenon of plastid polarity in which (i) division and migration of the plastid in prophase reflects establishment of the cytoplasmic domains of the future spore tetrad, and (ii) microtubules organized at the plastids contribute to both spindle and cytokinetic apparatus. Monoplastidic meiosis, which occurs in bryophytes (except hepatics) and the heterosporous vascular cryptograms *Isoetes* and *Selaginella*, has provided a wealth of information on the reorganization of the cytoplasm in preparation for division. Morphogenetic plastid migrations clearly mark changes in division polarity during sporogenesis.

The behaviour of the plastids in monoplastidic cell division mimics the behaviour of centrosomes in animal cells and is a valuable

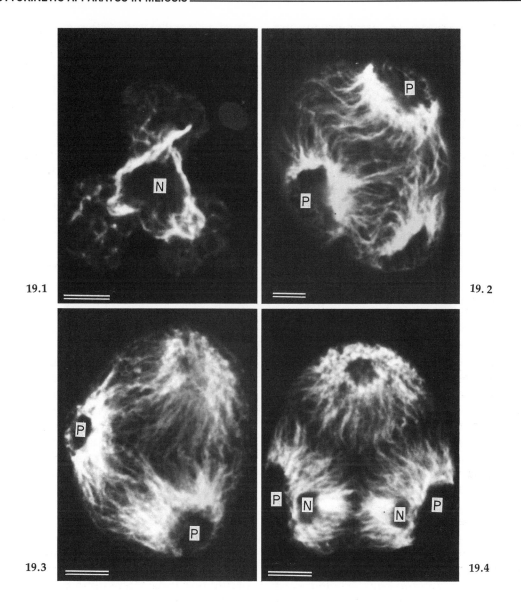

Figs 19.1–19.4. Microtubule systems associated with sporocyte quadripolarity in lower land plants. Confocal laser scanning micrographs of microtubules localized by indirect immunofluorescence.

Fig. 19.1. Meiotic prophase in *Symphyogyna brongniartii*, × 1250. In the polyplastidic hepatics, microtubules encaging the central nucleus (N) emanate from a centrosomal-like focus in each of the four cytoplasmic lobes. Scale bar = 10 μm.

Fig. 19.2. Meiotic prophase in the moss *Andreaea rothii*, × 3000. The QMS emanates from the four tetrahedrally arranged plastids (P). Scale bar = 3 μm.

Fig. 19.3. Megasporocyte of *Isoetes melanopoda*, × 1200. The plastid-based QMS in meiotic prophase. Plastids (P). Scale bar = 10 μm.

Fig. 19.4. Megasporocyte of *Isoetes melanopoda*, × 1200. Telophase II. Second division results in a nucleus (N) adjacent to a plastid (P) in each spore domain. Spindles are organized among microtubules emanating from the plastids. Scale bar = 10 μm.

source of information on organization and function of microtubules in sporogenesis. Whereas the material responsible for nucleation of microtubules in cells of higher plants is thought to be associated with the nuclear envelope (for reviews, Wick and Duniec, 1983; Lloyd, 1987), microtubules in monoplastidic cells clearly emanate from the plastid envelope (Brown and Lemmon, 1987a, b, 1988b, 1989a, 1990a; Busby and Gunning, 1988a, b). This suggests that the microtubule organizing centre (MTOC) responsible for nucleating microtubules is intimately associated with the plastid envelope. The observation that microtubules reappear at plastids in monoplastidic sporocytes recovering from drug-induced microtubule depolymerization strengthens the concept of the plastid MTOC (Busby and Gunning, 1989).

Plastid MTOCs result in unique microtubule systems that make up both the cytokinetic apparatus and spindle apparatus (Figs 19.2–19.4). A quadripolar microtubule system (QMS) emanating from four plastids (or the four tips of two dividing plastids) is associated with predictive positioning of the plastids and organization of the spindle in monoplastidic meiosis (Brown and Lemmon, 1990a). Two important variations of the QMS occur; in lobed, polyplastidic sporocytes of hepatics the QMS is not plastid-based (Fig. 19.1), and in microsporocytes of Selaginella a unique plastid-based microtubule system known as the procytokinetic plate predicts polarity of the first, rather than the second division (Brown and Lemmon, 1985).

The study of cell division in hornworts (Brown and Lemmon, 1988b) has provided information critical to understanding various aspects of the QMS, such as its development and organization, its role in plastid division and migration, its relationship to establishment of division planes, and its transformation into the meiotic spindle. In both meiosis and mitosis of hornworts, the single plastid divides and migrates in response to division polarity and serves to nucleate microtubules involved in nuclear division and cytokinesis. In preparation for mitosis, the plastid migrates to a position parallel to the future spindle axis. A unique axial microtubule system (AMS)

develops parallel to the isthmus of the dividing plastid and precisely intersects the division site which is girdled by a PPB. Since plastid migration and AMS development occur prior to PPB formation, both serve as reliable markers of polarity. As the plastid divides, the AMS elongates into two opposing sets of microtubules focused on tips of the daughter plastids. These microtubules contribute directly to development of the mitotic spindle.

A similar AMS is associated with successive plastid divisions that result in predictive positioning of a plastid at each of the future tetrad poles in prophase of meiosis (Brown and Lemmon, 1990b). An AMS arises in association with the first plastid division that gives rise to two plastids which elongate on either side of the nucleus. An AMS reappears at the midregion of each plastid as they simultaneously prepare for a second division. The two AMSs develop into the primary arrays which connect sister plastids after second plastid division. Additional microtubules radiate from the four plastid MTOCs to produce secondary arrays between non-sister plastids. Together the two systems comprise the mature QMS which interconnects the four tetrahedrally arranged plastids and encages the nucleus. In Isoetes and mosses, the plastids divide without a conspicuous AMS associated with the isthmus. However, microtubule arrays connecting pairs of sister plastids are often recognizable as primary arrays before the secondary arrays develop. This is especially obvious in Funaria where the second plastid division is delayed until after meiosis I, and plastid tips at the tetrad poles serve to organize microtubules of the QMS (Busby and Gunning, 1988a, b).

It is clear that plastid division and AMS formation are essentially the same in meiosis and mitosis. However, in meiosis, two rounds of plastid division occur and plastid positioning reflects the establishment of four spore domains rather than polarity of the dyad. Development of the first division spindle involves a partial merger of the four opposing arrays of the QMS and establishment of a division axis that terminates between non-sister plastids. In this way, the QMS of prophase is transformed into a meiotic spindle that is functionally bipolar

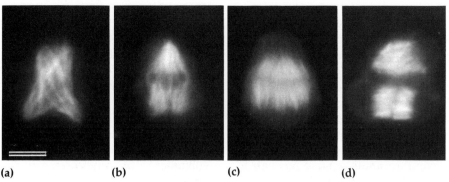

(a)   **(b)**   **(c)**   **(d)**

Fig. 19.5. Monoplastidic meiosis in the moss *Fissidens cristatus*. Epifluorescence micrographs of microtubules stained by indirect immunofluorescence, × 2100. a, Quadripolar origin of the meiotic spindle from the QMS in prometaphase; b, the meiotic spindle appears triangular because the lower pair of poles is seen side-by-side while the upper pair is superposed; c, the complex first division phragmoplast. No cell plate is laid down; d, the second division phragmoplasts lying at right angles to each other in the undivided cytoplasm. Note absence of microtubules in the first division site. Scale bar = 5 μm.

(Fig. 19.5), and the prophase II nuclei will each be positioned between pairs of plastids marking poles of the second meiotic division. Each telophase II nucleus will lie adjacent to a plastid (Fig. 19.4). The quadripolar origin of the meiotic spindle resulting from the merger of four cones of microtubules is unique. In the bryophyte type (Fig. 19.5), the plastids remain in tetrahedral arrangement throughout meiosis and the foci of microtubules migrate towards the spindle axis during development of the first division spindle. In the lycopsid type, the plastids are located at poles of both first and second division. In *Isoetes*, plastids with associated microtubules leave their prophasic tetrahedral positions to converge in pairs upon the first division spindle axis.

Microsporogenesis in *Selaginella* represents a special type in which plastid polarity sequentially establishes division polarity (Brown and Lemmon, 1985). Unlike the typical QMS, which establishes polarity for second meiotic division, the procytokinetic plate (PCP) of *Selaginella* establishes polarity for first nuclear division. It is not until after first nuclear division that the four plastids migrate to tetrahedral arrangement where they serve as poles for the second nuclear division.

Microtubule arrays are smoothly transformed from one configuration to another during sporogenesis with no evidence of total depolymerization between any of the transitions.

Although dynamic instability of microtubules as shown in animal cells (Cassimeris *et al.*, 1987) has yet to be documented for plant microtubules, this concept best explains the reorganization of microtubules in monoplastidic cell division. We assume that the dramatic reorganization of microtubules into stage specific arrays is a function of MTOC migration with turnover of microtubules. For example, during ontogeny of the meiotic spindle from the precursor QMS in *Isoetes* megasporogenesis, the plastids with associated microtubules converge on the spindle axis. In microsporogenesis of *Selaginella*, plastids with associated microtubules migrate to opposite sides of the nucleus during development of the PCP. These, and examples of microtubule systems that grow and recede from plastid MTOCs throughout the course of monoplastidic meiosis, are explained as resulting from movement of the MTOCs in response to changes in polarity, and to specific controls that regulate the growth of microtubules.

Postmeiotic determination of division planes

Cytokinetic planes appear to be determined after nuclear division in polyplastidic sporocytes of the majority of vascular plants and in one group of bryophytes, the Marchantiidae. In these plants there is no cytoplasmic lobing of the sporocyte or predictive prophasic micro-tubule system to indicate that spore domains have been established prior to division. Instead, microtubules radiating from the nuclei figure prominently in measuring spore domains. Thus the postmeiotic disposition of nuclei in the sporocyte is a key element in determination of the pattern of cleavage. It should be understood that any mechanism for orienting spindle axes, and thus arrangement of tetrad nuclei, would impose another level of control on the pattern of tetrad formation. Although such a mechanism is yet unknown, the precise arrangement of tetrads in many groups sug-gests the presence of such a mechanism. At another level, histological studies on certain cereal grains have established that the tetrads occur in a particular arrangement so that the distal aperture of each microspore is adjacent to the surrounding tapetum (Christensen and Horner, 1974). This would require strict, but as of yet inexplicable, control of spindle orientation at the tissue level.

Simultaneous versus successive cytokinesis
Cytokinesis in meiosis is categorized as either successive or simultaneous. In the successive type, a dyad is formed after first meiosis and in the simultaneous type, a four-nucleate coenocyte is quadripartitioned after meiosis. Successive cytokinesis is accomplished by the deposition of cell plates in association with phragmoplasts after each nuclear division. In simultaneous cytokinesis, quadripartitioning is typically initiated by infurrowing of cytoplasm along boundaries of spore domains defined by postmeiotic systems of nuclear-based micro-tubules. However, the distinction between successive and simultaneous cytokinesis is not absolute, and patterns that appear intermediate between the two major types have been reported. Sampson (1969) proposed that simul-taneous cytokinesis with no remnants of a cell plate is the most highly derived type. All major plant taxa, bryophytes, lycopsids, pteropsids, gymnosperms and angiosperms have examples of both types of cytokinesis. In *Isoetes*, cyto-kinesis is successive in microsporogenesis, but simultaneous in megasporogenesis. Among angiosperms, it is traditional to associate dicots

with simultaneous cytokinesis and the mono-cots with successive. However, sporocytes in two large groups of monocots, the Orchidales and certain of the Liliales such as the Iridaceae, divide simultaneously. Multiple modes for accomplishing cytokinesis are suspected in the dicots (Sampson, 1969), probably reflecting their polyphyletic evolution.

Common features of the cytokinetic apparatus in meiosis include the lack of a PPB and involvement of nuclear-based postdivision microtubules. In some cases, successive cyto-kinesis appears to differ from simultaneous cytokinesis only in that the spore domains are formed sequentially rather than simultaneously. For example, in certain hybrid orchids, a wall may or may not form after meiosis I (Brown and Lemmon, 1991). In addition, experimentally produced coenocytes of *Lilium* will sometimes attempt to divide after second meiotic division (Heslop-Harrison, 1971) suggesting that the determination of the spore domains is post-poned. Unfortunately, the nature of the micro-tubule systems was not determined, but this should now be possible using methods of indirect immunofluorescence. In the classical example of daylily, micronuclei after meiosis I nucleate microtubules and alter the pattern of dyad wall formation (Fig. 19.15). On the other hand, evidence for other levels of control comes from analysis of certain male sterile mutants of maize. In the meiotic mutant *dv*, multiple spindles result in multiple nuclei after meiosis I (Staiger and Cande, 1990). However, quite unlike the case in daylily discussed above, the micronuclei have no effect on cyto-kinesis and a dyad wall forms in the wild type position. One interpretation is that a division site is present even though not marked as in mitosis by a PPB. Alternatively, the micronuclei in this mutant may be, for reasons unknown, incapable of commanding a domain of cytoplasm. Such findings serve to emphasize the importance of sporogenesis to our under-standing of cytokinesis and the need for additional comparative and experimental studies.

Organelle band
In a wide variety of sporocytes that divide

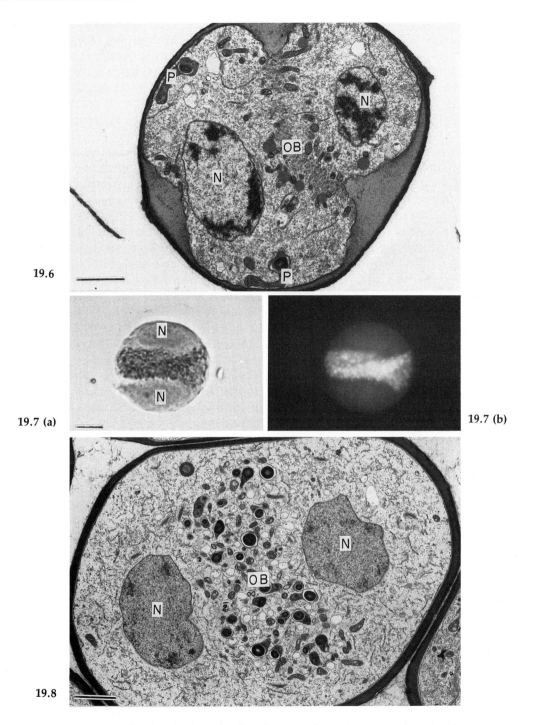

Figs 19.6–19.8. An organelle band (OB) divides the cytoplasm into two domains in sporocytes with simultaneous cytokinesis. Nucleus (N).

Fig. 19.6. Transmission electron micrograph (TEM) of prophase II in the moss *Entodon seductrix*, × 6500. In monoplastidic meiosis the plastids (P) are positioned in lobes and are not included in the organelle band. Scale bar = 2 μm.

Fig. 19.7. *Equisetum hyemale*, × 1500. a, DIC micrograph; b, organelle autofluorescence. Scale bar = 5 μm.

Fig. 19.8. TEM of prophase II in a hybrid *Phalaenopsis* orchid, × 3700. Scale bar = 3 μm.

simultaneously, the cytoplasm is divided into two functional domains after meiosis I with the first division site occupied by an embedded cell plate, encircling wall ingrowth or a polarized band of organelles.

Organelle bands have been reported in sporogenesis of a diverse assemblage of plants from bryophytes to orchids (Figs 19.6–19.8). The organelle band effectively divides the cytoplasm into two domains; it facilitates equal apportionment of membrane bound organelles and may prevent interaction of second division spindles (Brown and Lemmon, 1987b, 1988d; Rodkiewicz and Duda, 1988). In monoplastidic meiosis where the plastids are deployed at the tetrad poles (Fig. 19.6), the organelle band consists of mitochondria, oil bodies and vesicles, whereas in polyplastidic sporocytes, plastids are included in the organelle band, and may be especially conspicuous when they contain starch grains (Fig. 19.8). Following second nuclear division, the organelles either disperse or reorganize into patterns that reflect the cleavage planes for quadripartitioning.

Development of the organelle band can be traced to regular patterns of organelle distribution occurring in prophase (Brown and Lemmon, 1987a, b; Rodkiewicz and Duda, 1988). An exceptional case of organelle polarization has been documented in microsporogenesis of *Malva silvestris* where organelles form a dense layer coating the prophase nucleus. Division of the organelle population accompanies each successive nuclear division so that a like quantity of organelles surrounds each nucleus in the tetrad (Rodkiewicz and Duda, 1988). Thus, each spore domain receives a like quantity of organelles in a manner similar to sorting of organelles into cytoplasmic domains in algae (Goff and Coleman, 1987).

The role of the cytoskeleton in bringing about the polarized distribution of organelles is uncertain. The organelle band in mosses remains intact throughout the course of second division even though the phragmoplast microtubules disappear in inframeiotic interphase (Brown and Lemmon, 1987b, 1988d; Busby and Gunning, 1988a). A raft of actin, which may maintain the organelle band, has been reported in sporocytes of eggplant (Traas *et al.*, 1989).

However, in *Gasteria*, F-actin does not remain in the organelle band, but reorganizes into two domains around the nuclei (Van Lammeran *et al.*, 1989). Much remains to be learned about the role of F-actin in sporogenesis.

Spore domains and the control of division planes

Simultaneous cytokinesis provides the most convincing evidence for microtubule measured spore domains in the determination of quadripartitioning, and is especially informative of the basic mechanisms of cell division because it involves the development of walls between non-sister as well as sister nuclei. In simultaneous cleavage, the position of telophase II nuclei controls the pattern of cytokinesis which in turn determines the arrangement of the tetrad (Heslop-Harrison, 1971; Huynh, 1976) and influences the location of germinal pores (Heslop-Harrison, 1971; Dover, 1972; Sheldon and Dickinson, 1983, 1986). Cellularization during simultaneous cytokinesis involves formation of microtubules between non-sister nuclei as well as the usual early phragmoplasts between sister nuclei (Figs 19.9–19.14). The role of postdivision microtubules in meiotic cytokinesis has long been recognized (Timberlake, 1900). Farr (1916, 1918) observed that fibrils (microtubules) radiate from reformed nuclei subsequent to nuclear division and accurately described the formation of six 'spindles' interconnecting the nuclei in *Nicotiana* and a 'complex spindle' among nuclei of *Magnolia*. These post-division microtubule systems were referred to simply as spindles in the early literature and later as 'secondary spindles' (Waterkeyn, 1962; Heslop-Harrison, 1971). Recent work has documented the role of radial microtubules in determination of division plane. The primary and secondary (interzonal) spindles give rise to phragmoplasts which develop along boundaries of the spore domains. The study of simultaneous cytokinesis in the thalloid liverwort *Conocephalum conicum*, where patterns of cytokinesis are variable, has provided important clues to the assembly and function of microtubules in determination of spore domains (Brown and Lemmon, 1988c). In spite of variation in pattern of the tetrad

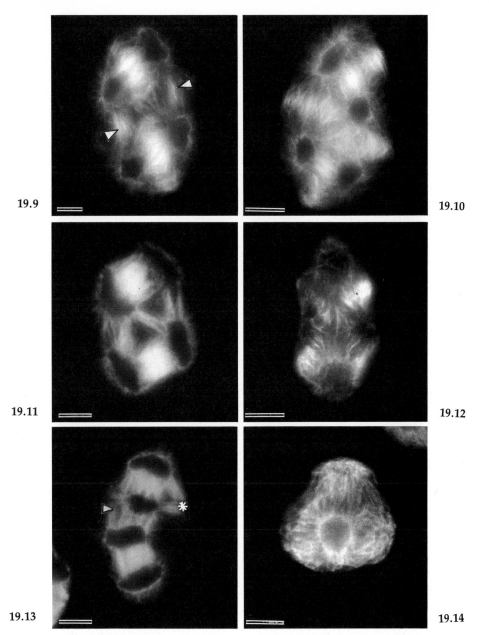

Figs 19.9–19.14. In simultaneous cytokinesis, microtubules form between both sister and non-sister nuclei. Arrangement of spore domains reflects position of nuclei. **Figs 19.9–19.10.** *Conocephalum conicum,* × 800. (After Brown and Lemmon, 1988c with permission of Wiley-Liss). **Figs 19.11–19.13.** The orchid *Phalaenopsis.*
Fig. 19.9. Primary phragmoplasts are initiated in interzonal regions and secondary phragmoplasts (arrowheads) form between non-sister nuclei. Scale bar = 10 μm.
Fig. 19.10. Phragmoplasts become confluent along all boundaries of cytoplasmic domains. Scale bar = 10 μm.
Fig. 19.11. Secondary spindle formation, × 3000. Scale bar = 3 μm.
Fig. 19.12. Phragmoplasts form along division planes defined by opposing nuclear-based microtubule arrays, × 1500. Scale bar = 8 μm.
Fig. 19.13. Secondary spindles form between a supernumerary nucleus (*) and the adjacent principal nuclei. Normally, only one secondary spindle (arrowhead) forms in a linear tetrad. × 3000. Scale bar = 3 μm.
Fig. 19.14. *Cassia alata,* × 1400. In the tetrahedral tetrads of most dicots, cytokinesis begins at the periphery of spore domains defined by the interaction of nuclear-based radial microtubules. Scale bar = 8 μm.

(tetrahedral, rhomboidal or linear), the spores produced are of approximately equal size and morphology. The mechanism for equal apportionment involves the development of postmeiotic microtubule arrays that radiate equally from the four nuclei which share the common cytoplasm. The opposing sets of microtubules divide the cytoplasm into four portions of approximately equal size. Thus, each nucleus appears to control the portion of the surrounding cytoplasm that will be included with it in the spore after cytokinesis.

The phragmoplasts form in areas of interaction between opposing sets of microtubules (Figs 19.9 and 19.10). Secondary phragmoplasts form between non-sister nuclei in addition to the primary phragmoplasts between sister nuclei. Phragmoplasts expand along all 'fronts' of interaction and become confluent with one another. The nucleation of phragmoplast microtubules appears to be triggered by the interaction of the opposing sets of microtubules. Cell plates which form along all planes occupied by the phragmoplasts fuse to cleave the cytoplasm simultaneously into four spores. The lack of a set arrangement of spores in the tetrads of *C. conicum* argues against the cause of secondary phragmoplast development somehow being related to predetermined division sites. Instead, it appears that the post-meiotic microtubules radiating equally from the four nuclei in the undivided sporocyte measure the cytoplasmic domains destined to become spores.

Further evidence for the postmeiotic selection of division planes is seen in examples such as *Impatiens* where postmeiotic growth of the sporocyte changes the configuration of the division planes (Brown and Lemmon, 1988d). In *Impatiens*, the young spores of the tetrad elongate after meiosis resulting in a packet of rod-like microspores. Cytokinesis progresses along extended fronts of newly organized microtubules that radiate from the elongating nuclei like bristles of a bottle brush.

In abnormal meiosis, the pattern of cytokinesis is altered when lagging chromosomes result in supernumerary nuclei. Chromosomes that fail to be incorporated into the telophase I nuclei frequently undergo second meiotic division on minispindles, and the resultant nuclei can claim cytoplasmic domains resulting in the cleavage of supernumerary spores (Fig. 19.13). Following meiosis, radial systems of microtubules measure cytoplasmic domains around each nucleus in the coenocyte. Walls are formed whenever opposing radial arrays interact and cytoplasm that is cleaved around micronuclei is proportionately smaller than that around the principal nuclei (Brown and Lemmon, 1989b). Since it is not logical to think that the cytoplasm could ancitipate the occasional meiotic mishap, the formation of supernumerary spores provides strong evidence for postmeiotic, rather than premeiotic, determination of division plane. Similar irregularities in successive cytokinesis emphasize the role of cytoplasmic domains in sporogenesis. If meiotic irregularities lead to micronuclei after the first division, the pattern of cytokinesis will be altered. Microtubules emanating from the micronuclei oppose those of the primary arrays and branching cell plates cut off small cells in addition to those of the dyad (Fig. 19.15). Such observations clearly indicate that nuclear-based microtubules are instrumental to apportionment of the cytoplasm in a coenocytic sporocyte.

Intersporal wall development

The actual process of wall formation in meiosis is variable and comparative data are woefully incomplete. In an early review, Farr (1916) concluded that cytokinesis in pollen mother cells may result from infurrowing rather than from cell plates, and later (1918) stated that, 'the furrow and the cell plate are not homologous structures though they accomplish the same end'. Longly and Waterkeyn (1979) recognized that meiotic cytokinesis is actually accomplished by deposition of the callosic special wall between tetrad members, whether or not a cell plate is formed. Thus, the cell plate is not the sole instrument of cytokinesis as it is in mitosis. In successive cytokinesis, a centrifugally expanding cell plate is formed after each nuclear division. These cell plates are subsequently covered by the callosic special wall that completes the intersporal septa.

In simultaneous cytokinesis, infurrowing due to deposition of callosic special wall material often marks the initiation of cyto-kinesis after second nuclear division and con-tinued infurrowing results in quadripartitioning. Cell plates that are 'embedded' or encased in callosic wall material have been described (Longly and Waterkeyn, 1979). The cell plates form simultaneously with infurrowing and are immediately encased in the advancing callosic sporocyte wall. Thus, there is clearly a basic difference in the role of the cell plate in meiotic and mitotic cytokinesis. In mitosis, the cell plate both contributes to wall formation and provides new plasma membrane for the daughter cells. In meiosis, intersporal wall deposition is a two-stage process (Longly and Waterkeyn, 1979; Van Went and Cresti, 1988). The cell plate serves as a template for the extracellular deposition of callose which results in massive walls separating the tetrad members.

In successive cytokinesis, the meiotic phrag-moplast consists of microtubules arising in the interzonal region between telophase nuclei and developing into opposing sets of micro-tubules which radiate from proximal surfaces of the reforming nuclei. Centrifugal expansion of the phragmoplast accompanies deposition of a cell plate that expands to partition the sporocyte into a dyad (Timberlake, 1900; Longly

and Waterkeyn, 1979; Sheldon and Dickinson, 1986; Hogan, 1987; Staiger and Cande, 1990). In simultaneous cytokinesis, the events follow-ing first meiotic division are similar, but no cell plate is laid down between domains. While most studies report a somewhat atypical phragmoplast in which microtubules extend from nuclei, a typical centrifugally expanding phragmoplast with free microtubule ends is seen following telophase I in the liverwort *C. conicum*. However, no cell plate is deposited (Brown and Lemmon, 1988c). This indicates that phragmoplast microtubule organization and vesicle coalescence leading to cell-plate formation can be uncoupled. Although a wall is not produced after first nuclear division in simultaneous cytokinesis, the cytoplasm is generally polarized into two domains separated by an organelle band.

Simultaneous quadripartitioning is often initiated after meiosis by infurrowing (Fig. 19.14) of the cytoplasm between the tetrad nuclei (Farr, 1916; Longly and Waterkeyn, 1979). Although infurrowing progresses centri-petally, the completion of cytokinesis is often accomplished by engulfment of embayments of wall resulting from irregular coalescence of vesicles along the division planes (e.g. Echlin

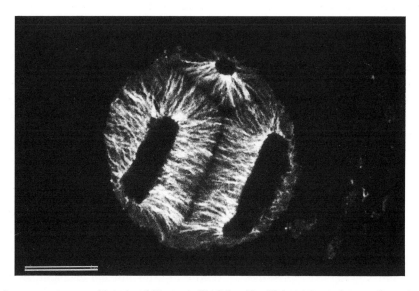

Fig. 19.15. In successive cytokinesis of *Hemerocallis fulva* (daylily) micronuclei resulting from irregularities in first meiotic division alter the pattern of cytokinesis. This demonstrates the importance of nuclear material in the organization of microtubules that determine the division plane. × 1800. Scale bar = 5 μm.

and Godwin, 1968; Pacini and Juniper, 1984; Van Went and Cresti, 1988). The latter stages of tetrad separation may appear ragged with cytoplasmic bridges often observed in equatorial regions between microspores (e.g. Buss *et al.*, 1969; Longly and Waterkeyn, 1979; Van Went and Cresti, 1988).

In simultaneous meiotic cytokinesis of orchids, there is no evidence of infurrowing whatsoever (Fig. 19.12), and cell-plate formation results from vesicle coalescence centrifugally along a phragmoplast complex defined by interaction among opposing nuclear-based microtubule arrays (Konta and Tsuji, 1982; Brown and Lemmon, 1989b). In *Gasteria*, another monocot with simultaneous cytokinesis, wall deposition occurs more or less evenly along equatorial regions defined by interaction of numerous microtubules radiating from the nuclei, and a thin layer of callose surrounds each newly formed microspore (Van Lammeren *et al.*, 1985). Recently, centrifugally expanding phragmoplasts have been reported in the primitive dicot *Laurelia* (Sampson, 1969). The author discusses other possible examples among little-studied primitive dicots.

A condition that appears to be intermediate between successive and simultaneous cytokinesis occurs in cycads (Rodkiewicz *et al.*, 1988) and certain primitive dicots (*Magnolia* (Farr, 1918) and *Asimina* (Locke, 1936)). In these cases, a much-thickened infurrowing develops at the first division equator. In cycads and *Asimina* the infurrowing grows until it contacts the interzonal spindle where a cell plate has formed. The infurrowing may stop at this point or, as in *Asimina*, the cell plate may break up and infurrowing continue until a dyad results. Similar infurrowings after second division result in a tetrad.

Conclusions

The cytokinetic apparatus in sporogenesis lacks the PPB component that faithfully predicts the division site in mitotic cells of vegetative tissues. Precise control of division plane in meiotic cells may occur at multiple levels and is reflected in the establishment of cytoplasmic domains, either before or after meiosis. Cytoplasmic lobing and complex cytoskeletal systems involved in the establishment and maintenance of division quadripolarity are early expressions of the cytokinetic apparatus peculiar to the sporocytes of lower plants. Spore domains are precociously established in anticipation of the disposition of nuclei resulting from two future nuclear divisions. This is contrary to most cases where cytoplasmic domains are organized around a nucleus, and indicates that the cytoplasm can respond to division signals in a way that is independent of the progress of nuclear division. In higher plants, there is no structural evidence of predivision determination of cleavage planes, and spore domains appear to be claimed after nuclear division by nuclear-based microtubule arrays. Thus, postmeiotically determined cleavage planes reflect the position of nuclei resulting from the orientation of spindle axes. Whereas orientation of the meiotic spindles in bryophytes and lycopsids is strictly controlled by the precursor QMS anchored in cytoplasmic lobes destined to become spores, the mechanisms responsible for spindle alignment in ferns and seed plants remain unknown.

Variation in the pattern of cytokinesis in sporogenesis provides some clues to the coordination of the various components of the cytokinetic apparatus. Cytokinesis can occur centripetally, centrifugally, or irregularly along the cytokinetic planes. It can be absent, incomplete, or complete after first nuclear division. Tetrad patterns can be precise or highly variable. This wide variation in the pattern of quadripartitioning during sporogenesis suggests that sporocytes are not under the control of a division site residing in the cortical cytoplasm, such as is thought to control the orderly progress of cell-plate formation in mitotic cytokinesis. In sporogenesis, premeiotic determination of division plane can be lost without affecting precise control of the division plane. We interpret this to mean that cytoplasmic domains, which can be determined before or after nuclear division, do not depend on predictive expressions of the cytokinetic apparatus; as for example the PPB in mitosis, or cytoplasmic lobing and the QMS in meiosis.

Ultimately, it appears that in some examples (the liverwort *Conocephalum* and the massulate orchids), strict control of spindle orientation can be lost and randomly arranged nuclei can claim equal domains of cytoplasm. This suggests that postmeiotic nuclei organize a radial microtubule system that measures a domain of cytoplasm and controls its apportionment. This lends support to the concept of the nucleus as an important centre for the organization of microtubules in cells of higher plants. However, in monoplastidic sporocytes, microtubules emanate from the plastids.

How microtubules define the extent of a cytoplasmic domain awaits further investigation. In all multinucleate coenobia, radial systems of microtubules appear to be an important component of the cytoskeleton. In microsporogenesis, microtubules radiating from the nuclei, even micronuclei resulting from lagging chromosomes, somehow serve to measure the spore domains with division occurring along the fronts of opposing arrays. Anucleate fragments of cytoplasm resulting from removal of free-nucleate endosperm from the ovary of *Haemanthus* develop phragmoplast-like structures where the ends of different microtubule arrays interact (Bajer and Mole-Bajer, 1986). Such observations suggest that the interaction of opposing sets of microtubules is a critical component in selection of the division plane. However, there are exceptions. For example, during embryogenesis in *Ephedra* and megagametogenesis of certain angiosperms, walls can be laid down around domains that are defined by an unopposed set of nuclear-based microtubules (see Timberlake, 1900, for discussion). Thus, the determining factors appear to be resident at the distal tips of microtubules.

Does the concept of cytoplasmic domains have any bearing on the control of division plane in mitosis? In *Drosophila* embryos, cytoplasmic domains (caps) become bilobed at the time of centriole division prior to mitosis, clearly indicating that the cytoplasm anticipates the cap domains in which the daughter nuclei will reside (Warn, 1986). In this case, the domains are organized around centrioles. Although mitotic cells of vegetative tissues in plants lack centrioles, and the possibility of detectable cytoplasmic lobing is precluded by the presence of an interconnecting system of rigid cell walls, there is nevertheless some reason to consider that the cytoplasm is reorganized into two domains while it still contains a single nucleus. Such polarity has been described in mitotic cells of the liverwort *Reboulia* where the first evidence of the establishment of division polarity is the appearance of opposing sets of microtubules emanating from centrosomes (Brown and Lemmon, 1990c). These astral-like microtubules radiating from the discrete centrosomes at opposite ends of the cell contribute to premitotic shaping of the nucleus, and to positioning of the PPB. Thus, the PPB seems to be a reflection of established polarity rather than the determinant. Experiments designed to analyse the cytokinetic apparatus in sporogenesis and to test the role of cytoplasmic domains in plant cell division may further advance the knowledge of the control of division plane, the nature of the division site, and the origin and function of the PPB.

Acknowledgements

This work was supported in part by NSF grant BSR 8610594. Professor J. Heslop-Harrison and Dr C. Staiger kindly read and commented on the manuscript.

References

Bajer, A.S. and Mole-Bajer, J. (1986). Reorganization of microtubules in endosperm cells and cell fragments of the higher plant *Haemanthus in vivo*. *J. Cell Biol.* **102**, 263–281.

Brown, R.C. and Lemmon, B.E. (1985). A cytoskeletal system predicts division plane in meiosis of *Selaginella*. *Protoplasma* **127**, 101–109.

Brown, R.C. and Lemmon, B.E. (1987a). Division polarity, development and configuration of microtubule arrays in bryophyte meiosis I. Meiotic prophase to metaphase I. *Protoplasma* **137**, 84–99.

Brown, R.C. and Lemmon, B.E. (1987b). Division polarity, development and configuration of microtubule arrays in bryophyte meiosis II. Anaphase I to the tetrad. *Protoplasma* **138**, 1–10.

Brown, R.C. and Lemmon, B.E. (1988a). Sporogenesis in bryophytes. *Adv. Bryol.* **3**, 159–223.

Brown, R.C. and Lemmon, B.E. (1988b). Prepro-phasic microtubule systems and development of the mitotic spindle in hornworts (Bryophyta). *Protoplasma* **143**, 11–21.

Brown, R.C. and Lemmon, B.E. (1988c). Cytokinesis occurs at boundaries of domains delimited by nuclear-based microtubules in sporocytes of *Conocephalum conicum* (Bryophyta). *Cell Motil. Cytoskel.* **11**, 139–146.

Brown, R.C. and Lemmon, B.E. (1988d). Microtubules associated with simultaneous cytokinesis of coenocytic microsporocytes. *Amer. J. Bot.* **75**, 1848–1856.

Brown, R.C. and Lemmon, B.E. (1989a). Morphogenetic plastid migration and microtubule organization during megasporogenesis in *Isoetes*. *Protoplasma* **152**, 136–147.

Brown, R.C. and Lemmon, B.E. (1989b). Minispindles and cytoplasmic domains in microsporogenesis in orchids. *Protoplasma* **148**, 26–32.

Brown, R.C. and Lemmon, B.E. (1990a). Monoplastidic cell division in lower land plants. *Amer. J. Bot.* **77**, 559–571.

Brown, R.C. and Lemmon, B.E. (1990b). The quadripolar microtubule system and meiotic spindle ontogeny in hornworts (Bryophyta: Anthocerotae). *Amer. J. Bot.* **77**, 1482–1490.

Brown, R.C. and Lemmon, B.E. (1990c). Polar organizers mark division axis prior to preprophase band formation in mitosis of the liverwort *Reboulia*. *Protoplasma* **156**, 74–81.

Brown, R.C. and Lemmon, B.E. (1991). Pollen development in orchids I. Cytoskeleton and the control of division plane in irregular patterns of cytokinesis. *Protoplasma*, in press.

Busby, C.H. and Gunning, B.E.S. (1988a). Establishment of plastid-based quadripolarity in spore mother cells of the moss *Funaria hygrometrica*. *J. Cell Sci.* **91**, 117–126.

Busby, C.H. and Gunning, B.E.S. (1988b). Development of the quadripolar meiotic cytoskeleton in spore mother cells of the moss *Funaria hygrometrica*. *J. Cell Sci.* **91**, 127–137.

Busby, C.H. and Gunning, B.E.S. (1989). Development of the quadripolar meiotic apparatus in *Funaria* spore mother cells: analysis by means of anti-microtubule drug treatments. *J. Cell Sci.* **93**, 267–277.

Buss, P.A., Jr, Galen, D.F. and Lersten, N.R. (1969). Pollen and tapetum development in *Desmodium glutinosum* and *D. illinoense* (Papilioneae: Leguminosae). *Amer. J. Bot.* **56**, 1203–1208.

Cassimeris, L., Walker, R.A., Pryer, N.K. and Salmon, E.D. (1987). Dynamic instability of microtubules. *BioEssays* **7**, 149–154.

Christensen, J.E. and Horner, H.T., Jr (1974). Pollen pore development and its spatial orientation during microsporogenesis in the grass *Sorghum bicolor*. *Amer. J. Bot.* **61**, 604–623.

Dover, G.A. (1972). The organization and polarity of pollen mother cells of *Triticum aestivum*. *J. Cell Biol.* **11**, 699–711.

Echlin, P. and Godwin, H. (1968). The ultrastructure and ontogeny of pollen in *Helleborus foetidus* L. II. Pollen grain development through the callose special wall stage. *J. Cell Sci.* **3**, 175–186.

Farr, C.H. (1916). Cytokinesis of the pollen-mother-cells of certain dicotyledons. *Mem. New York Bot. Gard.* **6**, 253–316.

Farr, C.H. (1918). Cell division by furrowing in *Magnolia*. *Amer. J. Bot.* 379–395.

Goff, L.J. and Coleman, A.W. (1987). The solution to the cytological paradox of isomorphy. *J. Cell Biol.* **104**, 739–748.

Gunning, B.E.S. (1982). The cytokinetic apparatus: Its development and spatial regulation. In *Cytoskeleton in Plant Growth and Development* (ed. C.W. Lloyd), pp. 229–292. Academic Press, London.

Heslop-Harrison, J. (1971). Wall pattern formation in angiosperm microsporogenesis. *Sym. Soc. Exp. Biol.* **25**, 277–300.

Hogan, C.J. (1987). Microtubule patterns during meiosis in two higher plant species. *Protoplasma* **138**, 126–136.

Huynh, K.L. (1976). Arrangement of some monosulcate, disulcate, trisulcate, dicolpate, and tricolpate pollen types in the tetrads and some aspects of evolution in the angiosperms. In *The Evolutionary Significance of the Exine* (eds I.K. Ferguson and J. Muller), pp. 101–124. Academic Press, London.

Konta, F. and Tsuji, M. (1982). The types of pollen tetrads and their formations observed in some species in the Orchidaceae in Japan. *Acta Phytotax. Geobot.* **33**, 206–217.

Lloyd, C.W. (1987). The plant cytoskeleton: The impact of fluorescence microscopy. *Ann. Rev. Plant Physiol.* **38**, 119–139.

Locke, J.F. (1936). Microsporogenesis and cytokinesis in *Asimina triloba*. *Bot. Gaz.* **98**, 159–169.

Longly, B. and Waterkeyn, L. (1979). Étude de la cytocinese III. Les cloisonnements simutanes et successifs des microsporocytes. *La Cellule* **73**, 65–80.

Menzel, D. (1986). Visualization of cytoskeletal changes through the life cycle in *Acetabularia*. *Protoplasma* **134**, 30–42.

Pacini, E. and Juniper, B. (1984). The ultrastructure of pollen grain development in *Lycopersicum peruvianum*. *Caryologia* **37**, 21–50.

Rodkiewicz, B. and Duda, E. (1988). Aggregations of organelles in meiotic cells of higher plants. *Acta Soc. Bot. Pol.* **57**, 637–654.

Rodkiewicz, B., Bednara, J., Kuras, M. and Mostowska, A. (1988). Organelles and cell walls of microsporocytes in a cycad *Stangeria* during meiosis I. *Phytomorphology* **38**, 99–110.

Sampson, F.B. (1969). Cytokinesis in pollen mother cells of angiosperms, with emphasis on *Laurelia novae-zelandiae* (Monimiaceae). *Cytologia* **34**, 627–634.

Sheldon, J.M. and Dickinson, H.G. (1983). Determination of patterning in the pollen wall of *Lilium henryi*. *J. Cell Sci.* **63**, 191–208.

Sheldon, J.M. and Dickinson, H.G. (1986). Pollen wall formation in *Lilium*: the effect of chaotropic agents, and the organization of the microtubular cytoskeleton during pattern development. *Planta* **168**, 11–23.

Staiger, C.J. and Cande, W.Z. (1990). Microtubule distribution in *dv*, a maize meiotic mutant defective in prophase to metaphase transition. *Dev. Biol.* **138**, 231–242.

Timberlake, H.G. (1900). The development and function of the cell plate in higher plants. *Bot. Gaz.* **30**, 73–170.

Traas, J.A., Burgain, S. and Dumas de Vaulx, R. (1989). The organization of the cytoskeleton during meiosis in eggplant (*Solanum melongena* L.): microtubules and F-actin are both necessary for coordinated meiotic division. *J. Cell Sci.* **92**, 541–550.

Van Lammeren, A.A.M., Keijzer, C.J., Willemse, M.T.M. and Kieft, H. (1985). Structure and function of the microtubular cytoskeleton during pollen development in *Gasteria verrucosa* (Mill.) H. Duval. *Planta* **165**, 1–11.

Van Lammeren, A.A.M., Bednara, J. and M.T.M. Willemse (1989). Organization of the actin cytoskeleton during pollen development in *Gasteria verrucosa* (Mill.) H. Duval visualized with rhodamine–phalloidin. *Planta* **178**, 531–539.

Van Went, J. and Cresti, M. (1988). Cytokinesis in microspore mother cells of *Impatiens sultani*. *Sex. Plant Reprod.* **1**, 228–233.

Warn, R.M. (1986). The cytoskeleton of the early *Drosophila* embryo. *J. Cell Sci. Suppl.* **5**, 311–328.

Waterkeyn, L. (1962). Les parois microsporocytaires de nature callosique chez *Helleborus* et *Tradescantia*. *Cellule* **62**, 225–255.

Wick, S.M. and Duniec, J. (1983). Immunofluorescence microscopy of tubulin and microtubule arrays in plant cells. I. Pre-prophase band development and concomitant appearance of nuclear envelope-associated tubulin. *J. Cell Biol.* **97**, 235–243.

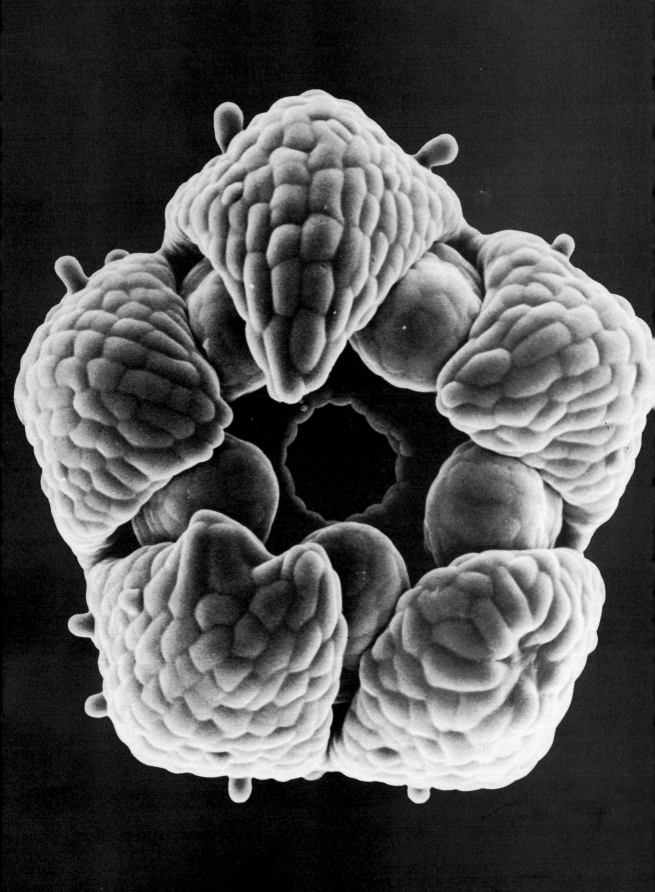

THE CYTOSKELETON IN PLANT DEVELOPMENT

The presence of a more or less rigid wall imposes limitations on the cytoplasmic behaviour of plants. For instance, plant cells cannot exchange partners as occurs so dramatically in animal development, regeneration and wound repair. The ways in which plants and animals are constructed from their basic units are therefore quite different. Although the ability of mature, thick-walled cells to change their shape is restricted, adaptive meristematic growth allows for new modules to be produced repetitively from the apex to add on to the increasingly inflexible earlier stages of growth. If we consider growth in organizationally simple cell files, the direction in which the apical cell extends, how it responds to environmental cues and how it divides can be seen to have a clear effect on the morphology of the file as a whole. To change the basic pattern of growth, subapical cells in filamentous organisms must divide in a plane different to that of the serial cross walls, and this occurs during side branch initiation.

The involvement of the cytoskeleton in filamentous growth is discussed in Chapters 20 and 21, which contain data on division plane determination that complement the earlier section on this theme. However, the alignment of the division plane is considered, here, more in the context of organismal development than as an isolated mechanism. Although the processes of directional cell expansion and cell plate orientation, as they occur in single cells, are known in very broad outline, it is far less clear how these processes as building blocks are co-ordinated in tissues to produce organized architecture. The way in which cells communicate with one another to mould development is only dimly perceived as is the genetic basis of cell organization. Filamentous plants, however, possess a simplified organization which may be useful as a model for events similar to those occurring in more complex tissues. Chapters 20 and 21 describe, for filamentous organisms, how endogenous plant growth substances and exogenous stimuli (gravity, light) that affect development, produce effects on the cytoskeleton. Establishing the links between stimulus and morphological response is an essential part of understanding how growth is co-ordinated. Understanding how the genetic programme unfolds is a much larger task, beyond the scope of this book, but developmental mutants are available for some mosses and it is obviously essential to accelerate the search for mutants that affect aspects of cytoskeletal structure and control.

Filamentous plants that advance by tip growth do not, however, model apical growth in higher plants for the reason that the expansion of many more cells by intercalary growth, their connectedness and the patterns that their divisions make, are considerably more complicated. Internal cells are compressed by neighbours, and outer cells will be under tension. Epigenetic, biophysical forces such as these are likely to be an intrinsic part of development, affecting the way in which cells pack together and, possibly, the way in which they divide. Such interactive forces cannot be studied in isolated cells or simple systems nor are they likely to be understood by reference to the genome alone. Biological information of this kind requires a different kind of analysis and in Chapter 22, Green and Selker discuss the possible relationship between the cytoskeleton, biophysical forces and apical growth.

20. THE CYTOSKELETON IN FERN PROTONEMATAL GROWTH IN RELATION TO PHOTOMORPHOGENESIS

Masamitsu Wada and Takashi Murata

*Department of Biology, Faculty of Science, Tokyo Metropolitan University,
Minami Ohsawa, Hachioji, Tokyo 192–03, Japan*

The function of cortical microtubules (MTs) is understood in general but the details are not clear. For instance, although the preprophase band (PPB) is involved in controlling the plane of cell division (Gunning, 1982), and cortical MTs are involved in controlling the arrangement of microfibrils (Robinson and Quader, 1982), we can only guess at precise mechanisms of MT behaviour. To study cytoskeletal function in morphogenesis, experimental systems in which cytoskeletal and/or morphogenetic changes can be predicted in advance are needed.

Fern protonemal cells are one of the best materials for studying plant morphogenesis at cellular and subcellular levels for the following reasons:

(1) First, since they are tip-growing filamentous cells when cultured appropriately, and they are not surrounded by any other tissues, they can be easily observed under a light microscope (Fig. 20.1) and cell manipulations can be performed easily.

(2) The development of fern protonemata is regulated by environmental light conditions in a clear-cut way such that the elementary processes of morphogenesis can be synchronously induced in many cases by a short pulse of light (apical swelling is an exception and needs continuous irradiation) (Miller, 1968; Furuya, 1983; Wada and Kadota, 1989).

(3) The phenomena are induced in a part of a cell, e.g. apical swelling occurs at the apical part of cells (Wada *et al.*, 1978), and photo-tropism is induced at the light irradiated side of a protonema (Kadota *et al.*, 1982).

(4) The photoreceptive sites of these pheno-nema are already known in most cases, e.g. the blue light receptor for cell division is localized in the nuclear region (Wada and Furuya, 1978) and phytochrome for the tropic response is localized in the subapical part of the cells (Wada *et al.*, 1981), so that these photoregulated phenomena can be induced by local irradiation at the photo-receptive site with a microbeam of a short pulse. These short and local irradiations can induce the phenomenon required without other side effects.

(5) The photoregulation by red light can produce long protonemal cells, or long and L-shaped protonemal cells, by changing the direction of the red light source (Fig. 20.1). In these long (and L-shaped) cells, the location of organelles can easily be changed by centrifugation (Wada *et al.*, 1983).

Cytoskeletal changes in fern protonemata during photomorphogenesis have been studied by electron microscopy (Stetler and DeMaggio, 1972; Wada and O'Brien, 1975; Wada *et al.*, 1980; Mineyuki and Furuya, 1986), but recently developed fluorescence microscopy has enabled us to study much more readily the overall distribution and arrangement of the cyto-skeleton. Fortunately, anti-chicken α tubulin and β tubulin antibodies cross-react with fern

The Cytoskeletal Basis of Plant Growth and Form   ISBN 0–12–453770–7

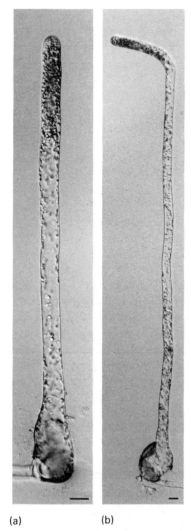

(a)   (b)

Fig. 20.1. *Adiantum* protonemata (with one proto-nemal cell and one rhizoid) cultured under continu-ous red light for 4 (a) or 6 (b) days. In (b), direction of red light was changed (from top to left in the panel) at 5.4 days. Protonemal cells grew towards the light source without cell division. Rhizoidal cells (growing from the base of protonemal cells) showed no phototropism, although only basal part of the rhizoidal cells is shown. Scale bars = 20 μm.

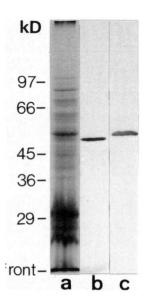

Fig. 20.2. Reactivities of anti-tubulin antibodies with proteins from *Adiantum* protonemata. Proto-nemata cultured under red light for 5 days were homogenized with the same weight of SDS buffer (0.25 M Tris-HCl (pH 6.8), 20% glycerol, 10% 2-mercaptoethanol, 4% SDS, 0.02% bromophenol blue) for 10 min on ice. The homogenate was boiled for 3 min in boiling water, and centrifuged at 15 000 × *g* for 30 min. The supernatant was analysed by SDS-PAGE (Laemmli's system) using 10% gel. The separated proteins were stained with Coomassie blue (a) or analysed by Western blotting procedure using anti-chicken α tubulin monoclonal antibody (N.356, Amersham) (b) or anti-chicken β tubulin monoclonal antibody (N.357, Amersham) (c). Molecular masses of marker proteins are shown in kilodaltons (kD).

The cytoskeleton in tip-growing protonemata under red light

Fern protonemal cells show, in general, typical tip growth and the growing zone is restricted mainly to the apical dome (Takahashi, 1961). Under continuous red light, protonemata grow linearly towards the light source with a low rate of cell division, resulting in long single cells (Wada and Furuya, 1970). The nucleus migrates in a cell towards the tip during the tip growth maintaining a constant distance (*c.* 60 μm in *Adiantum*) from the tip. Phyto-chrome is the photoreceptor for tip growth

tubulin (Fig. 20.2) and rhodamine-labelled phalloidin binds to fern F-actin. We have been studying changes of the cytoskeleton during light-induced cell growth, apical swelling, tropic responses, and cell division in *Adiantum* proto-nemata in order to establish the function of the cortical cytoskeleton in morphogenesis.

(Kadota and Furuya, 1977) as is the case for the tropic response, but its localization in a protonema is not known. The red light effect on cell growth is, however, somewhat different in some species. In *Pteris vittata*, a protonemal cell grows without cell division under red light, but a phototropic response could not be observed (Kadota *et al.*, 1989). In *Lygodium japonicum*, two-dimensional growth is induced even under red light (Raghavan, 1973), and filamentous growth occurs under far-red light or in the darkness (Raghavan, 1973; Takeno and Furuya, 1975).

The arrangement of cortical MTs around the apical part of a protonemal cell was first studied in the electron microscope by Stetler and DeMaggio (1972) in *Dryopteris filix-mas* protonemata. They reported a random arrangement of MTs around the subapical part and an arrangement parallel to the long axis in the cylindrical part. In *Adiantum* protonemata, a circular arrangement of MTs (MT-band) (Murata *et al.*, 1987; see Fig. 20.3b) was observed around the subapical part of the protonemal cells, although in tip growing rhizoids of the same species such a structure

has not been observed so far (Murata *et al.*, 1987). In the apical dome of a protonemal cell, MTs are rather scarce and run at random directions if any. MTs in the cylindrical part run mostly parallel to the long axis as in *Dryopteris* (Stetler and DeMaggio, 1972) but in some cases may be oblique. The existence of a MT-band at the subapical region has now been confirmed in all fern species investigated, such as *Anemia phyllitidis, Dryopteris erythrosora, Matteuccia struthiopteris, Onoclea sensibilis, Pteridium aquilinum, Pteris vittata*, even in *Dryopteris filix-mas* (Fig. 20.4). No such structure has been observed in other tip-growing cells such as moss protonemata, pollen tubes, root hairs, not even in a fern rhizoid of the same species. However, the general occurrence of the MT-band at the subapical part implies that it plays an essential role in tip-growing fern protonemata. However, the MT-band does disappear in some cases. When a protonema was centrifuged to take organelles down to basal part, away from the tip, the MT-band disappeared even though a thin layer of cortical cytoplasm (ectoplasm) still remained (Murata *et al.*, unpublished data). When protonemata were

(a) (b) (c)

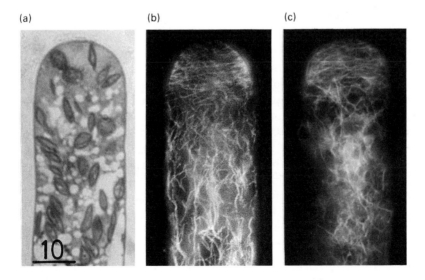

Fig. 20.3. Light micrographs of the apical part of *Adiantum* protonemata. (a) Longitudinal section. (b) Cortical MTs stained by indirect fluorescence with anti-tubulin antibodies. (c) Cortical microfilaments stained with rhodamine-labelled phalloidin. Note that transversely arranged cortical MTs (MT-band) and microfilaments can be seen at the subapical part of a protonema. Scale bar = 10 μm. (b, from Wada, *et al.*, 1990a, with permission; c, from Kadota and Wada, 1989a, with permission.)

(a)   (b)

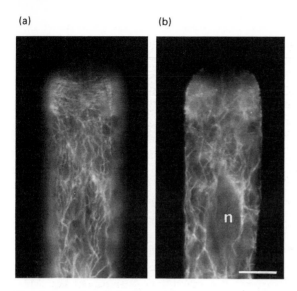

Fig. 20.4. Microtubule arrangement in a *Dryopteris filix-mas* protonema cultured under continuous red light. a, Cortical MTs. Note the existence of MT-band; b, endoplasmic MTs emerging from the nucleus (n). Scale bar = 10 μm.

transferred from red light to the dark, the MT-band remained during cell growth but disappeared when apical growth ceased (Wada *et al.*, 1980; Murata and Wada, 1989a). These results tend to suggest that the MT-band functions during protonemal growth.

When F-actin was stained with rhodamine–phalloidin (Kadota and Wada, 1989a) a similar circular arrangement of cortical microfilaments was observed around the subapical part of protonemal cells in both *Adiantum capillus-veneris* (Fig. 20.3c) and *Pteris vittata*. This establishes the co-existence of MTs and microfilaments in the same region. However, neither the molecular relationship between both filament systems nor their roles are known. In a growing rhizoid, a distinct polarity of F-actin distribution was observed, with the apex being stained densely. However, no circular arrangement of F-actin was observed.

The arrangement of microfibrils in the innermost layer of the cell wall was studied by electron microscopy (Murata and Wada, 1989c), and it was revealed that the arrangement of microfibrils is very similar to those of MTs throughout the protonemal cells (Fig. 20.5). In

the apical part of the cell, microfibrils are randomly oriented. In the subapical part, they are perpendicular to the growing axis and parallel with the MT-band. In the cylindrical part, they are parallel with the growing axis. Since the cellulose microfibrils are thought to be oriented by rosettes (membrane synthases) moving along tracks provided by underlying MTs (see Giddings and Staehelin, Chapter 7, this volume), the MT/microfibril parallelism has the appropriate arrangement in these cells. It is already known that rosettes are abundant in the tip region of the fern protonemal the cell where growth activity is prominent (Wada and Staehelin, 1981).

Microtubule bands and apical cell swelling

The direction of cell expansion in higher plants is determined by the physical properties of the cell wall (see Kutschera, Chapter 11, this volume). Tip-growing cells may also regulate their diameter during growth in some manner. Spherical expansion at the tip of protonemal cells can easily be induced by irradiation with blue light (Fig. 20.6). This is a good model for cell expansion because the phenomenon occurs at the restricted area under the control of blue light. When red-light-grown protonemata (Fig. 20.6a) are transferred to blue light, they continue to grow for a while without any change of cell shape and then the apical part of the cell starts to swell 1–2 h after the onset of blue-light irradiation at the area newly grown under blue-light condition (Fig. 20.6b, c). The cylindrical part grown under red light does not swell (Murata and Wada, 1989c). The blue-light receptor (chemical properties unknown) that mediated the cell swelling is believed to be localized close to, or on, the plasma membrane because of the dichroic effect of polarized blue light on the photoresponse (Wada *et al.*, 1978). First, the MT-bands around the subapical part of protonemata disappear, randomly oriented MTs remain (Fig. 20.7a), and then cell swelling becomes detectable. Microfibrils also change their arrangement from transverse to random at the subapical part of the cells before swelling is seen. These results clearly indicate that

(a) (b) (c)

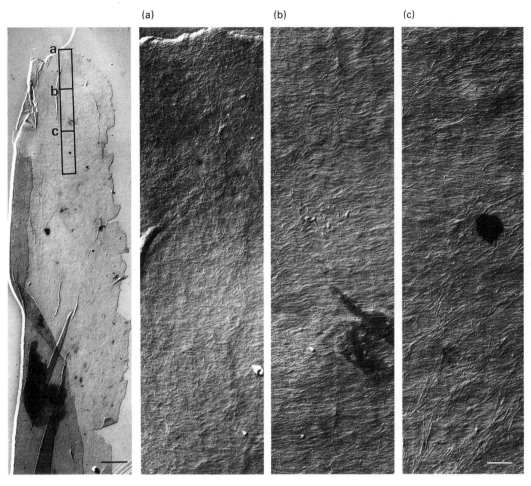

Fig. 20.5. Arrangement of microfibrils on the innermost layer of the cell wall of a protonemal cell of *Adiantum* growing under red light. The boxed regions (a, b, c) in a cell-wall sample (left-hand side panel) are magnified and shown in panels a, b, c, respectively. Scale bars = 5 μm (left) and 0.5 μm (right). (Adapted from Murata and Wada, 1989c, with permission.)

oriented cortical MTs are involved in regulating cell diameter in tip-growing protonemal cells, mediated by microfibrils. Similarly, cell swelling at the apex can be induced by treatments with colchicine and amiprophos-methyl, known anti-MT agents (Murata and Wada, 1989b) or sulphite, an air pollutant (Wada *et al.*, 1990b). These chemicals disrupted MT-bands and apical swelling subsequently occurred after a lag period. However, in protonemata treated with anti-MT agents, the pattern of microfibril deposition was similar to that in untreated cells (Murata and Wada, 1989b).

Rings of microfilaments at the subapical part

of a cell also disappeared preceding apical swelling (Kadota, personal communication). There is clear evidence, therefore, that the cytoskeleton is involved in regulating the diameter of the tip, although the relationship (if any) between actin filaments and MTs is unknown.

Microtubule bands and tropic responses

Polarotropism or phototropism is easily induced in *Adiantum* protonemata by irradiating the whole cell with polarized red light, or one side

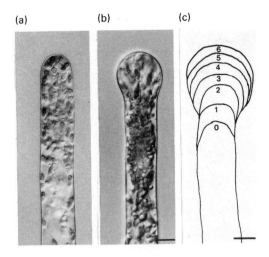

Fig. 20.6. Photomicrographs of *Adiantum* protonemal cells growing under red light (a) and irradiated with blue light for 4 h (b), and time-lapse sequence of apical swelling under blue light (c). Cell outlines were traced at 1-h intervals. Numbers are the time in hours after the onset of blue light. Scale bars = 10 μm. (From Murata and Wada, 1989c, with permission.)

of the subapical part of a protonema with a red microbeam. During the tropic response in a tip-growing cell, a new growing point is established and the former growing point ceases to function. These responses have been carefully analysed photobiologically (Wada and Kadota, 1989). The photoreceptor (that is phytochrome) is localized on, or very close to, the plasma membrane (Wada *et al.*, 1983) and the angle of the tropic response is controlled by the difference in the concentration of Pfr (far-red light absorbing form of phytochrome, considered to be the active form) on both sides of the subapical region of protonema (Iino *et al.*, 1990). Time-course studies of these responses revealed that the protonemal cell grew straight for about 1 h before it showed symptoms of tropic response, i.e. a slight distortion of the apical dome (Wada *et al.*, 1990a). The time lag of this response, as was also the case for apical cell swelling (Murata and Wada, 1989c; see also the previous section), may be due to rearrangements of shape-controlling structures (e.g. wall, cytoskeleton) that allow the effects of light signals to become manifest. The changes in MT and microfibril arrangements occur before the manifestation of the tropic responses (Wada *et al.*, 1990a; see Fig. 20.7b, c). Twenty minutes after inducing the tropic response (by turning the electrical vector of red polarized light) the MT-band began to change its orientation from perpendicular to oblique to the initial growth axis. This was before any changes in cell shape. After 30 min, the change in the orientation of the MT-band became clearer under 45° polarized light, but subsequently the MT-band began to disappear under 70° (Fig. 20.7c). The inconsistency of MT-band

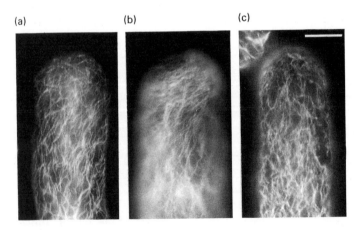

Fig. 20.7. Arrangement of cortical MTs in *Adiantum* protonemal cells under various light conditions. a, Blue light for 3 h; b, c, polarized red light, 1 h after turning the electrical vector 45° (b) or 70° (c). Scale bar = 10 μm. (a, from Murata and Wada, 1989c, with permission; b, c, from Wada *et al.*, 1990a, with permission.)

behaviour under different electrical vectors is hard to understand, but it may depend on local differences of Pfr concentration or its distribution in the apical dome. Actually, when local irradiation was employed to induce the tropic response (i.e. high Pfr concentration on the irradiated side and very low on the non-irradiated side), the MT-band on the irradiated side disappeared or became faint within 20 min, but did not disappear nor reorientate on the non-irradiated side.

The precise mechanism by which the MT-band changes its orientation or is disrupted after red-light irradiation is not known but it is obvious that a new growing point is established at the side with higher concentration of Pfr (Iino et al., 1990). Pfr may control the depolymerization or polymerization of MTs indirectly and one possible function of Pfr in this response could be to promote Ca influx (Serlin and Roux, 1984). A high concentration of Ca^{2+} ions may depolymerize MTs (Kakimoto and Shibaoka, 1986; Wang et al., 1989) and may reorientate the MT-band depending upon intracellular gradients. In turn, the changes in the MT-band influence microfibril orientation as was shown for apical swelling (Murata and Wada, 1989c). One hour after turning the electrical vector through 45° the transverse arrangement of the innermost layer of microfibrils in the subapical part of a protonema changed its orientation from perpendicular to the growing axis to oblique towards the direction of bending, as did the MT-band. This was seen in half of the cells tested but after 2 h the phenomenon became clearer in all protonemata examined. The changes in microfibril arrangement occur at about the same time as the appearance of the tropic response. This suggests that a new growing point may easily be formed, and cell protrusion may occur, where randomly deposited wall microfibrils overlie randomly oriented or depolymerized MTs.

The preprophase band of microtubules

The spatial regulation of cytokinesis is a prerequisite for the development of multicellular plants since cell movements cannot

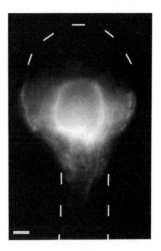

Fig. 20.8. A PPB formed in an *Adiantum* protonemal cell irradiated with white light for 13 h. A dashed white line indicates cell outline. Scale bar = 5 μm.

take place because of the presence of rigid cell walls. Cytoskeletal elements in the PPB and in the phragmosome are the only structures which predict or correlate with the site of the future cell plate, and hence are considered to play an important role in determining the division site. For this reason PPBs have been well studied in higher plant cells (see Chapters 17 and 18 this volume). However, the mechanism for regulating the actual site at which the PPB forms is at least as important, but has hardly been studied. For this purpose, a long cell and the technique of cell centrifugation can be very useful in analysing the roles of cytoplasm (including the nucleus) in regulating the site of PPB formation (Murata and Wada, 1991). Since fern protonemal cells have a PPB (Murata and Wada, 1989a; see Fig. 20.8), in contrast to moss protonemata (Doonan et al., 1985), we studied this structure in ferns.

Synchronous cell division was induced at the apical region of red-light-grown protonemal cells by transferring them to continuous blue or white light or to the dark. PPBs were found under all these conditions. At an early stage of PPB development the PPB – which was transverse to the cell's axis – coincided with an interphase cortical array of MTs which were random or parallel to the cell axis. The

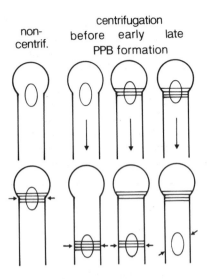

centrifugation

non-centrif.   before   early   late

PPB formation

Fig. 20.9. Schematic drawing of PPB formation (shown as parallel lines perpendicular to the cell axis) in protonemata centrifuged at various times. Oval shapes in protonemata indicate the nuclei. Long arrows in protonemata of upper row indicate the direction of nuclear displacement. Short arrows on both sides of protonemata of lower row indicate the sites and orientation of cell plate after cell division.

interphase cortical array disappeared thereafter. Whereas the width of the PPB became narrower during development under dark conditions, the width of the PPB did not change under blue light (Murata and Wada, 1989a).

Centrifugation of protonemata was performed at various times after, or just before, the induction of synchronous cell division by blue-light irradiation (Fig. 20.9). When protonemata were centrifuged before PPB formation, a PPB was formed in the region of the displaced nucleus in most cells; very few PPBs formed at the former nuclear region. When protonemata were centrifuged after or during PPB formation, PPBs which had already formed in the apical region were not disrupted but remained intact until the sedimented nucleus entered telophase or early interphase, even though newly formed PPB at the new nuclear region disappeared before metaphase. This means that the cytoplasm sedimented by centrifugation may include a factor(s) which disrupts PPBs just before metaphase. Two

PPBs could be observed in the same cell, one in the former and one in the new nuclear position, if centrifugation was performed at the appropriate time.

Formation of a new PPB at the sedimented nuclear region indicates that positioning of PPBs is influenced by the position of the nucleus (and/or other cytoplasm moved down with the nucleus) in *Adiantum* protonemata. The fact that two PPBs are induced in a cell by centrifugation indicates the following possibilities: if the nucleus induces PPB formation only once at the beginning of its development, but is not required thereafter, the centrifugation could re-set the clock so that the whole process is recapitulated at each new position. On the other hand the nucleus may retain its inductive activity throughout PPB development, in which case double PPB formation could represent a normal scheduled activity, but at two separate locations. Either way, it is clear that a nucleus can greatly influence PPB location and that this inductive activity, retained or re-set, may last until at least the latter stages of PPB development.

When protonemal cells were centrifuged at the late stage of, or after, establishment of the PPB (PPB maturation), the number of protonemata without PPBs at the new nuclear region increased and oblique cell plates were observed at high frequency (Murata and Wada, 1991). Cell division without a PPB was also observed in protonemata cultured as follows: when red-light-grown protonemata were transferred to the dark to induce cell division the protonemata continued to grow for one day and then cell growth ceased. When the cells were returned to red light again, the cells resumed growth and then cell division occurred. Under these red-light conditions the frequency of protonemata without PPBs and that of oblique cell division increased (Mineyuki *et al.*, in preparation). These results confirm that in *Adiantum* protonemata, formation of a PPB plays an essential role in cell plate orientation.

Fern gametophytes have long been an experimental model for studying the mechanisms of cell plate orientation in plant cells, because, at the process of differentiation from filamentous protonemata to the two-dimensional prothalli, the direction of cell division changes from transverse to longitudinal, i.e. from perpendicular, to parallel to the growing axis (Raghavan, 1989). Three hypotheses for the mechanisms determining the cell plate orientation have been proposed or supported using fern systems: genetic control (Hotta and Osawa, 1958; Raghavan, 1974), minimum area (Sobota and Partanen, 1966; Wada and Murata, 1988), and the anisotropic stress hypothesis (Miller, 1980). Despite the long dispute based on the experiments with fern gametophytes, the outcome has not, unfortunately, been fruitful. Perhaps the PPB should be analysed during two-dimensional differentiation in protonemata for the further advancement of this field.

The transition from cortical microtubules to the PPB

In higher plant cells, the pattern of disappearance of the interphase cortical MTs during the mitotic phase cannot be seen clearly, although MT disruption seems to occur gradually throughout the cell. In fern protonemal cells, the disruption pattern is unique (Murata and Wada, 1989a). The cortical array of MTs at interphase runs parallel with the cell axis in the cylindrical region. MTs begin to disappear at nearly the same time as PPB formation begins. Under blue light, the disruption of cortical MTs starts at approximately 150 μm from the tip (about 120 μm from the nucleus), and spreads towards the tip as far as the nuclear region and towards the base to an area approximately 300–400 μm from the tip (Fig. 20.10a, b). The cortical MTs far from the nucleus remain throughout the cell cycle. The rate of MT destruction is about 20–30 μm h^{-1} in both directions. The disappearance pattern of the cortical MTs between the protonemal tip and the nucleus could not be determined in normal cells, probably because the cell may not be sufficiently long in that region to observe the disruption pattern. Actually, if protonemata were centrifuged to displace the nucleus down towards the base, a similar disruption pattern was observed both from the tip and from the

base of the protonemata (Fig. 20.10c, d). Under dark conditions, the pattern of the disappearance of cortical MTs was somewhat different in many cells from that encountered upon exposure to blue light. Generally, a clear starting point of MT destruction was not observed, but MTs first reoriented from longitudinal to transverse, and then gradually disappeared (Murata and Wada, 1989a).

The mechanism of the disruption of interphase cortical MTs is not known. The area 150 μm below the tip where the disruption starts has a large vacuole and no distinct structure is observed there during interphase. Within the MT-disrupted area, no distinct structure or accumulation of organelles could be found under the light microscope. Displacement of the starting point of MT disruption by nuclear centrifugation suggests that some kind of information from the nuclear region may control the destruction point in long protonemata.

Concluding remarks

Fern protonemal cells are excellent material for cell biological studies. Their length and shape are controllable by light treatments. Cell division, and its timing, are also controlled by light with unparalleled precision. One of the disadvantages of fern protonemal cells is the wax layer surrounding the cell wall (Wada and Staehelin, 1981). This layer prevents the penetration of chemicals and antibodies into a cell. Protoplasts are difficult to obtain because enzymes such as cellulase cannot reach the cell wall. Nevertheless, these problems may yet be overcome by skill. Actually, penetration of antibodies has been performed by cutting the cells instead of using enzymes to digest the cell wall (Murata et al., 1987; Murata and Wada, 1989a, b, c), and protoplasts have been made using wax-less protonemata obtained by a modification of culture conditions (Kadota and Wada, 1989b). The use of fern protonemata, coupled with the techniques of photo-regulation and cell manipulation, permits unique studies which cannot be done using organizationally complicated higher plant cells.

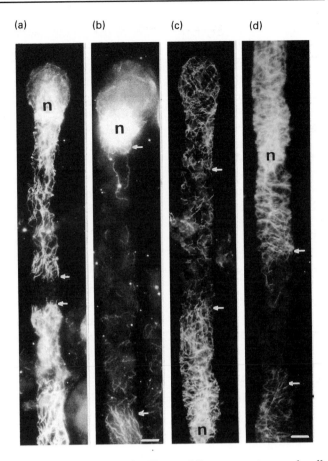

(a) (b) (c) (d)

Fig. 20.10. Disruption of interphase array of MTs in *Adiantum* protonemal cells. n, The nucleus. Small arrows indicate fronts of MT disruption. a, Early stage of disruption; b, late stage of disruption; c, d, disruption of the MT array in a centrifuged cell. Apical half and basal half of a single cell are shown in (c) and (d), respectively. Two regions of MT disruption in the cell are seen between the tip and the nucleus and between the nucleus and the base. Scale bars = 10 μm. (a, b, from Murata and Wada, 1989a, with permission; c, d, Murata and Wada, unpublished.)

References

Doonan, J.H., Cove, D.J. and Lloyd, C.W. (1985). Immunofluorescence microscopy of microtubules in intact cell lineages of the moss, *Physcomitrella patens* I. Normal and CIPC-treated tip cells. *J. Cell Sci.* **75**, 131–147.

Furuya, M. (1983). Photomorphogenesis in ferns. In *Encyclopedia of Plant Physiology, N.S. 16A, B. Photomorphogenesis* (eds W. Schropshire and H. Mohr), pp. 569–600. Springer-Verlag, Berlin.

Gunning, B.E.S. (1982). The cytokinetic apparatus: Its development and spatial regulation. In *The Cytoskeleton in Plant Growth and Development* (ed. C.W. Lloyd), pp. 229–292. Academic Press, London.

Hotta, Y. and Osawa, S. (1958). Control of differentiation in the fern gametophyte by amino acid analogs and 8-azaguanine. *Exp. Cell Res.* **15**, 85–94.

Iino, M., Shitanishi, K., Kadota, A. and Wada, M. (1990). Phytochrome-mediated phototropism in *Adiantum* protonemata – I. Phototropism as a function of the lateral Pfr gradient. *Photochem. Photobiol.* **51**, 469–476.

Kadota, A. and Furuya, M. (1977). Apical growth of protonemata in *Adiantum capillus-veneris*. I. Red far-red reversible effect on growth cessation in the dark. *Develop. Growth Differ.* **19**, 357–365.

Kadota, A. and Wada, M. (1989a). Circular arrangement of cortical F-actin around the subapical region of a tip-growing fern protonemal cell. *Plant Cell Physiol.* **30**, 1183–1186.

Kadota, A. and Wada, M. (1989b). Enzymatic isolation of protoplasts from fern protonemal cells stainable with a fluorescent brightener. *Plant Cell Physiol.* **30**, 1107–1113.

Kadota, A., Wada, M. and Furuya, M. (1982). Phytochrome-mediated phototropism and different dichroic orientation of Pr and Pfr in protonemata of the fern *Adiantum capillus-veneris* L. *Photochem. Photobiol.* **35**, 533–536.

Kadota, A., Kohyama, I. and Wada, M. (1989). Polarotropism and photomovement of chloroplasts in protonemata of the ferns *Pteris* and *Adiantum*: Evidence for the possible lack of dichroic phytochrome in *Pteris*. *Plant Cell Physiol.* **30**, 523–531.

Kakimoto, T. and Shibaoka, H. (1986). Calcium-sensitivity of cortical microtubules in the green alga *Mougeotia*. *Plant Cell Physiol.* **27**, 91–101.

Miller, J.H. (1968). Fern gametophytes as experimental material. *Bot. Rev.* **34** 361–440.

Miller, J.H. (1980). Orientation of the plane of cell division in fern gametophytes: The roles of cell shape and stress. *Amer. J. Bot.* **67**, 534–542.

Mineyuki, Y. and Furuya, M. (1986). Involvement of colchicine-sensitive cytoplasmic element in premitotic nuclear positioning of *Adiantum* protonemata. *Protoplasma* **130**, 83–90.

Murata, T. and Wada, M. (1989a). Re-organization of microtubules during preprophase band development in *Adiantum* protonemata. *Protoplasma* **151**, 73–80.

Murata, T. and Wada, M. (1989b). Effects of colchicine and amiprophos-methyl on microfibril arrangement and cell shape in *Adiantum* protonemal cells. *Protoplasma* **151**, 81–87.

Murata, T. and Wada, M. (1989c). Organization of cortical microtubules and microfibril deposition in response to blue-light-induced apical swelling in a tip-growing *Adiantum* protonema cell. *Planta* **178**, 334–341.

Murata, T. and Wada, M. (1991). Effects of centrifugation on preprophase band formation in *Adiantum* protonemata. *Planta*, **183**, 391–398

Murata, T., Kadota, A., Hogetsu, T. and Wada, M. (1987). Circular arrangement of cortical microtubules and the subapical part of a tip-growing fern protonema. *Protoplasma* **141**, 135–138.

Raghavan, V. (1973). Photomorphogenesis of the gametophytes of *Lygodium japonicum*. *Amer. J. Bot.* **60**, 313–321.

Raghavan, V. (1974). Control of differentiation in the fern gametophyte. *Amer. Sci.* **62**, 465–475.

Raghavan, V. (1989). *Developmental Biology of Fern Gametophytes*. Cambridge University Press, Cambridge.

Robinson, D.G. and Quader, H. (1982). The microtubule–microfibril syndrome. In *The Cytoskeleton in Plant Growth and Development* (ed. C.W. Lloyd), pp. 109–126. Academic Press, London.

Serlin, B.S. and Roux, S.J. (1984). Modulation of chloroplast movement in the green alga *Mougeotia* by the Ca^{2+} ionophore A23187 and by calmodulin antagonists. *Proc. Natl Acad. Sci. U.S.A.* **81**, 6368–6372.

Sobota, A.E. and Partanen, C.R. (1966). The growth and division of cells in relation to morphogenesis in fern gametophytes. I. Photomorphogenetic studies in *Pteridium aquilinum*. *Can. J. Bot.* **44**, 497–506.

Stetler, D.A. and DeMaggio, A.E. (1972). An ultrastructural study of fern gametophytes during one- to two-dimensional development. *Amer. J. Bot.* **59**, 1011–1017.

Takahashi, C. (1961). The growth of protonema cells and rhizoids in bracken. *Cytologia* **26**, 62–66.

Takeno, K. and Furuya, M. (1975). Bioassay of antheridiogen in *Lygodium japonicum*. *Develop. Growth Differ.* **17**, 9–18.

Wada, M. and Furuya, M. (1970). Photocontrol of the orientation of cell division in *Adiantum*. I. Effects of the dark and red periods in the apical cell of gametophytes. *Develop. Growth Differ.* **12**, 109–118.

Wada, M. and Furuya, M. (1978). Effects of narrow-beam irradiations with blue and far-red light on the timing of cell division in *Adiantum* gametophytes. *Planta* **138**, 85–90.

Wada, M. and Kadota, A. (1989). Photomorphogenesis in lower green plants. *Annu. Rev. Plant Physiol. Plant Mol. Biol.* **40**, 169–191.

Wada, M. and Murata, T. (1988). Photocontrol of the orientation of cell division in *Adiantum* IV. Light-induced cell flattening preceding two-dimensional growth. *Bot. Mag. Tokyo* **101**, 111–120.

Wada, M. and O'Brien, T.P. (1975). Observations on the structure of the protonema of *Adiantum capillus-veneris* L. undergoing cell division following white-light irradiation. *Planta* **126**, 213–227.

Wada, M. and Staehelin, L.A. (1981). Freeze-fracture observations on the plasma membrane, the cell wall and the cuticle of growing protonemata of *Adiantum capillus-veneris* L. *Planta* **151**, 462–468.

Wada, M., Kadota, A. and Furuya, M. (1978). Apical growth of protonemata in *Adiantum capillus-veneris* II. Action spectra for the induction of apical swelling and the intracellular photoreceptive site. *Bot. Mag. Tokyo* **91**, 113–120.

Wada, M., Mineyuki, Y., Kadota, A. and Furuya, M. (1980). The changes of nuclear position and distribution of circumferentially aligned cortical microtubules during the progression of cell cycle in *Adiantum* protonemata. *Bot. Mag. Tokyo* **93**, 237–245.

Wada, M., Kadota, A. and Furuya, M. (1981). Intracellular photoreceptive site for polarotropism in protonema of the fern *Adiantum capillus-veneris* L. *Plant Cell Physiol.* **22**, 1481–1488.

Wada, M., Kadota, A. and Furuya, M. (1983). Intracellular localization and dichroic orientation of phytochrome in plasma membrane and/or ectoplasm of a centrifuged protonema of fern *Adiantum capillus-veneris* L. *Plant Cell Physiol.* **24** 1441–1447.

Wada, M., Murata, T. and Shibata, M. (1990a). Changes in microtubules and microfibril arrangement during polarotropism in *Adiantum* protonemata. *Bot. Mag. Tokyo* **103**, 391–401.

Wada, M., Murata, T., Shimizu, H. and Kondo, N. (1990b). A model system to study the effect of SO_2 on plant cells. III. Effects of sulfite on the ultrastructure of fern protonemal cells. *Bot. Mag. Tokyo* **103**, 403–417.

Wang, H., Cutler, A.J., Saleem, M. and Fowke, L.C. (1989). Microtubules in maize protoplasts derived from cell suspension cultures: effect of calcium and magnesium ions. *Eur. J. Cell Biol.* **49**, 80–86.

21. THE CYTOSKELETON AND MOSS MORPHOGENESIS

John H. Doonan

Department of Cell Biology, John Innes Institute,
Colney Lane, Norwich NR4 7UH, UK

Moss – an experimental organism for cellular morphogenesis

Mosses, despite their small size, have made a significant contribution to our understanding of plant cellular morphogenesis, in particular the relationship between the cytoskeleton and cell shape. While the moss does not produce flowers nor set seed, it does possess a cell wall and therefore cellular morphogenesis occurs under similar constraints to those of higher plants. During their brief life cycle, these primitive land plants appear to anticipate many of the cellular processes which are required for morphogenesis of flowering plants, but because of their simple colony-like morphology and small size most of these developmental stages are directly accessible without requiring surgery. Their simple, sequential and yet highly ordered cellular development make moss an attractive experimental system in which to investigate the cellular and molecular basis of plant morphogenesis. The particular properties which make the moss so suitable for cellular studies can be listed as follows:

Spores and protoplasts provide a convenient supply of non-polarized cells which readily acquire polarity

In response to light, moisture and suitable temperature, the single-celled haploid spore germinates to form a tip-growing uniseriate filament. Filament emergence can be orientated by unidirectional or polarized light (Jenkins and Cove, 1983a). Protoplasts, which also lack intrinsic polarity, can readily be made from vegetative filaments of *Physcomitrella* and will regenerate at high frequency directly into filaments (Jenkins and Cove, 1983a). Moreover, the generation of asymmetry in response to environmental stimuli occurs in protoplasts much as it does in spores (Jenkins and Cove, 1983b).

Apical growth is indeterminate

Apical or tip growth is a common phenomenon amongst plant and fungal cells, being the primary mode of cell expansion in pollen tubes, root hairs, fungal hyphae, and in the filaments of many types of algae. Young *Physcomitrella* plants are composed of two types of filamentous, tip-growing cells referred to as chloronemata and caulonemata, each with radically different growth characteristics. Spores and protoplasts germinate to produce chloronema. After several cell cycles, the chloronemal apical cell can divide to produce a new type of apical cell, a caulonema. The transition is mediated by auxin-like plant growth regulators (Cove and Ashton, 1984) and the caulonemal cell thus produced has a highly polarized subcellular structure, increased division rate and altered responses to various environmental stimuli. The caulonemata represent the invasive phase of the life cycle, dividing faster and extending more quickly than the chloronemata. The actively growing cells are usually found at the periphery of the colony and by subculture can be maintained indefinitely. These cells are very accessible, both to observation and to experimentation: they can be easily grown on microscope slides, microinjected, and subjected to drug treatments or to environmental gradients. By blotting colonies directly to suitable substrates, the spatial arrangements of cells can be maintained through even complex procedures such as immunofluorescent staining (Doonan *et al.*, 1985).

The vector of apical growth responds to environmental stimuli

Plants respond to environmental gradients, often by altering the orientation of growth. The apical cells of moss caulonemata are very sensitive to unidirectional light and gravity, being readily reorientated by changes in these environmental stimuli (Cove *et al.*, 1978). Thus the response of single plant cells to environmental stimuli can be directly monitored while the cells remain within the intact plant.

Many cell divisions are asymmetric

Asymmetric divisions have important and long range consequences in plant morphogenesis, since daughter cells thus produced generally have different developmental fates; such divisions are often associated with the creation of new cell lineages. A peculiar feature of early moss development is the high frequency of asymmetric divisions – every cell division in caulonemal filaments is asymmetric. Apical cell division produces an apical and a subapical daughter cell. Partly due to its position, but also due to the large share of cytoplasm received, the apical daughter cell has the potential to divide directly whereas the subapical cell remains quiescent for some time. The differential timing of subsequent cell divisions in the apical daughter cell, and branch formation in its subapical sister, gives rise to the characteristic colony morphology. Branch formation also involves a highly asymmetric division, and the small but densely cytoplasmic branch initial thus produced has not only an altered growth polarity but also an increased developmental repertoire.

Developmental switches (i.e. between tip growth and meristematic growth) are dramatic and can be controlled easily

Developmental events in plants frequently involve changes in cell type which usually occur subsequent to an asymmetric cell division. The switch from chloronema to caulonema and from caulonema to buds both occur following unequal cell divisions. Application of auxin or cytokinin to competent cells will induce these events. Caulonemal cells, compared with chloronemata, have an increased developmental repertoire: whereas chloronema normally only give rise to chloronema or, in the presence of suitable levels of auxin, to caulonema, caulonemal subapical cells can form side branches whose developmental fate is variable (Doonan *et al.*, 1987). The course of side-branch development is largely determined by the environment and at least four distinct fates are recognizable – the side branch can (i) remain as an initial (with no further growth); (ii) form chloronemata; (iii) form caulonemata; or (iv) in presence of adequate light, auxin, and cytokinin, form buds. The buds develop into a simple meristem, producing a shoot with simple leaves and eventually bearing the sexual structures.

Other factors

An additional advantage of small size is the ability to treat the moss as a microbe and maintain it under defined *in vitro* culture conditions. As the plant is haploid the isolation of morphological mutants is relatively easy (Ashton and Cove, 1977; Courtice and Cove, 1983; Knight *et al.*, 1988). Analysis of developmentally abnormal mutants has elucidated the role of plant hormones in morphogenesis and shown that auxin and cytokinin are directly involved in cell differentiation (Cove and Ashton, 1984). Response to environmental stimuli is modified by mutations at several loci (Cove *et al.*, 1978; Jenkins and Cove, 1983c; Jenkins *et al.*, 1986). The position (Doonan, 1984) and development (Ashton and Cove, 1977) of branch initials and morphology of the leafy shoots (Courtice and Cove, 1983) can all be altered by mutation. Recently, DNA-mediated transformation has been developed, holding out the promise of isolation of the genes responsible for various aspects of development.

The two closely related species, *Physcomitrella patens* and *Funaria hygrometrica*, are particularly amenable to experimental manipulation and have been widely used for cellular and genetic analysis of morphogenesis. This chapter will concentrate on the role of the cytoskeleton in cellular morphogenesis of *Physcomitrella* and some of its close relatives in the Funariales and *Ceratodon*. Areas where the moss could be a useful experimental tool for elucidating basic cellular development in plants will be discussed.

Cell polarity and the cytoskeleton

Restriction of cell growth to an exclusive site on the cell cortex is characteristic of tip-growing cells. Materials necessary for growth, made in the main body of the cell, have to be delivered within Golgi-derived vesicles to the apical dome, where directed exocytosis occurs. Highly polarized cytoplasmic organization is very characteristic of tip-growing cells, but the molecular mechanisms by which tip growth is established, is maintained, or is modified in response to the environment is not understood. Recently, indirect immunofluorescence has provided new insight into how cellular architecture is organized at the site of cell growth, in the apical dome.

What is the cytoskeletal basis for tip growth in moss?

A cytoskeletal basis for protonemal polarity was suggested on the grounds that colchicine interferes with cell morphogenesis – at high concentrations this drug could induce subapical bulges (Schmiedel and Schnepf, 1980) and was later shown to cause irregular distribution of the membrane rosettes (Schnepf *et al.*, 1985) believed to be involved in cellulose synthesis (Reiss *et al.*, 1984). The direct involvement of microtubules in tip growth was in doubt because conventional EM failed to demonstrate their presence within the apical dome. Improved microtubule fixation protocols used for indirect immunofluorescent studies subsequently showed that microtubules did indeed extend into the apical dome (Fig. 21.1a) to the limits of the cytoplasm (Doonan *et al.*, 1985). Freeze substitution of moss protonemata (Doonan *et al.*, unpublished data) has confirmed this, showing microtubules abutting onto the plasma membrane of the apical dome.

Unlike most other plant cells, where cytoplasmic microtubules disappear during mitosis, the apical microtubules of moss protonemata are persistent. Time-lapse video microscopy has shown that the caulonemata continue growth throughout M-phase (C.D. Knight and D.J. Cove, unpublished data), and the persistence of apical microtubules would be important for maintaining normal apical growth.

However, the presence of a given structure in a particular location does not confirm its involvement in a suspected function, so the next question we asked was what would happen if microtubules were removed or if their organization was disturbed? The microtubule organizing centre (MTOC)-perturbing drug, isopropyl *N*-(3-chlorophenyl)-carbamate (CIPC), fragments and disorganizes microtubule arrays and concomitantly causes the cell to bend (Fig. 21.1b) or develop subapical bulges or new 'tips', similar to those reported by Schmiedel and Schnepf (1980) for colchicine treatment. The effects of CIPC drew our attention to the organization of microtubules within the apical dome, because the primary effect of this drug was to disorganize the foci of microtubules normally found there (Doonan *et al.*, 1985). The specific perturbation of these microtubule foci by a drug widely used to disrupt spindle pole organization in vertebrate cells, higher plants and algae suggests that these foci may have functional significance for tip growth in moss. Other unrelated antimicrotubule drugs, such as colchicine (Schmiedel and Schnepf, 1980), cremart (Doonan *et al.*, 1988) and oryzalin (Wacker *et al.*, 1988) can be used to disorganize microtubule arrays in the sense that they cause fragmentation, but only CIPC perturbs organization of the foci to produce such a variety of aberrant arrays within the apical dome. Any treatment which perturbs microtubule organization seems to cause some degree of filament bending and production of subapical protrusions. The idea that microtubule organization and/or dynamics may be important in tip growth and tip bending receives additional support from the effects of taxol when added to cremart-treated cells. This microtubule-stabilizing compound partly reverses the effects of cremart as far as

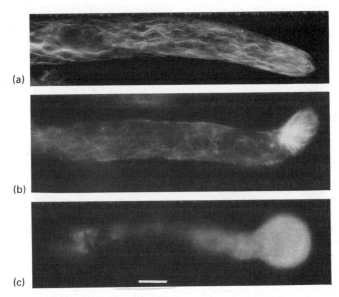

(a)

(b)

(c)

Fig. 21.1. The microtubules in caulonemal tip cells are organized into foci, which may be involved in directing cell elongation. a, Apical microtubules extend to the tip of the apical dome where they may form foci; b, treatment with CIPC disrupts the organization of the apical foci and causes bending and other aberrations in growth; c, cremart can totally remove microtubules and will cause cells to swell at the tip. Scale **bar = 10** μm. Reproduced in part from Doonan *et al.* (1985) and Doonan *et al.* (1988), with permission.

visualization of microtubules is concerned but actually intensifies the amount of bending and new tips formed (Doonan *et al.*, 1988).

The molecular nature of the microtubule organizer is unknown. For instance, it is not clear that there are any microtubule nucleation sites within the apical dome – microtubule regrowth experiments after cremart-induced disassembly suggest that such sites are mainly associated with the nucleus (Doonan, unpublished data). However, it may be that microtubule nucleation may be a separate function from microtubule organization in these cells.

Microtubules appear to direct polar growth

Complete removal of microtubules has rather different consequences for growth, compared with the disorganization or partial disassembly discussed above. Cremart is highly effective in disassembling plant microtubules and, unlike either CIPC or colchicine, cremart can reversibly eliminate all detectable microtubules (as assayed by indirect immunofluorescence (IIF))

from moss cells (Doonan *et al.*, 1988). In apical cells where microtubules are *completely* removed, no subapical swellings are found; instead the apical domes of such cells tend to swell to form spheres (Fig. 21.1c). Apparently, growth can take place in the absence of microtubules, albeit at a lower rate, but this growth is no longer polarized.

The microtubule system is involved in transduction of environmental signals into directional growth responses

Apical cells of moss protonemata respond to changes in several environmental stimuli, including light and gravity, by altering the orientation of growth. If microtubule organization does have a role in maintaining and modifying growth, then we might expect to observe alterations in microtubule organization in response to environmental stimuli, perhaps even prior to modification of growth. Response to gravity by *Ceratodon* apical cells (Schwuchow *et al.*, 1991) involves the formation of a microtubule 'enrichment' along the flank of the cell apex near to the gravity vector. This rearrangement of the apical dome microtubules precedes a change in growth. These microtubule enrichments demonstrate that microtubules within the apical dome are capable of reorganization in response to environmental stimuli. There are similarities in appearance between gravity-stimulated foci and some CIPC-induced microtubule foci in *Physcomitrella* (Doonan *et al.*, 1985), which are also associated with cell bending.

Antimicrotubule drugs are very potent in interfering with normal response to environmental stimuli. Graviresponse in *Ceratodon* apical cells may depend more than photo-response on the microtubule system; Hartmann (1984) observed that 1 mM colchicine could inhibit gravitropic bending, whereas photo-tropic bending was unaffected or even increased! Schwuchow *et al.* (1990) have shown that amiprophos-methyl (APM) and oryzalin will simultaneously disrupt the microtubule network and the ability of the cell to respond to gravity. Colchicine disrupts the ability of germinating *Physcomitrella* spores to respond to unidirectional light without affecting growth in a gross manner (Burgess and Linstead, 1981). The preservation of microtubules unfortunately was not monitored in these experiments, so one cannot be certain that the effects are solely or mainly due to the drug's effect on the microtubule system. The effect of other, more specific, antimicrotubule drugs on tropic bending will be informative.

The stage of the cell cycle can also modify the cell's ability to respond to gravity (Knight and Cove, 1988). Before M-phase, the bending not only ceases but appears to reverse. These authors suggest that the partial disassembly of microtubules known to occur during M might induce this response.

Although not proven, there is increasing evidence that microtubules play an important role in orientating the growth of caulonemal and perhaps all protonemal cells. There appears to be some form of microtubule-organizing capacity close to the tip which may be responsive to environmental signals and capable of transducing these signals into vectorial growth responses. Microfilaments are also present within the apical dome (Doonan *et al.*, 1988), and in chloronemal cells form a fine dense meshwork (Quader and Schnepf, 1989) but their interaction with microtubules has not been investigated. Nothing is known about the molecular basis of apical microtubule organization, except that this organization is at least as sensitive as mammalian MTOCs are to CIPC. The organization of apical MTOCs may be the key in dissecting how the cytoskeleton influences tip growth in moss.

The role of microfilaments in tip growth

Cytochalasin B, a drug commonly used to remove microfilaments, slowed the rate of growth but induced no abnormalities in *Funaria* caulonemal tip cells (Schmiedel and Schnepf, 1980). Using rhodamine–phalloidin staining, we found that cytochalasin D was effective in severely fragmenting the microfilament (MF) system in *Physcomitrella* apical cells (Doonan *et al.*, 1988). Such treatment prevents both normal growth and cremart-induced swelling, indicating that MFs are required for the process of cell extension. Cremart-treated cells which were swollen and presumably lacked microtubules contained an essentially complete microfilament system when stained with phalloidin-derivatives. Taken together, these results suggest that the MF system is required for growth *per se*, perhaps transporting precursors to the site of growth, but that MFs do not directly determine the exact site of growth in the apical cell.

However, during the formation of side branches in *Funaria* subapical cells, MFs do appear to be directly involved in determining the site of cell expansion. Recent improvements in MF fixation (Quader and Schnepf, 1989) have allowed the detailed description of a ring of MFs which is formed at the site of branch formation *prior* to any discernible cell expansion. The ring of MFs is, at first, compact with an internal diameter of 2 μm and covers an area of about 10 μm diameter. As the initial swell the ring spreads out, becoming more diffu

Actin and ion currents are involved in establishment and maintenance of apical growth

Another early manifestation of polar growth is the appearance of ionic currents associated with the point of outgrowth. Such currents can be measured in a variety of cells including the *Fucus* egg, where they anticipate the direction of growth. Indeed, growth of the *Fucus* egg and many types of spore can be orientated by placing them in an electrical field. Gradients of the Ca^{2+} ionophore A23187 will orientate *Funaria* sporelings (Chen and Jaffe, 1979), while electrical fields have been shown to do the same for *Physcomitrella* protoplasts (Burgess and Linstead, 1982). Calcium ions are implicated as one of the more important components of the current induced either artificially or by natural environmental stimuli (Saunders, 1986b).

The relative positions of ion currents is crucial for correct siting of side branches and MFs have been implicated in this process. Repositioning of these ion currents occurs prior to branch formation. Relocalization of membrane-bound calcium is the first manifestation of branch formation, predicting the site of cell division (Saunders and Hepler, 1981). Active ion pumps also relocate prior to cell division, but treatment of *Funaria* subapical cells with cytochalasin B or D (but not colchicine) disrupts both the repositioning of ion channels and results in the aberrant positioning of the side branch initial (Saunders, 1986a), implying that MFs are required for this process. Branch formation can be induced by artifical modification of calcium currents (Saunders and Hepler, 1982) or inhibited by blocking calcium fluxes (Saunders and Hepler, 1983; Saunders, 1986a).

The effects of Ca^{2+}-modulating drugs on the cytoskeleton is unclear. A large number of drugs, thought to act directly or indirectly on cellular calcium, have been used to probe cellular development in moss. Not surprisingly their effects are nearly as diverse as their pedigree. Application of A23187 on *Funaria* caulonemata changes the diameter of the cell but not its volume (Schmiedel and Schnepf, 1980). Thus a short fat cell is produced, but the normal polar distribution of organelles and cytoplasm is maintained. A variety of calcium channel blockers have been used to block cytokinin-induced side branch and bud formation (Saunders and Hepler, 1983) and to perturb tip cell growth in *Funaria* (Wacker and Schnepf, 1990). Nifedipine causes a dose-dependent decrease in the rate of cell extension, with about 90% inhibition at 10^{-5} μM and slight swelling of the apex. As judged by IIF, the microtubule cytoskeleton appears intact. The effect of Ca^{2+} channel blockers on *Physcomitrella* is rather similar to that observed in *Funaria* and the effects can be intensified by the addition of EGTA. Apical cells can become balloon-like and at least some of these still divide, apparently normally, whilst microtubule networks appear very well preserved. The effect of these substances on the MF system has not been investigated in either species.

Saunders (1986b) observed that the zone of maximal inward current in apical cells was positioned not at the tip, but over the nucleus. She suggested that the effect of ion flux might locally modify the structure of the cytoplasm, perhaps increasing microtubule polymerization around the nucleus. Microtubules are abundant in the nuclear region in the tip cells (Schmiedel and Schnepf, 1980; Doonan *et al.*, 1985; Wacker *et al.*, 1988). One possibility is that the ion channels may impose direction on the outgrowth of microtubules nucleated from the nucleus. Thus, during side branch formation, ion channels located at the branch initial may direct the unidirectional assembly of microtubules emanating from the migrating nucleus. During light-induced side branch formation in *Physcomitrella*, microtubules run from the migrating nucleus towards the division site, increasing dramatically in density as the nucleus moves (Doonan *et al.*, 1986), but it is not clear if ion currents are required for microtubule orientation.

Terminating tip growth

In higher plants there is much direct and circumstantial evidence that plant hormones

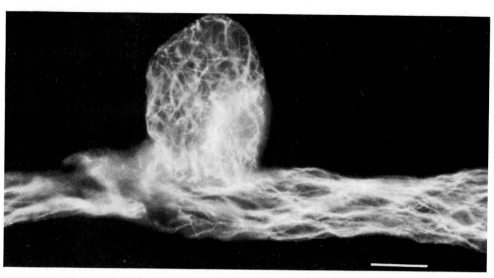

Fig. 21.2. Microtubule arrays in a young, developmentally plastic, side branch initial are abundant, but do not appear to terminate at the cell surface and are not conspicuously arranged into foci. Scale bar = 10 μm. Reproduced from Doonan *et al.* (1986).

affect cell shape through their effects on the cytoskeletal stability (reviewed by Shibaoka, Chapter 12, this volume). The case for plant hormones affecting microtubules in moss is unclear. Non-physiological concentrations of both auxin and cytokinin do indeed fragment microtubules. Ljubesic *et al.* (1989) found that high concentrations of exogenously applied auxin can simultaneously fragment microtubules and cause aberrant cell shape. Likewise, high levels of exogenous cytokinin in conjunction with auxin can cause microtubule disassembly and swelling in caulonemal tip cells of *Physcomitrella* (Doonan *et al.*, 1988). The latter observation was compared with the normal physiological action of lower cytokinin levels on side branch initials, which are induced by this hormone to swell and form buds. Compared with either uncommitted side branches (Fig. 21.2) or those committed to tip growth, IIF data suggest that interphase microtubules within buds are rather poorly preserved (see Fig. 21.3b). Although this may arise due to differences in cell permeability to fixation and/or antibody, it may be significant that EM studies never report the presence of cytoplasmic interphase microtubules in young buds. One inference is that the microtubules are indeed

less stable in cells responding to cytokinin. The effect of cytokinin on the shape of side branch initials can be mimicked by drugs which alter calcium fluxes (Saunders and Hepler, 1982) or by microtubule-destabilizing drugs (Doonan *et al.*, 1988). However, 'buds' induced by these drugs do not develop further so the case that cytokinin or auxin may affect microtubule organization or stability under more normal physiological conditions remains unclear and worthy of further investigation.

What is beyond doubt is that true tip growth ceases in cytokinin-treated side branches, and this correlates with the loss of microtubule foci within these cells. The bud initials lose an obvious apical dome and swell more evenly rather than locally. The cell cycle time may be shortened and the apical dominance is lost, allowing subapical cells to divide faster than they would in a caulonemal filament.

The transitional stages between tip growth and meristematic growth in moss are complex and occur over several cell generations. Cells which grow by localized extension at the tip contain a dramatically different cytoplasm and cytoskeletal organization to meristematic and leaf cells which grow by a more general expansion. Whole cell immunofluorescence

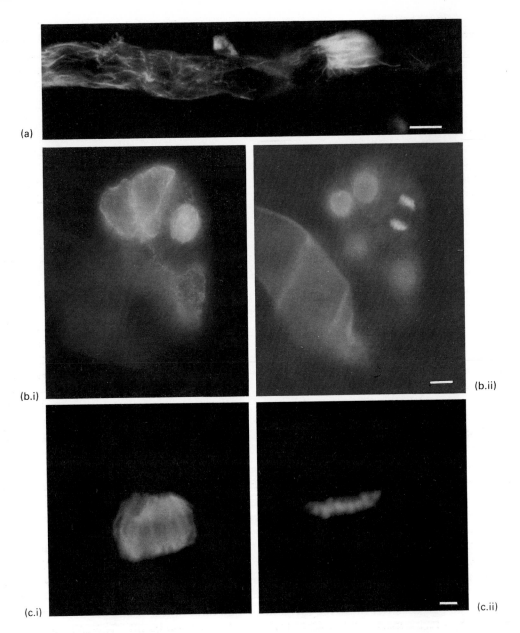

Fig. 21.3. The cytoplasmic microtubule arrays associated with the mitotic spindle suggests that microtubules may be involved in positioning the spindle in tip-growing cells and young buds, but not in older leafy shoots. a, Anaphase spindles from caulonemal cells are associated with many pole to cortex microtubules; b, similar cytoplasmic microtubules can be observed in multicellular buds prior to formation of leaves (i) microtubules, (ii) DAPI; c, spindles from leafy shoot cells completely lack associated astral microtubules (i) microtubules, (ii) DAPI. Scale bar = 10 μm. Reproduced from Doonan *et al.* (1987).

studies on this transition demonstrate that as the bud grows and forms leaves, cells develop a microtubule cell cycle very similar to that of higher plants, differing in several respects to the more 'primitive' cycle of the filament cells. When viewed by IIF, the filament cells contain a complex array of interphase microtubules which is both endoplasmic as well as cortical. EM studies on *Funaria* show that the endoplasmic and cortical microtubules form a continuous network. In cell squashes from developing buds, interphase microtubules are predominantly cortical and in expanding leaves, microtubules are organized as cortical helical arrays (Doonan *et al.*, 1987). Preprophase bands, completely lacking from the filamentous stages, are found in these cells, and mitotic spindles occur in the absence of associated cytoplasmic microtubules (Fig. 21.3). The cells within *Sphagnum* leaves also contain a higher plant-like arrangement of microtubules, although the postmitotic arrays associated with cytokinesis are somewhat different from those observed in most other cell types (Schnepf, 1984).

The origin of the preprophase band (PPB) is a matter of some interest both in evolutionary terms and in its appearance during each cell cycle. During moss development, the appearance of a PPB during the cell cycle coincides with the appearance of regularly organized cortical microtubules. This coincidence of regular cortical microtubule arrays with PPBs is observed in many other types of plant cells. Conversely, PPBs are absent from a wide variety of other plant cells which lack highly organized cortical arrays (i.e. pollen, endosperm, moss and fern gametogenic cells). This implies that cortical interphase microtubules may be one prerequisite (perhaps not the only one) for PPB formation. PPBs are themselves cortical arrays, and so the ability of a cell's microtubules to associated with the plasma membrane in a highly organized fashion may be an essential prerequisite for band formation. This idea is supported by the observation of a mitosis-related cortical band of microtubules in *Chlamydomonas* (Johnstone and Porter, 1968; Doonan and Grief, 1987), a unicellular algae which during interphase contains a set of

cortical microtubules. However, the idea that the PPB is simply due to a bunching up of interphase microtubules in G2 or early M has to be re-examined in the light of evidence that the PPB may be a nascent array (see Lloyd, Chapter 18, this volume).

Cell division and the cytoskeleton

In plant morphogenesis, asymmetrical cell divisions commonly produce daughters with different or potentially different developmental fates. Moss morphogenesis includes several examples of such unequal divisions: division of the caulonemal tip cell, side branch formation, and the early divisions of the young bud are the best characterized. The microtubule cytoskeleton has a conspicuous role in the physical separation of chromatin (spindle) and formation of a cross wall (phragmoplast). Both these structures appear to behave much as in other systems and will not be considered further here. However, both the precision with which the spindle and the phragmoplast in filament cells are positioned, and the sensitivity with which they are reorientated in response to environmental stimuli, are remarkable. In filamentous cells, cross walls are not only precisely aligned in the absence of a PPB of microtubules but the alignment is responsive to environmental stimuli. As with the reorientation of tip growth, microtubules may act to transduce vectorial environmental information into positional information for the cell division apparatus. Jensen (1981) has characterized the division of the apical cell in *Physcomitrium*, both in terms of how and when the cells divide and how environment influences the spatial aspects of that division and the succeeding division of the subapical daughter. Apical cells need to attain a minimum length before becoming competent to divide. Although mitosis may be delayed for some time after a cell becomes division competent, there is a stage at which the cell is committed to division and two spatial aspects of that division are decided. The overall length of the subapical daughter cell is fixed such that under constant conditions subapical cells will be of constant length. The orientation of the future cross wall is also fixed at this stage, such that the leading edge of the oblique cell plate is on the lit side of a unidirectionally illuminated filament. Changing the direction of light before commitment can alter the orientation of the cell plate, but thereafter has no effect. As pointed out by Jensen, the orientation of the cell plate defines the orientation of the branch formation by the subapical cell. When the subapical cell divides (also by unequal division) the nucleus migrates from its resting position on the mid flank of the cell to the site of branch formation, close to the leading edge of the anterior cell plate. The subsequent cell division therefore throws off a side branch initial towards the light source prevailing at the previous division. Thus transient environmental signals can influence the cellular structure of the filament system for at least two cell divisions after the perception and fixation of the stimulus.

The sequence of events by which the stimulus is perceived and the cell's response is executed is unclear, but an intact microtubule cytoskeleton is essential to ensure that the spatial decisions taken at each of these divisions are implemented with precision. The spindle poles of the metaphase spindle within apical cells are connected to the cell cortex via an extensive array of astral microtubules (Doonan et al., 1985). During anaphase the number of these cytoplasmic microtubules increases, particularly between the spindle and the immediate flanking wall (Fig. 21.3). Any treatment which perturbs or destroys the astral microtubules has severe effects on spindle orientation in Physcomitrella and Funaria (Doonan et al., 1985; Wacker et al., 1988). The effects of these compounds on light-induced reorientation of cross walls needs to be examined – it may be that low concentrations of drug would interfere with the light regulation of reorientation without affecting reorientation itself.

Nuclear positioning and migration

Not only is the orientation of the division plane under strict control, but the site of cell division is also regulated with a high degree of precision.

If not already positioned, then the premitotic nucleus will migrate into the division plane. Nuclear positioning and migration can be continuous (as in the growing apical cell) or can occur as a premitotic event as in side branch formation. In apical cells, nuclear positioning appears to be dependent on both actin and microtubules since the specific antagonists of microtubules and MFs have opposing effects (Schmiedel and Schnepf, 1979, 1980) – antimicrotubule drugs cause the nucleus to shift towards the basal end and cytochalasins cause a shift towards the apex.

Environmental signals (light in the case of Physcomitrella and cytokinin for Funaria) can induce nuclear migration in a competent subapical cell. The migration is dependent on an intact array of microtubules and is correlated with a major rearrangement of these microtubules in Physcomitrella (Doonan et al., 1986).

Summary

Although there are several indications that an intact cytoskeleton is essential for normal morphogenesis, it would appear that the cytoskeleton does not, of itself, 'control' development. Instead, the cytoskeleton may be part of a regulatory control loop, for there is evidence that the cytoskeleton is involved in transducing information, either from the environment or from other cells or even within the same cell, into a change in cell morphology. To understand this further, the molecules which mediate between the environment and the cytoskeleton need to be found. At least two routes are available to identify such molecules. Biochemical analysis of the moss cytoskeleton might provide the most direct route. Almost nothing is known about the composition of moss cytoskeletons, despite the fact they are easily isolated (Powell et al., 1980; Doonan and Duckett, 1988) from protoplasts. The remarkable differences in cytoskeletal organization in tip-growing and meristematic cells imply that the biochemical composition of the cytoskeleton may radically change during development. Do microtubule-associated proteins differ in the different cell types? Does auxin or cytokinin alter microtubule stability, and if so is this due to a different spectrum of associated proteins? Are there tip-specific cytoskeletal proteins as have been recently discovered in yeast shmoo cells (Gehrung and Snyder, 1990)?

Genetic analysis is another under-exploited approach. As experience in yeast, filamentous fungi, *Drosophila* and other systems have shown, even the most complex and intractable biological process can be dissected and understood by appropriate genetic analysis. Moss, being both small and haploid, are clearly amenable to this approach as extensive genetic analysis of plant hormones has demonstrated (Cove and Ashton, 1984). Mutants defective in various aspects of cellular morphogenesis would be immensely useful, but have not been extensively isolated from plants, perhaps because such mutations are usually either lethal or conditionally lethal and are thus perceived as being difficult to isolate and handle. However, certain characteristics of the moss can be used to advantage. Firstly, because the organism is haploid, recessive mutations can be readily isolated. Secondly, because the organism can be maintained vegetatively at almost any age, mutations which have their effect at a given stage can be rescued by growing an earlier developmental stage. An excellent example of this has been recently reported by Abel and co-workers (1989) where they have identified a chloroplast division defect which only becomes lethal in the bud stage. Processes such as bud formation and leaf morphogenesis have already been extensively analysed in this manner, and side branch formation and positioning are also good candidates for such experiments.

Well-characterized drug resistance mutants have been immensely useful in yeast, fungal and mammalian cell biology because they provide an internal control to assure that the effects observed are indeed due to the action of the drug on the suspected target molecule (e.g. That *et al.*, 1988). Moss cell biology studies depend a great deal on the use of drugs but drug-resistant mutants used as controls are conspicuous by their absence.

Mutants may also provide information on the molecules which communicate with, or regulate the function of, the cytoskeleton. Tropic mutants, which affect the cell's ability to alter its growth vector in response to environmental signals, have been isolated and partially characterized (Cove *et al.*, 1978; Jenkins *et al.*, 1986). These mutants represent the first step in dissecting how cell growth responds to environmental stimuli. Since *Physcomitrella* is amenable to DNA-modified transformation (Schaefer *et al.*, 1991), there is now the possibility of isolating these genes.

The control of cellular morphogenesis is exceedingly complex, as one might expect from a plant which must grow and respond to a variable environment. The moss, because of its simple structure, is highly attractive as a system to study cell morphogenesis during plant development, but its future potential for this lies in combining cell biology with both biochemical and genetical analysis.

References

Abel, W.O., Knebel, W., Koop, H.-U., Marienfeld, J.R., Quader, H., Reski, R. *et al.* (1989). A cytokinin-sensitive mutant of the moss *Physcomitrella patens*, defective in chloroplast division. *Protoplasma* **152**, 1–13.

Ashton, N.W. and Cove, D.J. (1977). The isolation and preliminary characterisation of auxotrophic and analogue resistant mutants of the moss, *Physcomitrella patens. Molec. Gen. Genet.* **154**, 87–95.

Burgess, J. and Linstead, P.J. (1981). Studies on the growth and development of protoplasts of the moss, *Physcomitrella patens*, and its control by light. *Planta* **151**, 331–338.

Burgess, J. and Linstead, P.J. (1982). Cell-wall differentiation during growth of electrically polarised protoplasts of *Physcomitrella. Planta* **156**, 241–248.

Chen, T.-H. and Jaffe, L.F. (1979). Forced calcium entry and polarized growth of *Funaria* spores. *Planta* **144**, 401–406.

Courtice, G.R.M. and Cove, D.J. (1983). Mutants of the moss *Physcomitrella patens* which produce leaves of altered morphology. *J. Bryol.* 12, 596–609.

Cove, D.J. and Ashton, N.W. (1984). The hormonal regulation of gametophytic development in bryophytes. In *The Experimental Biology of Bryophytes* (eds A.F. Dyer and J.G. Duckett), pp. 177–201. Academic Press, London.

Cove, D.J., Schild, A., Ashton, N.W. and Hartman, E. (1978). Genetic and physiological studies of the effect of light on the development of the moss. *Physcomitrella patens. Photochem. Photobiol.* **27**, 249–254.

Doonan, J.H. (1984). Morphogenetic role of the cytoskeleton in *Physcomitrella patens*. PhD thesis, University of Leeds.

Doonan, J.H. and Duckett, J.G. (1988). The bryophyte cytoskeleton: experimental and immunofluorescence studies of morphogenesis. *Adv. Bryol.* **3**, 1–31.

Doonan, J.H. and Grief, C. (1987). Microtubule cycle in *Chlamydomonas reinhardtii*: an immunofluorescence study. *Cell Motil. Cytoskel.* **7**, 381–392.

Doonan, J.H., Cove, D.J. and Lloyd, C.W. (1985). Immunofluorescence microscopy of microtubules in intact cell lineages of the moss, *Physcomitrella patens. J. Cell Sci.* **75**, 131–147.

Doonan, J.H., Jenkins, G.I., Cove, D.J. and Lloyd, C.W. (1986). Microtubules connect the migrating nucleus to the prospective division site during side branch formation in the moss, *Physcomitrella patens. Eur. J. Cell Biol.* **41**, 157–164.

Doonan, J.H., Cove, D.J., Corke, F.M.K. and Lloyd, C.W. (1987). Pre-prophase band of microtubules, absent from tip-growing moss filaments, arises in leafy shoots during transition to intercalary growth. *Cell Motil. Cytoskel.* **7**, 138–153.

Doonan, J.H., Cove, D.J. and Lloyd, C.W. (1988). Microtubules and microfilaments in tip growth: evidence that microtubules impose polarity on protonemal growth in *Physcomitrella patens J. Cell Sci.* **89**, 533–540.

Gehrung, S. and Snyder, M. (1990). The SPA2 gene of *Saccharomyces cerevisiae* is important for pheromone-induced morphogenesis and efficient mating. *J. Cell Biol.* **111**, 1451–1464.

Hartmann, E. (1984). Influence of light on phototropic bending of moss protonemata of *Ceratodon purpureus*. (Hedw.) *Brid. J. Hattori Bot. Lab.* **55**, 87–98.

Jenkins, G.I. and Cove, D.J. (1983a). Light requirements for regeneration of protoplasts of the moss *Physcomitrella patens. Planta* **157**, 39–45.

Jenkins, G.I. and Cove, D.J. (1983b). Phototropism and polarotropism of primary chloronemata of the moss *Physcomitrella patens*: responses of the wildtype. *Planta* **158**, 357–364.

Jenkins, G.I. and Cove, D.J. (1983c). Phototropism and polarotropism of primary chloronemata of the moss *Physcomitrella patens*: responses of mutant strains. *Planta* **159**, 432–438.

Jenkins, G.I., Courtice, G.R.M. and Cove, D.J.

(1986). Gravitropic responses of wild-type mutant strains of the moss *Physcomitrella patens. Pl. Cell and Environ.* **9**, 637–644.

Jensen, L.C.W. (1981). Division, growth, and branch formation in protonema of the moss *Physcomitrium turbinatum*: studies of sequential cytological changes in living cells. *Protoplasma* **107**, 301–317.

Johnstone, U.G. and Porter, K.R. (1968). Fine structure of cell divisions in *Chlamydomonas reinhardtii*. Basal bodies and microtubules. *J. Cell Biol.* **38**, 403–425.

Knight, C.D. and Cove, D.J. (1988). Time-lapse microscopy of gravitropism in the moss *Physcomitrella patens*. In *Methods in Bryology* (ed. J.M. Glime), pp. 127–129. Proc. Bryol. Meth. Workshop, Mainz. Hattori Bot. Lab., Nichinan.

Knight, C.D., Cove, D.J., Boyd, P.J. and Ashton, N.W. (1988). The isolation of biochemical and developmental mutants in *Physcomitrella patens*. In *Methods in Bryology* (ed. J.M. Glime), pp. 47–58. Proc. Bryol. Meth. Workshop, Mainz. Hattori Bot. Lab., Nichinan.

Ljubesic, N., Quader, H. and Schnepf, E. (1989). Correlation between protonema morphogenesis and the development of the microtubule system in *Funaria* spore germination under normal conditions and at high auxin concentrations: an immunofluorescence study. *Can. J. Bot.* **67**, 2227–2234.

Powell, A.J., Lloyd, C.W., Slabas, A.R. and Cove, D.J. (1980). Demonstration of the microtubular cytoskeleton of the moss, *Physcomitrella patens*, using antibodies against mammalian brain tubulin. *Pl. Sci. Letts.* **18**, 401–404.

Quader, H. and Schepf, E. (1989). Actin filament array during side branch initiation in protonema cells of the moss *Funaria hygrometrica*: an actin organization center at the plasma membrane. *Protoplasma* **151**, 167–170.

Reiss, H.-D., Schnepf, E. and Herth, W. (1984). The plasma membrane of the *Funaria* caulonema tip cell: morphology and distribution of particle rosettes, and the kinetics of cellulose synthesis. *Planta* **160**, 428–435.

Saunders, M.J. (1986a). Cytokinin activation and redistribution of plasma-membrane ion channels in *Funaria. Planta* **167**, 402–409.

Saunders, M.J. (1986b). Correlation of electrical current influx with nuclear position and division in *Funaria* caulonema tip cells. *Protoplasma* **132**, 32–37.

Saunders, M.J. and Hepler, P.K. (1981). Localization of membrane-associated calcium following cytokinin treatment in *Funaria* using chlorotetracycline. *Planta* **152**, 272–281.

Saunders, M.J. and Hepler, P.K. (1982). Calcium ionophore A23187 stimulates cytokinin-like mitosis in *Funaria*. *Science* **217**, 943–945.

Saunders, M.J. and Hepler, P.K. (1983). Calcium antagonists and calmodulin inhibitors block cytokinin-induced bud formation in *Funaria*. *Devel. Biol.* **99**, 41–49.

Schaefer, D., Zryd, J.-P., Knight, C.D. and Cove, D.J. (1991). Stable transformation of the moss *Physcomitrella patens*. *MGG*, in press.

Schmiedel, G. and Schnepf, E. (1979). Side branch formation and orientation in the caulonema of the moss, *Funaria hygrometrica*: experiments with inhibitors and with centrifugation. *Protoplasma* **101**, 47–59.

Schmiedel, G. and Schnepf, E. (1980). Polarity and growth of caulonema tip cells of the moss *Funaria hygrometrica*. *Planta* **147**, 405–413.

Schnepf, E. (1984). Pre- and postmitotic re-orientation of MT arrays in young *Sphagnum* leaflets: transitional stages and initiation sites. *Protoplasma* **120**, 100–112.

Schnepf, E., Witte, O., Rudolph, V., Deichgräber, G. and Reiss, H.-D. (1985). Tip cell growth and the frequency and distribution of particular rosettes in the plasmalemma: experimental studies in *Funaria* protonema cells. *Protoplasma* **127**, 222–229.

Schwuchow, J., Sack, F.D. and Hartmann, E. (1990). Microtubule disruption in gravitropic protonema of the moss *Ceratodon*. *Protoplasma* **159**, 60–69.

That, T.C.C.-T., Rossier, C., Barja, F., Turian, G. and Roos, U.-R. (1988). Induction of multiple germ tubes in *Neurospora crassa* by antitubulin agents. *Eur. J. Cell Biol.* **46**, 68–79.

Wacker, I. and Schnepf, E. (1990). Effects of nifedipine, verapamil, and diltiazem on tip growth in *Funaria hygrometrica*. *Planta* **180**, 492–501.

Wacker, I., Quader, H. and Schnepf, E. (1988). Influence of the herbicide oryzalin on cytoskeleton and growth of *Funaria hygrometrica* protonemata. *Protoplasma* **142**, 55–67.

22. MUTUAL ALIGNMENTS OF CELL WALLS, CELLULOSE, AND CYTOSKELETONS: THEIR ROLE IN MERISTEMS

Paul B. Green[1] and Jeanne M.L. Selker[2]

[1] Department of Biological Sciences, Stanford University,
Stanford, CA 94305, USA
[2] Department of Biology, University of Oregon,
Eugene, OR 97403–1210, USA

Cell patterns in meristems are constrained by physical considerations that come from continuity requirements. These exist because the tissue consists of adjacent turgid cells with strong cell walls. Symplastic tissue can 'flow' only in certain ways. There are also apparent constraints on the nature of stable cytoskeletal alignments in single cells and between adjacent cells. The two sets of constraints, physical and cytoskeletal, must be followed concurrently in the plant. It is only within these biophysical constraints that one can seek the mechanisms to account for (i) the consistent cell patterns seen in root sections, (ii) the origin of such patterns, and (iii) the geometrically more complex cyclic morphogenesis of the shoot. We discuss first the three-fold interactions between growth rate, growth direction, and morphology/cytology, under two special conditions: there is symplastic growth, and the geometry of the organ is constant (root) or cycles regularly (shoot) despite the fact that its elements, cells, are being displaced through it. Later, influences on the pertinent cytoskeletal behaviour will be analysed.

Biophysics of meristems

The freely growing plant cell, exemplified by the internode of *Nitella*, serves as a useful starting point for this chapter which concerns the biophysics of meristems. These are often considered to be bundles of such cylindrical cells.

The elongating cell and the strain rate cross

The *Nitella* cell's growth performance is shown in Fig. 22.1. Ignoring minor twisting aspects, growth is quantitatively described by a strain rate cross (defining an ellipse). The long arm gives the maximal rate of linear extension, the short arm gives the minimal. The rates are equivalent to rates of continuously compounded interest, e.g. a rate of 0.69 per day means a doubling of that dimension per day. The ellipse reflects the expansion of an initially circular patch of surface (Goodall and Green, 1986). The characteristic difference in rates, when iterated over time, accounts for the progressive elongation of the cell.

The immediate basis of the directionality is believed to be the continued presence of transverse microfibrils in the wall. This overcompensates the stress pattern of the turgid cylindrical cell, which would favour increase in cell girth to give overall lengthening. A deeper cause of this directionality is thought to be the persistent alignment of many microtubules in the transverse direction (Ledbetter and Porter, 1963; Gunning and Hardham, 1982). This stable orientation cannot be a simple

The Cytoskeletal Basis of Plant Growth and Form   ISBN 0–12–453770–7

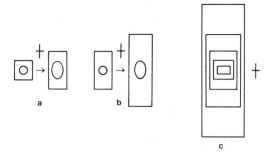

Fig. 22.1. Characterizing the growth of a cylindrical cell, as seen in profile. An initially square outline is changed to a rectangle over a short period of time (a). The originally circular patch is deformed to an ellipse. The fractional change along the major and minor axis is measured as $(\ln L_2 - \ln L_1)/(t_2 - t_1)$. L is length and t is time. This gives the relative rate, equivalent to a rate of continuously compounded interest. See Goodall and Green (1986) for details. The rates are displayed as the lengths of the arms of a cross. Repeated identical episodes of such growth (as in b) integrate to generate the successive outlines of the cell (c).

matter because microtubules are elongate and would be expected to be strain-aligned, by the stretching action of growth, into the longitudinal direction. A kind of a self-cinching mechanism could serve to maintain the transverse alignment (Lloyd and Barlow, 1982). Parallel sets of microtubules could form self-tightening circumferential arrays, to maintain their alignment.

Assuming that the above constitutes a plausible description of the performance of a roughly cylindrical cell, can a meristem be safely regarded as a simple aggregate of such cells? This chapter will argue that it cannot. There are major physical constraints on what can happen in a coherent tissue and there also appear to be constraints on how cytoskeletons in adjacent cells can be aligned. These two sets of constraints apparently must be obeyed simultaneously. They apply, in relatively simple fashion, in the subterminal parts of roots and in cylindrical organs in general. In certain root tips, and in shoot meristems, the geometrical

activity is more complex. None the less, the constraints confer coherence of meristematic activity in all cases. We will start with the root.

The root in section – and the growth tensor

A striking feature of the longitudinal section of many roots is the perpendicularity of the intersection of the walls (Fig. 22.2). This pattern is often made somewhat zig-zag by later adjustments (see Lloyd, Chapter 18, this volume). The pattern can be approximated by two intersecting sets of parabolas. Such curved configurations are the obvious precursors to the simple Cartesian (rectangular) grids seen in mature parts of the root. Hejnowicz (1989) has studied these patterns in terms of a growth tensor. He argues that such patterns are the inevitable consequence of symplastic growth and that they can reveal much about the kinetics of growth inside the meristems.

A major first observation he makes is that the orthogonal sets of lines seen in actual roots are not proper parabolas as are the $u + v$ lines in Fig. 22.2, but are actually mathematically more arbitrary curves as in Figs 22.3 and 22.4. The property of orthogonal intersection is preserved in these new curves. An important fact is that the strict parabolas (Fig. 22.2) require that the growth performance along the u set of curves be a function of position on the u lines only (and v of v). This simple condition is evidently not present in real roots. Extension in the two directions is interdependent. This was also the case in *Nitella* elongation. Hejnowicz has identified two paraboloid patterns. One is for a root with a prominent apical cell, the other is for a root with a quiescent centre (Figs 22.3 and 22.4). He emphasizes the fact that the patterns reflect trajectories of points. We will emphasize here the cell behaviour (growth rate + direction) which sums to give the trajectories.

To intepret the real patterns, it is useful to recognize the constraints on the strain rate crosses of neighbouring cells. Two constraints define symplastic growth. There is no slipping between cells. Also, gradients in extension rate in certain directions must be gentle. The common wall between cells has to have a strain

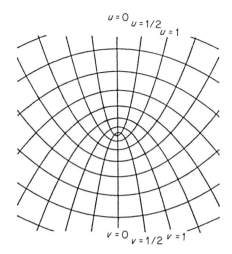

Fig. 22.2. Two families of intersecting parabolas (u and v). There is a common focus and the intersections are perpendicular. Such true parabolas are not encountered in sections of roots. From Hejnowicz (1989) with permission.

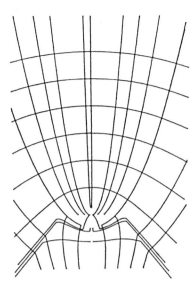

Fig. 22.4. Two families of perpendicularly intersecting curves. This pattern perpetuates itself only if there is growth rate *minimum* near the focus. From Hejnowicz (1989) with permission.

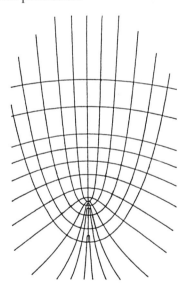

Fig. 22.3. Two families of related curves which retain the property of perpendicular intersection and resemble patterns seen in roots. For this patten to perpetuate itself there must be a *maximum* growth rate near the focus. From Hejnowicz (1989) with permission.

rate compatible with that of the cells on *both* sides. This occurs when the projection, onto the shared wall, is the same for the crosses of both cells. Various examples of possible and impossible strain patterns for adjacent cells are

shown in Fig. 22.5. Note that the minimal rate of one cross can be parallel to the maximal rate of the other cross, as long as both of these rates have the same projection on the common wall as in Fig. 22.5A(2). If there would be a discrepancy, the common wall must tear or wrinkle. Since neither effect is seen in most meristems, it can be concluded that juxtapositions of cells with incompatible expansion (that cannot be accommodated by curvature as in Fig. 22.5A(3)) are not normally produced. When they do occur, the tissue 'breaks up' as in the interior of some staminal filaments and in spongy mesophyll.

The orientation of the growth maximum has been worked out for the root patterns (Nakielski, 1990). The orientation generally follows one set of the paraboloidal lines. The orientation of the other set appears to follow the direction of the local direction of cellulose reinforcement. This combination is diagrammed in Fig. 22.6. The harmonious condition for the cellulose alignment throughout most of the section could be enhanced by any tendency for adjacent cells to have their cellulose orientation in parallel, as is often strikingly evident in secondary wall thicken-

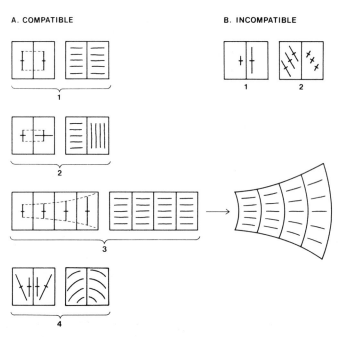

Fig. 22.5. A, Four groups of figures in which the left-hand pair shows compatible juxtaposition of growth rate crosses, and the right-hand pair shows the corresponding pattern of cellulose reinforcement and cortical microtubule array. B, Juxtaposed cells where continuity of growth is not maintained across the common wall. In such cases the common wall would wrinkle or tear.

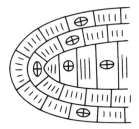

Fig. 22.6. Imaginary root longitudinal section where harmony of reinforcement is shown by continuity of lines over most districts, as in Fig. 22.5A(1), (2) and (3). Harmony of growth rate crosses also follows patterns in Fig. 22.5A(1), (2) and (3).

ings. Gunning (1982a) has shown parallel microtubule alignment in adjacent cells of the *Azolla* root; see also Wick (1985). Hence, for large districts, a root could be considered as an aggregate of *Nitella* cells, but with gradients in growth rate and pattern being mutually adjusted to meet the continuity requirements.

The perpetuation of such patterns is explained if old walls extend, and new walls are added, along the paraboloidal lines. This occurs automatically if cells divide perpendicular to, or parallel to, the growth trajectory lines. These are the directions free of shear as pointed out by Lintilhac (1984). Two patterns of addition of new cross walls in this fashion are given in Fig. 22.7.

One can conclude that there are two important orthogonal properties of symplastic growth in a 'solid' tissue. On the one hand are the wall outlines and cellulose reinforcement patterns which coincide with the local strain rate cross. On the other, and inside the cell, is the perpendicular relation between the spindle axis and the phragmosome. All of these 'crosses' appear to co-align during symplastic growth. This defines a type of biophysical 'harmony'. Within this state there is remarkable room for variation.

The analysis of Hejnowicz points out that the orthogonal grid prescribes the orientation of

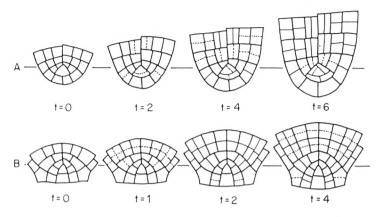

Fig. 22.7. Progressive cell patterns generated by adding new walls parallel to one or the other set of paraboloid lines in Figs 22.3 and 22.4. Each diagram is a simple symplastic field. A, Corresponding to Fig. 22.3 where there is a growth maximum near the focus. Note that the histological pattern is maintained even though most cells pass through the pattern. This resembles the root of *Azolla*; B, corresponding to Fig. 22.4 where there is a growth minimum at the focus (a quiescent centre). From Hejnowicz (1989) with permission.

the minimal and maximal extension rates for the cells in the grid; it does not specify which is which. This means that, within a harmonious orthogonal pattern there could be gradual shift from one orientation of the ellipse maximum, through the isotropic case (a circle), to the other. In fact, Nakielski (1990) shows that a shift from basically transverse, through isotropic, to longitudinal may occur for the growth maximum in tangential planes of a root.

There is also room for great variation in the local rate of increase in area. This is measured as the sum of the lengths of the arms of the growth rate cross. One of Hejnowicz's patterns has a growth rate *maximum* near the focus of the lines and corresponds to a root with a prominent apical cell (Fig. 22.3). This is the case in *Azolla* (Gunning, 1982a). The other has a rate *minimum* near the focus and corresponds to a root with a quiescent centre (Fig. 22.4). There is thus a three-way coupling between (i) gradients in the overall rate of growth, (ii) the local directionality of growth (as shown by Nakielski, 1990) and (iii) the pattern of histology. This situation persists when cell division, new reinforcement and strain rate crosses are all aligned with the curved coordinate system. The two root patterns may therefore each be a unique compromise to

combine, within symplastic growth, a characteristic gradient in overall growth rate with a general transverse reinforcement (tendency to elongate) of most of the cells.

The orthogonal grid perspective addresses well the many questions of self-perpetuation of histological patterns. It would also explain the origin of such patterns if the right angle relationships could persist down to infinitesimal dimensions at the point-like focus of the two sets of lines. There are two indications, however, that activity at the focus can be quite different from that in the more developed tissue. In the *Azolla* root tip the apical cell has three major 'cutting faces' which divide in turn (at 120°) to produce three developmental packages or merophytes. These merophytes would form a root comprised of only three files of cells except that each package undergoes a defined subroutine of divisions which yields the typical root histology. A fourth cutting face makes the root cap. This complex apical activity, while symplastic, is clearly not a back-extrapolation of the smooth growth of the cylindrical part of the root. This cyclic cell production at the tip might be expected to generate four filaments radiating in space or a spherical mass of adhering cells. The products of three of the cutting faces, however, show great physical coherence to yield the cylindrical structure. Symplastic growth thus greatly modifies the consequences of the four-directional activity at the focus.

Uniqueness at the focus is also found in the structure of the apical cell of the *Onoclea* shoot which also divides sequentially at 120° orientations. This cell has no preferred direction of cellulose reinforcement, but all its derivatives do (Lintilhac and Green, 1976). There are thus strong indications that activity at and near the focus is different from that in more established parts of a symplastic field. Such a difference is also evident in the shoot apex of some angiosperms.

The shoot in surface view – rhythmic morphogenesis

The surface of the apical dome often has a distinctive histology of anticlinal walls; it has a known cellulose reinforcement pattern in some cases (Green, 1985; Jesuthasan and Green, 1989). Microtubule arrays have also been analysed (Sakaguchi *et al.*, 1988b). Regular patterns are more evident for the cellulose, reflecting either that microtubule preservation is difficult, or that well-defined cellulose alignments can arise from microtubule patterns

which, at any given moment, appear relatively disorganized. We will emphasize the cellulose patterns relative to the epidermal cell pattern.

An orthogonal pattern of cells can be observed, particularly in plants with distichous (*Tradescantia*) and decussate (*Vinca*) phyllotaxis. The latter pattern has much in common with that of two opposing paraboloid patterns, fused together. In *Vinca*, a pattern of intersecting elliptical and paraboloid lines can be recognized (Fig. 22.8). The histologial pattern for 'harmony' is present. A surprise is that the cellulose pattern displays striking contrast at the centre. Cells with reinforcement at 90° are seen there (Jesuthasan and Green, 1989). Abrupt contrast is also evident at the tip of *Tradescantia* (Green, 1986). It appears that a perpendicular juxtaposition of microtubules, or at least cellulose synthesis, is compatible with a 'harmonious' orthogonal histology. See Fig. 22.5A(2). There is an apparent progression of three types of reasonable cytoskeleton juxtapositions for such shoots (Fig. 22.8B).

In a decussate apex, the perpendicular cytoskeletal alignments at the tip need to change by 90° during each plastochron because the new pair of leaf primordia arises at 90° to their

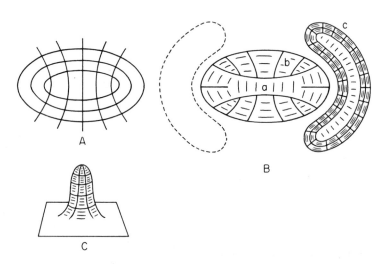

Fig. 22.8. A, Pattern of surface cells like that seen on the apical dome of the decussate plant *Vinca*, as seen from above (Green, 1985; Jesuthasan and Green, 1989). Perpendicularity is between elliptical and paraboloid lines; B, corresponding reinforcement (and microtubule ?) pattern. Compatibility, first at (a) is of the type in Fig. 22.5A(2). Later, in the dome periphery, it is of the type shown in Fig. 22.5A(1) and (3) as at (b). Formation of a leaf requires a second 'disjointed' pattern at (c). This ultimately smooths to the hoop-reinforced surface as in (c). Shoot development involves a succession of symplastic fields.

predecessors. In a distichous apex, the geometry at the tip may remain relatively constant. In both cases, as tissue is displaced towards the periphery, reinforcement appears to shift from one condition of intercellular harmony, 90° opposition of alignment, to another: parallel co-alignment. A circumferential arrangement of microtubules, corresponding to that for cellulose, has been established for the peripheral dome of *Vinca major* (Sakaguchi *et al.*, 1988b).

During the production of leaves, the harmonious pattern of reinforcement and microtubular co-alignment seen at the periphery of the dome has to be disrupted. It is topologically impossible to deform simply the circumferentially reinforced dome to produce hoop-reinforced appendages (Green, 1986). A suggestion to explain how such appendages do form, follows. The idea is that the surface buckles upward at sites where three conditions obtain:

(1) The radius is long (there is available space).
(2) The curvature of reinforcement in the surface plane is great and has arcs with bilateral symmetry facing the centre of the dome.
(3) There is a resistance, in the form of the arc-like crease pattern at the base of recently formed leaves, which is also bilaterally symmetrical with the concave side facing the centre of the dome.

This peripheral resistance combines with a presumed resistance at the dome centre (apparently slow growing and showing few mitoses) to provide the two necessary 'fixed' ends of the line along which the surface is to buckle. All three conditions appear to be cyclically regenerated in the three common types of phyllotaxis – distichous, decussate, and spiral (see Green, 1986; Lyndon, 1990).

The buckling in question is not elastic. Cuts in meristems gape, showing that the epidermis is in elastic tension ('wants to be shorter'). Tension cannot cause buckling. Instead, the buckling apparently reflects a growth potential which can only be realized, because of the two resistances, if the tissue along the line buckles upward. Relative to growth, the tunica is in compression (it 'wants to be longer'). A straight

growing coleoptile segment, its ends touching two fixed barriers, would soon also bow or buckle despite the fact that its epidermis is clearly in a state of elastic tension (Kutschera and Briggs, 1987; also see Kutschera, Chapter 11, this volume). The same may be true for the tunica: it can only buckle upward. The cellulose alignment and resistances apparently facilitate and localize the buckling which is thought to be similar to an aneurysm or a blister. This local upward buckling may bring on the well-known internal periclinal divisions commonly thought to be the first activity associated with appendage formation. Both buckling, and the divisions, take place in the 'peripheral' zone where mitoses are common (e.g. Bernier *et al.*, 1981). The buckling disrupts the previous harmony because the dome surface is no longer smooth.

The new organ is a small ridge. On its crest, it is reasonable to assume there are strong tensions perpendicular to the axis of the ridge. The resulting strains may lead to the 90° changes in birefringence seen there (Jesuthasan and Green, 1989). These changes in cellulose orientation break up the dome's symplastic field. They are essential to the initiation of hoop reinforcement in the appendage. Soon after, the new lateral organ is in the harmonious condition described earlier for the subapical part of the root (Green and Brooks, 1978; see Fig. 22.8C).

In Fig. 22.8 we review the biophysical transitions for a small group of cells on the shoot tip surface. A small piece of tunica tissue, originally near the centre of the dome, appears to be in an initial state of cytoskeletal harmony involving 90° confrontations of microtubule alignments with its neighbours. During subsequent expansion and division it will shift to a different condition of harmony: co-alignment of cellulose and microtubules with its neighbours in the dome periphery. This state is then altered by the 90° shifts associated with appendage formation. This gives a second configuration involving a 90° juxtaposition of cytoskeletal alignments. Finally, a second 'smoothing' leads to the co-aligned situation characteristic of an established plant axis. For the progression down the median section of the shoot interior, a single 90° shift in microtubule alignment has been described for *Vinca major* by Sakaguchi *et al.* (1988a). This is all that is needed in that plane.

To summarize, Hejnowicz has provided the useful concept of a symplastic field of cells. Within such a field there is remarkable room for variation, e.g. in the direction of the growth rate maximum. Certain such fields appear to have their origin in activity which is not a simple back extrapolation of the features of the established field (*Azolla* root apex). Remarkably disjointed cytoskeletal behaviour can be found at the origin of a symplastic field (apical dome of *Vinca*). Once a field is established, it is subject to being converted to an array of new fields (leaf formation in the dome periphery). It may thus be the developmental strategy of the shoot that physical consequences of the growth of appendages, themselves symplastic fields, feed back on the dome periphery – itself a very coherent field – to break it up periodically to produce new appendages.

From the above, it is clear that knowledge of all the influences on cellulose, and especially microtubule alignment, is of great importance. Because several factors may interact, and because the response may be statistical rather than absolute, the format for such studies is not obvious. It will be addressed in the following section. The question posed is, what factors correlate meaningfully with the orientation of microtubular arrays in a shoot meristem?

Potential cues for microtubule alignment

The mechanism for the alignment of cortical microtubules, and even their mode of origin, is not established. One can none the less seek to identify factors that influence the orientation mechanism, and perhaps help to clarify the issue. The geometrical features of apical tissue form a logical starting point for the search for factors which serve as cues for the alignment. The intrinsic character of microtubules can provide justification for this: microtubules are relatively long and stiff (Yamazaki *et al.*, 1982; Dustin, 1984; Bereiter-Hahn, 1987; Weiss *et al.*,

1987). This could make the orientation of their arrays sensitive to the flow associated with rapid cell shape change. Certain specific tissue features have been correlated with microtubule or cellulose orientation in a cell. These include (i) direction of apparent excessive tissue stretch (Green and Brooks, 1978; Green and Lang, 1980; Hardham *et al.*, 1980; Green, 1985; Jesuthasan and Green, 1989; Selker, 1990); (ii) orientation of division planes (Green and Brooks, 1978; Green and Lang, 1980; Hardham *et al.*, 1980; Green, 1984; Selker and Green, 1984; Jesuthasan and Green, 1989; Sylvester *et al.*, 1989; Selker, 1990); and (iii) orientation of microtubules in adjacent cells (Wick, 1985; Jesuthasan and Green, 1989; Marc and Hackett, 1989; Jung and Wernicke, 1990; Selker, 1990).

Features correlating with microtubule change during axis initiation

The above three features are interesting because of their apparent involvement in axis initiation from established axial organs. Green and Brooks (1978) used the succulent *Graptopetalum* to study how a radially symmetrical stem is produced in the middle of a residual meristem with initial bilateral symmetry. They found that remodelling proceeds by means of a regional change in cellulose orientation. Cellulose orientation was detected by carving off a thin paradermal tissue strip from the surface and viewing it in polarized light with a compensator of known orientation. The regional change in cellulose orientation in *Graptopetalum* is preceded by (i) insertion of cell walls at a new orientation and (ii) rapid surface extension, both (i) and (ii) oriented parallel to the new cellulose direction. Cellulose in a non-growing region is generally parallel to the most recently formed wall. Green and Lang (1980) demonstrated a dependence between cell shape and the cellulose-orienting activity of a cell wall inserted at a new orientation. Only when the cell wall is parallel to the daughter cell long axis will the orientation of cellulose in the daughter cell be changed relative to the mother cell. Surface extension parallel to divisions will increase the chance of the new wall being parallel to the long axis. Alternatively, frequent divisions in a single orientation will eventually produce daughter cells elongated parallel to the inserted walls. Hardham *et al.* (1980) showed that microtubules paralleled cellulose microfibrils during axis initiation in this residual meristem system. Selker and Green (1984) showed that periclinal divisions in the interior of the residual meristem precede changes in the microtubule orientation in these cells. Green (1984) assembled detailed data on cellulose orientations relative to division history in epidermal tissue from *Graptopetalum* stem, leaf, root and residual meristem. Cells with longitudinal cellulose are occasionally seen in established axes not in the process of initiating new branches. These cells with longitudinal cellulose have all recently divided longitudinally, indicating that the relationship between cellulose and division direction is a rule that can operate generally and not just under special conditions pertaining during axis initiation. In a slowly growing region of the residual meristem, brightness in polarized light with a compensator is proportional to the number of divisions parallel to the orientation of high refractive index for the compensator and inversely proportional to the number of divisions perpendicular to the compensator's direction of high refractive index. This means that relative amounts of cellulose in any particular orientation correspond to the number of divisions in that orientation.

Axis initiation has also been studied at the vegetative and floral apex (Green, 1985, 1988; Jesuthasan and Green, 1989). Shifts in cellulose orientation associated with appendage formation have been found in specific locations. In a decussate apex such as *Vinca major*, shifts occur in regions between the major and minor axes of the elliptical apex (Green, 1985; Jesuthasan and Green, 1989). Shifts are in general parallel to the direction of overall apex shape change, and they are often seen in cells which have recently divided in a new direction. Sometimes shifts smooth discontinuities in cellulose orientation between adjacent cells. As cells reach the periphery of the apical dome, they cease shifting and consolidate a circumferential alignment. Microtubule patterns have been found in several investigations to be fairly disorganized toward the centre of the apex and ordered,

both within cells and globally, towards the periphery (Sakaguchi *et al.*, 1988b; Marc and Hackett, 1989; Selker, 1990). In spite of involvement in shifting processes, most central cells have detectable order in cellulose alignment, as seen with polarized light (Jesuthasan and Green, 1989). Microtubules may have an average order over time which is sometimes difficult to detect. In all of the published views of apex microtubules (Sakaguchi *et al.*, 1988b; Marc and Hackett, 1989; Selker, 1990) there are some central cells which are ordered at the moment of observation. In apices of all the major types of phyllotaxis, the cellulose pattern over most of the apex (except the extreme periphery or, in some cases, the extreme centre) shows discontinuities along lines corresponding to the edges of leaf bases (Green, 1985, 1988; Jesuthasan and Green, 1989; see Fig. 22.8B). Each field of cellulose orientation (reinforcement field) is traceable to association with a specific existing leaf, and the cellulose is aligned parallel (tangential) to the base of the associated leaf. Alignment of microtubules is also strikingly tangential to leaf bases (Sakaguchi *et al.*, 1988b; Marc and Hackett, 1989; Selker, 1990). Because of this clarity of alignment adjacent to leaves, it appeared likely that the leaves were providing an orientation cue, perhaps by stretching the apical tissue adjacent to their bases. However, recent measurements (Green, unpublished) of apical strain in individual living plants has shown that strain tangential to leaf bases, although present, is probably not dramatic enough to make it a major signal for microtubule orientation shifts. The *Echeveria* floral apex provides a clear example of strain preceding a shift in cellulose orientation to adirection parallel to the strain (Green, 1988). Apparent strong strain parallel to radii of the apex aligns cellulose to make the sides of the stamen primordium, a new axis. The front and back of the stamen come from pre-existing circumferential reinforcement fields.

Other studies relating to strain and microtubule orientation

The relation between strain and microtubule orientation has also been examined in other types of systems. Microtubule change to parallel an excessive cytoplasmic flow (perhaps similar to a flow during tissue extension) has been seen in regions of *Nitella* cells in which chloroplasts have been caused to dislodge by bleaching with blue light irradiation (Kriskovich and Richmond, 1988). In these regions, microtubules were directly adjacent to the streaming cytoplasm from which they were normally shielded by the chloroplasts. This system should be useful for quantifying flow rates that can reorient microtubules.

It is clear that microtubules can maintain a transverse orientation during the course of continued longitudinal extension of cells in an established organ. This is seen wherever cell files have a consistent microtubule or cellulose synthesis alignment perpendicular to the long axis of the file (e.g. Green and Brooks, 1978; Sakaguchi *et al.*, 1988b; Wilms and Derksen, 1988; Marc and Hackett, 1989; Hush *et al.*, 1990). Experiments in which a *Nitella* internode cells was stretched showed that cellulose synthesis (and presumably also microtubule orientation) remains transverse even when an outside force is applied (Gertel and Green, 1977). The actual longitudinal strain in this experiment was similar in magnitude to that in control cells, but pulling the cell produced transverse contraction, thus changing the ratio of longitudinal to transverse extension. Strain seems to be part of a set of rules that operates when tissue is being remodelled, as for axis initiation. When cellulose is shifting, it can use the strain direction as an alignment cue, as seen in the examples of axis initiation described above. Whether or not strain direction correlates with a shift in cellulose or microtubule orientation may also depend on the strain rate (Jesuthasan and Green, 1989).

White and Overall (personal communication) have evidence that stretch can be a cue to align a microtubule array *perpendicular* to the stretch. They can control the future axis of elongation in *Mougeotia* protoplasts by physically compressing them into an elongate shape. The protoplasts, which would otherwise be equally likely to elongate in any direction, continue to elongate along the long axis they developed by the squashing treatment. This result indicates

that the microtubule array, known to precede cellulose synthesis parallel to itself and elongation perpendicular to itself, has become oriented perpendicular to the direction of stretching of the protoplast.

As cells get older and further from the growing point, their growth slows down. Several reports have shown that a decrease in longitudinal growth is correlated with a gradual change in the orientation of microtubules from transverse to longitudinal (Takeda and Shibaoka, 1981a; Traas *et al.*, 1984; Hogetsu and Oshima, 1986; Iwata and Hogetsu, 1988; Wasteneys and Williamson, 1987; Laskowski, 1990). Cells also elongate more slowly in the winter. Sakaguchi *et al.* (1990) reported that microtubules in the rib meristem are transverse when the tissue is growing and random in tissue collected in winter. Determination of the exact timing of the microtubule change relative to the decrease in longitudinal growth in the pea stem indicates that the decrease in growth rate begins at least 4 h before the beginning of the change in microtubule orientation (Laskowski, 1990). Thus microtubules are not responsible for initiating the decline in longitudinal growth rate through their ability to orientate cellulose deposition. The idea that a decrease in longitudinal strain could be a signal to microtubules to change their orientation is supported by an experiment of Gertel and Green (1977) in which a region of a *Nitella* cell prevented from growing by being held in a plexiglass box ceased transverse cellulose deposition and commenced random deposition. Similarly, osmotically induced reduction in growth rate shifted microtubules from transverse to oblique orientations (Roberts *et al.*, 1985).

Wasteneys and Williamson (1987) point out that microtubule control over the growth rate should lead to a change in the ratio of longitudinal to transverse extension, rather than to an eventual cessation of growth, as occurs in mature cells. Métraux and Taiz (1978) measured longitudinal and transverse extensibility of cell walls from old and young *Nitella* cells by applying tension to the walls. Young cell walls have a ratio of longitudinal to transverse extensibility two to four times greater than old cells, compatible with the microtubule data of Wasteneys and Williamson (1987). However, the transverse extensibility values are not correlated with growth (Métraux and Taiz, 1978). They conclude that metabolic processes other than cellulose microfibril deposition are influencing transverse growth. As Wasteneys and Williamson (1987) state: 'Any relation between strain and MT [micro-tubule] alignment could potentially operate in either direction'. A metabolically initiated decrease in growth rate could signal a change in microtubule orientation which in turn could affect the directionality of wall extensibility.

Laskowski (1990) has demonstrated that the magnitude of an apparent effect by strain rate on microtubule orientation is dependent on the specific conditions of the growth reduction. More pea stem cells with transverse micro-tubules remained after blue light irradiation was used to stop elongation than after the cells were allowed to progress naturally to the non-growing region of the stem.

An osmotically induced change in cell stret-ching seems to be necessary for a shift in microtubule orientation in developing guard cells of grasses and sedges (Palevitz, 1981, 1982). The microtubules shift from a pattern where they radiate from the centre of the stomatal pore site to a longitudinal alignment, parallel to the stem long axis and to the division which originally formed the two guard cells. Experimental interruption of the develop-mental process and comparison of different species pinpoints the critical event preceding the microtubule shift as a reconstriction of cells which had just transiently swelled. Both the swelling and reconstriction appear to be caused by active osmotic processes. A possible inter-pretation of the results in this system in relation to the above potential influences is that an increase in stretching (osmotically induced cell swelling) does not signal a microtubule orientation shift but a decrease in dimension does. The change is to the same direction as in cells that are stretching more slowly as they age.

Division plane and cell shape related to microtubule orientation

Sylvester et al. (1989) reported microtubule organization during growth of the leaf, the stem, and the transition zone between them in a plant growing at the base of a leaf of Graptopetalum. They found that cells of different shape tended to have microtubules arranged differently: cells widened transversely usually had transverse microtubules, square cells usually had longitudinal microtubules and irregularly shaped cells usually had randomly oriented microtubules. Among cells of similar shape, the orientation of microtubules was correlated with the orientation of the most recently formed cell wall, transverse micro-tubules being much more common than other orientations after transverse divisions, slightly more common after oblique divisions and not more common after longitudinal divisions. These results are in agreement with the papers on axis initiation mentioned above.

Flanders et al. (1989) assembled three-dimensional views of interphase microtubule arrays in cells of different shape in the stem and petiole epidermis of Datura stramonium. Elongated cells, found along the petiole, usually have ordered arrays of microtubules, while isodiametric cells on the stem have variable orientations of microtubules within individual cells. The three-dimensional reconstruction method allowed them to follow microtubule arrays around cell edges. They were able to verify that the arrays of elongated cells are continuous from one cell face to another. The isodiametric cells also have arrays that continue across cell edges. The microtubule arrays of the anticlinal cell surfaces are actually ordered, but since adjacent anticlinal walls are at angles much less than 180°, continuation of the anticlinal arrays across to the outer and inner periclinal surfaces produces a criss-crossed alignment of microtubules on these surfaces. Flanders et al. (1989) term these periclinal cell faces 'sacrificial' because the coherent alignment of microtubules present on the anticlinal cell faces cannot be maintained on the periclinal faces. The situation is analo-gous to that of a box with pieces on the side folded over to make the lid. If the sides have vertical lines, the lid will have lines that criss-cross.

A subsequent paper by the same authors (Flanders et al., 1990) explored the basis of differences between the elongated and the isodiametric cells. The two types of cells differ in their patterns of microtubule strands radiating from the nucleus prior to mitosis. Elongated cells have most strands oriented close to the transverse plane, while isodiametric cells have strands connecting the nucleus to the midpoint of each wall. These microtubule strand patterns correspond to the later location of the pre-prophase band. Elongated cells usually divide transversely, with a preprophase band formed at the surface of the region occupied by the transverse microtubule strands. Isodiametric cells can have a preprophase band in a single plane, at the surface of the plane occupied by a set of microtubule strands. Alternatively, the preprophase band can curve from a location at the surface of one set of strands to a location at the surface of a set of perpendicular strands. The cell wall will then form obliquely. Strands of actin microfilaments form similar patterns to the microtubule strands. They demonstrated that it is possible to model the behaviour of the strands with either bubbles or springs allowed to form spontaneous patterns within frame-works shaped like the different types of cells. They suggest therefore that the strands may be under tension. Their observations show how cell shape can influence division direction. It is then easy to imagine how division direction in turn influences interphase microtubule orienta-tion. For instance, an isodiametric cell which has just divided obliquely could have micro-tubules along intersecting perpendicular anti-clinal walls each parallel to their segment of the recent preprophase band. Then continuation of these (perpendicular) orientations across to the outer epidermal wall would produce a criss-cross array of microtubules.

Microtubule orientations in individual cells are not always predictable on the basis of previous division plane and cell shape, how-ever. Marc and Hackett (1989) present examples of recently divided cells in which interphase microtubule arrays have been set up either parallel to or perpendicular to the division plane, leading to the conclusion that there is no relation between division plane and microtubule orientation, even considering cell shape. However, some of the cases where microtubules are found perpendicular to a long recent wall are explainable by considering the effect of the probable stretch direction on the original cell shape: the cell originally, at the time of formation of the interphase microtubule array, would not have been elongated parallel to the recent wall.

In some cases, microtubule orientations have been found to change *before* the appearance of divisions in the new orientation. These changed microtubules then predict the subsequent division direction (Wilms and Derksen, 1988; Hush et al., 1990). Both of these examples involve tissue reorganizations after a cut. Wilms and Derksen (1988) found microtubules change from transverse to longitudinal and then to random in a region of a tobacco explant which later divides randomly and finally forms buds. The random microtubules develop only in explants cultured in high (10^{-6} M) benzyl-aminopurine and not in levels that fail to induce bud formation (10^{-7} M). Hush et al. (1990) cut a wedge of tissue from pea roots and observed that microtubules first change from transverse to longitudinal and then orient so as to parallel the edges of the wound. They had measured a large inward current preceding the time of the microtubule orientation shift, and also consider stress patterns at the wound a possible orienting influence for microtubules.

Sakaguchi et al. (1988a) point out that micro-tubules, in the layer of cells directly below the epidermis, are oriented differently from recent division planes in those cells. Based on the division patterns of surrounding cells, these cells (located at the flanks of the apical dome) may be about to divide parallel to their microtubule direction (parallel to the surface or obliquely). Cells with randomly arranged microtubules adjacent to their anticlinal walls have been seen in the epidermal layer of gymnosperm apices following normal anticlinal divisions (Sakaguchi et al., 1990). They point out that gymnosperms commonly have super-ficial cells dividing periclinally or obliquely.

Correspondence between microtubules in adjacent cells

When cells have microtubules different from their recent division planes, they may be subject to alternative influences. One example of a potentially competing influence comes from adjacent cells. Wick (1985) presented a dramatic example of consistency of microtubule orientation across cell boundaries in *Cyperus* epidermis. Jung and Wernicke (1990) have studied the formation of microtubule bands in wheat mesophyll cells and have shown that patterns of microtubule clustering are similar in adjacent cells. Marc and Hackett (1989) presented microtubule data for the apical epidermis of *Hedera helix* and reported that adjacent cells often have the same orientation, including cells along a single cell packet and cells contacting each other across cell packets. Some cells had 'V-strands' of microtubules, made from two oriented arrays of microtubules that were at different angles. Sometimes the two different orientations were observed to meet the arrays of two adjacent cells. Not only interphase microtubules but also preprophase bands and phragmoplast planes were aligned compatibly from cell to cell. Numerous examples can be found in apex epidermal data from Jesuthasan and Green (1989) and Selker (1990) showing cells with an orientation of cellulose or microtubules compatible with adjacent cells but not parallel to a recent division.

The adjacent cell influence is not universally seen. When the division rule is operating, it is common to see daughter cells of different dimensions having conflicting cellulose orientations (Green, 1984). Traas *et al.* (1984) have emphasized that neighbouring cells in the root cortex of *Raphanus sativus* can have completely different microtubule orientations. The pattern of cellulose in the epidermis of a variety of species shows striking discontinuities (Green, 1985; Jesuthasan and Green, 1989).

Other factors which influence microtubules

Some of the reported discrepancies in the relationship between anatomical features and microtubule orientation may be a consequence of different systems providing different con-texts. Additionally, any operation of rules involving anatomical features probably includes other, less obvious processes in its mechanisms of action. There are several examples in the literature of the types of elements which may be included in these contexts or mechanisms. As discussed by Shibaoka (Chapter 12, this volume), hormones have been demonstrated to affect microtubule orientation or stability, including auxin (Bergfeld *et al.*, 1988), gibberellic acid (Takeda and Shibaoka, 1981b; Simmonds *et al.*, 1983; Mita and Shibaoka, 1984; Akashi and Shibaoka, 1987), ethylene (Steen and Chadwick, 1981; Lang *et al.*, 1982; Roberts *et al.*, 1985), and abscisic acid (Rikin *et al.*, 1983; Sakiyama and Shibaoka, 1990). Electrical fields are able to specify the future axis of elongation of spherical protoplasts (White and Overall, 1989; White *et al.*, 1990). The axis of elongation has been shown to be under micro-tubular control (Galway and Hardham, 1986), so the electrical field must be able to influence the microtubules. Other cytoskeletal elements of the cytoplasm such as actin filaments (Seagull *et al.*, 1987) or intermediate filaments (Lloyd *et al.*, 1985) have sometimes been found to parallel microtubules and could potentially help organize microtubule arrays. Fine filaments of the same diameter as actin filaments have repeatedly been observed parallel to cortical interphase microtubules (Hardham *et al.*, 1980; Seagull and Heath, 1980; Lancelle *et al.*, 1986). Hepler *et al.* (1990) have observed an extensive cortical endoplasmic reticulum network which they suggest could function in stabilizing or anchoring the microtubule array. Akashi *et al.* (1990) have shown that presence of a cell wall, and specifically the wall component extensin, stabilize microtubules against cold treatment.

Assessing multiple potential influences on microtubule alignment

Given the information that microtubules have been found at various times oriented predictably in association with certain tissue and cellular features, we now need a method for analysing whether these apparent correlations actually deviate from random. The method must include an evaluation of every cell, have a statistical basis, and deal with the problem that the potential influences are often found not in concert. Such a method has been applied to microtubule orientation in *Vinca minor* considering three anatomical features as potential cues (Selker, 1990). The remaining portion of the chapter will present the method and some results from applying it to available data on microtubules in the epidermis of plant apices.

Three studies of microtubules in shoot apical meristems

The sources of data are studies by Sakaguchi *et al.* (1988b), Marc and Hackett (1989), Selker (1990) and Selker and Bradley (unpublished). Sakaguchi *et al.* (1988b) prepared cryosections of fixed shoot apices of *Vinca major*. The sections were then thawed and treated successively with detergent and cell wall degrading enzymes. Staining of the sections was with a mixture of α- and β-tubulin-specific antibodies followed by a fluorescein-conjugated second antibody. They obtained detailed images of microtubule arrays over extensive regions of the apex and presented sections from early and mid-plastochron stages. At both stages it is common to find centrally located cells with randomly oriented microtubules. Further from the centre most but not all cells have microtubules oriented perpendicular to radii of the apex. The average orientation of cellulose, checked with polarized light, corresponded with the microtubule orientation, and with the direction of growth of the apex.

Marc and Hackett (1989) used cyanoacrylic glue to anchor fixed apical meristems of *Hedera helix* onto slides. After incubation in cell-wall degrading enzymes, interior cells were brushed away, leaving the epidermal layer isolated on the slide. This tissue was then stained for tubulin using indirect immunofluorescence. Microtubules were visible in extremely large regions of tissue. Data were presented from two focal planes, the cell centre and the inner epidermal surface. The inner surface view offers the most detail. The central focal plane viewed microtubules perpendicular to their long axis, as they went up the anticlinal walls, connecting arrays on the inner epidermal cell

surface. This central plane showed the average microtubule orientation because staining was brightest along anticlinal walls with the most microtubules. The cells in the centre of the meristem from a mature-phase plant with distichous phyllotaxis often had arrays of microtubules with 'mixed' orientations, and groups of cells could have conflicting orientations, so that the microtubules of the central tissue formed a complex network. As cells attained increasing distances from the centre, they developed parallel circumferential arrays which were arranged parallel to the topographic contours of the meristem. This process of global polarization appeared to be promoted by gibberellic acid, since an apex treated with GA₃ had a smaller central region with the complex network of microtubules than did the mature apex without hormone treatment. Cell outlines were traceable only for the untreated apex, so only this apex was analysed as part of the analysis presented here.

Selker (1990) serially sectioned a mid-plastochron apex of *Vinca minor* fixed with glutaraldehyde followed by osmium, and photographed microtubule patterns seen in epidermal cells viewed with transmission electron microscopy. Unpublished data of Selker and Bradley from early and late plastochron apices fixed the same way and from one freeze-substituted apex have also been included in the analysis.

Deciding whether ordered microtubule arrays are correlated with anatomical features

To apply the method for analysing orientation correlations, each cell's array of microtubules is first classified as ordered or unordered. Second, the microtubule orientation of each 'ordered cell' (cell with an ordered microtubule array) is compared with the orientation of each of the three features: (i) overall direction of recent tissue stretch, (ii) recent division planes involved in the production of the cell, and (iii) the orientation of microtubule arrays in adjacent cells. Orientations within 22.5° of each other were considered co-aligned. A microtubule array parallel to either of the two most recent divisions was counted as being parallel to a division plane, and an array parallel to any one of the arrays in the several adjacent cells was counted as positive for adjacent cell alignment. This part of the analysis gives each cell a number from 0 to 3, indicating the number of features with which the cell's microtubule array lines up. A correlation between the orientation of the features and the microtubules would be indicated by finding high numbers to be more common than low numbers. Cells from the above studies were scored. The result was that, out of 407 cells, 208 had microtubule arrays aligned with three features, 164 with two, 33 with one, and 2 with zero.

It is possible to calculate the number of arrays which would be expected to be parallel to all three features if the arrays were oriented randomly. The calculation is based on knowing the number of cells that do have a way for all three cues to be co-aligned; these cells will be termed '3p' cells, meaning 'score of 3 is possible'. The interaction between the features has a cell-specific character since the recent division planes and the set of adjacent cells are specific to each cell. Some, but not all, cells will have a recent division parallel both to inferred stretch and to microtubules in one of the adjacent cells; these are the 3p cells. Figure 22.9 presents examples of two 3p cells and two cells where there is no orientation possible for the cell's array which would be compatible with all the features at once. For a 3p cell, the probability of having microtubules in the cell in the same orientation as all three features is computed from the number of ways for microtubules to be parallel to the three features divided by the total number of ways the microtubules can be oriented. If there has been one direction of stretch in the recent past, there will be only one microtubule alignment that agrees with the three features. If there have been two stretch directions, then there are two possibilities for microtubule orientations that would give the cell a score of 3. Since orientations within a 22.5° range are considered co-aligned, the total number of possible orientations is 180/22.5 = 8. Random placement of the microtubule orientation would give agreement with three features 1/8 of the time, a probability of 0.125, if the tissue has had one

direction of stretch. It would be 2/8, a probability of 0.25, if there had been two directions of stretch. Table 22.1 summarizes the data for each species and presents the calculated expected values for each observed number. A χ^2 test performed on the data indicates that the microtubule arrays are aligned parallel to the direction compatible with all three features significantly more frequently than would be expected from alignment of arrays at random orientations. Ordered arrays thus have their orientation correlated with sets of three features. Similar calculations were made for expected scores of cells where the three features were less coherent than for the 3p cells. The general trend observed was that scores of 3 and 2 were seen more frequently than expected, while scores of 1 and 0 were seen less frequently than expected (data not shown). Thus the selected features appear to act as meaningful cues.

Relationship between order in microtubule arrays and a unified signal from the anatomical features

A question independent from whether the anatomical features serve as cues for ordered microtubule arrays is: why are some cells unordered? Unordered cells could be in transition between ordered states or could simply be insensitive to the orientation cues, for example because of their stage in the cell cycle. One reason for cells to undergo a transition in microtubule orientation could be if the direction of one of the oriented features changed. Such a change could erase a cell's 3p status. This means that cells with coherent cues (3p cells) should be more likely to be ordered than unordered, if the unordered state is primarily a transition phenomenon. Among all the 3p cells from the same cells used above, 256 were ordered and 22 were unordered. For cells with features less coherent than those of 3p cells, there was a trend toward a lower ratio of ordered to unordered cells (data not shown). These results support the idea that a coherent signal from the potential cues is correlated with an ordered microtubule array.

In summary, a statistical approach to understanding the orientation of microtubule arrays

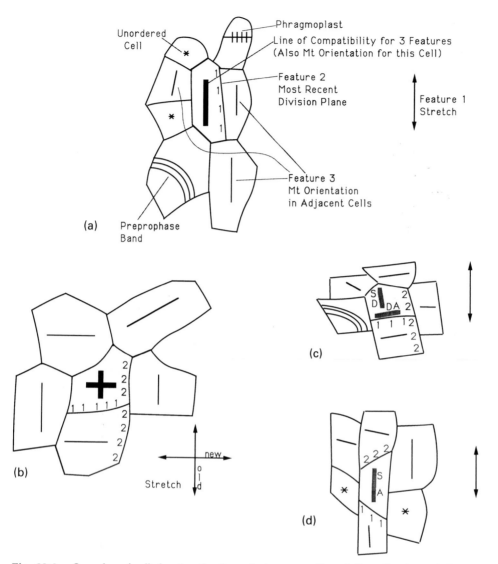

Fig. 22.9. Samples of cells having the three features co-aligned (3p cells, (a) and (b)) and cells with the three features not sharing a common alignment (c) and (d). In each case, the three features can be seen: numbers adjacent to a cell outline indicate recent division planes, microtubule alignments in adjacent cells are indicated with lines (ordered cells) or asterisks (unordered cells) and the recent direction(s) of stretch are shown by arrows to the side of the cell pattern. Stretch direction was inferred from elliptical apex shape and the straightness of cell file boundaries. In (a) and (b), the large bar(s) in the central cell indicate a common orientation for three features. In (c) and (d), initials of the coincident features that are parallel to each bar are written next to the bar (S, stretch; D, division; A, adjacent cell). Preprophase bands and a phragmoplast have been drawn in where appropriate. a, Cells from a mid-plastochron *Vinca major* apex (Selker, 1990, her fig. 6a). The central cell is a 3p cell because the stretch direction (vertical) is parallel to the most recent division (labelled '1') and three adjacent cells have microtubule orientations in this same direction. The cell's own microtubules were also vertical. b, Cells from a late plastochron *Vinca minor* apex (Selker and Bradley, unpublished). The central cell has recently been involved in divisions parallel to each of the two stretch directions, and is surrounded by adjacent cells among which microtubule alignments parallel to both of the stretch directions are found. This cell therefore has two possible ways to be a 3p cell, indicated by the two bars. The cell actually had microtubules parallel to the vertical line. c, Cells from the central region of a *Hedera helix* apex with spiral phyllotaxis (Marc and Hackett, 1989, their fig. 2, immediately above star). The stretch

Table 22.1. Number of ordered 3p cells observed to have microtubule orientations agreeing with all three features compared with the number of cells expected to have microtubules aligned with all three features if the ordered microtubule array were placed randomly in the cell

Species	Number of ordered 3p cells	Number of cells with microtubule arrays co-aligned with the three features	
		Observed	Expected*
Hedera helix	113	87	28.2
Vinca major	28	25	6.6
Vinca minor	116	96	26.4

* Expected if ordered MT arrays took on orientations at random.
The χ^2 statistic computed for these data is 358. (Critical $\chi^2 =$ 5.99 for $\alpha = 0.05$ with 2 degrees of freedom.)

has led to two tentative conclusions. First, ordered arrays are found lined up with all three features (potential cues), more frequently than would be expected if microtubule alignment were occurring randomly. Second, unordered cells are rare among the cells that have three co-aligned features (3p cells). The picture of the microtubule-orientation process supported by these findings is one where a cell is sensitive to multiple inputs and makes a decision based on some kind of integration of the information from different sources. Refinement of the analysis will need to come from adding more detailed information about the recent stretch history of each cell, information about how cells respond to stretch, and data

from more apices representing more species and stages of the plastochron.

Acknowledgements
We thank Z. Hejnowicz and G. Steucek for contributing comments on drafts of the manuscript. Work in the authors' laboratories has been supported by grants from NSF and the USDA competitive grants programme.

References

Akashi, T. and Shibaoka, H. (1987). Effect of gibberellin on the arrangement and the cold stability of cortical microtubules in the epidermal cells of pea internodes. *Plant Cell Physiol.* **28**, 339–348.

Akashi, T., Kawasaki, S. and Shibaoka, H. (1990). Stabilization of cortical microtubules by the cell wall in cultured tobacco cells. Effects of extensin on the cold-stability of cortical microtubules. *Planta* **182**, 363–369.

Bereiter-Hahn, J. (1987). Mechanical principles of architecture of eukaryotic cells. In *Cytomechanics. The Mechanical Basis of Cell Form and Structure* (eds J. Bereiter-Hahn, O.R. Anderson and W.E. Reif), pp. 3–30. Springer-Verlag, London.

Bergfeld, R., Speth, V. and Schopfer, P. (1988). Reorientation of microfibrils and microtubules at the outer epidermal wall of maize coleoptiles during auxin-mediated growth. *Bot. Acta.* **101**, 57–67.

Bernier, G., Kinet, J.-M. and Sachs, R.M. (1981). *The Physiology of Flowering*, vol. II. CRC Press, Boca Raton, FL.

Dustin, P. (1984). *Microtubules*, 2nd edn. Springer-Verlag, Berlin.

Flanders, D.J., Rawlins, D.J., Shaw, P.J. and Lloyd, C.W. (1989). Computer-aided 3-D reconstruction of interphase microtubules in epidermal cells of *Datura stramonium* reveals principles of array assembly. *Development* **106**, 531–541.

Fig. 22.9. *cont.*

direction is superimposable over the orientation of the second most recent division of the cell, while the adjacent cells have microtubule orientations co-aligned with only the other, most recent division plane. The central cell thus has two lines (SD and DA) where sets of two features agree and no line where three features coincide. Microtubules were parallel to DA; d, cells from a mid-plastochron *Vinca major* apex (Sakaguchi *et al.*, 1988, their fig. 3, second file from right). The cells are in long files, not having recently divided parallel to the stretch direction. There are two cells with microtubules parallel to the stretch direction, giving the central cell a line of compatibility only for stretch and adjacent cells (SA). The cell had an unordered array of microtubules.

Flanders, D.J., Rawlins, D.J., Shaw, P.J. and Lloyd, C.W. (1990). Nucleus-associated microtubules help determine the division plane of plant epidermal cells: avoidance of four-way junctions and the role of cell geometry. *J. Cell Biol.* **110**, 1111–1122.

Galway, M.E. and Hardham, A.R. (1986). Microtubule reorganization, cell wall synthesis and establishment of the axis of elongation in regenerating protoplasts of the alga *Mougeotia*. *Protoplasma* **135**, 130–143.

Gertel, E.T. and Green, P.B. (1977). Cell growth pattern and wall microfibrillar arrangement. Experiments with *Nitella*. *Plant Physiol.* **60**, 247–254.

Goodall, C.R. and Green, P.B. (1986). Quantitative analysis of surface growth. *Bot. Gaz.* **147**, 1–15.

Green, P.B. (1980). Organogenesis – a biophysical view. *Annu. Rev. Plant Physiol.* **31**, 51–82.

Green, P.B. (1984). Shifts in plant cell axiality: histogenetic influences on cellulose orientation in the succulent, *Graptopetalum*. *Dev. Biol.* **103**, 18–27.

Green, P.B. (1985). Surface of the shoot apex: a reinforcement field theory for phyllotaxis. In *The Cell Surface in Plant Growth and Development* (eds K. Roberts, A.W.B. Johnstone, C.W. Lloyd, P. Shaw and H.W. Woolhouse), pp. 181–201. *J. Cell Sci. Suppl.* **2**, 1985.

Green, P.B. (1986). Plasticity at the stem apex: a biophysical view. In *Plasticity in Plants* (eds D.H. Jennings and A.J. Trewavas), pp. 211–232. Company of Biologists, Cambridge, UK.

Green, P.B. (1988). A theory for inflorescence development and flower formation based on morphological and biophysical analysis in *Echeveria*. *Planta* **175**, 153–169.

Green, P.B. and Brooks, K.E. (1978). Stem formation from a succulent leaf: its bearing on theories of axiation. *Amer. J. Bot.* **65**, 13–26.

Green, P.B. and Lang, J.M. (1980). Toward a biophysical theory of organogenesis: birefringence observations on regenerating leaves in the succulent, *Graptopetalum paraguayense* E. Walther. *Planta* **151**, 413–426.

Gunning, B.E.S. (1982a). The root of the water fern *Azolla*: cellular basis of development and multiple roles for cortical microtubules. In *Developmental Order: Its Origin and Regulation* (eds S. Subtelny and P.B. Green), pp. 379–421. (Symp. 40 of Society for Developmental Biology). Alan R. Liss, New York.

Gunning, B.E.S. (1982b). The cytokinetic apparatus: its development and spatial regulation. In *The Cytoskeleton in Plant Growth and Development* (ed. C.W. Lloyd), pp. 229–294. Academic Press, London.

Gunning, B.E.S. and Hardham, A.H. (1982). Microtubules. *Ann. Rev. Pl. Physiol.* **33**, 651–698.

Hardham, A.R., Green, P.B. and Lang, J.M. (1980). Reorganization of cortical microtubules and cellulose deposition during leaf formation in *Graptopetalum paraguayense*. *Planta* **149**, 181–195.

Hejnowicz, Z. (1989). Differential growth resulting in the specification of different types of cellular architecture in root meristems. *Env. Exp. Bot.* **29**, 85–93.

Hepler, P.K., Palevitz, B.A., Lancelle, S.A., McCauley, M.M. and Lichtscheidl, I. (1990). Cortical endoplasmic reticulum in plants. *J. Cell Sci.* **96**, 355–373.

Hogetsu, T. and Oshima, Y. (1986). Immunofluorescence microscopy of microtubule arrangment in root cells of *Pisum sativum* L. var. Alaska. *Plant Cell Physiol.* **27**, 939–945.

Hush, J.M., Hawes, C.R. and Overall, R.L. (1990). Interphase microtubule re-orientation predicts a new cell polarity in wounded pea roots. *J. Cell Sci.* **96**, 47–61.

Iwata, K. and Hogetsu, T. (1988). Arrangement of cortical microtubules in *Avena* coleoptiles and mesocotyls and *Pisum* epicotyls. *Plant Cell Physiol.* **29**, 307–315.

Jesuthasan, S. and Green, P.B. (1989). On the mechanism of decussate phyllotaxis: biological studies on the tunica layer of *Vinca major*. *Amer. J. Bot.* **76**, 1152–1166.

Jung, G. and Wernicke, W. (1990). Cell shaping and microtubules in developing mesophyll of wheat (*Triticum aestivum* L.). *Protoplasma* **153**, 141–148.

Kriskovich, M.D. and Richmond, P.A. (1988). Flow-alignment of microtubules in *Nitella*. *J. Cell Biol.* **107**, 29a.

Kutschera, U. and Briggs, W.R. (1987). Differential effect of auxin on *in vivo* extensibility of cortical cylinder and epidermis in pea internodes. *Plant Physiol.* **84**, 1361–1366.

Lancelle, S.A., Callaham, D.A. and Hepler, P.K. (1986). A method for rapid freeze fixation of plant cells. *Protoplasma* **140**, 141–150.

Lang, J.M., Eisinger, W.R. and Green, P.B. (1982). Effects of ethylene on the orientation of microtubules and cellulose microfibrils of pea epicotyl cells with polylamellate cell walls. *Protoplasma* **110**, 5–14.

Laskowski, M.J. (1990). Microtubule orientation in pea stem cells: A change in orientation follows the initiation of growth rate decline. *Planta* **181**, 44–52.

Ledbetter, M. and Porter, K.R. (1963). A 'microtubule' in plant cell fine structure. *J. Cell Biol.* **19**, 239–250.

Lintilhac, P.M. (1984). Stress induced alignment of plant tissues grown *in vivo*. *Nature* **307**, 363–364.

Lintilhac, P.M. and Green, P.B. (1976). Patterns of microfibrillar order in a dormant fern apex. *Amer. J. Bot.* **63**, 726–728.

Lintilhac, P.M. and Vesecky, T.B. (1980). Mechanical stress and cell wall orientation in plants I. Photoelastic derivation of principal stresses with a discussion of the concept of axillarity and the significance of the 'arcuate shell zone'. *Amer. J. Bot.* **67**, 1477–1483.

Lintilhac, P.M. and Vesecky, T.B. (1981). Mechanical stress and cell wall orientation in plants II. The application of controlled directional stress to growing plants; with a discussion of the nature of the wound reaction. *Amer. J. Bot.* **68**, 1222–1230.

Lloyd, C.W. and P.W. Barlow (1982). The coordination of cell division and elongation: the role of the cytoskeleton. In *The Cytoskeleton in Plant Growth and Development* (ed. C.W. Lloyd), pp. 203–228. Academic Press, London.

Lloyd, C.W., Clayton, L., Dawson, P.J., Doonan, J.H., Hulme, J.S., Roberts, I.N. *et al.* (1985). The cytoskeleton underlying side walls and cross walls in plants: molecules and macromolecular assemblies. *J. Cell Sci. Suppl.* **2**, 143–155.

Lyndon, R.F. (1990). *Plant Development. The Cellular Basis*. Unwin Hyman, London.

Malik, H.J. and Mullen, K. (1973). *A First Course in Probability and Statistics*. Addison-Wesley, Menlo Park, CA, USA.

Marc, J. and Hackett, W.P. (1989). A new method for immunofluorescent localization of microtubules in surface cell layers: application to the shoot apical meristem of *Hedera*. *Protoplasma* **148**, 70–79.

Métraux, J.-P. and Taiz, L. (1978). Transverse viscoelastic extension in *Nitella*. I. Relationship to growth rate. *Plant Physiol.* **61**, 135–138.

Mita, T. and Shibaoka, H. (1984). Effects of S-3307, an inhibitor of gibberellin biosynthesis, on swelling of leaf sheath cells and on the arrangement of cortical microtubules in onion seedlings. *Plant Cell Physiol.* **25**, 1531–1539.

Nakielski, J. (1990). Distribution of relative elemental rates of growth in length in root meristems. *Acta Soc. Bot. Pol.* **59**, in press.

Palevitz, B.A. (1981). The structure and development of stomatal cells. In *Stomatal Physiology* (eds P.G. Jarvis and T.A. Mansfield), pp. 1–203. Cambridge University Press, Cambridge.

Palevitz, B.A. (1982). The stomatal complex as a model of cytoskeletal participation in cell differentiation. In *The Cytoskeleton in Plant Growth and Development* (ed. C.W. Lloyd), pp. 345–376.

Academic Press, London.

Rikin, A., Atsmon, D. and Gitler, C. (1983). Quantitation of chill-induced release of a tubulin-like factor and its prevention by abscisic acid in *Gossypium hirsutum* L. *Plant Physiol.* **71**, 747–748.

Roberts, I.N., Lloyd, C.W. and Roberts, K. (1985). Ethylene-induced microtubule reorientations: mediation by helical arrays. *Planta* **164**, 439–447.

Sakaguchi, S., Hogetsu, T. and Hara, N. (1988a). Arrangement of cortical microtubules in the shoot apex of *Vinca major* L. *Planta* **175**, 403–411.

Sakaguchi, S., Hogetsu, T. and Hara, N. (1988b). Arrangement of cortical microtubules at the surface of the shoot apex in *Vinca major* L.: observations by immunofluorescence microscopy. *Bot. Mag. Tokyo* **101**, 497–507.

Sakaguchi, S., Hogetsu, T. and Hara, N. (1990). Specific arrangements of cortical microtubules are correlated with the architecture of meristems in shoot apices of angiosperms and gymnosperms. *Bot. Mag. Tokyo* **103**, 143–163.

Sakiyama, M. and Shibaoka, H. (1990). Effects of abscisic acid on the orientation and cold stability of cortical microtubules in epicotyl cells of the dwarf pea. *Protoplasma* **157**, 165–171.

Seagull, R.W. and Heath, I.B. (1980). The organization of cortical microtubule arrays in the radish root hair. *Protoplasma* **103**, 205–229.

Seagull, R.W., Falconer, M.M. and Weerdenburg, C.A. (1987). Microfilaments: dynamic arrays in higher plant cells. *J. Cell Biol.* **104**, 995–1004.

Selker, J.M.L. (1990). Microtubule patterning in apical epidermal cells of *Vinca minor* preceding leaf emergence. *Protoplasma* **158**, 95–108.

Selker, J.M.L. and Green, P.B. (1984). Organogenesis in *Graptopetalum paraguayense* E. Walther: shifts in orientation of cortical microtubule arrays are associated with periclinal divisions. *Planta* **160**, 289–297.

Simmonds, D., Setterfield, G. and Brown, D.L. (1983). Organization of microtubules in dividing and elongating cells of *Vicia hajastana* Grossh. in suspension culture. *Eur. J. Cell Biol.* **32**, 59–66.

Sokal, R.R. and Rohlf, F.J. (1969). *Biometry; the Principles and Practice of Statistics in Biological Research*. Freeman, San Francisco.

Steen, D.A. and Chadwick, A.V. (1981). Ethylene effects in pea stem tissue. Evidence for microtubule mediation. *Plant Physiol.* **67**, 460–466.

Sylvester, A.W., Williams, M.H. and Green, P.B. (1989). Orientation of cortical microtubules correlates with cell shape and division direction: Immunofluorescence of intact epidermis during development of *Graptopetalum paraguayensis*. *Proto-*

plasma **153**, 91–103.

Takeda, K. and Shibaoka, H. (1981a). Changes in microfibril arrangement on the inner surface of the epidermal cell walls in the epicotyl of *Vigna angularis* Ohwi et Ohashi during cell growth. *Planta* **151**, 385–392.

Takeda, K. and Shibaoka, H. (1981b). Effects of gibberellin and colchicine on microfibril arrangement in epidermal cell walls of *Vigna angularis* Ohwi et Ohashi epicotyls. *Planta* **151**, 393–398.

Traas, J.A., Braat, P. and Derksen, J.W. (1984). Changes in microtubule arrays during the differentiation of cortical root cells of *Raphanus sativus*. *Eur. J. Cell Biol.* **34**, 229–238.

Wasteneys, G.O. and Williamson, R.E. (1987). Microtubule orientation in developing internodal cells of *Nitella*: a quantitative analysis. *Eur. J. Cell Biol.* **43**, 14–22.

Weiss, D.G., Langford, G.M. and Allen, R.D. (1987). Implications of microtubules in cytomechanics: static and motile aspects. In *Cytomechanics. The Mechanical Basis of Cell Form and Structure.* (eds J. Bereiter-Hahn, O.R. Anderson and W.E. Reif),

pp. 100–113. Springer-Verlag, London.

White, R.G. and Overall, R.L. (1989). Elongation of initially non-polar protoplasts is oriented by electric fields. *Biol. Bull.* **176**, 145–149.

White, R.G., Hyde, G.J. and Overall, R.L. (1990). Microtubule arrays in regenerating *Mougeotia* protoplasts may be oriented by electric fields. *Protoplasma* **158**, 73–85.

Wick, S.M. (1985). Immunofluorescence microscopy of tubulin and microtubule arrays in plants cells. 3. Transition between mitotic/cytokinetic and interphase microtubule arrays. *Cell Biol. Intl Rep.* **9**, 357–372.

Wilms, F.H.A. and Derksen, J. (1988). Reorganization of cortical microtubules during cell differentiation in tobacco explants. *Protoplasma* **146**, 127–132.

Yamazaki, S., Maeda, T. and Miki-Noumura, T. (1982). Flexural rigidity of singlet microtubules estimated from statistical analysis of fluctuating images. In *Biological Functions of Microtubules and Related Structures* (eds H. Sakai, H. Mohri and G.G. Borisy), pp. 41–48. Academic Press, London.

INDEX

B